The chemistry of
acyl halides

THE CHEMISTRY OF FUNCTIONAL GROUPS

A series of advanced treatises under the general editorship of
Professor Saul Patai

The chemistry of alkenes (published in 2 volumes)
The chemistry of the carbonyl group (published in 2 volumes)
The chemistry of the ether linkage (published)
The chemistry of the amino group (published)
The chemistry of the nitro and nitroso group (published in 2 parts)
The chemistry of carboxylic acids and esters (published)
The chemistry of the carbon–nitrogen double bond (published)
The chemistry of amides (published)
The chemistry of the cyano group (published)
The chemistry of the hydroxyl group (published in 2 parts)
The chemistry of the azido group (published)
The chemistry of acyl halides (published)

The chemistry of
acyl halides

Edited by

SAUL PATAI

The Hebrew University, Jerusalem

1972

INTERSCIENCE PUBLISHERS
a division of John Wiley & Sons

LONDON — NEW YORK — SYDNEY — TORONTO

Library of Congress Catalog Card No. 70–37114
ISBN 0 471 66936 9

Printed in Great Britain by
John Wright & Sons Ltd., at the Stonebridge Press, Bristol

Contributing authors

M. F. Ansell — Queen Mary College, University of London, England.

D. V. Banthorpe — University College, London, England.

P. Beltrame — University of Cagliari, Italy.

N. J. Bunce — Guelph University, Guelph, Ontario, Canada.

S. Cohen — Israel Institute for Biological Research, Ness-Ziona, Israel.

H. Egger — Sandoz Forschungsinstitut, Vienna, Austria.

P. H. Gore — Brunel University, Uxbridge, Middlesex, England.

D. N. Kevill — Northern Illinois University, Dekalb, Illinois, U.S.A.

A. Kivinen — University of Helsinki, Finland.

K. T. Potts — Rensselaer Polytechnic Institute, Troy, New York, U.S.A.

C. Sapino — Rensselaer Polytechnic Institute, Troy, New York, U.S.A.

D. P. N. Satchell — King's College, University of London, England.

R. S. Satchell — Queen Elizabeth College, University of London, England.

U. Schmidt — Organisch-chemisches Institut, University of Vienna, Austria.

M. Simonetta — University of Milan, Italy.

B. V. Smith — Chelsea College of Science and Technology, University of London, England.

D. D. Tanner — University of Alberta, Edmonton, Alberta, Canada.

H. Weiler-Feilchenfeld — Hebrew University of Jerusalem, Israel.

O. H. Wheeler — Omni Research Inc., Mayaguez, Puerto Rico, U.S.A.

Foreword

As all the volumes of the series 'The Chemistry of Functional Groups', the present volume is again organized according to the same general plan described in the Preface to the series, printed on the following pages.

The series as a whole, considering volumes already published as well as those in active preparation, is now well advanced. I would like to use this opportunity to express again my appreciation and thanks to Dr. Arnold Weissberger, whose initiative and help were instrumental in carrying out this project.

The original plan of the present volume included sixteen chapters, but unfortunately three of these, on 'Decarbonylation and Fragmentation', on 'Syntheses and Uses of Isotopically Labelled Acyl Halides' and on 'Synthetic Uses of Acyl Halides', did not materialize.

Jerusalem, October 1971   SAUL PATAI

The Chemistry of Functional Groups
Preface to the series

The series 'The Chemistry of Functional Groups' is planned to cover in each volume all aspects of the chemistry of one of the important functional groups in organic chemistry. The emphasis is laid on the functional group treated and on the effects which it exerts on the chemical and physical properties, primarily in the immediate vicinity of the group in question, and secondarily on the behaviour of the whole molecule. For instance, the volume *The Chemistry of the Ether Linkage* deals with reactions in which the C—O—C group is involved, as well as with the effects of the C—O—C group on the reactions of alkyl or aryl groups connected to the ether oxygen. It is the purpose of the volume to give a complete coverage of all properties and reactions of ethers in as far as these depend on the presence of the ether group, but the primary subject matter is not the whole molecule, but the C—O—C functional group.

A further restriction in the treatment of the various functional groups in these volumes is that material included in easily and generally available secondary or tertiary sources, such as Chemical Reviews, Quarterly Reviews, Organic Reactions, various 'Advances' and 'Progress' series as well as textbooks (i.e. in books which are usually found in the chemical libraries of universities and research institutes) should not, as a rule, be repeated in detail, unless it is necessary for the balanced treatment of the subject. Therefore each of the authors is asked *not* to give an encyclopaedic coverage of his subject, but to concentrate on the most important recent developments and mainly on material that has not been adequately covered by reviews or other secondary sources by the time of writing of the chapter, and to address himself to a reader who is assumed to be at a fairly advanced post-graduate level.

With these restrictions, it is realized that no plan can be devised for a volume that would give a *complete* coverage of the subject with *no* overlap between chapters, while at the same time preserving the readability of the text. The Editor set himself the goal of attaining *reasonable* coverage with *moderate* overlap, with a minimum of cross-references between the chapters of each volume. In this manner, sufficient freedom is given to each author to produce readable quasi-monographic chapters.

The general plan of each volume includes the following main sections:

(a) An introductory chapter dealing with the general and theoretical aspects of the group.

(b) One or more chapters dealing with the formation of the functional group in question, either from groups present in the molecule, or by introducing the new group directly or indirectly.

(c) Chapters describing the characterization and characteristics of the functional groups, i.e. a chapter dealing with qualitative and quantitative methods of determination including chemical and physical methods, ultraviolet, infrared, nuclear magnetic resonance, and mass spectra; a chapter dealing with activating and directive effects exerted by the group and/or a chapter on the basicity, acidity or complex-forming ability of the group (if applicable).

(d) Chapters on the reactions, transformations and rearrangements which the functional group can undergo, either alone or in conjunction with other reagents.

(e) Special topics which do not fit any of the above sections, such as photochemistry, radiation chemistry, biochemical formations and reactions. Depending on the nature of each functional group treated, these special topics may include short monographs on related functional groups on which no separate volume is planned (e.g. a chapter on 'Thioketones' is included in the volume *The Chemistry of the Carbonyl Group*, and a chapter on 'Ketenes' is included in the volume *The Chemistry of Alkenes*). In other cases, certain compounds, though containing only the functional group of the title, may have special features so as to be best treated in a separate chapter, as e.g. 'Polyethers' in *The Chemistry of the Ether Linkage*, or 'Tetraaminoethylenes' in *The Chemistry of the Amino Group*.

This plan entails that the breadth, depth and thought-provoking nature of each chapter will differ with the views and inclinations of the author and the presentation will necessarily be somewhat uneven. Moreover, a serious problem is caused by authors who deliver their manuscript late or not at all. In order to overcome this problem at least to some extent, it was decided to publish certain volumes in several parts, without giving consideration to the originally planned logical order of the chapters. If after the appearance of the originally planned parts of a volume it is found that either owing to non-delivery of chapters, or to new developments in the subject, sufficient material has accumulated for publication of an additional part, this will be done as soon as possible.

The overall plan of the volumes in the series 'The Chemistry of Functional Groups' includes the titles listed below:

The Chemistry of Alkenes (*published in two volumes*)
The Chemistry of the Carbonyl Group (*published in two volumes*)
The Chemistry of the Ether Linkage (*published*)
The Chemistry of the Amino Group (*published*)
The Chemistry of the Nitro and the Nitroso Group (*published in two parts*)
The Chemistry of Carboxylic Acids and Esters (*published*)
The Chemistry of the Carbon–Nitrogen Double Bond (*published*)
The Chemistry of the Cyano Group (*published*)
The Chemistry of Amides (*published*)
The Chemistry of the Hydroxyl Group (*published in two parts*)
The Chemistry of the Azido Group (*published*)
The Chemistry of Acyl Halides (*published*)
The Chemistry of Carbon–Halogen Bond (*in preparation*)
The Chemistry of the Quinonoid Compounds (*in preparation*)
The Chemistry of the Thiol Group (*in preparation*)
The Chemistry of the Carbon–Carbon Triple Bond
The Chemistry of Imidoates and Amidines
The Chemistry of the Hydrazo, Azo and Azoxy Groups
The Chemistry of the SO, $-SO_2$, $-SO_2H$ and $-SO_2H$ Groups
The Chemistry of the $-OCN$, $-NCO$ and $-SCN$ Groups
The Chemistry of the $-PO_3H_2$ and Related Groups

Advice or criticism regarding the plan and execution of this series will be welcomed by the Editor.

The publication of this series would never have started, let alone continued, without the support of many persons. First and foremost among these is Dr. Arnold Weissberger, whose reassurance and trust encouraged me to tackle this task, and who continues to help and advise me. The efficient and patient cooperation of several staff-members of the Publisher also rendered me invaluable aid (but unfortunately their code of ethics does not allow me to thank them by name). Many of my friends and colleagues in Israel and overseas helped me in the solution of various major and minor matters, and my thanks are due to all of them, especially to Professor Z. Rappoport. Carrying out such a long-range project would be quite impossible without the non-professional but none the less essential participation and partnership of my wife.

The Hebrew University, SAUL PATAI
Jerusalem, ISRAEL

Contents

Contents

CHAPTER **1**

General and theoretical aspects

MASSIMO SIMONETTA

Istituto di Chimica fisica, Università, Milano, Italy

and

PAOLO BELTRAME

Istituto Chimico, Università, Cagliari, Italy

I. INTRODUCTION

In the present chapter, general and theoretical aspects of saturated, unsaturated and aromatic mono- and di-carboxylic acid halides are considered.

Some physico-chemical properties relevant to the understanding of the chemistry of the COX group will be discussed, such as geometry and

1

dipole moments, thermodynamic behaviour, interactions with the electromagnetic field, mass spectra and the electronic structure as obtained by theoretical calculations.

This chapter is not intended as a comprehensive review of the subject and the number of examples given is limited to those required by the discussion of the various properties. The reactivity of acyl halides is not discussed here; this subject will be dealt with in later chapters.

II. GEOMETRY

A. X-ray Diffraction

X-ray data for molecules containing the COX group are extremely scanty. A determination of the geometry of phosgene in the solid phase at liquid nitrogen temperature leads to the following results[1]: C—O, 1.15 ± 0.02 Å; C—Cl, 1.74 ± 0.02 Å; Cl—C—Cl angle, $111.0 \pm 1.5°$. The molecule was found to be planar, within experimental error, and planarity was also found for oxalyl bromide[2, 3]. In the crystals of the latter the molecules are arranged in non-planar sheets; it is suggested that every molecule is linked to four neighbours by weak charge-transfer O···Br bonds, of length 3.27 Å (Figure 1). The following molecular parameters were found[2, 3]: C—O, 1.17 Å; C—Br, 1.84 Å; O—C—Br angle, $128.3°$.

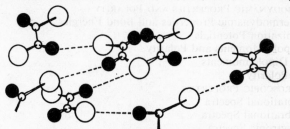

FIGURE 1. Intermolecular interactions in oxalyl bromide crystals[3] (○, carbon; ●, oxygen; ○, bromine).

B. Electron Diffraction

Most of the available data were collected before 1960, and refer to simple molecules. Results are shown in Table 1. For phosgene the geometry does not appear to be significantly different from that found in the solid state. In acetyl derivatives the C—C bond length was consistently found to be very close to 1.50 Å[6, 8, 9]: this is the expected value for a C_{sp^3}—C_{sp^2} single bond. The higher value (1.534 Å) reported for the C_{sp^2}—C_{sp^2} single bond in oxalyl chloride[11] may be justified by electrostatic repulsion between the polarized COCl groups.

It has always been found or assumed that the COX group is co-planar with the atom to which it is bound.

The carbonyl C—O bond length is independent of the halogen, within experimental uncertainty (Table 1). On the average, this distance is

TABLE 1. Geometry of the COX group from electron diffraction

Halogen	Compound	C—O (Å)	C—X (Å)	OCX angle (degrees)	Reference
F	Formyl fluoride	$1 \cdot 19 \pm 0 \cdot 01$	$1 \cdot 35 \pm 0 \cdot 01$	$121 \cdot 9 \pm 0 \cdot 9$	4
	Carbonyl fluoride	$1 \cdot 17 \pm 0 \cdot 02$	$1 \cdot 32 \pm 0 \cdot 02$	124	5
	Acetyl fluoride	$1 \cdot 16 \pm 0 \cdot 02$	$1 \cdot 37 \pm 0 \cdot 02$	125	6
Cl	Phosgene	$1 \cdot 18 \pm 0 \cdot 03$	$1 \cdot 74 \pm 0 \cdot 02$	124	7
	Acetyl chloride	$1 \cdot 17 \pm 0 \cdot 04$	$1 \cdot 82 \pm 0 \cdot 02$	$122 \cdot 5 \pm 2 \cdot 5$	6
		$1 \cdot 22 \pm 0 \cdot 04$	$1 \cdot 77 \pm 0 \cdot 02$	123 ± 3	8
	Chloroacetyl chloride	$1 \cdot 21 \pm 0 \cdot 04$	$1 \cdot 74 \pm 0 \cdot 03$	122 ± 3	9
	Acrylyl chloride	$1 \cdot 20 \pm 0 \cdot 02$	$1 \cdot 74 \pm 0 \cdot 02$	125 ± 2	10
	Oxalyl chloride	$1 \cdot 189 \pm 0 \cdot 002$	$1 \cdot 749 \pm 0 \cdot 003$	$123 \cdot 5$	11
Br	Acetyl bromide	$1 \cdot 17^{a}$	$2 \cdot 00 \pm 0 \cdot 04$	125 ± 5	6
I	Acetyl iodide	$1 \cdot 18^{b}$	$2 \cdot 21 \pm 0 \cdot 04$	125 ± 5	6

[a] Most probable value from radial distribution curve.
[b] Assumed.

shorter in acyl halides than in acids or esters[12], indicating more double-bond character.

As to carbon–halogen bond lengths, it may be noted from Table 1 that they are generally longer than is usually reported for bonds between olefinic carbon and halogens, that is[13]: C_{ol}—F, 1·333; C_{ol}—Cl, 1·719; C_{ol}—Br, 1·89; C_{ol}—I, 2·09 Å. This is in agreement with the pronounced double-bond character of the C—O bonds.

C. Microwave Spectroscopy

Only for small molecules is there a sufficient number of isotopic species available for the complete determination of the structure from microwave spectra, independently of other data. Results from the literature are collected in Table 2.

The reported geometry for formyl fluoride is in complete agreement with previous results[20] in which the most reliable electron diffraction data[4]

TABLE 2. Geometry of the COX group from microwave spectroscopy

Compound	C—O (Å)	C—X (Å)	$\widehat{OCX}$ (degrees)	$\widehat{XCZ}$[a] (degrees)	Reference
Formyl fluoride	1·181 ± 0·005	1·338 ± 0·005	122·8 ± 0·5		14
Carbonyl fluoride	1·17₄ ± 0·01	1·31₂ ± 0·01	126·0 ± 0·5	(Z = F) 108·0 ± 0·5	15
Acetyl fluoride[b]	1·181 ± 0·01	1·348 ± 0·015	121·4 ± 1	(Z = C) 110·3 ± 1	16
Carbonyl chloride fluoride[c]	(X = F) 1·162 ± 0·007	1·304 ± 0·006	130·6 ± 0·04	111·9 ± 0·3	17
	(X = Cl)	1·751 ± 0·003	117·5 ± 0·3		
Phosgene	1·166 ± 0·002	1·746 ± 0·004	124·4 ± 0·1	(Z = Cl) 111·3 ± 0·1	18
Acetyl chloride[b]	1·192 ± 0·010	1·789 ± 0·005	120·2 ± 0·5	(Z = C) 112·7 ± 0·5	19

[a] Z is the third atom to which the carbonyl C is bound, besides O and X.

[b] Assuming a symmetric methyl group.

[c] Results based also on data for similar molecules (COCl$_2$ and COF$_2$), used to establish lower and upper limits to bond lengths and angles.

were incorporated. In carbonyl fluoride, both the C—O and C—F distances are shorter than the corresponding ones in formyl fluoride, indicating a general trend with increasing fluorine substitution. Also the relatively small value of the FCF angle (108·0 ± 0·5°) in COF$_2$ should be noticed. The experimental results for acetyl fluoride were treated in two different ways, with or without the requirement of methyl group symmetry; figures in Table 2 refer to the former treatment and are in remarkable agreement with those for formyl fluoride. The CCF angle in acetyl fluoride has a value reasonably consistent with a bent bond description of the CO double bond[16]. This bond appears to be shorter in formyl and acetyl fluorides than in formaldehyde (1·21 ± 0·01 Å)[21] and acetaldehyde (1·216 ± 0·002)[22], respectively.

Microwave spectra together with a number of reasonable assumptions allowed the determination[17] of the structural parameters for COClF given in Table 2.

Variations of CO and CX distances, and of OCX angles, on going from $COCl_2$ to CH_3COCl or from COF_2 to CH_3COF are comparable. Furthermore, the most stable conformation for acetyl chloride was found to be the same as for acetaldehyde, acetyl fluoride and acetyl cyanide: that is, a methyl hydrogen eclipses the oxygen atom[19]. In the case of fluoroacetyl fluoride, the most stable rotamer has all the heavy atoms in a plane, with the CH_2F fluorine eclipsing the oxygen atom[23].

III. THERMODYNAMIC PROPERTIES AND POLARITY

A. Thermodynamic Properties and Bond Energies

A critical examination of the scanty data found in the literature for the heats of formation of acyl halides has led to the figures shown in Table 3.

TABLE 3. Heats of formation ($-\Delta H_{f,298}$) of acyl halides (kcal/mole) from elements in their standard states

Halogen	Compound	State		Reference
		Liquid	Gas	
F	Carbonyl fluoride		153	24
	Acetyl fluoride	111·4	105·2	24
Cl	Phosgene		52·8	24
	Acetyl chloride	65·5	58·3	24
	Benzoyl chloride	39·3	(28·9)[a]	25
	Oxalyl chloride	85·6	78·0	26
Br	Acetyl bromide	53·2	(45·9)[a]	24
	Benzoyl bromide	25·5	(14·7)[a]	25
I	Acetyl iodide	39·0	30·3	24, 27
	Benzoyl iodide	12·55	(1·3)[a]	25

[a] Estimated value (see text).

The $\Delta H_{f,298}$ values for the gas given in brackets were obtained from the corresponding values for the liquid by applying the Trouton rule to estimate the heat of vaporization.

From the heats of formation of gaseous acetyl halides (Table 3), of the acetyl radical ($\Delta H_{f,298}^0 = -11$ kcal/mole)[28], and of halogen atoms[29], the C—X bond dissociation energies in acetyl halides were calculated by considering the following sequence of reactions:

$$(CH_3COX)_g \longrightarrow 2C + \tfrac{3}{2}H_2 + \tfrac{1}{2}O_2 + \tfrac{1}{2}X_2$$
$$2C + \tfrac{3}{2}H_2 + \tfrac{1}{2}O_2 \longrightarrow (CH_3CO^{\bullet})_g$$
$$\tfrac{1}{2}X_2 \longrightarrow (X^{\bullet})_g$$

$$\overline{(CH_3COX)_g \longrightarrow (CH_3CO^{\bullet})_g + (X^{\bullet})_g}$$

Results are (in kcal/mole): 112·7 (C—F), 76·2 (C—Cl), 61·6 (C—Br), and 44·8 (C—I). An analogous calculation for gaseous benzoyl halides ($\Delta H_{f,298}^0$ for benzoyl radical $= 15\cdot2$ kcal/mole)[28], gave slightly different values: 73·0 (C—Cl), 56·6 (C—Br) and 42·0 (C—I). There is substantial agreement between these results and the bond energy terms for C–halogen bonds discussed by Cottrell[29].

For the three benzoyl derivatives there are also experimental determinations of the C–halogen bond dissociation energy by the kinetic method, assuming that the activation energy for the recombination of radicals is zero. The following values (in kcal/mole) were obtained: 73·6 for benzoyl chloride[30], 57·0 for benzoyl bromide[31] and 43·9 for benzoyl iodide[32]. Agreement with the thermochemical results previously given is good.

A rough calculation of the CO double bond energy is possible, using the heat of formation of carbonyl and acetyl halides from monoatomic gaseous elements and the bond energy terms for CC, CH and CX single bonds. A value close to 190 kcal/mole is obtained, indicating the C=O bond to be particularly strong in this series of compounds.

Thermodynamic functions for a few compounds containing the COX group have been tabulated. In Table 4 a list of these compounds is given, together with the temperature range and references.

For $COCl_2$ and COF_2 accurate calorimetric determinations of the standard entropy gave results in complete agreement with the statistical values from spectroscopic data[36, 39], provided that the stable form of solid phosgene is considered.

B. Ionization Potentials

The most common method for the measurement of ionization potentials is by electron impact in a mass spectrometer. Ionization with and without fragmentation may be obtained:

$$R - X + e^- \longrightarrow R - X^+ + 2e^- \qquad (1)$$
$$R - X + e^- \longrightarrow R^+ + X + 2e^- \qquad (2)$$

Process (1) gives the radical ion with the same molecular weight as the parent molecule, and allows the measurement of the ionization potential of R−X, I (RX). Processes of type (2) give rise to a radical ion R^+ and a

TABLE 4. Tables available in the literature for the thermodynamic functions of acyl halides (heat capacity, entropy, heat content function, Gibbs energy function for ideal gas state at 1 atm. pressure)

Halogen	Compound	Temperature range (K)	Reference
F	HCOF	200–1500	33
	DCOF	200–1500	33
	COF_2	273·16–1500	34, 24
F, Cl	COFCl	273·16–1500	34
Cl	$COCl_2$	15–4000	35, 36, 24
	CH_3COCl	298–1000	24
	ClCO—COCl	100–1000	37
Cl, Br	COClBr	298·15–1000	38
Br	$COBr_2$	298·15–1000	38

neutral radical X; the minimum energy of the bombarding electrons at which R^+ appears gives its appearance potential $A(R^+)$.

A list of values of ionization and appearance potentials is given in Table 5. The sequence of values for $A(CH_3CO^+)$ shows the expected

TABLE 5. Ionization and appearance potentials (eV)[a]

Compound (RX)	I (RX)	A (R[+])	Reference
HCOF	11·4	—	40
COF_2	13·2 ± 0·1	—	41
$COCl_2$	11·77 ± 0·04	—	40
	11·7₅[b]	—	42
CH_3COF	—	12·3	43
CH_3COCl	11·08 ± 0·06	—	44
	—	11·2	43
CH_3COBr	—	10·6	43
PhCOF	10·6	11·5	45
	—	11·18	46
PhCOCl	10·6	10·5	45
	9·70 ± 0·01	—	44
PhCOBr	—	10·0	45
p-MeOC$_6$H$_4$COCl	8·87 ± 0·05	—	44
p-MeC$_6$H$_4$COCl	9·37 ± 0·01	—	44
p-ClC$_6$H$_4$COCl	9·58 ± 0·03	—	44
p-NO$_2$C$_6$H$_4$COCl	10·66 ± 0·01	—	44

[a] From mass spectrometry, unless otherwise stated.
[b] From u.v. spectrum.

relationship with the carbon–halogen bond dissociation energies of the corresponding acetyl halides. Analogous behaviour is shown by benzoyl halides. The ionization potentials of unsubstituted and *para*-substituted benzoyl chlorides from reference 44 were correlated with σ^+ substituent constants (Figure 2).

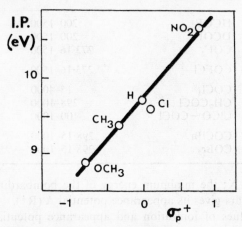

FIGURE 2. Correlation[44] of ionization potential for p-RC$_6$H$_4$COCl with $\sigma_p^+(R)$.

C. Dipole Moments and Polarity

I. Dipole moments

Electric dipole moments of a number of simple molecules were measured from the Stark effect on microwave spectra. For other compounds the usual dielectric constant determination, mostly in solution, was used. A sample of values is shown in Table 6.

From the sequence of dipole moment values for formaldehyde (2·34 D)[53], formyl fluoride (2·0 D) and carbonyl fluoride (0·95 D) it has been shown that even in COF$_2$ the moment is directed toward the carbonyl group[15].

For fluoroacetyl fluoride, the dipole moments of two rotamers have been obtained, that is of the most stable one (*trans* isomer) and of the *cis* isomer[23]. Figure 3 shows the two conformations and the direction of the respective moments.

A microwave study of acrylyl fluoride has shown that the molecule is planar, and the two isomers *s-cis* and *s-trans* have about the same energy and the same dipole moment[49]. The trend observed when comparing the dipole moments of formyl, acetyl and acrylyl fluoride (Table 6) shows the increasing polarizability of the H, CH$_3$ and CH=CH$_2$ groups.

TABLE 6. Dipole moments of acyl halides

Compound	μ (Debye)	Conditions	Reference
	From microwave spectroscopy		
HCOF	$\begin{cases} 2 \cdot 02 \pm 0 \cdot 02 \\ 1 \cdot 99 \pm 0 \cdot 03 \end{cases}$	gas gas	20 47
COF_2	$0 \cdot 951 \pm 0 \cdot 010$	gas	15
COClF	$1 \cdot 23 \pm 0 \cdot 01$	gas	48
CH_3COF	$2 \cdot 96 \pm 0 \cdot 03$	gas	16
FCH_2COF	$\begin{cases} 2 \cdot 67 \quad (trans) \\ 2 \cdot 05 \quad (cis) \end{cases}$	gas	23
$CH_2 = CHCOF$	$\begin{cases} 3 \cdot 26 \pm 0 \cdot 05 (s\text{-}trans) \\ 3 \cdot 21 \pm 0 \cdot 05 (s\text{-}cis) \end{cases}$	gas	49
	From dielectric constants[a]		
$COCl_2$	$1 \cdot 179 \pm 0 \cdot 015$	gas, 20–180°	51
CH_3COCl	$\begin{cases} 2 \cdot 71 \\ 2 \cdot 47 \end{cases}$	gas, 50–110° benzene, 20°	
CH_3COBr	$2 \cdot 45$	benzene, 20°	
$BrCH_2COCl$	$2 \cdot 14$	CCl_4, 25°	
$BrCH_2COBr$	$2 \cdot 07$	CCl_4, 25°	
$ClCH_2COCl$	$\begin{cases} 2 \cdot 20 \\ 2 \cdot 24 \\ 2 \cdot 17 \end{cases}$	gas, 85–255° benzene, 20° CCl_4, 25°	
$Cl_2CHCOCl$	$1 \cdot 58$	CCl_4, 25°	
Cl_3CCOCl	$\begin{cases} 1 \cdot 12 \\ 1 \cdot 20 \end{cases}$	CCl_4, 25° benzene, 20°	
CH_3CH_2COCl	$2 \cdot 63$	benzene, 20°	
$C_6H_5CH_2COCl$	$2 \cdot 56$	benzene, 20°	
ClCOCOCl	$0 \cdot 93$	benzene, 20°	
$ClCOCH_2COCl$	$2 \cdot 81$	benzene, 20°	
$ClCO(CH_2)_2COCl$	$3 \cdot 03$	benzene, 20°	
C_6H_5COCl	$3 \cdot 42$	benzene, 25°	52
C_6H_5COBr	$3 \cdot 40$	benzene, 20°	
$p\text{-}MeC_6H_4COCl$	$3 \cdot 84$	benzene, 20°	
$p\text{-}ClC_6H_4COCl$	$2 \cdot 02$	benzene, 20°	
$p\text{-}BrC_6H_4COCl$	$2 \cdot 05$	benzene, 20°	
$p\text{-}NO_2C_6H_4COCl$	$1 \cdot 12$	benzene, 20°	

[a] From reference 50 unless otherwise specified.

The dipole moment of formyl fluoride has been measured both for the ground and for the excited $n–\pi^*$ singlet states[54]. For the last state, $\mu \simeq 2 \cdot 60$ D. The fact that the dipole moment increases during the $n–\pi^*$ transition is surprising for a carbonyl compound.

The dependence of dipole moments on the nature of the halogen in the COX group is illustrated by μ-values for acetyl, bromoacetyl and benzoyl halides (Table 6): while values for X = Cl and Br are almost equal, there seems to be an increase on going to X = F. Benzoyl derivatives have larger

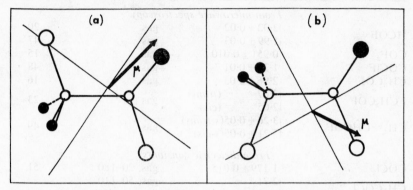

FIGURE 3. *Trans* (a) and *cis* (b) forms of CH_2FCOF. Principal axes and dipole moments directions are shown[23].

μ-values than the corresponding acetyl derivatives, due to the larger polarizability of the phenyl group. The influence of substituents on these dipole moments shows normal trends; for example, progressive α-substitution of hydrogen by chlorine in acetyl chloride causes a regular decrease in dipole moment.

While oxalyl chloride has been found to exist in a nearly *trans* form in the solid[3] the dipole moment found in benzene solution (0·93 D) is far from zero. A calculation based on the use of C=O and C—Cl bond moments, sp^2 hybridization at the carbon atoms and free rotation around the C—C bond gives a value coincident with the experimental one. This suggests also the possibility that the molecule is present in two forms, *cis* and *trans*, at least in benzene solution. The most recent results on i.r. and Raman spectra of oxalyl chloride in the liquid/gaseous states have been explained in terms of a mixture of *cis* and *trans* conformations, in which the *cis* isomer is present in 15–20% concentration[55].

2. Polarity

The dielectric constant of acetyl chloride is fairly high (15·9 at 20°)[56]; it possesses a moderate solvolysing power and intervenes in the formation of complexes between bases and Lewis acids. This has been interpreted[57] as evidence of a weak ionization, according to the equation:

$$CH_3COCl \rightleftharpoons CH_3CO^+ + Cl^-$$

Further support derives from measurements of electric conductivity[58]. Similar observations were made with benzoyl chloride[56, 59].

The behaviour of acetyl and benzoyl chlorides can be interpreted also in terms of their 'donicity': that is, the ability to make an electron pair available to form a co-ordinate bond[60].

IV. SPECTROSCOPIC PROPERTIES

A. Rotational Spectra

The most convenient technique for the study of rotational levels of molecules is microwave spectroscopy. Usually one obtains information on moments of inertia, on dipole moments and on nuclear quadrupole constants. Some of these topics are treated in other parts of this chapter.

When internal rotation is present in the molecule under study, in special circumstances the rotational energy barrier can be evaluated and, moreover, possible rotational isomers can be identified and their difference in energy established. A theory has been developed for cases in which the molecule can be treated as a rigid symmetric rotor attached to a rigid asymmetric frame. While the spectrum of a rigid molecule is made up of single lines, the theory predicts for the above molecules a series of doublets; the splitting in each doublet is a sensitive function of the height of the barrier hindering internal rotation[22].

Acetyl halides belong to this class of compounds, and have been studied from this viewpoint. Since the methyl group has a threefold axis, the hindering potential may be approximately written as

$$V(\alpha) = \tfrac{1}{2}V_3(1 - \cos 3\alpha) \tag{3}$$

where α is the rotation angle, measured from a minimum position, and V_3 is the barrier height. Values of V_3 for these and related compounds are shown in Table 7. The height of the barrier is of the order of 1 kcal/mole

TABLE 7. Internal rotation barriers for acetyl halides and related compounds

Compound	CH_3COF	CH_3COCl	CH_3COBr	CH_3COI
V_3 (cal/mole)	1041 ± 20	1296 ± 30	1305 ± 30	1301 ± 30
Reference	16	19	61	62

Compound	CH_3CHO	CH_3COCN	CF_3COF
V_3 (cal/mole)	1150 ± 30	1270 ± 30	1390 ± 210^a
Reference	22	63	64

^{a} From i.r. spectrum.

in all cases, and does not seem to be very sensitive either to the nature of X (= H, F, Cl, Br, I, CN) in the COX group or to substitution of hydrogen with fluorine atoms in the methyl group.

Replacement of one hydrogen by a fluorine atom in acetyl fluoride has a drastic effect on the shape of the potential function for internal rotation: this function (Figure 4) has two minima, differing by 910 ± 100 cal/mole[23]

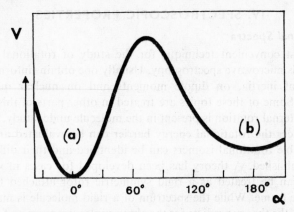

FIGURE 4. A reasonable form for the potential energy curve for internal rotation in CH_2FCOF[23]; (a) and (b) have the same meaning as in Figure 3.

and separated by a barrier higher than 4 kcal/mole when measured from the most stable isomer, where the two fluorine atoms are *trans* to one another. The two rotamers have been shown in Figure 3.

The microwave spectrum of COF_2, examined under high resolution, has allowed the determination of ^{19}F nuclear spin–nuclear spin and ^{19}F spin–rotation interaction constants[65]. The microwave spectra of two vibrationally excited states have also been identified[66].

The analysis of the i.r. spectra of HCOF and DCOF, recorded by means of a high-resolution spectrometer, has allowed the determination of rotational constants for these molecules[67].

B. Vibrational Spectra

I.r. and Raman spectra can be used for various purposes, the most prominent of which are the assignment of the fundamental frequencies to the normal vibrational modes of the molecules, the determination of the force constants in appropriate potential functions, the detection of rotational isomers with the evaluation of their difference in energy, and the determination of rotational barriers. When data for a series of analogous molecules are available, frequencies of fundamental vibrations, for instance the $C=O$ stretching, can be correlated with physico-chemical properties of the atoms or groups varying along the series, for instance the electronegativity of halogens in the series RCOX.

Carbonyl and formyl halides, being four-atom molecules, are well suited for both normal vibration assignments and force constant calculations. Planar COX_2 and $COXY$ molecules have five fundamental vibrations in the molecular plane and one out-of-plane. Molecules COX_2 have C_{2v} symmetry: three vibrations (ν_1, ν_2, ν_3) are totally symmetric, that is belong to species A_1, vibrations ν_4 and ν_5 are symmetric only with respect to the molecular plane and belong to species B_1; the out-of-plane vibration ν_6 belongs to species B_2, that is the vibration is antisymmetric with respect to the plane of the molecule. These normal vibrations[68] are shown in Figure 5. The in-plane vibrations can be described as follows: ν_1 is the in-phase stretching of the C—X bonds, ν_2 is the $C=O$ stretching, ν_3 is the X—C—X bending, ν_4 is the out-of-phase C—X stretching, ν_5 is the rocking of the XCX triangle (sometimes described as X—C—O bending).

Molecules COXY have C_s symmetry: the five in-plane vibrations belong to A' species, and the out-of-plane vibration to A''. The description is the same as for the previous case.

The fundamental frequencies and assignments for a number of such molecules are shown in Table 8.

TABLE 8. Fundamental frequencies (cm^{-1}) for simple acyl halides

Species	A_1			B_1		B_2	Reference
Vibration	ν_1	ν_2	ν_3	ν_4	ν_5	ν_6	
COF_2	965	1928	584	1249	626	774	69
$COCl_2$	567	1827	285	849	440	580	70, 38
$COBr_2$	425	1828	181	757	350	512	38

Species	A'					A''	Reference
Vibration	ν_1	ν_2	ν_3	ν_4	ν_5	ν_6	
HCOF	2976	1834	1344	1064	661		71
COClF	776	1868	501	1095	415	667	72
COBrF	721	1874	398	1068	335	620	73
COClBr	517	1828	240	806	372	547	38

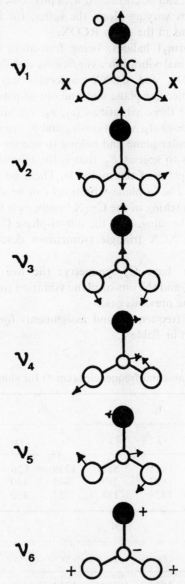

FIGURE 5. Normal vibrations for a molecule COX_2: ν_1–ν_5, in-plane vibrations; ν_6, out-of-plane vibration.

The vibrational energy for a N-atomic molecule can be expressed by means of a potential energy function, involving the internal displacement co-ordinates. Assuming that only quadratic terms need be considered, a general force field is obtained, of the form:

$$2V = \sum_{i,j=1}^{3N-6} f_{ij} S_i S_j \qquad (4)$$

where S_i and S_j are internal co-ordinates, and f_{ij}s are force constants. Owing to the fact that even for small symmetric molecules the number of independent force constants in (4) is large and greater than the number of observed frequencies, simplified force fields are commonly preferred. The most generally used is the Urey–Bradley force field[73a], in which the potential energy changes as a function of changes in bond lengths (Δr_i), bond angles ($\Delta \alpha_j$) and distances between non-bonded atoms (Δq_k):

$$2V = \sum_i [2K_i' r_{i0} \Delta r_i + K_i(\Delta r_i)^2]$$
$$+ \sum_j [2H_j' r_{j0}^2 \Delta \alpha_j + H_j(r_{j0} \Delta \alpha_j)^2]$$
$$+ \sum_k [2F_k' q_{k0} \Delta q_k + F_k(\Delta q_k)^2] \qquad (5)$$

In (5) r_{i0} and q_{k0} are equilibrium interatomic distances for bonded and non-bonded atoms, respectively, r_{j0} is the geometric mean of the equilibrium lengths of the bonds that define the bond angle α_j; K_i, H_j and F_k are the stretching, bending and repulsive force constants, respectively. Since Δr_i, $\Delta \alpha_j$ and Δq_k are not independent of one another, the presence of first-order terms in (5) is required, and the new force constants K_i', H_i' and F_k' come in. Equilibrium distances r_{i0}, r_{j0} and q_{k0} are introduced in order to make all the force constants dimensionally homogeneous. Removal of redundant co-ordinates, introduction of the equilibrium condition and further simplifications[74] allow reduction of the independent parameters to K_i, H_j and F_k force constants. These are used, in linear combinations, to build up the elements of the force constant matrix $\mathbf{F}$, for the computation of vibrational frequencies according to the method of the $\mathbf{GF}$ matrix[75]. In Table 9 Urey–Bradley force constant values for symmetric carbonyl halides are shown, limited to in-plane deformations. By transferring these values or interpolating between them [in the case of K(C—O), H(XCY) and F(X$\cdots$Y)], force constants for mixed carbonyl halides COXY were obtained, and used to calculate vibrational frequencies for COClF, COBrF and COBrCl[69]. Good agreement with observed frequencies was obtained, as appears from Table 10.

For HCOF force constants of in-plane deformations were calculated in two different approximations. In one case[71] normal co-ordinates were used

TABLE 9. Urey–Bradley force constants (md/Å) for in-plane deformations of symmetric carbonyl halides[69]

Molecule	Stretching		Bending		Repulsive	
	K(C—O)	K(C—X)	H(XCX)	H(OCX)	F(O···X)	F(X···X)
COF_2	12·85	4·528	0·450[a]	0·261	1·578	1·120
$COCl_2$	12·61	1·99	0·11[a]	0·145	0·860	0·523
$COBr_2$	12·83	1·565	0·060[a]	0·104	0·631	0·528

[a] Assumed value, transferred from the corresponding perhaloethylene.

TABLE 10. Calculated and observed in-plane vibrational frequencies (cm^{-1}) of COXY molecules

Assignment	COClF		COBrF		COClBr	
	calc.[69]	obs.[72]	calc.[69]	obs.[73]	calc.[69]	obs.[38]
ν_1	762	776	731	721	519	517
ν_2	1877	1868	1876	1874	1828	1828
ν_3	485	501	384	398	236	240
ν_4	1117	1095	1099	1068	806	806
ν_5	425	415	374	335	371	372

and only diagonal terms were retained in the potential function. In the second calculation[76] changes of bond lengths and bond angles were taken as variables, and all interaction terms were included. A summary of results is given in Table 11. Both sets of force constants reproduce the observed frequencies (Table 8) precisely.

TABLE 11. Force constants for planar vibrations of HCOF (md/Å)

Vibration	Reference 71	Reference 76
C—H stretch	4·80	4·78
C=O stretch	11·34	13·66
C—H bend	0·26	—
HCO bend	—	1·09
C—F stretch	4·76	5·43
FCO bend	1·43	2·13

The different force fields used as approximations of the general force field (4) have been criticized[77] because of the introduction of arbitrary assumptions, required by the insufficient amount of experimental data. To overcome this difficulty, it has been suggested to apply a minimum of very mild constraints to the interaction constants, and to calculate the range of acceptable solutions for the general force field problem. This method has been applied to $COCl_2$ and $COBr_2$, where five vibrational frequencies are available for the determination of nine force constants. The constraints were assumptions about some signs and reasonable ranges for the interaction constants. The calculated range of the $C=O$ stretching constant, for instance, is 12·5–13·9 md/Å for $COCl_2$ and 12·9–14·2 for $COBr_2$; the corresponding values for the $C-Cl$ and $C-Br$ stretching force constants are 2·94–3·27 and 2·27–2·63, respectively. It is gratifying that no overlap occurs between the last two ranges, indicating that $C-Cl$ bonds are tighter than $C-Br$ bonds, as expected.

The i.r. spectra of acetyl halides have been obtained, both in gas and in liquid phase, and frequencies have been assigned to the fundamental vibrations[78]. Vibrational assignments of prominent interest to the study of the COX group are reported in Table 12. Vibrations involving the halogen

TABLE 12. Vibrational frequencies (cm⁻¹) for the COX group in the i.r. spectra of acetyl halides[78] and of trifluoroacetyl chloride[79]

Vibration	Species	CH_3COX				CF_3COX
		X = F	X = Cl	X = Br	X = I	X = Cl
$C=O$ stretch	A'	1867	1820	1826	1808	1811
$C-C$ stretch	A'	1185	1106	1089	1075	937
$C-X$ stretch	A'	832	605	563	542	750
XCO bend	A'	598	435	335	285	511
XCC bend	A'	473	346	305	264	334
$C=O$ o.o.p. bend	A''	569	514	490	476	390
$C-C$ torsion	A''	208ᵃ	260ᵃ	—	—	—

ᵃ From Raman spectra.

atom show the most remarkable changes in frequency on going from X = F to X = I, as expected, while other vibrations are less affected. For purpose of comparison, data for trifluoroacetyl chloride[79] are also given in Table 12.

Vibrational spectra were used to investigate rotational isomerism in a number of α-substituted acetyl halides, as CH_2DCOCl[80], CH_2BrCOX[81],

CH_2FCOX^{82}, $CH_2ClCOX^{81, 83}$, $CHCl_2COCl^{84}$ and $Me_2CBrCOBr^{85}$. Two sets of bands are observed in such cases in the fluid states, while only one of them is usually present in the spectrum of the solid.

For chloroacetyl chloride electron diffraction[9] proved that the most stable form has the halogens *trans* to each other; the same conclusion was drawn for fluoroacetyl fluoride from microwave spectra (section II, C and Figures 3, 4). It is generally agreed that the vibrational spectrum that persists in the solid state of monohalogeno-acetyl halides pertains to the *trans* conformation (Figure 6, a), although a *gauche* conformation (Figure 6, b) has also been proposed[85]. The conformation of the other rotamer

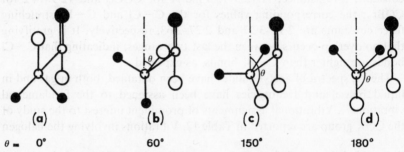

$\theta =$ $0°$ $60°$ $150°$ $180°$

FIGURE 6. Schematic representation of some possible conformations of mono-halogenoacetyl halides (θ = azimuthal angle).

has been studied by calculating vibrational frequencies by a Urey–Bradley force field at different values of the azimuthal angle θ (Figure 6) and comparing calculated and observed values. An additional criterion is the 'product rule'[81], which states that the ratio of the products of the normal frequencies of rotational isomers depends on the **G**-matrix values according to equation (6).

$$\frac{\Pi\nu \, (\theta \text{ isomer})}{\Pi\nu \, (0° \text{ isomer})} = \left(\frac{|\mathbf{G}(\theta)|}{|\mathbf{G}(0°)|}\right)^{\frac{1}{2}} \tag{6}$$

(The calculation of the **G**-matrix involves the geometry of the molecule and therefore depends on the azimuthal angle θ.) A *gauche* form with $\theta = 150°$ has been proposed for $CH_2ClCOCl$ and $CH_2BrCOCl^{81}$ (Figure 6, c), while a form close to *cis* is favoured by other authors in the case of CH_2ClCOX (X = F, Cl, Br)[83] (Figure 6, d).

By studying intensities of absorption at specific bands of the two rotamers as a function of temperature, and applying the van t'Hoff equation, the energy difference between rotational isomers was evaluated in some cases[81, 82, 83]. Results are (in kcal/mole): $CH_2BrCOCl$, $1·0 \pm 0·1$;

$CH_2BrCOBr$, $1\cdot9\pm0\cdot3$; CH_2FCOCl, $1\cdot0\pm0\cdot1$; CH_2ClCOF, $0\cdot8\pm0\cdot2$; $CH_2ClCOCl$, $1\cdot0\pm0\cdot2$; $CH_2ClCOBr$, $1\cdot9\pm0\cdot3$.

Dichloroacetyl chloride appears to be present in fluid states in two forms of different polarity, as evaluated from the effect of solvents on the intensity of one group of bands with respect to the other[84]. Dipole moments were calculated by the vector addition of bond moments, normal vibrations were calculated on the basis of the Urey–Bradley potential function, and the 'product rule' was also applied. The less polar rotamer was so identified as the form with the hydrogen of the $CHCl_2$ group eclipsing the oxygen atom. For the more polar rotamer, the conformation with the $CHCl_2$ group rotated by an angle of 90° is favoured. Only this isomer persists in the solid state. On the other hand, in the gas phase the less polar form is slightly more stable (by ca. $0\cdot2$ kcal/mole) than the more polar form[84].

A less polar and a more polar form are also described in the case of 2-bromo-2-methyl-propionyl bromide, and two *gauche* conformations are suggested, having azimuthal angles θ of 60° (more polar) and 120° (less polar)[85].

I.r. spectra are reported of a series of saturated acyl halides[86]. The presence of a number of C—Cl or C—Br stretching vibrations proves that conformational differences, not only at the α- but also at the β-carbon atom, affect the carbon–halogen stretching frequency by intramolecular non-bonded interaction.

The study of the vibrational spectra of oxalyl halides has led to controversial interpretations. While the most recent discussion of oxalyl chloride spectra is in favour of the existence of both a *cis* and a *trans* form in fluid states[55], no indication of a *cis* conformation was found for oxalyl fluoride and oxalyl fluoride chloride[87].

The vibrational spectra of acrylyl chloride[88] have been discussed in terms of an equilibrium between *trans* and *cis* rotamers in the fluid states. From the temperature dependence of the absorbances of the two C—Cl stretching vibrations in the vapour spectrum, an energy difference of ca. $0\cdot6$ kcal/mole between the rotational isomers was calculated, the *trans* form being the more stable. Also i.r. spectra of methacrylyl, crotonyl and fumaryl chlorides are reported and discussed[88].

The i.r. spectra of the four benzoyl halides in the liquid state have been recorded and the fundamental vibrations assigned[89]. Frequencies assigned to vibrations of the COX group are collected in Table 13, where the stretching frequencies depend on the atomic weight of the halogen in the usual way.

I.r. spectra of substituted benzoyl chlorides and benzoyl bromides have been reported[90].

TABLE 13. Typical vibrational frequencies (cm^{-1}) for the COX group in the i.r. spectra of benzoyl halides[89]

Vibration	X = F	X = Cl	X = Br	X = I
C=O stretch	1802	1771	1765	1753
C—C stretch	1260	1209	1199	1198
C—X stretch	1245	873	675	603
XCO bend	648	—	—	—

Relations have been found between the electronegativity of halogens and the stretching frequency of the carbonyl group in acyl halides[89, 91]. An analogous correlation has been suggested for the out-of-plane C=O bending frequency[92]. Force constants for carbonyl out-of-plane deformation in a series of compounds of the type COXY were found to be linearly related (Figure 7) to Taft's inductive and resonance substituent constants, σ_I and σ_R respectively, according to the equation[93]:

$$k = 0.233 + 0.199 \sum_{X,Y} \sigma_I - 0.268 \sum_{X,Y} \sigma_R \qquad (7)$$

It appears that k is increased by electron withdrawal due to the inductive effect of groups X and Y, as well as by electron release from the same groups into the π-system.

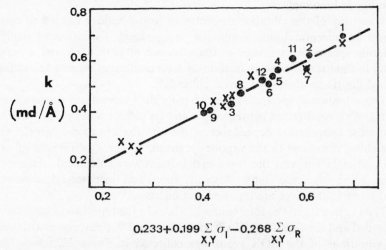

FIGURE 7. Correlation between o.o.p. bending force constants k for COXY molecules and Taft's parameters[93]. Full circles (No., X, Y): 1, F, F; 2, F, Cl; 3, F, D; 4, Cl, Cl; 5, Cl, Br; 6, Br, Br; 7, F, Br; 8, F, CH$_3$; 9, Cl, CH$_3$; 10, Br, CH$_3$; 11, Cl, OCH$_3$; 12, Cl, SCH$_3$. Crosses: X and Y other than halogens. The straight line represents equation (7).

C. Electronic Spectra

The near u.v. spectrum of formyl fluoride presents an absorption band with origin at 267 nm ($\varepsilon = 0{\cdot}01 \, 1 \, cm^{-1} \, mole^{-1}$) and $\lambda_{max} = 210$–220 nm ($\varepsilon = 50$)[94, 95, 96]. The band is assigned to a $n \rightarrow \pi^*$ singlet–singlet electronic transition. Its detailed analysis allows to specify that the transition is from a planar ground state to a pyramidal excited state, characterized by a longer CO bond distance, a smaller FCO bond angle and by an out-of-plane angle between the CO axis and the HCF plane in the range 30–35°.

The electronic spectrum of carbonyl fluoride has been observed down to 121·5 nm. Four transitions occur in this spectral region, and have been interpreted, in order of increasing energy, as two $n \rightarrow \pi^*$, one $n \rightarrow \sigma^*$ and one $\pi \rightarrow \pi^*$ type[41].

The electronic spectrum of phosgene has been recorded over the regions 315–210 nm[97] and 200–60 nm[42]. In the former region an $n_O \rightarrow \pi^*$ transition was identified, with origin near 299 nm. A high-dispersion study of this band has led to the conclusion that in the corresponding excited state the molecule has undergone a deformation similar to that of HCOF. In the far-u.v. region six electronic transitions were identified and tentatively assigned as follows, in order of increasing energy: $n_O \rightarrow \sigma^*$, $n_{Cl} \rightarrow \sigma^*$, $\pi \rightarrow \pi^*$, $n_O \rightarrow R_O$, $n_{Cl} \rightarrow R'_{Cl}$, $n_{Cl} \rightarrow R''_{Cl}$, where R_O and R_{Cl} orbitals are atomic-like (Rydberg) orbitals.

The near u.v. spectrum of carbonyl fluoride chloride presents an $n \rightarrow \pi^*$ absorption band with origin near 259 nm and $\lambda_{max} = 203$ nm ($\varepsilon = 35 \, 1 \, cm^{-1} \, mole^{-1}$)[98].

The $n \rightarrow \pi^*$ absorption band of acetyl derivatives presents its maximum intensity at the following wavelengths: CH_3COF, 205 nm ($\varepsilon = 37$; in heptane)[99]; CH_3COCl, 235 nm ($\varepsilon = 53$; in hexane)[100], 241 nm ($\varepsilon = 38$; in heptane)[99]; CH_3COBr, 250 nm ($\varepsilon = 94$; in heptane)[99]; CF_3COCl, 253 nm ($\varepsilon = 17$; gas)[79]. For CF_3COF, the spectral region from 250 to 115 nm was explored[101]; the recorded spectrum is shown in Figure 8. The four bands

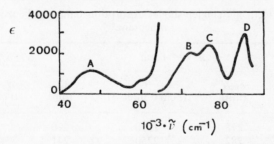

FIGURE 8. Electronic spectrum of trifluoroacetyl fluoride in the gas phase[101]. The curve containing band A represents $\varepsilon \times 20$.

labelled A, B, C and D were assigned to $n \to \pi^*$, Rydberg, $\pi \to \pi^*$ and Rydberg transitions, respectively.

The near u.v. spectra of some oxalyl halides have been observed[102]. For each molecule, two electronic transitions were identified, one singlet–singlet and the other singlet–triplet, both resulting from $n \to \pi^*$ electron promotion. In the case of oxalyl chloride the vibrational structure of both bands has been carefully analysed. The observed spectra are consistent with transitions between planar, *trans* electronic states[103, 104].

The origins of the $n \to \pi^*$ absorption bands for some simple acyl halides are collected in Table 14. Within homogeneous series of compounds a

TABLE 14. Lowest energy $n \to \pi^*$ singlet–singlet absorption bands in simple acyl halides

Compound	Electronic origin of the band (cm^{-1})	Reference
HCOF	37,488	96
COF$_2$	42,084	41
COFCl	38,658	98
COCl$_2$	33,500	97
FCO—COF	32,445	102
FCO—COCl	28,724	102
ClCO—COCl	27,189	103
BrCO—COBr	25,371	102

dependence of the frequency on the electronegativity of α-atoms is apparent[98, 99, 105] and is regarded as mainly due to stabilization of the ground state n-electrons by electronegative adjacent atoms. A linear correlation is shown in Figure 9 for derivatives of the type COXY.

The u.v. absorption maxima of some benzoyl halides are reported in Table 15. The absorption spectrum in the singlet–triplet region and the

TABLE 15. Electronic singlet–singlet absorption bands of benzoyl halides in n-heptane[99]

Compound	$n \to \pi^*$ band		$\pi \to \pi^*$ band	
	λ_{max} (nm)	ε (l cm^{-1} mole^{-1})	λ_{max} (nm)	ε (l cm^{-1} mole^{-1})
PhCOF	278	1910	230	18,400
PhCOCl	282	2250	241	27,200
PhCOBr	282	1520	245	16,500

phosphorescence emission spectrum of benzoyl bromide have been measured in ether–toluene glass at 77 K[107]; the transition has been assigned as a $\pi \rightarrow \pi^*$ type.

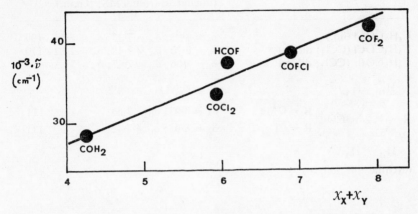

FIGURE 9. Correlation between wavenumber at the origin of the $n \rightarrow \pi^*$ absorption band for COXY molecules and the Gordy[106] electronegativity sum for X and Y.

D. N.m.r. Spectra

The COX group contributes to nuclear magnetic spectra through the nuclei of ^{13}C (nuclear spin $I = \frac{1}{2}$), of ^{17}O ($I = \frac{5}{2}$) and of the halogen atoms. Among the latter nuclei, the most commonly investigated is ^{19}F ($I = \frac{1}{2}$) while little is known about the other halogen atoms[108].

Furthermore, COX substituents influence the chemical shifts of other active nuclei in a molecule. Some data of this kind are presented in Table 16 for the active nuclei ^{1}H and ^{19}F. Recalling that the proton chemical shifts downfield from TMS are 7·26 p.p.m. in benzene, 7·40 (δ_α) and 6·30 p.p.m. (δ_β) in furan[112], and that the ^{19}F chemical shift upfield from $CFCl_3$ is 163·0 p.p.m. in C_6F_6[116], it appears from the table that COX groups are electron-withdrawing and act mainly on *ortho* and *para* positions. Comparison with other electronegative substituents reveals that the effect of COX groups is one of the strongest[110].

Table 17 shows that in the COF group the ^{19}F chemical shift is to low field from $CFCl_3$, indicating the strong electron-withdrawing character of the carbonyl. A few ^{19}F chemical shifts in acyl halides are shown in Figure 10 together with some reference compounds.

Within the COF group the spin–spin coupling constants J_{CF} between ^{13}C and ^{19}F have been determined in a few cases. Values given in Table

Massimo Simonetta and Paolo Beltrame

TABLE 16. Effect of COX groups on ^1H and ^{19}F magnetic resonance spectra

Compound (solvent)	Chemical shifts in ^1H spectra (downfield from TMS; p.p.m.)	Ref.
C_6H_5COCl (liq.)	$\delta_o = 7\cdot93$; $\delta_m = 7\cdot35$; $\delta_p = 7\cdot54$	109
C_6H_5COCl (CCl_4; dil. ∞)	$\delta_o = 8\cdot10$; $\delta_m = 7\cdot48$; $\delta_p = 7\cdot62$	110
C_6H_5COBr (CCl_4; dil. ∞)	$\delta_o = 8\cdot06$; $\delta_m = 7\cdot47$; $\delta_p = 7\cdot63$	110

| | 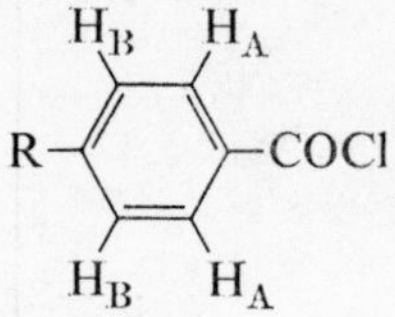 R = OMe | $\delta_A = 8\cdot07$; $\delta_B = 7\cdot63$ | 111 |
| | R = Cl | $\delta_A = 8\cdot03$; $\delta_B = 7\cdot42$ | 111 |

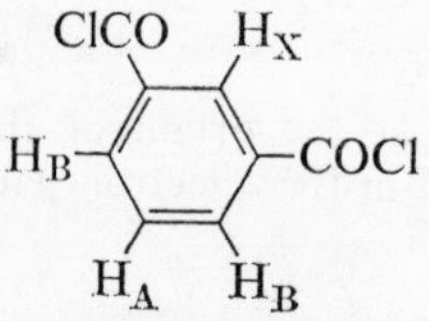

(cyclohexane; dil. ∞)

$\delta_X = 8\cdot90$; $\delta_A = 7\cdot57$; $\delta_B = 8\cdot35$ 111

(cyclohexane; dil. ∞)

| | 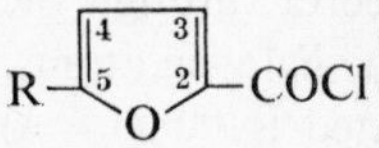 R = H (CCl_4) | $\delta_3 = 7\cdot50$; $\delta_4 = 6\cdot69$; $\delta_5 = 7\cdot86$ | 112 |
| | R = Me (CCl_4; dil. ∞) | $\delta_3 = 7\cdot31$; $\delta_4 = 6\cdot21$ | 113 |

| Maleyl fluoride ($CFCl_3$) | $\delta = 6\cdot73$ | 114 |
| Fumaryl fluoride ($CFCl_3$) | $\delta = 6\cdot94$ | 114 |

	Chemical shifts in ^{19}F spectra (p.p.m.)	
m-FC_6H_4COF (several solvents)	$\delta_{\text{ring-F}} = 2\cdot15 \pm 0\cdot08^a$	115
p-FC_6H_4COF (cyclohexane)	$\delta_{\text{ring-F}} = 11\cdot15^a$	115
p-FC_6H_4COCl (cyclohexane)	$\delta_{\text{ring-F}} = 11\cdot20^a$	115
C_6F_5COCl (benzene)	$\delta_o = 130\cdot8^b$; $\delta_m = 160\cdot0^b$; $\delta_p = 146\cdot1^b$	116

^{a} Shift to low field from C_6H_5F.
^{b} Shift to high field from $CFCl_3$; for data in other solvents see references 117 and 118.

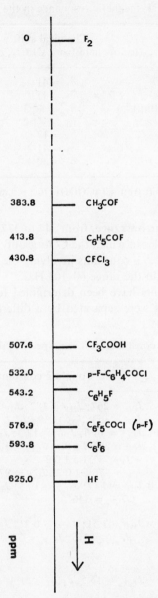

FIGURE 10. ^{19}F Chemical shifts for reference and typical compounds, upfield from F_2 (reference 122, besides those already given).

TABLE 17. ^{19}F magnetic resonance in the COF group

Compound (solvent)	Chemical shift (to low field from $CFCl_3$; p.p.m.)[a]	Reference
HCOF (liq.)	$\delta_F = 41$	119
CH_3COF (liq.)	$\delta_F = 47$	119
$CF_2Br-CHBr-COF_y$ (liq.)	$\delta_{F_y} = 40\cdot3$	120
$CF_2=CF-COF_y$ (liq.)	$\delta_{F_y} = 21\cdot2$	121
C_6H_5COF (liq.)	$\delta_F = 17$	119
Maleyl fluoride ($CFCl_3$)	$\delta_F = 41\cdot2$	114
Fumaryl fluoride ($CFCl_3$)	$\delta_F = 30\cdot4$	114

[a] Shift from $CFCl_3$ = shift from $CF_3COOH - 76\cdot8$ p.p.m[120].

18 are in a relatively narrow range, from 344 to 377 Hz. The lowest values were found for molecules in which the COF group is linked to an olefinic or an aromatic system. J_{CF} constants involving the carbon atom adjacent to the COF group fall in the range 60–103 Hz.

J_{FF} coupling constants have been determined for molecules in which the two fluorine atoms were separated by a different number of bonds,

TABLE 18. Coupling constants in n.m.r. spectra (absolute values; Hz)

Compound	Coupling constants	Reference
HCOF	$J_{CH} = 267; J_{HF} = 182; J_{CF} = 369$	123, 119
C_yH_3COF	$J_{CF} = 353; J_{OF} = 39 \pm 6; J_{C_yF} = 59\cdot7$	119, 124
$CF_xF_yBr-CHBr-COF$	$J_{FF_x} = 14\cdot6; J_{FF_y} = 12\cdot3; J_{HF} = 3\cdot1$	120
F_yC_yO-COF	$J_{CF} = 365\cdot9 \pm 0\cdot4; J_{CF_y} = 103\cdot2 \pm 0\cdot4;$ $J_{FF_y} = 51\cdot5 \pm 0\cdot2$	125
ClC_yO-COF	$J_{CF} = 376\cdot5 \pm 0\cdot2; J_{C_yF} = 97\cdot0 \pm 0\cdot2$	125
$H_y-C-C_yOF_y$ $\parallel$ $H-C-COF$	$J_{HF} = 4\cdot17; J_{H_yF} = 1\cdot83; J_{FF_y} = 4\cdot78;$ $J_{C_yF_y} = 346$	114
$F_yOC_y-C_x-H_y$ $\parallel$ $H-C-COF$	$J_{HF} = 7\cdot51; J_{H_yF} = 0\cdot21; J_{FF_y} = 0\cdot22;$ $J_{C_yF_y} = 345; J_{C_xF_y} = 72$	114
$\begin{matrix} F_a \\ \diagdown \\ F_b \diagup \end{matrix} C=C \begin{matrix} F_x \\ \diagdown \\ COF_y \end{matrix}$	$J_{F_aF_y} = 15\cdot5; J_{F_bF_y} = 51\cdot4; J_{F_xF_y} = 32\cdot0$	121
⬡$-C_yOF$	$J_{C_yF} = 344; J_{C_xF} = 61\cdot0$	119

from 3 to 5 (Table 18). The amount of coupling is not solely determined by the 'through-bond' distance, indicating the presence of a 'through-space' coupling. The lowest value in the table ($J_{FF} = 0.22$ Hz for fumaryl fluoride) occurs when both the distance in space and the number of intervening bonds are large.

The chemical shifts of ^{13}C in acetyl halides CH_3COX are as follows (downfield with respect to benzene): 41.1 p.p.m. (X = Cl), 40.9 (X = Br), 29.3 (X = I)[126]. These values have been correlated with the π-bond polarity in the carbonyl group[127], within the theory of carbon chemical shifts[128].

The ^{17}O chemical shift and the J_{OF} coupling constant in acetyl fluoride have been determined[124]: $\delta = 374$ p.p.m. downfield from $H_2^{17}O$; $J_{OF} = 39 \pm 6$ Hz. For two series of aliphatic acyl halides the ^{17}O chemical shifts (on the same scale) are available[124, 129].

$$CH_3COX: \quad 374 \ (X = F); \ 507 \ (X = Cl); \ 536 \ (X = Br)$$
$$CH_3CH_2COX: \quad 373 \ (X = F); \ 498 \ (X = Cl); \ 526 \ (X = Br)$$

E. Mass Spectra

The available information on the mass spectra of acyl halides indicates a facile cleavage according to the following scheme:

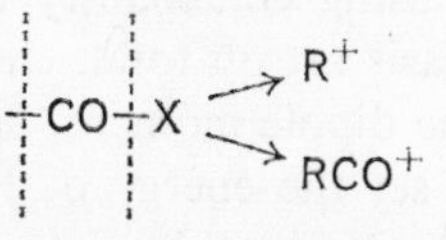

$$R \dashv CO \dashv X \begin{array}{c} \nearrow R^+ \\ \searrow RCO^+ \end{array}$$

Acylium ions appear as base peaks in the spectra of acetyl chloride ($m/e = 43$)[130] and of nitrobenzoyl chlorides ($m/e = 150$)[131]. On the other hand, loss of COX gives the next most abundant ion in the case of acetyl chloride ($m/e = 15$)[130], and the base peak in the positive ion mass spectrum of CF_3COF ($m/e = 69$)[132]. Loss of COX is also favoured when it gives rise to a benzyl- or tropylium-like cation, as is the case for $C_6F_5CH_2COCl$, that gives $C_7F_5H_2^+$ as the most abundant ion[133], and for $C_6H_5CHBrCOCl$, that gives an intense peak at $m/e = 169-171$[134]. In the last case, however, the base peak has been found at $m/e = 125-127$ (M − Br − CO) and it has been interpreted as due to loss of Br followed by migration of Cl towards the cationic centre, accompanied by loss of CO:

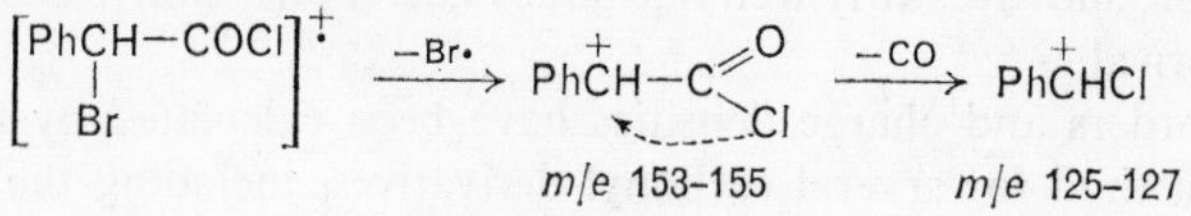

$$\begin{bmatrix} PhCH-COCl \\ | \\ Br \end{bmatrix}^{+} \xrightarrow{-Br\cdot} PhCH-C{\overset{\displaystyle O}{\underset{\displaystyle Cl}{\diagdown}}} \xrightarrow{-CO} Ph\overset{+}{C}HCl$$

$$m/e \ 153-155 \qquad\qquad m/e \ 125-127$$

It should be noted that the behaviour of acyl halides in the mass spectrometer, as summarized above, is not different from that of other carbonyl derivatives, of formula R—CO—X with X other than halogen[135].

V. THEORETICAL ASPECTS

As might be expected, the most simple acyl halide molecule, namely HCOF, has been the most thoroughly studied by theoretical methods. The Pople–Pariser–Parr method[136] has been used to justify the blue shift of the first $n \to \pi^*$ transition on going from formaldehyde to formyl fluoride[137]. The dipole moment of the same molecule has been calculated by three different semi-empirical molecular orbital methods: extended Hückel (EH)[138], iterative extended Hückel (IEH)[139], and CNDO/2 method[140]. The results are[141]: $\mu = 6.32$, 1.78 and 1.98 D, respectively, while the experimental value is about 2.0 D (Table 6). As for a number of other molecules, while the EH method overestimates the charge densities and hence the dipole moment, both IEH and CNDO/2 methods give a satisfactory result.

Several physical properties of HCOF in its ground state were calculated by a non-empirical quantum-mechanical method[142]. Molecular wave functions were obtained as linear combinations of atomic orbitals by the usual SCF procedure[143], using Gaussian-type functions[144]. The influence of the variation of the basis set on total, electronic, binding and orbital energies, as well as on the dipole moment, has been investigated. With a reasonably limited basis set the energy of HCOF was determined as a function of bond lengths, OCH angle and o.o.p. distortion. The experimental geometry was well reproduced, while stretching force constants were predicted one order of magnitude too large[142].

A more recent non-empirical SCF calculation is based on a set of Gaussian orbitals, in which two basis functions are used to represent each atomic orbital[101]. Each basis function is a linear combination of Gaussian functions. The results were used to predict ground-state properties, as well as the energy and intensity of $n \to \pi^*$ and $\pi \to \pi^*$ transitions, in HCOF, HCONH$_2$ and HCOOH. For formyl fluoride the calculated dipole moment is $\mu = 2.80$ D, the ionization potentials are 14.15 and 15.35 eV for n and π electrons, respectively, and the net charge on fluorine (-0.305) is more negative than on oxygen (-0.278). Spectral properties of the three mentioned molecules are fairly well reproduced, as far as relative energy values are concerned.

Bond orders and charge densities have been calculated by the simple Hückel method for several carbonyl derivatives, including the following acyl halides[145]: HCOF, COF_2, COFCl, $COCl_2$, $COBr_2$, CH_3COCl, CCl_3COCl, CF_3COX (X = F, Cl, Br, I). Both the CO bond order and the charge density on oxygen linearly correlate with the carbonyl-stretching frequency observed in the vapour state.

EH molecular orbital calculations were performed for a number of acetyl and benzoyl systems[99]. Energies of highest occupied and lowest vacant molecular orbitals were obtained and used to evaluate the shift of $n \rightarrow \pi^*$ and $\pi \rightarrow \pi^*$ bands in the electronic spectra with respect to the corresponding bands in acetone and acetophenone (Table 19). The

TABLE 19. Shifts in $n \rightarrow \pi^*$ and $\pi \rightarrow \pi^*$ transitions due to substitution of a methyl group by a halogen[99]

Compound	$\Delta E_{n \rightarrow \pi^*}$ (eV)		$\Delta E_{\pi \rightarrow \pi^*}$ (eV)	
	calc.	obs.	calc.	obs.
CH_3COCH_3[a]	0·0	0·0	0·0	0·0
CH_3COF	−1·72	−1·59	−0·69	—
CH_3COCl	−0·60	−0·70	−0·11	+0·20
CH_3COBr	−0·17	−0·51	+1·07	+0·30
$PhCOCH_3$[a]	0·0	0·0	0·0	0·0
PhCOF	−0·42	−0·54	−0·22	−0·13
PhCOCl	−0·10	−0·53	−0·13	+0·12
PhCOBr	−0·08	−0·52	+0·02	+0·20

[a] Reference compound.

calculated $n \rightarrow \pi^*$ band shifts are approximately linearly related to the electronegativity of the atom linked to the carbonyl carbon; observed shifts satisfy the same relationship only for acetyl derivatives.

A semi-empirical molecular orbital calculation, including the evaluation of spin–orbit interaction, has been performed[146] to study the triplet–singlet $\pi^* \rightarrow n$ transition in COX_2 molecules (X = hydrogen or halogens). It has been shown that, owing to the mixing of the n-orbital on oxygen with carbon and halogen p-orbitals, oscillator strength and phosphorescence lifetime depend on the atomic number of the halogen. Lifetimes ranging from $\approx 10^{-2}$ s (COF_2) to $\approx 10^{-5}$ s (COI_2) were obtained. A computation along similar lines was made for XCO—COX molecules (X = H, Cl or Br)[147]. The inclusion of heavy atoms is predicted to produce an increased intensity of the singlet–triplet absorption band, which was confirmed by experiment.

VI. REFERENCES

1. B. Zaslow, M. Atoji and W. N. Lipscomb, *Acta Cryst.*, **5**, 833 (1952).
2. P. Groth and O. Hassel, *Proc. Chem. Soc.*, 343 (1961).
3. P. Groth and O. Hassel, *Acta Chem. Scand*, **16**, 2311 (1962).
4. M. E. Jones, K. Hedberg and V. Schomaker, *J. Am. Chem. Soc.*, **77**, 5278 (1955).
5. T. T. Broun and R. L. Livingston, *J. Am. Chem. Soc.*, **74**, 6084 (1952).
6. P. W. Allen and L. E. Sutton, *Trans. Faraday Soc.*, **47**, 236 (1951).
7. V. Schomaker, D. P. Stevenson and J. E. Luvalle, unpublished results cited by V. Schomaker *et al.*, *J. Am. Chem. Soc.*, **72**, 4222 (1950).
8. Y. Morino, K. Kuchitsu, M. Iwasaki, K. Arakawa and A. Takahashi, *J. Chem. Soc. Japan, Pure Chem. Sect.*, **75**, 647 (1954); *Chem. Abstr.*, **48**, 8644 (1954).
9. Y. Morino, K. Kuchitsu and M. Sugiura, *J. Chem. Soc. Japan, Pure Chem. Sect.*, **75**, 721 (1954); *Chem. Abstr.*, **48**, 11128 (1954).
10. T. Ukaji, *Bull. Chem. Soc. Japan*, **30**, 737 (1957); *Chem. Abstr.*, **52**, 5906 (1958).
11. K. E. Hjortaas, *Acta Chem. Scand.*, **21**, 1379 (1967).
12. M. Simonetta and S. Carrá, *The Chemistry of Carboxylic Acids and Esters* (Ed. S. Patai), Wiley, New York, 1969, Ch. 1.
13. L. E. Sutton, Ed., 'Tables of Interatomic Distances, Supplement', *Chem. Soc. Special Publ.*, No. 18 (1965).
14. R. F. Miller and R. F. Curl, Jr., *J. Chem. Phys.*, **34**, 1847 (1961).
15. V. W. Laurie, D. T. Pence and R. H. Jackson, *J. Chem. Phys.*, **37**, 2995 (1962).
16. L. Pierce and L. C. Krisher, *J. Chem. Phys.*, **31**, 875 (1959).
17. A. M. Mirri, A. Guarnieri, P. Favero and G. Zuliani, *Nuovo Cimento*, **25**, 265 (1962).
18. G. W. Robinson, *J. Chem Phys.*, **21**, 1741 (1953).
19. K. M. Sinnott, *J. Chem. Phys.*, **34**, 851 (1961).
20. O. H. LeBlanc, Jr., V. W. Laurie and W. D. Gwinn, *J. Chem. Phys.*, **33**, 598 (1960).
21. G. Glockler, *J. Phys. Chem.*, **62**, 1049 (1958).
22. E. B. Wilson, Jr., C. C. Lin and D. R. Lide, Jr., *J. Chem. Phys.*, **23**, 136 (1955); R. W. Kilb, C. C. Lin and E. B. Wilson, Jr., *J. Chem. Phys.*, **26**, 1695 (1957).
23. E. Saegebarth and E. B. Wilson, Jr., *J. Chem. Phys.*, **46**, 3088 (1967).
24. D. R. Stull, E. F. Westrum, Jr. and G. C. Sinke, *The Chemical Thermodynamics of Organic Compounds*, Wiley, New York, 1969.
25. A. S. Carson, H. O. Pritchard and H. A. Skinner, *J. Chem. Soc.*, 656 (1950).
26. L. C. Walker and H. Prophet, *Trans. Faraday Soc.*, **63**, 879 (1967).
27. R. Walsh and S. W. Benson, *J. Phys. Chem.*, **70**, 3751 (1966).
28. P. Gray and A. Williams, *Chem. Rev.*, **59**, 239 (1959).
29. T. L. Cottrell, *The Strengths of Chemical Bonds*, 2nd edn., Butterworths, London, 1958.
30. M. Szwarc and J. W. Taylor, *J. Chem. Phys.*, **22**, 270 (1954).
31. M. Ladacki, C. Leigh and M. Szwarc, *Proc. Roy. Soc.*, **A214**, 273 (1952).
32. E. T. Butler and M. Polanyi, *Trans. Faraday Soc.*, **39**, 19 (1943).

33. K. Venkateswarlu, S. Jagatheesan and K. V. Rajalakshmi, *Proc. Indian Acad. Sci. Sect. A*, **58**, 373 (1963); *Chem. Abstr.*, **60**, 12785 (1964).
34. R. J. Lovell, C. V. Stephenson and E. A. Jones, *J. Chem. Phys.*, **22**, 1953 (1954).
35. J. S. Gordon and D. Goland, *J. Chem. Phys.*, **27**, 1223 (1957).
36. W. F. Giauque and J. B. Ott, *J. Am. Chem. Soc.*, **82**, 2689 (1960).
37. J. S. Ziomek, A. G. Meister, F. F. Cleveland and C. E. Decker, *J. Chem. Phys.*, **21**, 90 (1953).
38. J. Overend and J. C. Evans, *Trans. Faraday Soc.*, **55**, 1817 (1959).
39. E. L. Pace and M. A. Reno, *J. Chem. Phys.*, **48**, 1231 (1968).
40. V. I. Vedeneyev, L. V. Gurvich, V. N. Kondrat'yev, V. A. Medvedev and Ye. L. Frankevich, *Bond Energies, Ionization Potentials and Electron Affinities*, Arnold, London, 1966.
41. G. L. Workman and A. B. F. Duncan, *J. Chem. Phys.*, **52**, 3204 (1970).
42. S. R. La Paglia and A. B. F. Duncan, *J. Chem. Phys.*, **34**, 125 (1961).
43. J. R. Majer, C. R. Patrick and J. C. Robb, *Trans. Faraday Soc.*, **57**, 14 (1961).
44. A. Foffani, S. Pignataro, B. Cantone and F. Grasso, *Z. Physik. Chem. (Frankfurt)*, **42**, 221 (1964).
45. J. R. Majer and C. R. Patrick, *Trans. Faraday Soc.*, **59**, 1274 (1963).
46. R. I. Reed and M. B. Thornley, *Trans. Faraday Soc.*, **54**, 949 (1958).
47. P. Favero and J. G. Baker, *Nuovo Cimento*, **17**, 734 (1960).
48. A. Guarnieri and A. M. Mirri, *Atti Accad. Naz. Lincei, Rend.*, **40**, 837 (1966).
49. J. J. Keirns and R. F. Curl, Jr., *J. Chem. Phys.*, **48**, 3773 (1968).
50. A. L. McClellan, *Tables of Experimental Dipole Moments*, Freeman, San Francisco, 1963.
51. D. W. Davidson, *Can. J. Chem.*, **40**, 1721 (1962).
52. C. Pigenet, J.-P. Morizur, Y. Pascal and H. Lumbroso, *Bull. Soc. Chim. France*, 361 (1969).
53. J. N. Shoolery and A. H. Sharbaugh, *Phys. Rev.*, **82**, 95 (1951).
54. J. R. Lombardi, D. Campbell and W. Klemperer, *J. Chem. Phys.*, **46**, 3482 (1967).
55. J. R. Durig and S. E. Hannum, *J. Chem. Phys.*, **52**, 6089 (1970).
56. R. C. Paul and S. S. Sandhu, in *The Chemistry of Nonaqueous Solvents* (Ed. J. J. Lagowski), Vol. III, Academic, New York, 1970, p. 187.
57. R. C. Paul and S. S. Sandhu, *Proc. Chem. Soc.*, 262 (1957).
58. M. Maunaye and J. Lang, *Compt. Rend.*, **261**, 3381 (1965); *Chem. Abstr.*, **64**, 4337 (1966).
59. R. C. Paul and G. Singh, *Current Sci. (India)*, **26**, 391 (1957); *Chem. Abstr.*, **52**, 15203 (1958).
60. V. Gutmann, *Chem. Brit.*, **7**, 102 (1971).
61. L. C. Krisher, *J. Chem. Phys.*, **33**, 1237 (1960).
62. M. J. Moloney and L. C. Krisher, *J. Chem. Phys.*, **45**, 3277 (1966).
63. L. C. Krisher and E. B. Wilson, Jr., *J. Chem. Phys.*, **31**, 882 (1959).
64. K. R. Loos and R. C. Lord, *Spectrochim. Acta*, **A21**, 119 (1965).
65. M.-K. Lo, V. W. Weiss and W. H. Flygare, *J. Chem. Phys.*, **45**, 2439 (1966).
66. V. W. Laurie and D. T. Pence, *J. Mol. Spectry.*, **10**, 155 (1963).
67. R. F. Stratton and A. H. Nielsen, *J. Mol. Spectry.*, **4**, 373 (1960).

68. G. Herzberg, *Molecular Spectra and Molecular Structure. II. Infrared and Raman Spectra of Polyatomic Molecules*, Van Nostrand, New York, 1945.

69. J. Overend and J. R. Scherer, *J. Chem. Phys.*, **32**, 1289, 1296 (1960).

70. E. Catalano and K. S. Pitzer, *J. Am. Chem. Soc.*, **80**, 1054 (1958).

71. H. W. Morgan, P. A. Staats and J. H. Goldstein, *J. Chem. Phys.*, **25**, 337 (1956).

72. A. H. Nielsen, T. G. Burke, P. J. H. Woltz and E. A. Jones, *J. Chem. Phys.*, **20**, 596 (1952).

73. R. R. Patty and R. T. Lagemann, *Spectrochim. Acta*, **A15**, 60 (1959).

73a. H. C. Urey and C. A. Bradley, *Phys. Rev.*, **38**, 1969 (1931).

74. T. Shimanouchi, *J. Chem. Phys.*, **17**, 245 (1949); Y. Morino, K. Kuchitsu and T. Shimanouchi, *J. Chem. Phys.*, **20**, 726 (1952).

75. E. B. Wilson, Jr., *J. Chem. Phys.*, **7**, 1047 (1939); **9**, 76 (1941).

76. H. Johansen, *Z. Physik. Chem. (Leipzig)*, **227**, 305 (1964).

77. W. J. Lehmann, L. Beckmann and L. Gutjahr, *J. Chem. Phys.*, **44**, 1654 (1966).

78. J. A. Ramsey and J. A. Ladd, *J. Chem. Soc.* B, 118 (1968).

79. C. V. Berney, *Spectrochim. Acta*, **A20**, 1437 (1964).

80. J. Overend, R. A. Nyquist, J. C. Evans and W. J. Potts, *Spectrochim. Acta*, **A17**, 1205 (1961).

81. I. Nakagawa, I. Ichishima, K. Kuratani, T. Miyazawa, T. Shimanouchi and S. Mizushima, *J. Chem. Phys.*, **20**, 1720 (1952).

82. J. E. F. Jenkins and J. A. Ladd, *J. Chem. Soc.* B, 1237 (1968).

83. A. Y. Khan and N. Johathan, *J. Chem. Phys.*, **50**, 1801 (1969).

84. A. Miyake, I. Nakagawa, T. Miyazawa, I. Ichishima, T. Shimanouchi and S. Mizushima, *Spectrochim. Acta*, **A13**, 161 (1958).

85. G. A. Crowder and F. Northam, *J. Mol. Spectry.*, **26**, 98 (1968).

86. J. E. Katon and W. R. Feairheller, Jr., *J. Chem. Phys.*, **44**, 144 (1966).

87. J. L. Hencher and G. W. King, *J. Mol. Spectry.*, **16**, 158, 168 (1965).

88. J. E. Katon and W. R. Feairheller, Jr., *J. Chem. Phys.*, **47**, 1248 (1967).

89. P. Delorme, V. Lorenzelli and A. Alemagna, *J. Chim. Phys.*, **62**, 3 (1965).

90. C. N. R. Rao and R. Venkataraghavan, *Spectrochim. Acta*, **A18**, 273 (1962).

91. R. E. Kagarise, *J. Am. Chem. Soc.*, **77**, 1377 (1955).

92. K. Shimizu and H. Shingu, *Spectrochim. Acta*, **A22**, 1528 (1966).

93. J. C. Evans and J. Overend, *Spectrochim. Acta*, **A19**, 701 (1963).

94. A. Foffani, I. Zanon, G. Giacometti, U. Mazzucato, G. Favaro and G. Semerano, *Nuovo Cimento*, **16**, 861 (1960).

95. L. E. Giddings, Jr. and K. K. Innes, *J. Mol. Spectry.*, **6**, 528 (1961).

96. G. Fischer, *J. Mol. Spectry.*, **29**, 37 (1969).

97. L. E. Giddings, Jr. and K. K. Innes, *J. Mol. Spectry.*, **8**, 328 (1962).

98. I. Zanon, G. Giacometti and D. Picciol, *Spectrochim. Acta*, **A19**, 301 (1963).

99. K. Yates, S. L. Klemenko and I. G. Csizmadia, *Spectrochim. Acta*, **A25**, 765 (1969).

100. H. H. Jaffé and M. Orchin, *Theory and Applications of Ultraviolet Spectroscopy*, Wiley, New York, 1962, p. 180.

101. H. Basch, M. B. Robin and N. A. Kuebler, *J. Chem. Phys.*, **49**, 5007 (1968).

102. W. J. Balfour and G. W. King, *J. Mol. Spectry.*, **25**, 130 (1968).

103. W. J. Balfour and G. W. King, *J. Mol. Spectry.*, **26**, 384 (1968).

104. W. J. Balfour and G. W. King, *J. Mol. Spectry.*, **27**, 432 (1968).

105. P. Borrell, *Nature*, **184**, 1932 (1959).
106. W. Gordy, *J. Chem. Phys.*, **14**, 305 (1946).
107. R. F. Borkman and D. R. Kearns, *J. Chem. Phys.*, **46**, 2333 (1967).
108. C. Hall, *Quart. Rev.*, **25**, 87 (1971).
109. K. Hayamizu and O. Yamamoto, *J. Mol. Spectry.*, **25**, 422 (1968).
110. K. Hayamizu and O. Yamamoto, *J. Mol. Spectry.*, **28**, 89 (1968).
111. J. Martin and B. P. Dailey, *J. Chem. Phys.*, **37**, 2594 (1962).
112. Y. Pascal, J.-P. Morizur and J. Wiemann, *Bull. Soc. Chim. France*, 2211 (1965).
113. J.-P. Morizur, Y. Pascal and F. Vernier, *Bull. Soc. Chim. France*, 2296 (1966).
114. T. P. Vasileff and D. F. Koster, *J. Org. Chem.*, **35**, 2461 (1970).
115. R. W. Taft, Jr., E. Price, I. R. Fox, I. C. Lewis, K. K. Andersen and G. T. Davis, *J. Am. Chem. Soc.*, **85**, 709, 3146 (1963).
116. M. G. Hogben and W. A. G. Graham, *J. Am. Chem. Soc.*, **91**, 283 (1969).
117. R. Fields, J. Lee and D. J. Mowthorpe, *J. Chem. Soc. B*, 308 (1968).
118. M. I. Bruce, *J. Chem. Soc. A*, 1459 (1968).
119. N. Muller and D. T. Carr, *J. Phys. Chem.*, **67**, 112 (1963).
120. R. R. Dean and J. Lee, *Trans. Faraday Soc.*, **64**, 1409 (1968).
121. K. C. Ramey and W. S. Brey, Jr., *J. Chem. Phys.*, **40**, 2349 (1964).
122. J. A. Pople, W. G. Schneider and H. J. Bernstein, *High-resolution Nuclear Magnetic Resonance*, McGraw-Hill, New York, 1959, p. 91.
123. N. Muller, *J. Chem. Phys.*, **36**, 359 (1962).
124. J. Reuben and S. Brownstein, *J. Mol. Spectry.*, **23**, 96 (1967).
125. J. Bacon and R. J. Gillespie, *J. Chem. Phys.*, **38**, 781 (1963).
126. G. B. Savitsky, R. M. Pearson and K. Namikawa, *J. Phys. Chem.*, **69**, 1425 (1965).
127. G. E. Maciel, *J. Chem. Phys.*, **42**, 2746 (1965).
128. J. A. Pople, *Mol. Phys.*, **7**, 301 (1964).
129. H. A. Christ, P. Diehl, H. R. Schneider and H. Dahn, *Helv. Chim. Acta*, **44**, 865 (1961).
130. H. Gutbier and H. G. Plust, *Chem. Ber.*, **88**, 1777 (1955).
131. R. H. Shapiro and J. W. Serum, *Org. Mass Spectrom.*, **2**, 533 (1969).
132. E. M. Chait, W. B. Askew and C. B. Matthews, *Org. Mass Spectrom.*, **2**, 1135 (1969).
133. M. I. Bruce and M. A. Thomas, *Org. Mass Spectrom.*, **1**, 417 (1968).
134. R. G. Cooks, J. Ronayne and D. H. Williams, *J. Chem. Soc. C*, 2601 (1967).
135. H. Budzikiewicz, C. Djerassi and D. H. Williams, *Interpretation of Mass Spectra of Organic Compounds*, Holden-Day, San Francisco, 1964.
136. A. Pariser and R. G. Parr, *J. Chem. Phys.*, **21**, 466, 767 (1953); J. A. Pople, *Trans. Faraday Soc.*, **49**, 1375 (1953).
137. G. Giacometti, G. Rigatti and G. Semerano, *Nuovo Cimento*, **16**, 939 (1960).
138. R. Hoffmann, *J. Chem. Phys.*, **39**, 1397 (1963).
139. A. Pullman, E. Kochanski, M. Gilbert and A. Denis, *Theor. Chim. Acta*, **10**, 231 (1968).
140. J. A. Pople and G. A. Segal, *J. Chem. Phys.*, **44**, 3289 (1965).
141. J. E. Bloor, B. R. Gilson and F. P. Billingsley II, *Theor. Chim. Acta*, **12**, 360 (1968).

142. I. G. Csizmadia, M. C. Harrison and B. T. Sutcliffe, *Theor. Chim. Acta*, **6**, 217 (1966).
143. C. C. J. Roothaan, *Rev. Mod. Phys.*, **23**, 69 (1951).
144. C. M. Reeves, *J. Chem. Phys.*, **39**, 1 (1963).
145. S. Forsén, *Spectrochim. Acta*, **A18**, 595 (1962).
146. D. G. Carroll, L. G. Vanquickenborne and S. P. McGlynn, *J. Chem. Phys.*, **45**, 2777 (1966).
147. H. Shimada and Y. Kanda, *Bull. Chem. Soc. Japan*, **40**, 2742 (1967).

CHAPTER 2

Preparation of acyl halides

Martin F. Ansell

Queen Mary College, (University of London), England

I. INTRODUCTION

Acyl halides are most commonly prepared from the parent carboxylic acid, although other carboxylic acid derivatives may be used as precursors and the direct introduction of the halocarbonyl group is possible. The various methods described in the sequel are in principle all applicable to the preparation of fluorides, chlorides, bromides and iodides. However, the method employed for a particular class of compounds varies according to the availability of the reagent. The choice of method in an individual case

35

must also depend on the functional groups in the rest of the molecule. In this review the reactions are classified on the basis of the carbonyl halide precursors, and for each class of precursors the methods available for its conversion into the related acyl halide are discussed. This article concentrates on the more recent developments in this field, the literature up to 1950 being well summarized in Houben-Weyl[1]. The preparation of carbonyl halides has also been reviewed by Buehler and Pearson[2], and by Sonntag[3].

II. FROM CARBOXYLIC ACIDS AND ANHYDRIDES

A. With Thionyl Chloride

Almost certainly the most widely used method of preparing acyl halides is the reaction between carboxylic acids and thionyl chloride. Thionyl fluoride[4] and thionyl bromide (obtainable from thionyl chloride and potassium bromide[5]) undergo analogous reactions, but have not been extensively used.

Since acyl chlorides react with carboxylic acids to give acid anhydrides, then irrespective of their detailed mechanism the following reactions occur:

$$RCOCl + RCO_2H \longrightarrow (RCO)_2O + HCl$$
$$(RCO)_2O + SOCl_2 \longrightarrow 2 RCOCl + SO_2$$

Thus acid anhydrides may replace carboxylic acids in this and similar preparations of acid halides.

The earlier literature on the use of thionyl chloride has been reviewed[6] and some examples of its use are given in Table 1.

Although thionyl chloride will usually react with carboxylic acids in the absence of a catalyst the use of the latter is sometimes recommended. Small amounts of pyridine have been used as in the preparation[18] of the acid chloride of geranylacetic acid. However, in the preparation of crotonyl chloride the addition of pyridine results in extensive charring (neither dimethyl formamide nor sodium chloride (see further) is an effective catalyst) and the acid chloride is best obtained[19] from the uncatalysed reaction at 0°. However, pyridine is an essential catalyst[20] for the conversion of phthalic or succinic acids into the corresponding diacid chloride. In the absence of the catalyst the anhydride is obtained. Anhydrous zinc chloride[21] is an alternative catalyst for this reaction. Pyridine is also necessary[22] for the conversion of aromatic hydroxy acids into the corresponding acyl chlorides.

The use of N,N-dimethylformamide as a catalyst was introduced by Bosshard and co-workers[23], the catalytic effect being considered due to the formation of an intermediate imidoyl chloride (see section II, G).

TABLE 1. Preparation of carbonyl chlorides with thionyl chloride

Acid	Yield of chloride (%)	Reference
$CH_3(CH_2)_2CO_2H$	85	7
$MeO_2C(CH_2)_2CO_2H$	90–93	8
Me_2CHCO_2H	90	9
$Me_2\overset{+}{N}CH_2CO_2H\ Cl^-$	98–100	10
$C_{17}H_{35}CO_2H$	97–99	11
$CH_2(COOH)_2$	72–85	12
$PhN=NC_6H_4CO_2H$	89	13
$m\text{-}NO_2C_6H_4CO_2H$	90–98	14
$C_6H_5CH(OAc)CO_2H$	85	15
$CH_3(CH_2)_7CH=CH(CH_2)_7CO_2H^a$	97–98	16

OCH₂CO₂H structure: 85–87, Reference 17

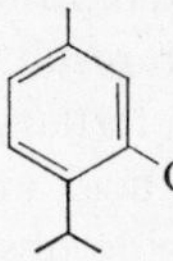 structure OCH$_2$CO$_2$H	85–87	17

^{a} Continuous process described.

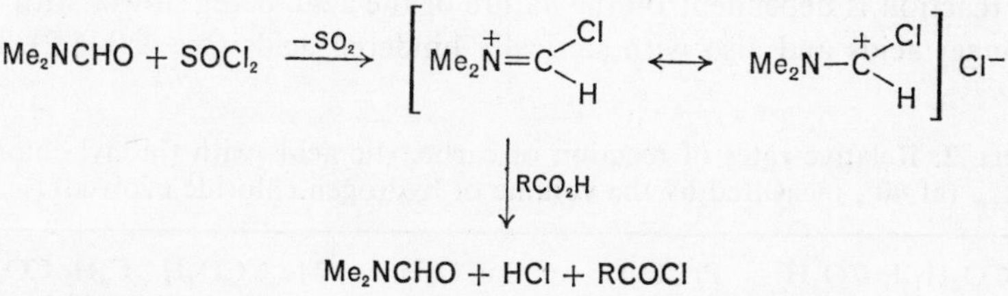

$$Me_2NCHO + SOCl_2 \xrightarrow{-SO_2} \left[Me_2\overset{+}{N}=C\overset{Cl}{\underset{H}{\diagdown}} \longleftrightarrow Me_2N-\overset{+}{C}\overset{Cl}{\underset{H}{\diagdown}} \right] Cl^-$$

$$\downarrow RCO_2H$$

$$Me_2NCHO + HCl + RCOCl$$

The effectiveness of dimethylformamide as a catalyst is illustrated by the following reaction:

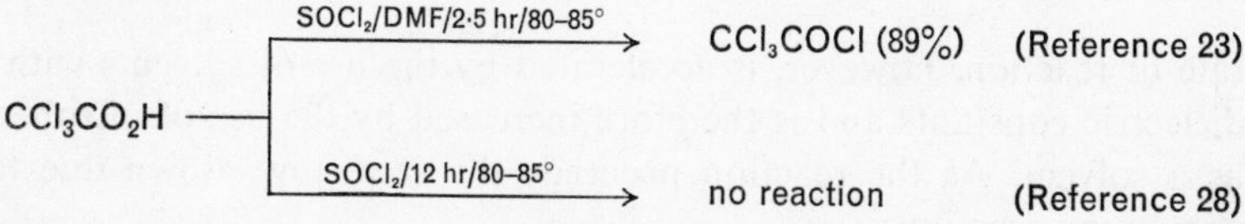

$$CCl_3CO_2H \begin{cases} \xrightarrow{SOCl_2/DMF/2\cdot5\ hr/80-85°} & CCl_3COCl\ (89\%)\quad \text{(Reference 23)} \\ \xrightarrow{SOCl_2/12\ hr/80-85°} & \text{no reaction}\quad\quad \text{(Reference 28)} \end{cases}$$

More recently the use of hexamethylphosphoramide and thionyl chloride has been recommended[24] as an effective reagent for converting acids to acid chlorides. It is said to be preferable to the pyridine-catalysed reaction, and particularly advantageous with α,β-unsaturated acids as the solvent traps the evolved hydrogen chloride and addition to the double bond does not occur, e.g.

$$CH_3CH{=}CHCO_2H \xrightarrow[\substack{HMPT \\ -20°}]{SOCl_2} CH_3CH{=}CHCOCl \quad (80\%)$$

However, as pointed out above[19], this reaction can be effected in the absence of any catalyst but the yield is only 56%.

Other catalysts which have been employed in these reactions are iodine, which is stated[25] to raise the yield of *p*-nitrobenzoyl chloride from 84 to 99%, and neutral chlorides[26]. The use of the latter is shown by their effect on the conversion of trichloroacetic acid into acid chloride. This reaction is slow and the acid is mainly unchanged after 12 h in boiling thionyl chloride. However, addition of an alkali metal chloride (e.g. 2 g KCl/mole $SOCl_2$) brought about a thirty-fold increase in the rate of the reaction and after 7 h the chloride was obtained in 87% yield. Similar results were obtained by the addition of quaternary ammonium salts or a tertiary base. The effects with weaker acids such as monochloroacetic acid are less dramatic, only a 2·3-fold increase in the rate of reaction being observed.

The kinetics of the reaction between thionyl chloride and carboxylic acids have been studied[27] and the reaction is shown to be second-order overall and first-order with respect to each reactant. The rate of the reaction is dependent on the nature of the acid, being slower with the stronger acids and also with sterically hindered acids (see Table 2). The

TABLE 2. Relative rates of reaction of carboxylic acids with thionyl chloride (at 40°, measured by the volume of hydrogen chloride evolved)

$MeCO_2H$	$EtCO_2H$	$PrCO_2H$	$BuCO_2H$	Me_3CCO_2H	$C_5H_{11}CO_2H$
1·0	0·67	0·69	0·67	0·30	0·71
$MeCO_2H$	ICH_2CO_2H	$BrCH_2CO_2H$	$ClCH_2CO_2H$	CCl_3CO_2H	
1·0	0·30	0·25	0·14	no reaction	

rate of reaction, however, is accelerated by the use of solvents with high dielectric constants and is therefore increased by the use of nitrobenzene as a solvent. As the reaction proceeds the rate slows down due to the competing reaction:

$$RCOCl + RCO_2H \longrightarrow (RCO)_2O + HCl$$

The authors[27] suggest that the initial reaction is the formation of the chlorosulphite, as suggested earlier by Gerrard and Thrush[28], which is then converted into the acyl chloride. The latter step may be considered as proceeding by the usual addition–elimination mechanism for nucleophilic substitution of acyl halides:

$$RCO_2H + SOCl_2 \longrightarrow RC\overset{\overset{O}{\|}}{}OSOCl + HCl$$

$$R\overset{\overset{O}{\|}}{C}\!-\!OSOCl \longrightarrow R\overset{\overset{O}{\|}}{C}\!-\!Cl + SO_2 + Cl^-$$

The catalytic effect of added neutral chlorides is not confined to Step 2, as it catalyses the reaction with trichloroacetic acid in which without catalyst Step 1 does not occur. As the catalytic effect is more pronounced with the stronger acids it is suggested[26] that the equilibrium involved is:

$$Cl_3CCO_2H + MCl \rightleftharpoons Cl_3CCO_2M + HCl$$

Many of the reactions of thionyl chloride are reported to proceed in high yield and often side reactions occur only with molecules containing acid-sensitive groups. However, the reaction of 4-nitro-2,5-dimethoxyphenyl-acetic acid with thionyl chloride in the presence of a small amount of a tertiary amine does not yield the acid chloride but the α-chloro-α-chloro-sulphenyl derivative (1) is obtained in 75% yield[29]. This might be attributed

(1)

to the presence of the highly reactive benzylic position. However, oxidation of the non-benzylic α-position occurs in the reaction of β-phenylpropionic acid[30] when it reacts with excess thionyl chloride at reflux temperature for 14 h, the main product being the sulphenyl halide and the minor product the acyl chloride. At higher temperature, a more complex mixture of products is obtained as is shown in Scheme I.

An attempt[31] to prepare acetylenecarbonyl chloride using thionyl chloride and dimethylformamide, pyridine or triethylamine led only to dichloromaleic anhydride, a product of *cis* chlorination.

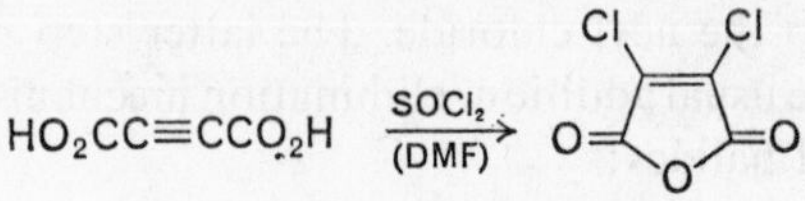

$$HO_2CC\equiv CCO_2H \xrightarrow[\text{(DMF)}]{SOCl_2} \text{(structure with Cl, Cl and O=, =O)}$$

The authors consider that the catalyst complexes with the thionyl chloride and the complex reacts with the triple bond possibly as illustrated:

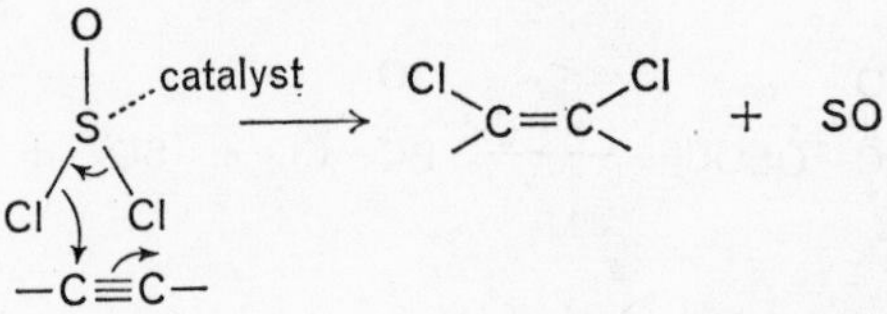

$$\text{(reaction scheme)} \quad \xrightarrow{\text{...catalyst}} \quad \begin{array}{c} Cl \\ \diagdown \\ C=C \\ \diagup \\ \end{array}\begin{array}{c} Cl \\ \diagup \\ \diagdown \end{array} \quad + \quad SO$$

$$2\,SO \longrightarrow SO_2 + S$$

This suggestion also accounts for the formation of sulphur.

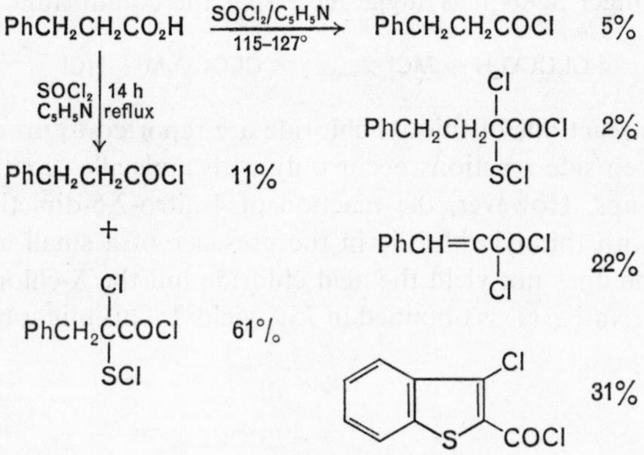

$$PhCH_2CH_2CO_2H \xrightarrow[115-127°]{SOCl_2/C_5H_5N} PhCH_2CH_2COCl \quad 5\%$$

$$\downarrow \begin{array}{c} SOCl_2 \\ C_5H_5N \end{array} \begin{array}{c} 14\,h \\ reflux \end{array}$$

$$PhCH_2CH_2\underset{\underset{SCl}{|}}{\overset{\overset{Cl}{|}}{C}}COCl \quad 2\%$$

$$PhCH_2CH_2COCl \quad 11\%$$

$$+$$

$$PhCH=\underset{\underset{Cl}{|}}{\overset{\overset{COCl}{}}{C}} \quad 22\%$$

$$PhCH_2\underset{\underset{SCl}{|}}{\overset{\overset{Cl}{|}}{C}}COCl \quad 61\%$$

$$\text{(benzothiophene structure with Cl, S, COCl)} \quad 31\%$$

Scheme 1

Although thionyl chloride is a valuable reagent for the preparation of acid chlorides, it does appear that the reaction conditions are often critical. From the reactions reported it is not possible to draw any generalization as to the best conditions. It is possible that purity of the thionyl chloride is an important factor (see section II, D). Thionyl chloride may be readily purified by careful distillation from triphenyl phosphite[32].

B. With Phosphorus Halides

The phosphorus chlorides PCl_3, PCl_5 and $POCl_3$ have been used to prepare acyl chlorides from carboxylic acids. Phosphorus bromides can be used similarly, but the related iodides and fluorides do not appear to have been utilized in similar reactions.

Phosphorus trichloride has been used extensively under a variety of conditions and a detailed investigation of the reaction[33] shows that 25–100% excess of the reagent is required, based on the equation:

$$3\,RCO_2H + PCl_3 \longrightarrow RCOCl + H_3PO_3$$

This simple equation does not account for the fact that all such reactions evolve hydrogen chloride. An explanation put forward by Cade and Gerrard[34] is that the reaction proceeds by the stepwise hydroxylation of phosphorus trichloride producing the intermediates $HPOCl_2$ and $HP(OH)Cl$ which may react with other molecules of carboxylic acid or degrade to hydrogen chloride and condensed phosphorus acids. The yield of the reaction is very dependent on the conditions of the reaction. Thus the conversion of oleic acid into its acid chloride is reported[35] to occur in 42% yield using 30% excess reagent, whereas under milder conditions and with only a 20% excess reagent an 85% yield of non-isomerized product may be obtained[33]. In some reactions phosphorus trichloride may be the preferred reagent. Thus[36] with 2-ethylmercaptopropionic acid it yields the colourless acid chloride in 64% yield whereas thionyl chloride gives a coloured product. Its use, in about 60% excess, for the conversion of phenylacetic acid into phenylacetyl chloride has been reported[37].

Phosphorus pentachloride has been used extensively for the preparation of acid chlorides. Only one of the five chlorines is normally utilized, one being lost as hydrogen chloride and the other three being retained as the relatively unreactive phosphorus oxychloride. Examples of its use are given below:

Reference 38 $\qquad$ $NCCH_2CO_2H \xrightarrow[\text{benzene}]{PCl_5} NCCH_2COCl$ $\qquad$ > 70%

Reference 39 $\quad AcOCH_2(CHOAc)_4CO_2H \xrightarrow[\text{ether}]{PCl_5} AcOCH_2(CHOAc)_4COCl$ $\quad$ 80–92%

Reference 40 $\qquad CH_2{=}CCO_2H \xrightarrow{2\ PCl_5} CH_2{=}CCOCl$ $\qquad$ 60–66%
$\qquad\qquad\qquad$ | $\qquad\qquad\qquad\qquad$ |
$\qquad\qquad\quad CH_2CO_2H \qquad\qquad\qquad CH_2COCl$

Reference 41 $\xrightarrow{PCl_5}$ $\xrightarrow[\text{(2) 3 H}_2\text{O}]{\text{(1) AlCl}_3}$

Reference 42 $\quad CH_3CHCHCO_2H \xrightarrow[\text{4 h below 20}^\circ]{PCl_5/CH_3COCl} (CH_3)_2CHCHCOCl$ $\quad$ 80%
$\qquad\qquad\qquad$ | $\qquad\qquad\qquad\qquad\qquad\qquad$ |
$\qquad\qquad\quad NH_2 \qquad\qquad\qquad\qquad\qquad\qquad NH_2$

High-molecular weight acyl chlorides may be prepared[43] by the action of phosphorus pentachloride on a solution of the acid dissolved in an inert solvent such as benzene, petroleum or carbon tetrachloride. The acidic by-products of the reaction can be removed, without substantial hydrolysis of the acid chloride, by washing the reaction mixture with water.

An attempt[44] to prepare acetylenedicarbonyl chloride from acetylene dicarboxylic acid and phosphorus pentachloride led only to chlorofumaryl chloride; the required acetylene dicarbonyl chloride was only obtainable from the acid chloride of the adduct of anthracene and acetylene dicarboxylic acid by displacement with maleic anhydride as shown in Scheme 2.

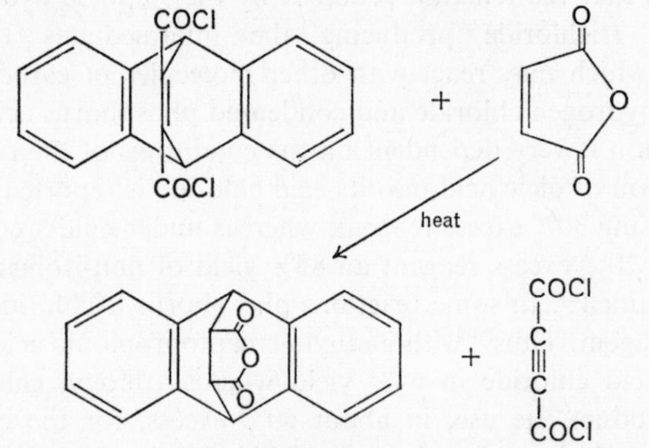

Scheme 2

For some reactions phosphorus pentachloride is the preferred reagent. Thus N-acetylacridan-9-carboxylic acid is smoothly converted into its acid chloride with phosphorus pentachloride, whereas thionyl chloride gives a difficultly separable mixture of products[45]. It is possible that with

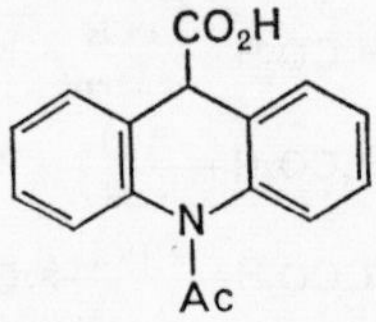

thionyl chloride, reaction occurs at the 9-position as discussed in section II, A.

Phosphorus oxychloride has not been extensively used in the preparation of acyl chlorides as it is not very reactive towards carboxylic acids. It has been suggested[46] that amines or amides might be effective catalysts for such reactions. One example of its use is[47] the conversion of the pyrimidine

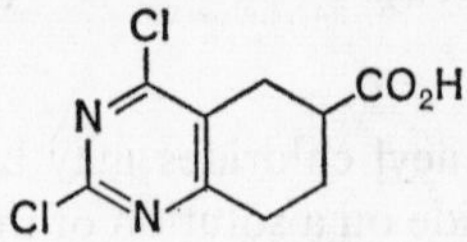

into its acid chloride in about 60% yield.

C. With Modified Phosphorus Halides

As an alternative to the phosphorus halides catechylphosphotrichloride[48] (**2**; X = Cl) or tribromide[49] (**2**; X = Br) may be used to convert carboxylic acids or anhydrides into the corresponding acyl halide.

The trichloride, which is a solid m.p. 61–62° (b.p. 132°/13 mm), is commercially available, but may be easily prepared from catechol and phosphorus pentachloride. It reacts readily with both carboxylic acids and anhydrides at room temperature or on slight warming to give the carbonyl chloride (70–80% yield) which may be distilled off from the catechylphospho-oxychloride (b.p. 121°/13 mm).

For the preparation of the bromides the following procedure is given[49]: a solution of equimolar quantities of the acid (or anhydride) and catechylphosphomonobromide in methylene chloride is treated at 0° with an equimolar amount of bromine, then warmed to 100° and distilled. The carbonyl bromides reported prepared by this method were obtained in yield greater than 80%.

Triphenylphosphine dichloride[50] and dibromide[51], obtained by the addition of the appropriate halogen to triphenylphosphine in an inert solvent (benzene or chlorobenzene), react with acids (or anhydrides) to yield the acid chloride which may be isolated by distillation or concentration and extraction.

$$Ph_3P + X_2 \longrightarrow Ph_3PX_2$$

$$Ph_3PX_2 + RCO_2H \longrightarrow RCOX + Ph_3PO + HX$$

$$Ph_3PX_2 + (RCO)_2O \longrightarrow 2\,RCOX + Ph_3PO$$

The acyl halides are obtained in 50–80% yield and the reaction is applicable to unsaturated acids such as cinnamic and 3,3-dimethylacrylic acid. Carboxylic acids react rapidly with triphenylphosphine in carbon tetrachloride to give acyl halides in good yield[52]. In the absence of hydrolysis the acid is rapidly converted into the neutral halide without the generation of acidic material. The exact mechanism of the reaction is not known but the following scheme has been suggested:

$$Ph_3P + CCl_4 \xrightarrow[\text{steps}]{\text{via radical}} [Ph_3PCCl_3]^+ \ Cl^-$$

$$\downarrow \begin{array}{l} +RCO_2H \\ -CHCl_3 \end{array}$$

$$RCOCl + Ph_3PO \longleftarrow [Ph_3POOCR]^+ \ Cl^-$$

The scope and limitations of this reaction do not seem to have been fully explored. The original work quotes good yields and reports the conversion of a number of alkanoic acids into the corresponding chlorides. However, although this method involves mild neutral conditions, when applied[19] to cis-crotonic acid, a low yield of product was obtained which was 85% trans-crotonyl chloride.

One example of a modified phosphorus halide is methyl tetrafluorophosphorane, which is reported[53] to react with acid anhydrides to produce carbonyl fluorides in high (up to 80%) yield

$$MePF_4 + (RCO)_2O \xrightarrow{80°} MePOF_2 + 2 RCOF$$

D. With Carbonyl and Sulphonyl Halides

When an acyl halide is added to a carboxylic acid the following equilibrium is set up:

$$R'COCl + RCO_2H \rightleftharpoons R'CO_2H + RCOCl$$

The general value of this reaction for preparing a lower-boiling acyl halide RCOCl from a higher-boiling acyl halide R'COCl was first pointed out by Brown[54]. Continuous removal of the lower-boiling product displaces the equilibrium to the right. In this way, for example, acetyl and propionyl chloride are obtained in over 80% yield from the corresponding acid and benzoyl chloride. This method may be extended to higher acids by the use of a higher-boiling acid chloride such as phthaloyl chloride. Thus with this reagent acetic acid is converted into acetyl chloride in 97% yield[55] and maleic anhydride into fumaroyl chloride[56]. The method is applicable to unsaturated acids[57], benzoyl chloride converting acrylic acid into the chloride in a 70% yield in contrast to the poor yields that have been obtained with thionyl chloride and phosphorus oxychloride. The conversion[58] of α-methylacrylic acid into α-methylacryloyl chloride (86%) on treatment with 20% excess of benzenesulphonyl chloride and an equimolar quantity of pyridine is a variation of this method. By the use of chloro- and fluorosulphonic acids[59] the related acyl halides may be obtained. The reaction proceeds readily at room temperature and is applicable to both aliphatic (yields 90–95%) and aromatic (yields 62–75%) anhydrides.

$$(RCO)_2O + ClSO_3H \longrightarrow RCOOSO_3H + RCOCl$$

As already described, the method discussed in this section is only applicable to the preparation of relatively low-boiling acid halides. An alternative way of displacing the equilibrium to the right is to remove the acid by-product, $R'CO_2H$. This is achieved by using oxalyl chloride or carbonyl chloride (phosgene). With both these reagents, as shown in the following equations, the acid by-product is unstable and decomposes to gaseous products which are readily removed from the reaction.

$$RCO_2H + COCl_2 \rightleftharpoons RCOCl + [ClCO_2H]$$

$$\downarrow$$

$$HCl + CO_2$$

$$RCO_2H + (COCl)_2 \rightleftharpoons RCOCl + [HO_2CCOCl]$$

$$\downarrow$$

$$CO + CO_2 + HCl$$

Oxalyl chloride will convert lauric acid into lauroyl chloride essentially quantitatively[3, 60], it is also a valuable reagent with unsaturated acids. The product from oxalyl chloride and oleic acid (*cis*) contained[60] no elaidoyl chloride. Oxalyl chloride has often been used as the preferred reagent for the conversion of a sensitive acid into its acid chloride, in some cases with the prior conversion of the acid to its sodium salt, and the use of the latter suspended in benzene containing pyridine. This is illustrated by the conversion[61] of the acid (3) via its sodium salt into the corresponding acid chloride, a conversion which could not be effected satisfactorily with thionyl chloride. Similarly[62] the conversion of the acid (4) into its chloride

(3)   (4)

was effected in high yield by this method. In contrast however, the conversions of the acids (5)[63] and (6)[64] into the acid chlorides were found to

(5)   (6)

be unsatisfactory when the sodium salts were treated with oxalyl chloride
in the presence of pyridine, and in both cases better results were obtained
using just oxalyl chloride in benzene. These results clearly illustrate the
difficulty in generalizing about the best method for converting carboxylic
acids into carbonyl chlorides. Especially when it is reported[65] that the
usual methods available for the preparation of acyl chlorides were not
applicable to the conversion of tubaric acid into tubaroyl chloride, due to
the lability of the side-chain double bond, but that the action of oxalyl
chloride on the potassium salt suspended in benzene in the *absence* of
pyridine readily produced the required compound.

   In general it does appear that oxalyl chloride is a much milder reagent
than thionyl chloride. The action of thionyl chloride on the isomeric half-
esters of acid such as α-ethyl-α-butylglutaric acid can lead to rearrangement, the derivatives obtained from the ester acid chlorides being mixtures
derived from both isomers[66]. This type of reaction was investigated by
Stenhagen[67] who studied the reaction of the two enantiomers of the acid
$MeO_2CCH_2CHMeCH_2CO_2H$, where rearrangement is readily detected
since it leads to racemization. With oxalyl chloride in benzene no rearrangement occurs with pure thionyl chloride if the reaction temperature
is kept below 30° and excess thionyl chloride removed below 50°. With
less pure thionyl chloride rearrangement occurs under these conditions.

   The following examples illustrate the use of carbonyl chloride (phosgene).
Whereas thionyl chloride with 7-oxooctadecanoic acid at 0° in benzene
gives only a 48% yield of the derived acid chloride, an 87% yield is
obtained[68] using phosgene under the same conditions. Phosgene bubbled
through a carboxylic acid (lauric or oleic) held at 140–160° gives[69] the acid
chloride in 70–90% yield. The acid chlorides can be prepared[70] without
the evolution of hydrogen chloride if the acid anhydride (not applicable
to cyclic anhydride) is used in the presence of a carboxyamide such as
formamide or dimethylformamide, as a catalyst

$$(RCO)_2O + COCl_2 \longrightarrow 2\,RCOCl + CO_2$$

The selective preparation of the N-benzyl-dl-aspartic acid mono acid chloride has been achieved[71] using carbonyl chloride:

$$\begin{array}{c} CH_2CO_2H \\ | \\ CHCO_2H \\ | \\ NHCH_2C_6H_5 \end{array} + COCl_2 \longrightarrow \begin{array}{c} CH_2CO_2H \\ | \\ CHCOCl \\ | \\ NHCH_2C_6H_5 \end{array} \quad 84\%$$

A possible useful reagent which has not yet been widely exploited is 2,2,4-trimethyl-3-oxo-valeryl chloride[72]. It is stable at elevated temperatures and it may be used in the preparation of chlorides of other acids by an exchange reaction which is forced to completion by the decarboxylation of the β-keto acid formed:

$$RCO_2H + Me_2CHCOCMe_2COCl$$

$$RCOCl + Me_2CHCOCMe_2CO_2H \longrightarrow CO_2 + Me_2CHCOCMe_2$$

These reactions are not restricted to the preparation of acyl chlorides. Thus acetyl bromide and acetyl fluoride have been prepared from acetic acid and an appropriate acid halide as illustrated by the following reactions:

Reference 73

$$CH_3{}^{14}CO_2H + C_6H_5COBr \longrightarrow C_6H_5CO_2H + CH_3{}^{14}COBr \quad (94\%)$$

Reference 74

$$CD_3CO_2D + C_6H_5COF \longrightarrow C_6H_5CO_2D + CD_3COF \quad (77\%)$$

Reference 58

$$(CH_3CO)_2 + FSO_3H \longrightarrow CH_3COOSO_3H + CH_3COF \quad (90\%)$$

Reference 58

$$(C_6H_5CO)_2 + FSO_3H \longrightarrow C_6H_5COOSO_3H + C_6H_5COF \quad (75\%)$$

E. With Chloroethers

Carboxylic acids react with α-chloroethers and with α-chlorovinyl ethers to form acyl chlorides. The advantage of this method is that no acidic by-products are evolved and therefore in principle it is well suited to the preparation of acyl chlorides containing acid-sensitive groups. These reactions have not been widely exploited as the α-chlorovinyl ethers are not readily available.

An early use of α,β-dichlorovinyl ether[75] does not record any yield of acid chlorides obtained but reports two competing reactions:

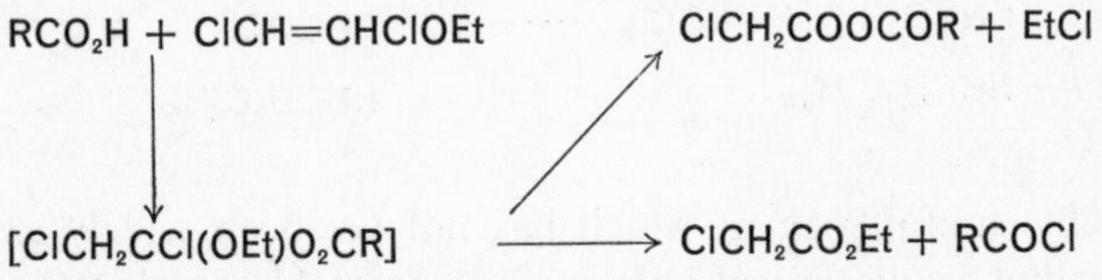

$$RCO_2H + ClCH=CHClOEt \longrightarrow ClCH_2COOCOR + EtCl$$

$$[ClCH_2CCl(OEt)O_2CR] \longrightarrow ClCH_2CO_2Et + RCOCl$$

From acetic, benzoic and chloroacetic acids, the acid chlorides were obtained pure, and succinic acid gave the anhydride. Similar reactions have been reported[76] to occur with β-bromo-α,β-dichlorovinyl ethyl ether, but the yields are variable.

$$RCO_2H + BrClC=CHClOEt \longrightarrow RCOCl + BrClCHCO_2Et$$

However, high yields of acetyl (83%), chloroacetyl (96%) and benzoyl (97%) chloride have been obtained[77] from the corresponding acid and 2-chlorodioxene at between 40° and 120°. However, with p-nitrobenzoic acid the yield fell to 38%.

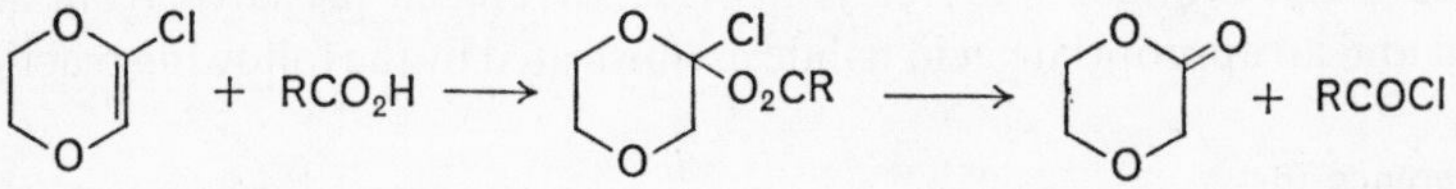

Instead of chlorovinyl ethers α,α-dihalo ethers may be used. Thus[78] α,α-dichloro- or α,α,β-trichloroethyl ether react with carboxylic acids to yield the derived acid chloride in yields of 70–90%.

$$XCH_2CCl_2OEt + RCO_2H \longrightarrow RCOCl + XCH_2CO_2Et + HCl$$

Such reactions are applicable to acids, their sodium salts and anhydrides, as illustrated below for the preparation of benzoyl chloride using α,α-dichlorodimethyl ether[79].

$$(PhCO)_2O + Cl_2CHOCH_3 \longrightarrow HCO_2CH_3 + PhCOCl \quad (98\%)$$

$$PhCO_2Na + Cl_2CHOCH_3 \longrightarrow HCO_2CH_3 + NaCl + PhCOCl \quad (86\%)$$

$$PhCO_2H + Cl_2CHOCH_3 \longrightarrow HCO_2CH_3 + HCl + PhCOCl \quad (96\%)$$

With succinic acid, succinic anhydride is produced. An example with a more complex acid is[80] the conversion of penta[O]acetyl-D-galactonic acid into the acid chloride (92%) by just warming it to 70° with excess α,α-dichlorodimethyl ether followed by evaporation of the excess reagent.

Acyl bromides may[50] be similarly prepared using α,α-dibromodimethyl methyl ether which is obtained from methyl formate and catechylphosphotribromide.

Methyl trichloromethyl ether is a highly reactive compound and in some of its reactions it behaves similarly to benzotrichloride. It undergoes hydrolysis to methyl chlorocarbonate and reacts with benzoic acid to give benzoyl chloride[81].

$$MeOCCl_3 + H_2O \longrightarrow MeOCOCl + 2\ HCl$$

$$MeOCCl_3 + C_6H_5CO_2H \longrightarrow MeOCOCl + C_6H_5COCl\ (46\%) + HCl$$

F. With Halogen Acids and Metallic Halides

Although the reaction between carboxylic acids and halogen acids does probably occur, the equilibrium normally lies on the side of the starting materials rather than the products. It would be difficult to displace the equilibrium left to right as the halogen acid is the most volatile constituent of the reaction mixture:

$$RCO_2H + HX \rightleftharpoons RCOCl + H_2O$$

However with acid anhydrides the reaction:

$$(RCO)_2O + HX \rightleftharpoons RCOX + RCO_2H$$

which is known to be reversible, can be used to prepare acid chlorides.

One example of this reaction is[82] the reaction at 360° between acetic anhydride and carbon tetrachloride in the presence of cobalt chloride and a cyclic hydrocarbon (cyclohexane, tetralin or decalin) when acetyl chloride is formed in 80–90% yield. The authors consider the reaction to involve the intermediate formation of hydrogen chloride which then reacts with the acetic anhydride (hexachloroethane is produced in the absence of the hydrocarbon). The reaction of hydrogen chloride with acetic anhydride was first reported by Colson[83] in 1897. Acetic anhydride reacts with the appropriate calcium salt in the presence of boron trifluoride etherate to give[84] acetyl bromide or acetyl chloride.

Although these reactions are not important for the preparation of acyl chlorides or bromides they are more widely exploited for the preparation of acyl fluorides. The reaction of acid anhydrides with hydrogen fluoride has been investigated by Olah and Kuhn[85]. Acetic, propionic and benzoic anhydrides react with hydrogen fluoride at reflux temperature, but with the higher acid anhydrides, higher temperatures are required and the reactions must therefore be carried out under pressure. The reaction is applicable to mixed anhydrides and in this way formyl fluoride can be prepared from formic–acetic anhydride. By continuously removing the low-boiling formyl fluoride the reaction can be forced to completion.

$$CH_3COOOCH + HF \rightleftharpoons CH_3CO_2H + HCOF$$

G. With Imidoyl Halides

The reaction of imidoyl chlorides[86] ($RCCl{=}NR'$) with carboxylic acids to yield acid chlorides was first reported[83] in 1897 and it has been suggested[23] that the imidoyl chloride, $[ClCH{=}NR_2]^+Cl^-$, derived from dimethylformamide, is responsible for the catalytic effect of dimethylformamide on the reaction between thionyl chloride and carboxylic acids (see section II, A). This compound (m.p. ca. 140°) has been prepared from dimethylformamide and carbonyl chloride[23, 87] or thionyl chloride[23] and has been shown to react with carboxylic acids[23, 88] to yield acyl chlorides.

The related fluoro compound obtained[89] from the chloro compound by the action of hydrogen fluoride is a relatively low-boiling liquid (80°/30 mm) and is considered to exist mainly as the covalent form (α,α-difluorotrimethylamine) rather than the ionic form

$$Me_2NCHF_2 \rightleftharpoons [Me_2N{=}CH{-}F]^+F^-$$

The α,α-difluorotrimethylamine (or its trihydrofluoride) reacts, at 0°, with carboxylic acids to give acyl fluorides. Thus benzoic acid yields benzoyl fluoride in 90% yield.

N-Phenyl-trimethylacetimidoyl chloride (b.p. 112°/13 mm) which is conveniently prepared by the action of phosphorus pentachloride on N-phenyl-trimethylacetamide,

$$Me_3CCONHPh + PCl_5 \longrightarrow \underset{\underset{Cl}{|}}{Me_3CC}{=}NPh + POCl_3 + HCl$$

reacts[90] with carboxylic acids usually in a solvent solution to give the acyl chlorides in good yields as shown in Table 3.

TABLE 3. Preparation of acyl chlorides with *N*-phenyl-trimethylacetimidoyl chloride

Acid	Solvent	Reaction temperature (°)	Isolated as	Yield (%)
Acetic	Ether	36	Anilide	69
Acetic	Ether	36	Titration Cl⁻	61–66
Benzoic	Benzene	80	Chloride	90
Benzoic	Benzene	80	Anilide	80
Salicylic	Dimethylformamide	80–90	Anilide	15
Phenylacetic	Dimethylformamide	80–90	Anilide	60
Propionic	—	90	Chloride	80
Chloroacetic	—	90	Chloride	75

H. With Benzotrichloride

There are a few reported examples of the use of the reaction between benzotrichloride and a carboxylic acid. Thus with acetic acid a mixture of acetyl and benzoyl chlorides is obtained[91].

$$PhCCl_3 + CH_3CO_2H \longrightarrow PhCOCl + CH_3COCl + HCl$$

Clearly the reaction has the disadvantage of giving a mixture of acid chlorides, except in the reaction between benzotrichloride and benzoic acid which leads to benzoyl chloride in 89% yield[92]. However, provided that the boiling points of the resulting acid chlorides differ considerably the method is satisfactory as illustrated by the preparation of fluoroacetyl chloride[93]:

$$C_6H_5CCl_3 + CH_2FCO_2H \longrightarrow HCl + \underset{\text{b.p. 197°}}{C_6H_5COCl} + \underset{\text{b.p. 72°}}{CH_2FCOCl}$$

I. Miscellaneous

Sulphur tetrafluoride[94] reacts with carboxylic acids in two stages, giving first the acyl fluoride and then the trifluoromethyl derivative:

$$C_6H_5CO_2H + SF_4 \longrightarrow C_6H_5COF + HF + SOF_2$$

$$C_6H_5COF + SF_4 \longrightarrow C_6H_5CF_3 + SOF_2$$

with monobasic acids it does not appear possible to stop the reaction at the first stage, but with polybasic acids the degree of fluorination can be

3

controlled by the amount of sulphur tetrafluoride used, as illustrated with 1,10-decanedioic acid.

$$HO_2C(CH_2)_8CO_2H \xrightarrow{\text{6 SF}_4} CF_3(CH_2)_8CF_3$$

$$\downarrow \text{3 SF}_4$$

$$CF_3(CH_2)_8COF + CF_3(CH_2)_8CF_3 + FOC(CH_2)_8COF$$
$$\quad\;45\% \qquad\qquad\quad 27\% \qquad\qquad\;\; 21\%$$

Anhydrides give products like those obtained from acids, but higher temperatures are required

Silicon tetrachloride reacts with isobutyric acid at 50° for 1 h to give isobutyroyl chloride in 85% yield[95].

III. FROM ACYL DERIVATIVES

A. From Acyl Halides by Halogen Exchange

Acyl bromides, iodides and fluorides have all been prepared from acyl chlorides by halogen exchange. Thus acyl chlorides react with hydrogen iodide[96, 97] to produce the acyl iodide and with hydrogen bromide to give the acyl bromide (aluminium chloride is reported[98] to be an effective catalyst for this reaction) and with hydrogen fluoride[85] the acyl fluoride is obtained.

$$RCOCl + HX \longrightarrow RCOX + HCl$$

Sodium iodide[99, 100] converts acyl chlorides into acyl iodides.

These are the most convenient routes to acyl bromides and iodides but they have not been widely used since these derivatives usually have no advantage over the more readily accessible acyl chlorides. However, this route has been widely exploited for the preparation of acyl fluorides.

A wide variety of fluorides has been used for converting acid chlorides into acid fluorides. The use of hydrogen fluoride is illustrated by the preparation of benzoyl fluoride[101]. The most widely used fluoride is potassium hydrogen fluoride which may be used[102] directly on the acid chloride, or for low-boiling acid fluorides, on the free acid in the presence of benzoyl chloride:

$$C_4H_9CO_2H + C_6H_5COCl \rightleftharpoons C_4H_9COCl + C_6H_5CO_2H$$

$$C_4H_9COCl + KHF_2 \rightleftharpoons KF + KCl + C_4H_9COF$$

Among the other reagents reported are sodium fluoride[103] Na_2SiF_6[104] and potassium fluorosulphinate[105]. The latter is a mild fluorinating agent which is effective at ca. 80° (heating is continued until SO_2 evolution ceases) and does not attack sensitive groups, e.g. it can be used to prepare cinnamoyl fluoride from the chloride. Thallous fluoride[106] in 24 h at room temperature or 0·5 h at 100° will convert s-butyl chloroformate to the fluoride (64%). Other fluorides used are triethylammonium fluoride[107] and the hexafluoroacetone–potassium fluoride adduct will[108] convert perfluoro-acyl chlorides into the corresponding fluorides.

$$R^FCOCl + (CF_3)_2CF-\overline{O}K^+ \longrightarrow R^FCOF + KCl + (CH_3)_2CO$$

Potassium fluoride may[109] be used to replace chlorine by fluorine in acyl halides and an interesting example of its basic properties is shown in the conversion[110] of 4-chlorobutyroyl chloride into the fluoride of cyclopropanecarboxylic acid

Analogous reactions were not observed with 5- and 6-halogenoacyl halides.

B. From Acyl Hydrazides and Imidazolides

The conversion of acyl hydrazides into acyl chlorides was first reported by Carpino[111]. Treatment of the hydrazides in nitromethane or methylene chloride with hydrogen chloride followed by chlorine gave the acyl chloride in 60–70% yield. This method has been applied[111] to the preparation of the acid chlorides of nitroacetic acid and α-nitro-isobutyric acid. Nitroacetyl chloride cannot be obtained from the parent acid and thionyl chloride or phosphorus pentachloride; it was previously[112] prepared by addition of nitryl chloride to ketene[113].

Hydrazides may[114] also be converted into acyl chlorides by the action of thionyl chloride (via sulphinyl hydrazides) or sulphuryl chloride:

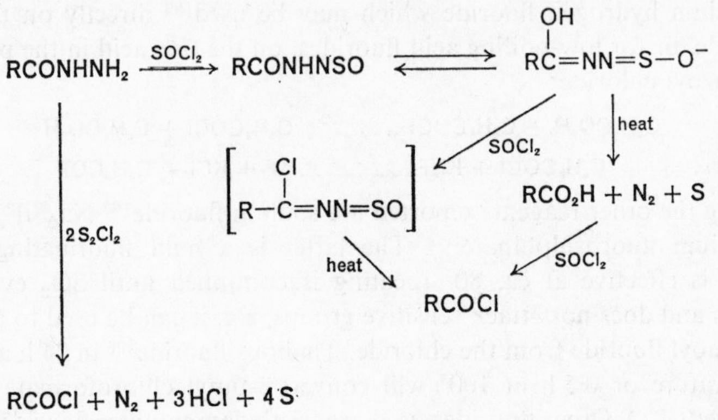

Related to the hydrazides are the imidazolides. These compounds, which are readily available from the room temperature reaction of N,N'-carbonyldi-imidazole and the carboxylic acid, react in chloroform or methylene chloride solution with hydrogen chloride to give the acyl chloride[115]:

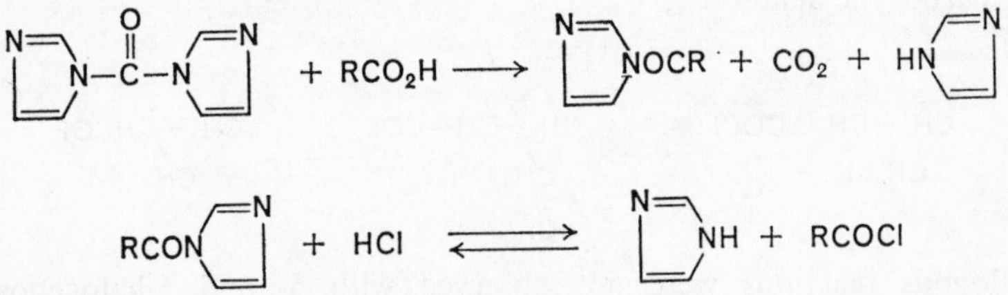

Although the reaction is reversible the yields are essentially quantitative since imidazolinium chloride precipitates from the solution. Using this method, formyl chloride has been prepared at $-60°$. It is stable at $-60°$ for about 1 h and is mostly decomposed when warmed to $-40°$. This method appears to be widely applicable and the following preparations have been reported (yields in parentheses): benzoyl chloride (81%); p-methoxybenzoyl chloride (91%); p-nitrobenzoyl chloride (94%); acetyl chloride (57%); capronyl chloride (73%); palmitoyl chloride (68%); 2,2-diphenylcyclopropanecarbonyl chloride (93%); Δ^2-2,2-diphenylcyclopropene carbonyl chloride (94%).

C. From Esters

Few examples are reported of the use of saturated esters as precursors of acyl halides. Catechylphosphotribromide (see section II, C) will convert butyl benzoate into benzoyl chloride, but methyl formate is converted by this reagent into dibromomethyl methyl ether[50]. However, isopropenyl esters have been successfully used as acyl halide precursors.

Hull reported[116] that isopropenyl acetate could be converted to acetyl chloride, bromide or iodide by passage of the appropriate hydrogen halide:

$$CH_3COOC\overset{\nearrow CH_2}{\underset{\searrow CH_3}{}} + HX \longrightarrow CH_3COX + O=C\overset{\nearrow CH_3}{\underset{\searrow CH_3}{}}$$

It has been shown more recently[117] that this reaction can also be used for the preparation of acetyl fluoride and is general for isopropenyl esters. It has been applied to the preparation of stearoyl chloride and fluoride, octanoyl and azelaoyl fluorides (yields 50 and 63% respectively) from the appropriate isopropenyl esters.

IV. FROM ALDEHYDES

The replacement of the proton in an aldehyde group leads to an acyl chloride. Such a reaction was effected with benzaldehyde to give benzoyl chloride by Liebig and Wöhler[118] in 1832. More recently it has been reported[119] that the action of dry chlorine on o-chlorobenzaldehyde at 140–160° gives o-chlorobenzoyl chloride in 82% yield. This reaction proceeds by a free-radical mechanism and can be effected with most reagents considered as radical halogen sources. Thus ethyl hypochlorite[120] or t-butyl hypochlorite[121] convert benzaldehyde into benzoyl chloride. With the latter reagent the yield is greater than 90% and the reaction is applicable to p-tolualdehyde, o- and m-chlorobenzaldehyde, but nitro-substituted aldehydes do not react, and with hydroxy-, methoxy- or dimethylamino-substituted aldehydes, chlorination of the nucleus occurred. N-Chlorosuccinimide[122] is also effective in converting benzaldehyde to benzoyl chloride in 65% yield, N-bromosuccinimide[123] appears less effective with benzaldehyde, but p-chloro and p-nitrobenzaldehyde are converted into the related benzoyl bromides in yields greater than 50% (as estimated by isolation of the derived amides and acids). Similarly[124] the N-halo-hydantoins (e.g. 1,3-dichloro- and 1,3-dibromo-5,5-dimethylhydantoins effectively (60–70%) convert benzaldehyde and m- and p-nitrobenzaldehyde into the related acyl halides. Halogenation may also be effected with sulphuryl chloride[125], a reaction which has been studied by Mannosuke Arai[126] who carried out the reaction in carbon tetrachloride solution using benzoyl peroxide as the catalyst, obtaining the following results:

Phenyl substituent	H	m-NO$_2$	p-NO$_2$	o-Cl	p-Cl	p-MeO
Yield of aroyl halide as anilide (%)	67	43	47	87	83	52

Vanillin and p-hydroxybenzaldehyde were chlorinated in the ring and salicylaldehyde and o-nitrobenzaldehyde failed to react.

An alternative method of converting aldehydes into acyl chlorides is their high-temperature reaction with carbon tetrahalides[127]. Aromatic aldehydes with carbon tetrachloride at 160–205° yield aroyl chlorides, and with bromotrichloromethane they yield aroyl bromides: aryl aldehydes with alkyl side-chains give the halo-alkyl derivatives.

$$X = Cl, 94\%$$
$$X = Br, 88\%$$

The reaction of the aliphatic aldehydes isovaleraldehyde and β-phenyl-valeraldehyde with carbon tetrachloride in the presence of benzoyl peroxide gives[128] the related acyl chlorides in about 60% yield. This is a chain-reaction of the type:

$$RCHO \xrightarrow{-H\cdot} R\dot{C}O$$
$$R\dot{C}O + CCl_4 \longrightarrow RCOCl + \dot{C}Cl_3$$
$$RCHO + \dot{C}Cl_3 \longrightarrow R\dot{C}O + CHCl_3$$

It is stated[129] that chlorination of acetaldehyde in acetic acid solution at room temperature yields acetyl chloride ($>80\%$) and vapour-phase photochemical-catalysed bromination of acetaldehyde yields[130] acetyl bromide. Similarly[131] photochemically catalysed bromination of β,β-dichloroacraldehyde gives the acid bromide whereas in the absence of light addition to the double bond occurs:

$$CCl_2BrCHBrCHO \xleftarrow{Br_2/dark} CCl_2{=}CHCHO \xrightarrow[50°]{Br_2/h\nu} CCl_2{=}CHCOBr$$

The free-radical nature of this type of reaction is confirmed by the observation[132] that reaction of α-chloroisobutyraldehyde with chlorine to give α-chloroisobutyroyl chloride is catalysed by benzoyl peroxide or u.v.-radiation.

V. DIRECT INTRODUCTION OF HALOCARBONYL GROUP

A. With Oxalyl Chloride or Carbonyl Chloride (Phosgene)

The photolysis of oxalyl chloride in the presence of alkanes or cycloalkanes leads[133] to the direct introduction of the chlorocarbonyl group. This is a free-radical reaction and suffers from the lack of specificity typical of such reactions, but from cyclohexane the acid chloride of cyclohexanecarboxylic acid was obtained in about 30% yield. The reaction will proceed in the dark in the presence of peroxides.

In the absence of any catalyst oxalyl chloride reacts[134] with 1,1-diphenylethylene, styrene α-methylstyrene and 1-methylcyclohexene to give respectively, β-phenylcinnamoyl chloride, cinnamoyl chloride, β-methylcinnamoyl chloride and the chloride of 1-methylcyclohexene 2-carboxylic acid. The yields vary from 50% for 1,1-diphenylethylenes to 6% for 1-methylcyclohexene. The simple alkenes (octene, cyclohexene, etc.) and symmetrical diaryl alkenes (stilbene) do not react. The effects of catalysts such as aluminium chloride or boron trifluoride were not investigated, but the authors suggest they might improve the yields. A similar reaction has been reported[135] with oxalyl bromide whereby isobutylene has been converted into β,β-dimethylacrylic acid (via the acid bromide) in 12% yield. The mechanism of such reactions is suggested to be:

Phosgene may also be used to prepare acyl chlorides from alkenes[136]. Thus propylene reacts as follows:

$$CH_3CH=CH_2 + COCl_2 \xrightarrow[\substack{100-300° \\ 10-40 \text{ Atm.}}]{AlCl_3} \overset{\overset{\displaystyle CH_3}{\displaystyle |}}{ClCH_2CHCOCl}$$

The reactions of aromatic compounds with oxalyl chloride and carbonyl chloride have been reviewed[137]. The reaction of carbonyl chloride with aromatic compounds in the presence of aluminium chloride is difficult to stop at the acid chloride stage for the latter will react further with the aromatic substrate to yield benzophenone derivatives. However, it is reported[138] that in the case of *m*-xylene, dimethylbenzoic acid is obtained in good yield, following hydrolyses of the first-formed acid chloride. The reaction of oxalyl chloride with aromatic compounds to give acyl chlorides appears preferable to that of phosgene and also to that of oxalyl bromide for the latter is more reactive and, for example with benzene, a considerable amount of benzil (PhCOCOPh) is formed. The reaction of oxalyl chloride with mesitylene gives mesitoic acid in good yield[139], the reaction steps being:

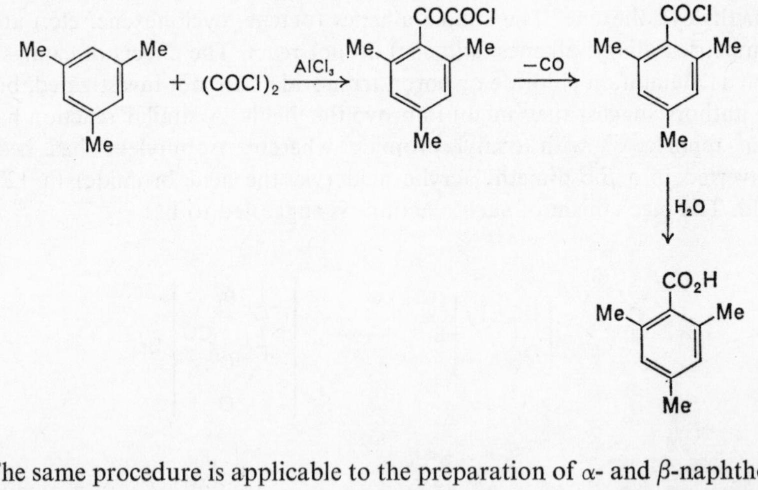

The same procedure is applicable to the preparation of α- and β-naphthoic acids, cumene carboxylic acid, 2,5-dimethylbenzoic acid and durene carboxylic acid.

The reaction of anthracene with oxalyl chloride[140] (in nitrobenzene solution) or with oxalyl bromide[135] to yield 9-anthroic acid (50–60%) via the appropriate anthroyl halide does not require a catalyst.

Although the reactions described in this section lead to acid chlorides, they have not been isolated and in the reactions involving aluminium chloride the acid chlorides are of course complexed with the aluminium chloride.

B. With Carbon Monoxide

Acyl chlorides may be obtained from alkanes and cycloalkanes by a radical chain reaction utilizing carbon tetrachloride and carbon monoxide in which the following reactions occur:

$$CCl_4 \longrightarrow Cl^{\cdot} + {}^{\cdot}CCl_3$$

$$RH + {}^{\cdot}CCl_3 \longrightarrow R^{\cdot} + CHCl_3$$

$$R^{\cdot} + CCl_4 \longrightarrow RCl + {}^{\cdot}CCl_3$$

$$R^{\cdot} + CO \longrightarrow R\overset{\cdot}{C}O$$

$$R\overset{\cdot}{C}O + CCl_4 \longrightarrow RCOCl + \overset{\cdot}{C}Cl_3$$

Initiation of the reaction may be by either peroxide addition or γ-irradiation[141]. Clearly the reaction may lead to mixtures of products. The complexity of the mixture is reduced when the starting material is symmetrical, e.g. cyclohexane.

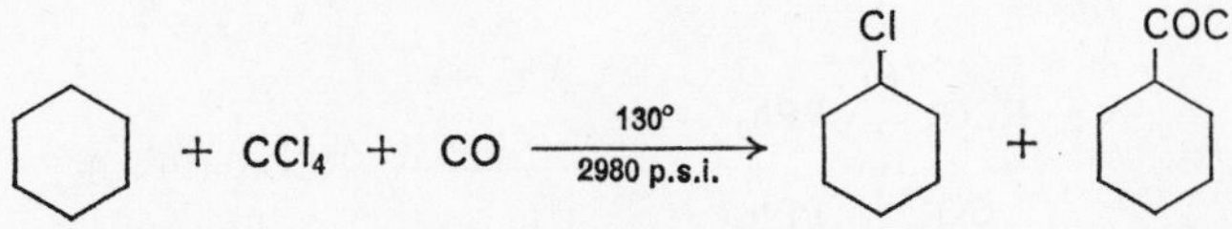

Tertiary alkyl halides react with carbon monoxide in the presence of boron trifluoride or bismuth trichloride at low temperature to give acid chlorides[142].

$$Me_3CCl + CO + BF_3 \xrightarrow[1\cdot25\ h]{0°/700\ atm.} Me_3COCl \quad (49\%)$$

The reactions between carbon monoxide and both aliphatic and aromatic halides are catalysed by transition metals. Thus carbonylation of allylic chlorides proceeds in the presence of π-allylic palladium complexes to give high yields of alk-3-enoyl chlorides[143]:

$$CO + CH_2{=}CHCH_2Cl \xrightarrow[110°/45\ min]{(\pi\text{-allyl PdCl})_2} CH_2{=}CHCH_2COCl \quad (93\%)$$

Under these conditions the two monomethyl allyl chlorides shown below give rise to the same product, pent-3-enoyl chloride, thus indicating the intervention of π-allylic palladium complexes:

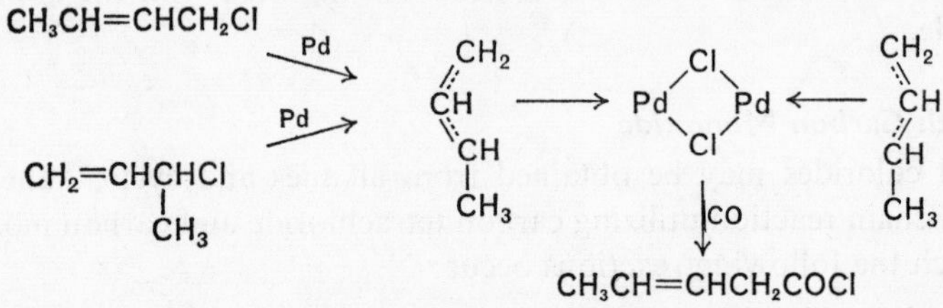

$$CH_3CH=CHCH_2Cl$$

$$CH_3CH=CHCH_2COCl$$

Under these conditions alkyl halides did not react and benzyl chloride was converted to a tar. However[144], in the presence of bistriphenyl-phosphine rhodium carbonyl chloride benzyl chloride was converted into phenylacetyl chloride. Allyl bromide with this catalyst gave crotonyl bromide (isolated as the ethyl ester). The following reaction scheme is proposed[144]:

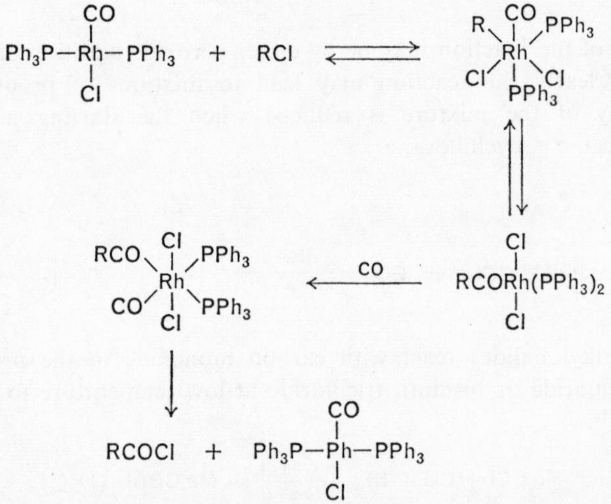

Aroyl halides are obtained[145] from aryl halides in the presence of palladium halides. In this way chlorobenzene may be converted into benzoyl chloride. Benzoyl bromide may be similarly prepared.

$$PhCl + CO \xrightarrow[\text{8 atm.}]{\text{PdCl}_2/160°} PhCOCl \quad (80\%)$$

Aliphatic and alicyclic compounds can be prepared[146] from acetylenic and ethylenic compounds by reaction with carbon monoxide and hydrogen chloride under pressure in the presence of palladium catalyst. Thus acetylene, carbon monoxide and hydrogen chloride react in the presence of bis(triphenylphosphine) palladium (II) dichloride at 80° (200 atm.) to give a mixture of 3-chloropropionyl chloride and succinoyl chloride.

A very different type of reaction is that of carbon monoxide (or of some source of carbon monoxide) with aromatic diazonium tetrafluoroborate, to yield aroyl fluorides[147].

$$\text{MeOH (sat. HCl)} + \text{Ni(CO)}_4 + \text{PhN}_2{}^+\text{BF}_4{}^- \xrightarrow[\text{pressure}]{\text{atmospheric}} \text{C}_6\text{H}_5\text{COF} \quad (32\%)$$

VI. FROM KETENES

The addition of compounds of the type Y-halogen across the carbon–carbon double bond of ketene can lead to derivatives of acetyl halides. Thus chlorine will react with ketene to give initially chloroacetyl chloride which then reacts further to yield dichloro- and trichloracetyl chloride: a mixture of all three products is usually obtained[148].

$$\text{CH}_2{=}\text{C}{=}\text{O} + \text{Cl}_2 \longrightarrow \text{CH}_2\text{ClCOCl}$$

$$\xrightarrow[+\text{Cl}_2]{-\text{HCl}} \text{CHCl}_2\text{COCl}$$

$$\xrightarrow[+\text{Cl}_2]{-\text{HCl}} \text{CCl}_3\text{COCl}$$

With halogen acids[149] such as hydrogen fluoride or chloride the appropriate acetyl halide is obtained:

$$\text{CH}_2{=}\text{C}{=}\text{O} + \text{HF} \longrightarrow \text{CH}_3\text{COF}$$

Alkyl and aryl halides do not normally react with ketene unless they are particularly reactive such as α-chloroethers, benzylic and allylic halides.

Reference 150 $\text{Ph}_3\text{CCl} + \text{CH}_2{=}\text{C}{=}\text{O} \longrightarrow \text{Ph}_3\text{CCH}_2\text{COCl}$

Reference 151

Reference 152

The addition of carbonyl halides to ketene yields acetyl carbonyl halides most of which are thermally labile, and they are not often isolated. Also since the primary product is a carbonyl halide it can undergo further reaction with another molecule of ketene. These features of the reaction are illustrated by the reaction of trichloroacetyl chloride and ketene[153].

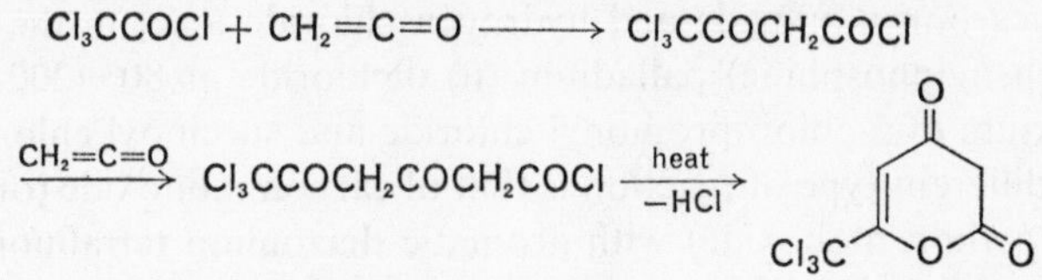

$$Cl_3CCOCl + CH_2{=}C{=}O \longrightarrow Cl_3CCOCH_2COCl$$

$$\xrightarrow{CH_2{=}C{=}O} Cl_3CCOCH_2COCH_2COCl \xrightarrow[-HCl]{heat}$$

Trifluoroacetyl chloride reacts with ketene in difluorodichloromethane at −30° to yield[154] (77%) fairly stable trifluoroacetyl–acetyl chloride F_3CCOCH_2COCl, b.p. 8°/6 mm.

Diacid chlorides may be used in this reaction. Thus oxalyl chloride reacts with diphenylketene, the initially formed α-keto-acid chloride losing carbon monoxide to form diphenylmalonyl chloride[155] (80–85%). The same compound is formed[156] from phosgene and diphenylketene

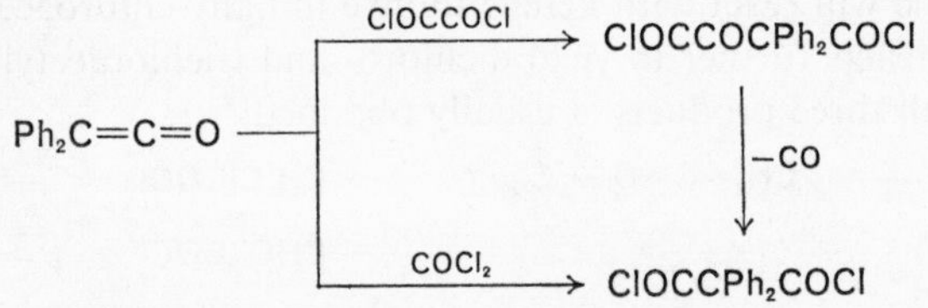

Similarly monoalkyl ketenes yield monoalkylmalonyl chlorides.

The β-keto carbonyl chlorides are also obtained from diketene by the action of hydrogen halides or halogens.

Reference 157

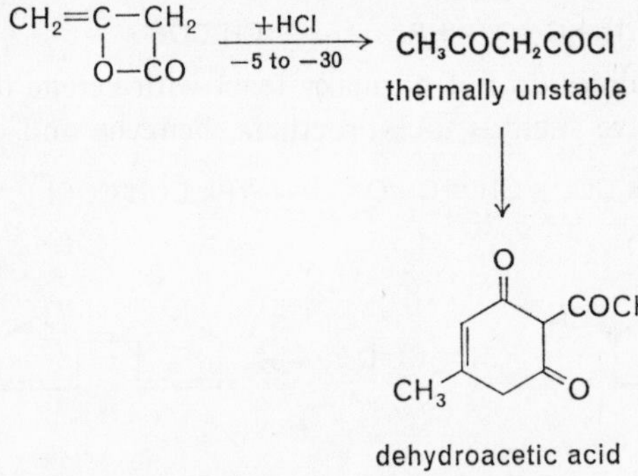

CH₃COCH₂COCl
thermally unstable

dehydroacetic acid

Reference 158

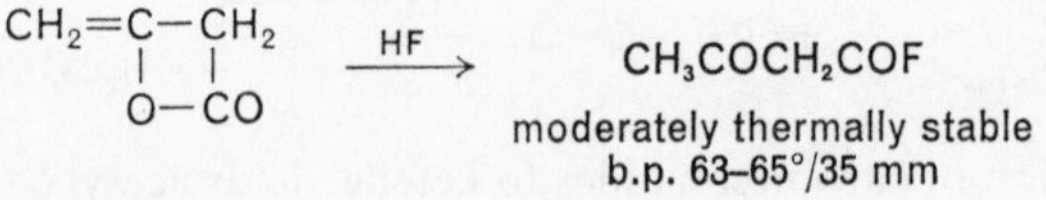

CH₃COCH₂COF
moderately thermally stable
b.p. 63–65°/35 mm

Reference 159

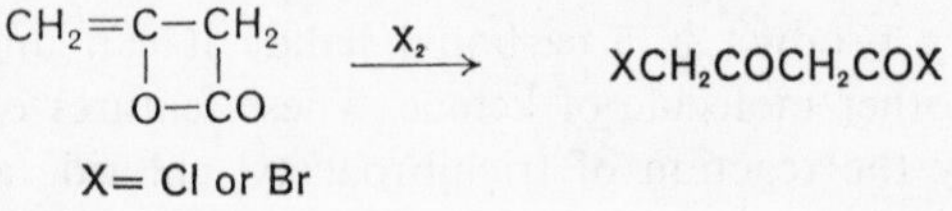

XCH₂COCH₂COX

X= Cl or Br

Related to the last reaction is the addition of hydrogen chloride to 3-hydroxy-2,2,4-trimethylpent-3-enoic acid β-lactone which yields[72] the thermally stable 2,2,4-trimethyl-3-oxovaleryl chloride which may be utilized in the preparation of other acid chlorides (see section II, D).

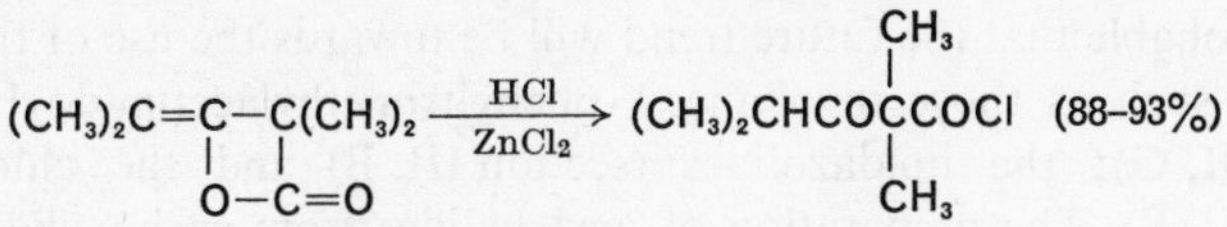

$$(CH_3)_2C=C-C(CH_3)_2 \xrightarrow[ZnCl_2]{HCl} (CH_3)_2CHCOCCOCl \quad (88-93\%)$$

with substituents indicated: O—C=O on the left structure, and CH₃ groups (top and bottom) on CCOCl of the product.

VII. MISCELLANEOUS METHODS

Acid chlorides are formed[160], among other products, on the oxidation of chloro olefins. The mechanism for the oxidation seems to be satisfactorily represented by the following free-radical mechanism:

$$Cl_2 + h\nu \longrightarrow 2Cl^{\bullet}$$
$$Cl^{\bullet} + CCl_2=CCl_2 \longrightarrow CCl_3\overset{\bullet}{C}Cl_2$$
$$CCl_3\overset{\bullet}{C}Cl_2 + O_2 \longrightarrow CCl_3CCl_2OO^{\bullet}$$
$$CCl_3CCl_2OO^{\bullet} \left| \begin{array}{l} \longrightarrow CCl_3COCl + ClO^{\bullet} \\ \longrightarrow COCl_2 + Cl^{\bullet} \end{array} \right.$$
$$CCl_2=CCl_2 + ClO^{\bullet} \longrightarrow CCl_3COCl + Cl^{\bullet}$$

Similarly oxidation of fluorinated alkenes may give rise to carbonyl fluorides[161].

$$CF_2=CFCl \xrightarrow[25-50°]{O_2} ClF_2CCOF \quad (43\%)$$

Cleavage of perfluoro ethers with aluminium chloride can lead to perfluoroacyl chlorides[162]

$$(n\text{-}C_6F_{13})_2O \xrightarrow[250°]{AlCl_3} n\text{-}C_5F_{11}COCl$$

Trichloromethyl aromatic compounds[163] react at high temperatures with metallic oxides such as TiO_2 or V_2O_5 to produce acid chlorides as illustrated by the following reactions:

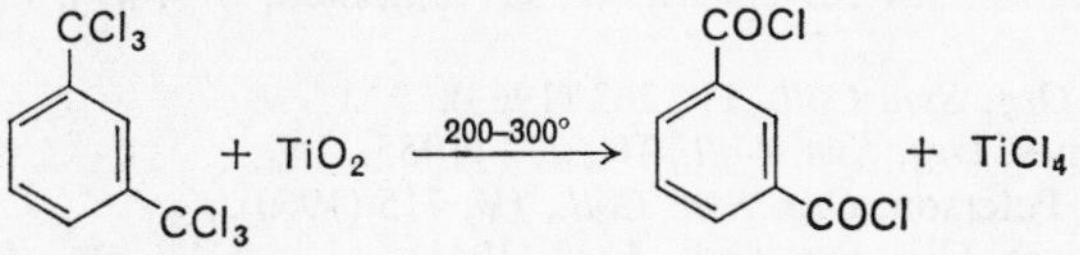

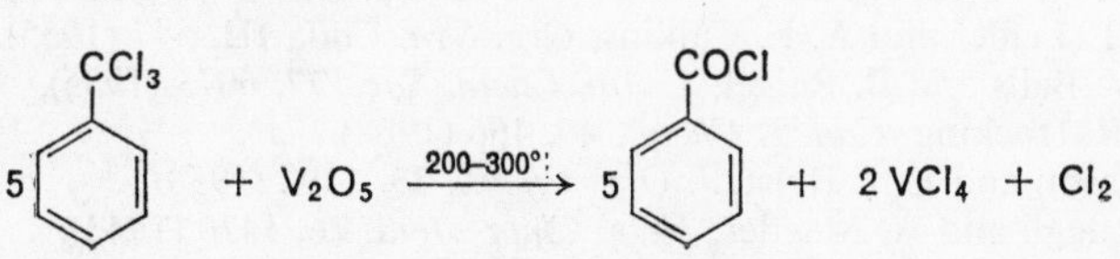

CCl₃ / CCl₃ (benzene ring) + TiO₂ $\xrightarrow{200-300°}$ COCl / COCl (benzene ring) + TiCl₄

5 CCl₃ (benzene ring) + V₂O₅ $\xrightarrow{200-300°}$ 5 COCl (benzene ring) + 2 VCl₄ + Cl₂

VIII. SUMMARY

From the foregoing review it is not possible to draw any conclusion as to the preferred reagent for the preparation of acyl halides. Many of the more recently introduced reagents have not yet been widely exploited. It seems probable that the future trend will be towards the use of the milder methods such as those involving N-phenyltrimethylacetimidoyl chloride (section II, G); the imidazolides (section III, B) and the chloroethers (section II, E). The preparation of acyl halides from carboxylic acids by the action of sulphonyl halides or carbonyl halides is also a convenient and generally satisfactory method. In view of the ease with which thionyl chloride produces unwanted side reactions it is surprising it has been so widely used.

The wide range of reagents and routes available for the preparation of acyl halides ensures that by experimentation a suitable method of preparation for a particular acyl chloride can always be found.

IX. REFERENCES

1. *Methoden der Organischen Chemie*, Vol. VIII, Georg Thieme Verlag, Stuttgart, 1952, p. 463.
2. C. A. Buehler and D. E. Pearson, *Survey of Organic Syntheses*, Wiley-Interscience, New York, 1970.
3. N. O. V. Sonntag in *Fatty Acids, their Chemistry, Properties, Production and Use* (Ed. K. S. Markley), Interscience, New York, 1961, pp. 1127–1142.
4. H. C. Roberts, *Brit. Pat.*, 908,177, (1962); *Chem. Abstr.*, **58**, 5579 (1963).
5. M. J. Frazer and W. Gerrard, *Chem. Ind.*, 280 (1954); cf. H. Hibbert and J. C. Pullman, *Inorg. Synth.*, **1**, 113 (1939).
6. G. Machell, *Chem. Prod.*, **19**, 307, 356 (1956).
7. B. Helferich and W. Schaefer, *Org. Syn. Coll.*, **I**, 147 (1941).
8. J. Cason, *Org. Syn. Coll.*, **III**, 169 (1955).
9. R. E. Kent and S. M. McElvin, *Org. Syn. Coll.*, **III**, 490 (1955).
10. B. Vassel and W. G. Skelly, *Org. Syn. Coll.*, **IV**, 154 (1963).
11. C. F. H. Allen, J. R. Byers and W. J. Humphlett, *Org. Syn. Coll.*, **IV**, 739 (1963).
12. C. Raha, *Org. Syn. Coll.*, **IV**, 263 (1963).
13. R. P. Barnes, *Org. Syn. Coll.*, **III**, 712 (1955).
14. J. Munch-Peterson, *Org. Syn. Coll.*, **IV**, 715 (1963).
15. F. K. Thayer, *Org. Syn. Coll.*, **I**, 13 (1941).
16. C. F. H. Allen, J. R. Byers and W. J. Humphlett, *Org. Syn.*, **37**, 66 (1957).
17. M. T. Leffler and A. E. Calkins, *Org. Syn. Coll.*, **III**, 547 (1955).
18. J. W. Balls and B. Reigel, *J. Am. Chem. Soc.*, **77**, 6073 (1955).
19. M. B. Hocking, *Can. J. Chem.*, **46**, 466 (1968).
20. J. Cason and E. J. Reist, *J. Org. Chem.*, **23**, 1492 (1958).
21. P. Ruggli and A. Maeder, *Helv. Chim. Acta*, **26**, 1476 (1943).

22. E. H. Wilson, *U.S. Pat.*, 2,899,458 (1959); *Chem. Abstr.*, **54**, 428 (1960).
23. H. M. Bosshard, R. Moray, M. Schmid and H. Zollinger, *Helv. Chim. Acta*, **62**, 1633 (1959).
24. J. F. Normant, J. P. Foulon and H. Deshayes, *Compt. Rend.*, **269**, 1325 (1969).
25. V. I. Zaïonts, *Zhur. Priklad. Khim.*, **33**, 711 (1960); *Chem. Abstr.*, 20965 (1960).
26. M. Ya. Kraft and V. V. Katyshkina, *Doklady. Akad. Nauk, U.S.S.R.*, **109**, 312 (1956).
27. M. A. Beg and H. N. Singh, *Z. Physik. Chemie*, **237**, 129 (1968); **227**, 272 (1964); *Fette und Seifen*, **70**, 640 (1968).
28. W. Gerrard and A. M. Thrush, *J. Chem. Soc.*, 2117 (1953).
29. M. S. Simon, J. B. Rogers, W. Saenger and J. Z. Gougoutas, *J. Am. Chem. Soc.*, **89**, 5838 (1967).
30. A. J. Krubsack and T. Higa, *Tetrahedron Letters*, 5149 (1968).
31. R. N. McDonald and R. A. Krueger, *J. Org. Chem.*, **28**, 2542 (1963).
32. L. F. Fieser and M. Fieser, *Reagents for Organic Synthesis*, Wiley, New York, 1967, p. 1158.
33. A. R. Galbraith, P. Hale and J. E. Robertson, *J. Am. Oil Chemist Soc.*, **41**, 104 (1964).
34. J. A. Cade and W. Gerrard, *J. Chem. Soc.*, 2030 (1954).
35. N. O. V. Sontag, J. R. Trowbridge and I. J. Krems, *J. Am. Oil Chemist Soc.*, **31**, 151 (1954).
36. A. Mooradian, C. J. Cavallito, A. J. Bergman, E. J. Lawson and C. M. Suter, *J. Am. Chem. Soc.*, **71**, 3372 (1949).
37. C. F. H. Allen and W. E. Barker, *Org. Syn. Coll.*, **I**, 156 (1943).
38. R. E. Ireland and M. Chaykovsky, *Org. Syn.*, **41**, 5 (1961).
39. C. E. Braun and C. D. Cook, *Org. Syn.*, **41**, 79 (1961).
40. H. Feuer and S. M. Pier, *Org. Syn.*, *Coll.*, **IV**, 554 (1963).
41. E. Ott, *Org. Syn.*, *Coll.*, **II**, 528 (1943).
42. H. Zinner and G. Brossman, *J. Prakt. Chem.*, **5**, 91 (1957).
43. C. G. Young, A. Epp, B. M. Craig and H. R. Sallans, *J. Am. Oil Chemist Soc.*, **34**, 107 (1957).
44. O. Diels and W. E. Thiele, *Ber.*, **71**, 1173 (1938).
45. L. J. Sargent, *J. Org. Chem.*, **21**, 1286 (1956).
46. R. Kostka and J. Ribka, *Ger. Pat.*, 1,245,958 (1967); *Chem. Abstr.*, **67**, 99872 (1967).
47. J. De Graw, L. Goodman, R. Koehler and B. R. Baker, *J. Org. Chem.*, **24**, 1632 (1959).
48. H. Gross and J. Gloede, *Chem. Ber.*, **96**, 1378 (1963).
49. L. Horner, H. Oediger and H. Hoffmann, *Annalen*, **626**, 26 (1959).
50. J. Gloede and H. Gross, *Chem. Ber.*, **100**, 1770 (1967).
51. H. J. Bestman and L. Mott, *Annalen*, **693**, 132 (1966).
52. J. B. Lee, *J. Am. Chem. Soc.*, **88**, 3440 (1966).
53. V. V. Lysenko, I. D. Shelakova, K. V. Karavanov and S. Z. Irvin, *Zhur. Obshchei Khim.*, **36**, 1507 (1966).
54. H. C. Brown, *J. Am. Chem. Soc.*, **60**, 1325 (1938).
55. M. Bubner and L. H. Schmidt, *Pharmazie*, **18**, 668 (1963).
56. L. P. Kyrides, *Org. Syn.*, *Col.*, **III**, 422 (1955).

57. G. H. Stempel, R. P. Cross and R. P. Marella, *J. Am. Chem. Soc.*, **72**, 2298 (1950).
58. J. Heyboer and A. J. Staverman, *Rec. Trav. Chim.*, **69**, 787 (1950).
59. M. Schmidt and K. E. Pichl, *Chem. Ber.*, **98**, 1003 (1965).
60. N. O. V. Sonntag, J. R. Trowbridge and I. J. Krems, *J. Am. Oil Chemist Soc.*, **31**, 151 (1954).
61. A. L. Wilds and C. H. Schunk, *J. Am. Chem. Soc.*, **72**, 2388 (1950).
62. G. Stork and F. H. Clarke, *J. Am. Chem. Soc.*, **83**, 3114 (1961).
63. F. Reber, A. Lardon and T. Reichstein, *Helv. Chim. Acta*, **37**, 45 (1954).
64. Ch. R. Engel and G. Just, *Can. J. Chem.*, **33**, 1515 (1955).
65. M. Miyano, *J. Am. Chem. Soc.*, **87**, 3958 (1965).
66. J. Cason, *J. Am. Chem. Soc.*, **69**, 1548 (1947).
67. S. Ställberg-Stenhagen, *J. Am. Chem. Soc.*, **69**, 2568 (1947).
68. R. H. Jaeger and Sir R. Robinson, *Tetrahedron*, **14**, 320 (1961).
69. J. Prat and A. Etienne, *Bull Soc. Chim. France*, **11**, 30 (1944).
70. F. J. Christoph, S. H. Parker and R. L. Seagrave, *U.S. Pat.*, 3,318,950 (1967); *Chem. Abstr.*, **67**, 43437 (1967).
71. L. and V. Arsenijevic and A. F. Damanski, *Compt. Rend.*, **256**, 4039 (1963).
72. E. U. Elam, P. G. Gott and R. H. Hasek, *Org. Syn.*, **48**, 126 (1968).
73. C. Heidelberger and R. Hurlbert, *J. Am. Chem. Soc.*, **72**, 4704 (1950).
74. G. A. Olah, S. J. Kuhn, W. S. Tolgyesi and E. S. Baker, *J. Am. Chem. Soc.*, **84**, 2733 (1962).
75. H. C. Crompton and P. L. Vanderstichele, *J. Chem. Soc.*, 691 (1920).
76. R. Rix and J. Arens, *Proc. Koninkl. Ned. Akad. Wetenschap*, **136**, 372 (1953).
77. M. J. Astle and J. D. Welks, *J. Org. Chem.*, **26**, 4325 (1961).
78. L. Heslinga, G. J. Katerberg and J. F. Arens, *Rec. Trav. Chim.*, **76**, 969 (1957).
79. A. Rieche and H. Gross, *Chem. Ber.*, **92**, 83 (1959).
80. R. Bognar, I. Farkas, I. F. Szabo and G. D. Szabo, *Magy. Kem. Folyoirat*, **69**, 450 (1963); *Chem. Abstr.*, **60**, 643 (1964).
81. J. B. Douglas and G. B. Warner, *J. Am. Chem. Soc.*, **78**, 6070 (1956).
82. V. N. Dubchenko and V. I. Kovaleko, *Zhur. Prikl. Khim.*, **41**, 157 (1968).
83. A. Colson, *Bull. Soc. Chim.*, *France*, **17**, 59 (1897).
84. E. Sorkin and J. Gmunder, *Helv. Chim. Acta*, **36**, 2021 (1953).
85. G. A. Olah and S. J. Kuhn, *J. Org. Chem.*, **26**, 237 (1961).
86. R. Bonnett, 'The Imidoyl Halides' in *The Chemistry of the Carbon Nitrogen Double Bond* (Ed. S. Patai), Interscience, New York, 1970, p. 597.
87. Z. Arnold, *Coll. Czech. Chem. Comm.*, **52**, 2013 (1958).
88. M. Zaoral and Z. Arnold, *Tetrahedron Letters* **14**, 9 (1960).
89. Z. Arnold, *Coll. Czech. Chem. Comm.*, **28**, 2047 (1963).
90. F. Cramer and K. Baer, *Chem. Ber.*, **93**, 1231 (1960).
91. A. N. Nesmejanov and E. J. Kahn, *Ber.*, **37**, 370 (1934).
92. I. K. Kacker, H. Dakshinamurty, S. G. Sibhu and I. S. Ahmed, *Indian Pat.*, 96,421 (1966); *Chem. Abstr.*, **70**, 28667 (1969).
93. E. Gryszkiewicz-Trochimowski, A. Sporzyński and J. Wanuk, *Rec. Trav. Chim.*, **66**, 419 (1947).
94. W. R. Hasek, W. C. Smith and V. A. Engelhardt, *J. Am. Chem. Soc.*, **82**, 543 (1960).

95. I. N. Nazarov and I. L. Kotlyarevskii, *J. Gen. Chem.*, *U.S.S.R.*, **20**, 1501 (1950).
96. H. Staudinger and E. Anthes, *Ber.*, **46**, 1417, 1426 (1913).
97. E. L. Gustus and P. G. Stevens, *J. Am. Chem. Soc.*, **55**, 374 (1933).
98. D. E. Lake and A. A. Asadorian, *U.S. Pat.*, 2,553,518 (1951); *Chem. Abstr.*, **46**, 2561 (1952).
99. D. W. Theobald and E. D. Bergman, *Chem. Ind.*, 1007 (1958).
100. J. Blum, H. Rosenman and E. D. Bergman, *J. Org. Chem.*, **33**, 1928 (1968).
101. G. A. Olah and S. J. Kuhn, *Org. Syn.*, **45**, 3 (1965).
102. G. Olah, S. Kuhn and S. Beke, *Chem. Ber.*, **89**, 862 (1956).
103. G. W. Tullock and D. D. Coffman, *J. Org. Chem.*, **25**, 2016 (1960).
104. J. Dahlmos, *Angew. Chem.*, **71**, 274 (1959).
105. F. Seel and J. Langer, *Chem. Ber.*, **91**, 2553 (1958).
106. S. Nakanishi, T. C. Myers and E. V. Jensen, *J. Am. Chem. Soc.*, **77**, 3099 (1955).
107. I. L. Knunyants, Yu. A. Cheburkov and M. D. Bargamova, *Izvest. Akad. Nauk S.S.S.R. Ser. Khim.*, 1393 (1963); *Chem. Abstr.*, **59**, 15175b (1963).
108. A. G. Pittman and D. L. Sharp, *J. Org. Chem.*, **31**, 2316 (1966).
109. M. Hanack and H. Eggensperger, *Chem. Ber.*, **96**, 1341 (1963).
110. R. E. A. Dear and E. E. Gilbert, *J. Org. Chem.*, **33**, 1690 (1968).
111. L. A. Carpino, *J. Am. Chem. Soc.*, **79**, 96 (1957); *Chem. Ind.*, **123** (1956).
112. L. W. Kissinger and H. E. Ungnade, *J. Org. Chem.*, **24**, 1244 (1959).
113. W. Steinkopff and M. Kuhnel, *Ber.*, **75**, 1323 (1942).
114. P. Hope and L. A. Wiles, *J. Chem. Soc.*, 5386, 5838 (1965).
115. H. A. Staab and A. P. Dutta, *Angew Chem. Int. Edn.*, **3**, 132 (1964); H. A. Staab, K. Wendel and A. P. Dutta, *Annalen*, **649**, 78 (1966).
116. D. C. Hull, *U.S. Pat.*, 2,475,966 (1949); *Chem. Abstr.*, **43**, 7954 (1949).
117. E. S. Rothman, G. G. Moore and S. Serota, *J. Org. Chem.*, **34**, 2486 (1969).
118. J. Liebig and F. Wöhler, *Annalen*, **3**, 262 (1832).
119. H. T. Clarke and E. R. Taylor, *Org. Syn.*, *Coll.*, **I**, 155 (1947).
120. S. Goldschmidt, R. Endress and R. Durch, *Ber.*, **58**, 576 (1925).
121. D. Ginsburg, *J. Am. Chem. Soc.*, **73**, 702 (1951).
122. M. F. Hebbelynck and R. H. Martin, *Bull. Soc. Chim. Belg.*, **60**, 54 (1951).
123. M. Yamaguchi and T. Adachi, *Nippon Kagaku. Zasshi*, **79**, 487 (1958); *Chem. Abstr.*, **54**, 4480b (1960).
124. O. A. Orio and J. D. Bonafede, *An. Asoc. Quim. Argentina*, **54**, 129 (1966).
125. T. H. Durrans, *J. Chem. Soc.*, 45 (1922).
126. Mannosuke Arai, *Bull. Chem. Soc., Japan*, **37**, 1280 (1964); **38**, 252 (1965).
127. Yu. A. Ol'dep and A. M. Kalinina, *Zhur. Obshchei Khim.*, **34**, 3473 (1964).
128. S. Winstein and F. H. Senbold, *J. Am. Chem. Soc.*, **69**, 2917 (1947).
129. Y. Kato, *Japan. Pat.*, 153, 599; *Chem. Abstr.*, **49**, 3027 (1949); **44**, 3319 (1950).
130. E. Buckley and E. Whittle, *Can. J. Chem.*, **40**, 1611 (1962).
131. M. and E. Levas, *Compt. Rend.*, **250**, 2819 (1960).
132. Y. Ohshiro, S. Hattori, T. Takaoka and S. Komori, *Kogyo. Kagaku. Zasshig*, **70**, 1262 (1967); *Chem. Abstr.*, **68**, 29221 (1968).
133. M. S. Kharasch and H. C. Brown, *J. Am. Chem. Soc.*, **64**, 329 (1942).

134. M. S. Kharasch, S. S. Kane and H. C. Brown, *J. Am. Chem. Soc.*, **64**, 333 (1942); F. Bergmann, M. Weizman, E. Dimant, S. Patai and J. Szmuskowics, *J. Am. Chem. Soc.*, **70**, 1612 (1948).
135. W. Treibs and H. Orttmann, *Chem. Ber.*, **93**, 545 (1960).
136. E. E. Reid, *U.S. Pat.*, 2,028,012; *Chem. Abstr.*, **30**, 1387 (1936).
137. G. A. and J. A. Olah, *The Friedel–Crafts Reaction* (Ed. G. A. Olah), Vol. III, Interscience, New York, 1964, p. 1257 *et seq.*
138. E. Ador and F. Meir, *Ber.*, **12**, 1968 (1879).
139. P. E. Sokol, *Org. Syn.*, **44**, 69 (1964).
140. H. G. Latham, E. L. May and E. Mosettig, *J. Am. Chem. Soc.*, **70**, 1079 (1948).
141. W. A. Thaler, *J. Am. Chem. Soc.*, **89**, 1902 (1967); **88**, 4278 (1966). cf. L. Schmerling, *U.S. Pat.*, 3,367,953 (1968); *Chem. Abstr.*, **69**, 35482 (1968); *Neth. Application*, 6,611,634; *Chem. Abstr.*, **67**, 53779 (1967).
142. R. E. Brook and I. D. Webb, *U.S. Pat.*, 2, 580,070, (1951); *Chem. Abstr.*, **46**, 6667 (1952).
143. W. T. Dent, R. Long and G. H. Whitfield, *J. Chem. Soc.*, 1588 (1964).
144. K. Ohno and J. Tsuji, *J. Am. Chem. Soc.*, **90**, 99 (1968).
145. *Neth. Application*, 6,614,185; *Chem. Abstr.*, **67**, 64066s (1967).
146. N. von Kutepow, K. Bittler, D. Neubauer and H. Reis, *Ger. Pat.*, 1,237,116 (1967); *Chem. Abstr.*, **68**, 21550y (1968).
147. R. G. Linville, *U.S. Pat.*, 2,517,898; *Chem. Abstr.*, **45**, 2505 (1951).
148. R. N. Lacey and E. E. Conolly, *Brit. Pat.*, 715,896,735,902. *Chem. Abstr.*, **49**, 13290 (1955); **50**, 7846 (1956).
149. F. Chick and N. T. M. Wilsmore, *J. Chem. Soc.*, **24**, 77 (1908).
150. A. T. Blomquist, R. W. Holley and O. J. Sweeting, *J. Am. Chem. Soc.*, **69**, 2356 (1947).
151. C. D. Hurd and A. J. Richardson, *J. Org. Chem.*, **32**, 3516 (1967).
152. M. Kratochvil, J. Uhdeova and J. Jonas, *Coll. Czech. Chem. Comm.*, **30**, 3205 (1965).
153. J. Berànek, J. Smrt and F. Sorm, *Chem. Listy*, **48**, 679 (1954); **49**, 73 (1955).
154. *Brit. Pat.*, 931,689 (1969); *Chem. Abstr.*, **60**, 2788f (1964).
155. P. D. Bartlett and L. B. Gortler, *J. Am. Chem. Soc.*, **85**, 1864 (1963).
156. H. Staudinger, C. Göhring and M. Schöller, *Ber.*, **47**, 40 (1914).
157. C. D. Hurd and C. D. Kelso, *J. Am. Chem. Soc.*, **62**, 1548 (1940).
158. G. A. Olah and S. J. Kuhn, *J. Org. Chem.*, **26**, 225 (1961).
159. F. Chick and N. T. M. Wilsmore, *J. Chem. Soc.*, 1978 (1910).
160. W. T. Miller and A. L. Dittman, *J. Am. Chem. Soc.*, **78**, 2793 (1956) and papers there cited.
161. V. R. Hurka, *U.S. Pat.*, 2,676,983 (1954); *Chem. Abstr.*, **49**, 5510 (1955).
162. G. V. D. Tiers, *J. Am. Chem. Soc.*, **77**, 6703 (1955).
163. R. C. Schreyer, *J. Am. Chem. Soc.*, **80**, 3483 (1958).

CHAPTER **3**

Detection, determination and characterization

HANNAH WEILER-FEILCHENFELD

Department of Organic Chemistry, The Hebrew University of Jerusalem, Israel

69

Acyl halides play a fundamental role in preparative, analytical and industrial organic chemistry. It is therefore of utmost importance to have reliable analytical tools for dealing with these compounds. Detection and determination of acyl halides are made difficult by the ease of their conversion to carboxylic acids and hydrogen halides on the one hand, and by the similarity of their reactions to those of anhydrides and other carboxylic acid derivatives on the other. Analysis of acyl halides is usually not carried out on the halides themselves but rather on the corresponding acids or their derivatives. Much useful information as well as detailed descriptions of the actual procedures can be found in most organic analytical textbooks. Only methods specifically referring to acyl halides will be described here.

I. CHEMICAL METHODS

A. Detection

I. Hydrolysis

Hydrolysis of acyl halides by water to the corresponding carboxylic acids and hydrogen halides is easy and fast.

$$RCOX + H_2O \longrightarrow RCOOH + HX$$

The reaction of anhydrides is much slower, and sometimes even needs the addition of a catalyst[1]; however, for definite proof that the substance undergoing hydrolysis is actually an acyl halide, the presence of hydrogen halide in the aqueous solution must be shown, for instance by argentimetry.

2. Hydroxamic acid test

Reaction of acyl halides with hydroxylamine yields hydroxamic acids; addition of ferric chloride then gives deep red-coloured complexes[2, 3].

$$RCOX + NH_2OH \longrightarrow RCONHOH + HX$$

$$3\,RCONHOH + FeCl_3 \longrightarrow \left[RC \underset{O \rightarrow}{\overset{NHO}{\diagup}} \right]_3 Fe + 3HCl$$

Similar reactions take place with anhydrides and esters. Buckles and Thelen[3] describe a further procedure to distinguish between esters on one side and anhydrides and halides on the other; no such test has been devised to differentiate between anhydrides and halides.

It should be noted that other classes of organic compounds, such as phenols, react directly with ferric chloride to give similarly coloured complexes. It is therefore essential, as a preliminary step, to test the substance to be studied directly with ferric chloride in the absence of hydroxylamine[4].

3. Amide formation

Acyl halides may be detected by their reactions with aniline, p-toluidine or benzylamine which yield almost insoluble substituted amides[5-9], benzylamine being more specifically useful for the detection of aroyl halides[10].

$$C_6H_5NH_2 + RCOX \longrightarrow C_6H_5NHCOR + HX$$

Again similar reactions take place with acid anhydrides and with esters, so that the positive result of the test is not unequivocal.

4. Microdetection

a. Hydroxamic acid test. Feigl[11] has developed a spot-test based on the reaction of the compounds with hydroxylamine to give hydroxamic acids, which in turn form red-coloured complexes with ferric ion (see section I, A, 2). Although in Feigl's original publication this test is applied to acid anhydrides, it can be used equally well for acyl halides.

b. Reaction with 4-(4-nitrobenzyl)pyridine. Acyl halides react with 4-(4-nitrobenzyl)pyridine according to the following scheme[12–14]:

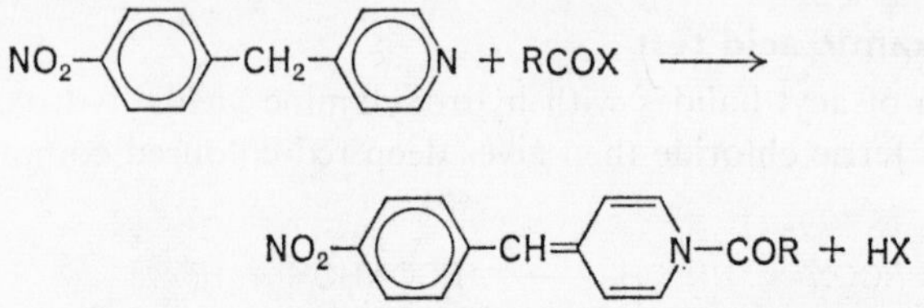

The acyl derivatives obtained have an absorption band between 445 and 475 nm and paper impregnated with 4-(4-nitrobenzyl)pyridine will show a strong red or orange stain in the presence of acyl halides; sometimes it is necessary to expose the paper to triethylamine (which acts as a colour stabilizer)[13].

Chloroformates give the same reaction; carbamoyl chlorides show some colour formation but more weakly and only after heating. Acid anhydrides do not react with 4-(4-nitrobenzyl)pyridine; hence this test is useful to distinguish between acid anhydrides and acyl halides. The reagent gives a similar reaction product

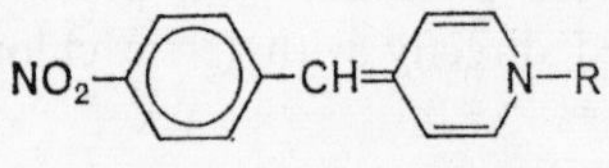

with alkyl halides, but the absorption maximum then lies between 545 and 575 nm and the colour obtained is different.

Alternative procedures use 4-pyridinecarboxaldehyde-2-benzothiazolyl-hydrazone or 4-acetylpyridine-2-benzothiazolylhydrazone in acetophenone, with addition of triethylamine, reaching detection limits as low as $0.1–0.5\ \mu g$[13].

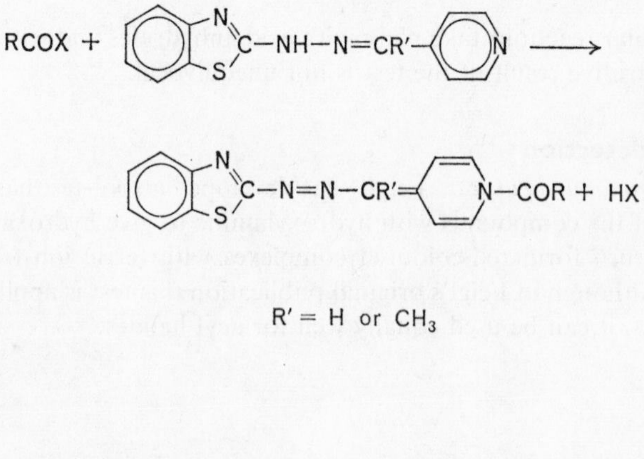

R′ = H or CH₃

B. Separation

The isolation of acyl halides is usually carried out after transformation into the corresponding carboxylic acids or into some of their insoluble derivatives.

I. Separation as carboxylic acids

The acyl halide is first converted by hydrolysis into the corresponding carboxylic acid; the acid can then be isolated by any of the many procedures available for carboxylic acids.

If the acid is dissolved in an aqueous solvent, it can be precipitated from the neutralized solution as a lead, silver or mercury salt, then regenerated by the action of hydrogen sulphide on this salt. From organic solvents immiscible with water, the acid can be extracted by an aqueous solution of a base and subsequent acidification of the aqueous layer. Countercurrent extraction techniques, based on the different solubilities of carboxylic acids in water and in organic solvents immiscible with water, can also be useful.

Distillation or steam distillation can be used for volatile acids and a variety of adsorption techniques are available for the less volatile ones. These include adsorption on a basic ion-exchange resin followed by washing out with an acid solution[15-17], sometimes on a microscale[18]; also used are column chromatography on silica gel[19, 20] or on large-surface polystyrene[21], liquid–liquid chromatography with water adsorbed on silica as the immobile phase and with an eluent made progressively more polar (for instance, chloroform to which increasing quantities of n-butanol are added)[22]; thin-layer and paper chromatography may also be applied to separation, but are better suited to the identification of the compounds (see section I, C, 3).

2. Separation as insoluble derivatives

The acyl halides can be separated in the form of insoluble or only slightly soluble derivatives. They can be converted into amides, anilides or p-toluidides by reaction with ammonia in water, or with aniline or p-toluidine in benzene[8, 9]. The amides or aryl amides can then be easily separated by precipitation or extraction and recrystallized from water or ethanol.

The acids may be precipitated from their solutions as a variety of other solid derivatives such as S-alkylisothiouronium salts or phenacyl esters. This has the added advantage of allowing further characterization of the acid, if necessary, as will be described later (see section I, C, 2).

C. Characterization

Identification procedures for acyl halides are not different from those used in the characterization of carboxylic acids or their derivatives. Three different approaches have been applied: determination of the molecular weight of the compound, preparation of solid derivatives of known melting point and use of chromatographic techniques.

I. Determination of the molecular weight by titration

The most direct method of measuring the molecular weight of an acyl halide is hydrolysis of a weighed quantity of the compound by water, followed by titration with sodium hydroxide. Two equivalents of the base are used up in the titration and the required molecular weight can easily be deduced.

$$RCOX + H_2O \longrightarrow RCOOH + HX$$

$$RCOOH + HX + 2\,NaOH \longrightarrow RCOONa + NaX + 2\,H_2O$$

The acyl halide itself can also be titrated directly with sodium methylate in non-aqueous solvents[23].

$$RCOX + CH_3ONa \longrightarrow RCOOCH_3 + NaX$$

The S-benzylthiouronium salt of an acyl halide (see section I, C, 2, a) can be titrated with perchloric acid in acetic acid[24], either by a potentiometric method, or by using crystal violet as a visual indicator; the titration yields the equivalent weight of the parent carboxylic acid and thus allows its identification; this can be checked by determining the melting point of the salt.

2. Preparation of solid derivatives

a. S-Alkylisothiouronium salts. S-Alkylisothiouronium salts have often been used for the identification of acyl halides[25-31]. Usually the alkyl group is benzyl[24, 28] or *para*-substituted benzyl (*p*-chloro, *p*-bromo or *p*-nitro)[24, 28-30]; naphthylmethylisothiouronium salts have also been cited[31]. The acyl halides are first converted to the sodium salts of the corresponding carboxylic acids, then the isothiouronium salts are prepared according to Vogel[32] or Friediger and Pedersen[33]:

$$\left[C_6H_5CH_2SC\begin{array}{c}{}^{NH_2}\\{}_{NH_2}\end{array}\right]^+Cl^- + RCOONa \longrightarrow$$

$$\left[C_6H_5CH_2SC\begin{array}{c}{}^{NH_2}\\{}_{NH_2}\end{array}\right]^+RCOO^- + \cdot NaCl$$

In the basic medium, the salts usually precipitate; however, the reaction mixtures must not be too alkaline, since the S-benzylisothiouronium bases then tend to decompose to mercaptans. For the same reason recrystallization should take place in neutral or slightly acid solvents. The salts can be identified by their melting points[24, 28-31, 34, 35]; however, these seem to be strongly dependent on the rate of heating[24] and care must be taken on this account.

b. *Esters*. Solid esters of the carboxylic acids, as for instance their phenacyl[36-39], p-chlorophenacyl[36-38], p-bromophenacyl[36-42], p-phenylphenacyl[43-45] or p-nitrobenzyl esters[46-49], can be used in the same manner for identification purposes by determination of their melting point.

c. *Anilides*, p-*toluidides* and N-*benzamides*. These derivatives have the advantage that for their preparation (see section I, A, 3) no prior conversion of the acyl halides into the sodium salts of the carboxylic acids is necessary. N-Benzylamides are more easily isolated than anilides or toluidides[10]. The preparation of N-(β-morpholinoethyl)amides, or, if these are liquid, of their quaternary methiodides, has also been suggested[50]. The relevant melting points can be found in the literature[5-10, 50].

d. *Other solid derivatives*. The preparation of hydrazides for the identification of acyl halides was recommended by Buu-Hoï and co-workers[51]; the reaction of the hydrazides with aryl isocyanates or aryl thioisocyanates yields 1-acyl-4-arylsemicarbazides or 1-acyl-4-arylthiosemicarbazides, substances with even higher melting points and lower solubilities than the hydrazides themselves.

$$RCONHNH_2 + ArNCO \longrightarrow RCONHNHCONHAr$$

or

$$RCONHNH_2 + ArNCS \longrightarrow RCONHNHCSNHAr$$

2,4-Dinitrophenylhydrazides were used by Cheronis[52] for the microidentification of carboxylic acids. A large number of additional solid derivatives have been suggested by different authors for the identification of acyl halides[53].

3. Chromatographic identification

In recent years, a number of investigations have dealt with the chromatographic separation and characterization of various carboxylic acids. Different techniques can be used, such as column chromatography on silica gel[19, 20], paper chromatography with dimethyl sulphoxide or other solvents[54-57] and reversed phase chromatography, for instance with a paraffin–aqueous methanol system[58]. The R_f values of a large number of acids have been published for different solvent mixtures[54-57]. Dibasic acids[54] can be separated and identified in this way even in the presence of monocarboxylic acids. Low molecular weight acids which are too volatile for paper chromatography are often separated in the form of salts, but these are liable to be hydrolysed by the humidity of air and large amounts of the acids may be lost during chromatography[59].

D. Determination

The quantitative analysis of acyl halides has been and still is the subject of a large amount of research work, and no wholly satisfying procedure has yet been developed. Titrimetric methods are hampered by the free acids present as impurities, while carboxylic acid derivatives other than acyl halides often interfere with colorimetric determinations. Gas chromatographic separation seems to be a promising technique, although it has been little used yet.

I. Determination by titration

Acyl halides are usually contaminated by small amounts of free carboxylic acids and hydrogen halides; this seriously complicates the quantitative analysis of these compounds. The presence of hydrogen halides renders useless, for instance, a gravimetric determination by argentimetry, or a direct acid–base titration of the mixture to be analysed. Early methods, such as titration by silver nitrate in pyridine–water mixture[60] or by sodium methylate in organic solvents (benzene, benzene–methanol or butyl-amine)[61], yielded results which were too high because free acids were titrated together with the acyl halides. In fact, the determination of acyl halides involves the simultaneous determination of the carboxylic acids and hydrogen halides. This makes the procedures cumbersome at best, and very often inaccurate as well, because at least one of the constituents is usually determined indirectly by subtraction.

a. Conversion to amides. Pesez and Willemart[62] were the first to attempt a complete determination. By reacting the acyl halide with aniline in dioxan

$$RCOX + C_6H_5NH_2 \longrightarrow RCONHC_6H_5 + HX$$

and by performing two separate titrations with sodium hydroxide and one with silver nitrate, they measured the acyl halide, carboxylic acid and hydrogen halide contents. All three values, however, were obtained indirectly and the results were of relatively low accuracy.

Other methods, also using the reaction of acyl halides with amines, attempted more direct determinations. Ackley and Tesoro[63] were able to determine the carboxylic acid present as impurity by reacting the sample to be analysed with ammonia; the acyl halides yielded amides, the carboxylic acids were converted into ammonium salts:

$$RCOX + 2 NH_3 \longrightarrow RCONH_2 + NH_4X$$

$$RCOOH + NH_3 \longrightarrow RCOONH_4$$

Subsequent acidification of the reaction mixture had no effect on the amides, but regenerated the free carboxylic acids from their ammonium salts; the acids were then extracted by ether and titrated. Bauer[64] used a similar procedure with aniline in place of ammonia.

Stahl and Siggia[65] improved Pesez's method by replacing aniline with *m*-chloroaniline and by using a potentiometric method for the sodium hydroxide titration. They obtained two breaks in the potential, one indicating completion of the titration of the free carboxylic acid, the other that of the hydrogen halide and the amine hydrohalide obtained in the reaction with *m*-chloroaniline. The free hydrogen halide originally present was determined by a separate potentiometric titration with tripropylamine in chlorobenzene–ether mixture.

Lohr[66] also used the reaction of an amine with acyl halides, directly titrating the mixture to be analysed with cyclohexylamine in tetrahydrofuran. In this procedure the free carboxylic acids present do not interfere with the reaction and are not determined; hydrogen halides are titrated together with the acyl halides, and a correction must be made by direct titration of a separate sample with tripropylamine[65].

b. *Hydrolysis in the presence of pyridine.* Mitchell and Smith[67] reacted the acyl halide with pyridine and a weighed quantity of water in dioxan or toluene:

$$C_5H_5N + RCOX + H_2O \longrightarrow RCOOH + C_5H_5N \cdot HX$$

The excess of water was then determined by a Karl Fischer titration; the free acids present did not interfere with the analysis. In a similar method Hennart and Vieillet[68] hydrolysed the mixture to be analysed in the presence of pyridine, then eliminated the excess of water by addition of an acid anhydride, and titrated the pyridine hydrohalide with perchloric acid; the originally present hydrogen halide was determined separately and the amount deducted from the result of the perchloric acid titration; carboxylic acids were not titrated.

Klamann[69] hydrolysed the mixture in a pyridine–water solution, then titrated with silver nitrate for total hydrogen halide content; the originally present hydrogen halide was determined by titration of the mixture to be analysed with silver nitrate in acetone, while the carboxylic acid content was obtained indirectly from a total sodium hydroxide titration.

   c. Esterification. Burger and Schulek[70] treat their mixture with ethanol, obtaining the ethyl ester from the acyl halide while the carboxylic acid

$$RCOX + C_2H_5OH \longrightarrow RCOOC_2H_5 + HX$$

does not react. If the carboxylic acid is weak, the total hydrogen halide content (produced by esterification and present originally) is determined by sodium hydroxide titration with methyl-red as indicator; the carboxylic acid content is then obtained by further sodium hydroxide titration using phenolphthalein as indicator; the ester is not titrated. If the carboxylic acid is strong it is titrated by sodium hydroxide together with the hydrogen halide and the hydrogen halide content must be determined by argentimetry. The actual acyl halide content is then determined by the quantity of alkali consumed in the saponification of the ester.

   d. Other methods. Simonyi and Kekesy[71], in a completely different approach, convert acyl chlorides into acyl iodides by reaction with sodium iodide in acetone.

$$RCOCl + NaI \longrightarrow RCOI + NaCl$$

The sodium chloride formed is insoluble in acetone and precipitates. It can be filtered off and titrated by Volhard's method. Free hydrogen chloride does not interfere with the determination, but other halogens which are easily exchanged by iodine are determined together with the acyl chloride.

   Stuerzer[72] analyses a mixture of acyl chloride and acid anhydride by potentiometric titration with silver acetate in pyridine, the hydrogen halide present as impurity being determined separately by tripropylamine titration.

   Finally one must mention the method developed by Patchornik and Rogozinski[73] for the micro-determination, in non-aqueous media, of the individual components of any mixture of hydrogen halide, carboxylic acid, acyl halide, acid anhydride and alkyl halide. This complex procedure involves several different titrations by sodium methylate and tributylamine, esterifications, reactions with amines and a titration with trimethylbenzyl-ammonium hydroxide. Patchornik does not recommend the use of this method if all five of the above components are simultaneously present; in this case equilibria such as

$$(RCO)_2O + HCl \rightleftharpoons RCOOH + RCOCl$$

are known to exist, and the proportions of the substances to be analysed may change during the course of the determination.

2. Colorimetric methods

a. Hydroxamic acid reaction. The hydroxamic acid test commonly used for the detection of acyl halides (see section I, A, 2) was applied by Goddu and co-workers[74] to the quantitative determination of esters and anhydrides. The analysis is made by measuring the optical density of the hydroxamic acid/ferric ion complex at the maximum of absorption (530 nm for aliphatic compounds, 550–560 nm for aromatic derivatives). Buzlanova and co-workers[75] adapted the procedure to the colorimetric determination of acyl halides, with a precision usually better than 5%. Both esters and anhydrides, if present, interfere with the measurement in alkaline medium; anhydrides interfere also in neutral solution. The carboxylic acids and the hydrogen halides resulting from partial hydrolysis do not contribute to the colour formation.

b. Reaction with 4-(4-*nitrobenzyl*)*pyridine.* The reaction used for the micro-detection of acyl halides (see section I, A, 4, b) was applied by Agree and Meeker[14] to their quantitative micro- or semimicro-determination in aprotic media such as methyl ethyl ketone or acetophenone. On addition of 4-(4-nitrobenzyl)pyridine the colour forms rapidly in most cases (sometimes a little heating is necessary) but a colour stabilizing agent such as triethylamine must be added. The acyl halide is then determined colorimetrically, molar absorbances being of the order of 10,000 to 25,000 l/(mol cm). The absorption band of the reaction product (situated between 445 and 475 nm) lies on the tail of the longest wavelength absorption band of 4-(4-nitrobenzyl)pyridine itself; by use of a baseline technique, Agree was able to avoid complications due to this factor. Mixtures of two acyl halides can also be determined in this way, using the difference in the rates of reaction of the two halides with the reagent[76]. As already mentioned, only chloroformates give the same reaction; alkyl or aryl halides have their maximum of absorption at much longer wavelengths, while carbamoyl halides interfere only slightly and acid anhydrides not at all.

c. Reaction with α-*oxo-aldoximes.* In aqueous alkaline medium acyl halides react with monoisonitrosoacetone

$$RCOX + CH_3COCHNOH + H_2O \longrightarrow RCOOH + CH_3COOH + HX + HCN$$

to yield hydrogen cyanide in quantity equivalent to the acyl halide to be determined[77]. The hydrogen cyanide can be converted into cyanogen chloride by reaction with chloramine T; subsequent addition of bis-(3-methyl-1-phenyl-5-pyrazolone) in pyridine yields an intense blue colour permitting colorimetric determination of microgram quantities. Acid anhydrides and chloroformates give the same reaction and their presence interferes with the determination.

3. Determination by gas chromatography

Direct quantitative gas chromatography of acyl halides is almost impossible because of hydrolysis by the humidity of air or decomposition on the column. However, if before the analysis the acyl halide is converted into a more stable derivative, while the carboxylic acid, present as impurity, remains unchanged, the gas chromatographic technique becomes applicable.

Hishta and Bomstein[78] thus reacted a mixture of several acyl halides with diethylamine in chloroform to obtain amides. These could then easily be separated and determined quantitatively. Carboxylic acids present in the original mixture did not interfere with the procedure. The same method can be applied to the determination of chloroformates.

Simonaitis and Guvernator[79] esterified the acyl halides with methanol under conditions such that carboxylic acids present in the mixture did not react. Subsequent gas chromatography allowed separation between the free acids and their methyl esters and quantitative determination of each of them.

II. PHYSICAL METHODS

A. Infrared Spectroscopy

The vibrational spectra of acyl halides are of exceptional interest, and are among those most frequently and extensively investigated. Measurements were made by i.r. and Raman spectroscopy, and the results reported here are derived from both techniques. Among the most often studied substances are the carbonyl halides themselves (COF_2, $COCl_2$, $COBr_2$, COFCl, COFBr, COClBr)[80-85], formyl fluoride (which is the only formyl halide stable enough to be examined under normal conditions)[86, 87], acetyl and halogenated acetyl halides[88-107], and some higher molecular weight halides[88, 90-94]. Complete vibrational assignments of the i.r. and Raman spectra of many of these compounds have been given.

1. The C=O stretching frequency (Tables 1 and 2)

a. Identification. The most striking feature of the spectra of acyl halides is their exceptionally high carbonyl stretching frequency. Phosgene has a frequency of 1827 cm^{-1} (as compared with 1745 cm^{-1} for formaldehyde) and the value of 1928 cm^{-1} found for COF_2 is probably the highest known. Acetyl halides have carbonyl stretching frequencies somewhat higher than 1800 cm^{-1}. Replacement of the CH_3 group by a heavier alkyl substituent only slightly decreases this frequency; Kahovec and Kohlrausch[90] and Seewann-Albert[91, 92] systematically examined a large number of acyl

TABLE 1. Carbonyl stretching frequencies of carbonyl halides

Compound	$\nu_{C=O}$ (cm^{-1})	State	Reference
HCOF	1837	vapour	87
	1834	vapour	86
	1809[a]	liquid	86
	1806	solid	86
COF$_2$	1928	vapour	82
COCl$_2$	1827	vapour	82
COBr$_2$	1828	vapour	85
COFCl	1868	vapour	82
COFBr	1874	vapour	84
COClBr	1828	vapour	85

[a] Value obtained by Raman spectroscopy.

chlorides ($C_nH_{2n+1}COCl$) and acyl bromides ($C_nH_{2n+1}COBr$) and found that, when n increases from 3 to 8, the carbonyl stretching frequencies are practically constant (about 1795 cm^{-1} for the chlorides, 1805 cm^{-1} for the bromides). The carbonyl stretching vibration of phenylacetyl chloride was also found at the same frequency[89]. Conjugation, withdrawing electrons from the C=O bond and diminishing its double bond character, lowers the carbonyl frequency; this can be seen clearly (Table 2) from a comparison of $\nu_{C=O}$ of CH_3COX and of C_6H_5COX or $CH_2=CHCOCl$ (although the case of benzoyl chloride is complicated by Fermi resonance).

These carbonyl stretching frequencies, higher by about 100 cm^{-1} than those of the corresponding aldehydes and ketones, are characteristic of acyl halides and provide the best way to their detection or identification. The only other organic compounds showing similar frequencies are the anhydrides and peroxides; in these cases, however, vibrational coupling between the two neighbouring carbonyl groups results in a splitting of their stretching frequencies into symmetrical and asymmetrical vibrations, so that these substances are characterized by the presence of *two* carbonyl stretching frequencies, separated by about 60 cm^{-1} for anhydrides and by 30 cm^{-1} for peroxides. Acyl halides, with some exceptions due to rotational isomerism or Fermi splitting (see sections II, A, 3 and II, A, 4), show only a single carbonyl stretching absorption.

b. Theoretical aspects. There is a great deal of uncertainty about the origin of the high carbonyl stretching frequency; this problem was recently discussed by Bellamy[108] in the framework of a general study of carbonyl frequencies.

One of the possible explanations is direct vibrational coupling of the carbonyl stretching vibration with the stretching or bending vibrations

TABLE 2. Carbonyl stretching frequencies of some acyl halides

Compound	$\nu_{C=O}$ (cm^{-1})	State	Technique[a]	Reference
CH$_3$COF	1870	vapour	i.r.	97, 98, 103, 105
	1841	liquid	i.r.	105
CH$_3$COCl	1822	vapour	i.r.	88, 98, 103, 120, 121
	1806	CCl$_4$ sol, liquid	i.r., R	98, 120, 121
CH$_3$COBr	1827	vapour	i.r.	98, 103, 105
	1814	solid	i.r.	95
	1802	liquid	i.r.	105
CH$_3$COI	1808	vapour	i.r.	105
	1799	liquid	i.r.	105
C$_2$H$_5$COCl	1786		R	90
n-C$_3$H$_7$COCl	1791		R	90
n-C$_5$H$_{11}$COCl	1794		R	90
C$_6$H$_5$CH$_2$COCl	1795	CCl$_4$ sol	i.r.	89
CH$_2$=CHCOCl	1761	CCl$_4$ sol	i.r.	107
C$_6$H$_5$COF	1820		i.r.	109
C$_6$H$_5$COCl	1778, 1735	CCl$_4$ sol	i.r., R	89, 131, 136, 138
p-ClC$_6$H$_4$COCl	1772, 1732	CCl$_4$ sol	R	90
C$_{10}$H$_7$COCl	1760	CCl$_4$ sol	i.r.	140
C$_6$H$_5$COBr	1778		i.r.	109
CF$_3$COF	1898	vapour	i.r.	99, 104
CF$_3$COCl	1811	vapour	i.r.	99, 100
	1799	liquid	R	100
CF$_3$COBr	1838	vapour	i.r.	99
CF$_3$COI	1812	vapour	i.r.	99
CCl$_3$COCl	1815	vapour	i.r.	121
	1803	CCl$_4$ sol	i.r.	121

[a] i.r.: infrared spectroscopy; R: Raman spectroscopy.

of neighbouring bonds[109-111]. Such a coupling would depend on three factors: the masses of the X and Y groups adjacent to the C=O bond, the angles between the C—X, C—Y and C=O bonds and the strengths of the C—X and C—Y bonds. Direct mass effects of the substituents X or Y in XCOY molecules were found to be negligible[112, 113] if the masses of X and Y are larger than 12. The bond angles in acyl halides should be practically constant, as the carbonyl halide group is sterically unstrained; the variation of $\nu_{C=O}$ from one halide to the other can hardly be explained by an angle effect. The bond strengths, on the other hand, vary strongly from one substituent to the other, and mechanical coupling of the $C=O$ bond with different $C-X$ or $C-Y$ bonds should lead to very different $C=O$ stretching frequencies. Overend and Scherer[109, 110] assumed that in acyl halide molecules the force constant K_{CO} of the carbonyl bond is independent of its neighbours, and calculated the shifts produced by interaction of the stretching force constants K_{CX}, K_{CY} and the bending force constants H_{XCY}, H_{XCO} and H_{YCO}[114] with K_{CO}. Their results show that such shifts can be very important, and that the very large difference between $\nu_{C=O}$ in $COCl_2$ and in COF_2 could be caused by the difference in strength between the $C-Cl$ and the $C-F$ bonds. However, the carbonyl frequencies predicted only on the basis of vibrational coupling with neighbouring bonds should decrease in the order $F > CH_3 > Cl > Br$, while the observed sequence is $F > Cl > Br > CH_3$[108]. Some coupling undoubtedly takes place, but this, and other experimental evidence, suggests that a different mechanism is responsible for the high value of $\nu_{C=O}$ or at least contributes to it.

The most probable explanation is that an inductive effect directly affects the $C=O$ bond strength and therefore the $C=O$ stretching frequency[115]. The halogen attracts electrons from the carbon which in turn attracts electrons from the oxygen, thereby decreasing the negative charge on the latter. This process reduces the polarity of the $C=O$ bond, giving it more of a covalent, double-bond character.

$$\text{>}C^+\text{—}O^- \longleftrightarrow \text{>}C=O$$

The bond length decreases with decreasing polarity of the bond; the relation between bond length and bond stretching frequency is well known and the high $\nu_{C=O}$ frequency in acyl halides is easily explained in this way. Kagarise[116] found a very good linear relationship between the sum of the electronegativities of X and Y in XCOY and the experimental $C=O$ stretching frequencies, strongly supporting this type of explanation; the above-mentioned sequence of frequencies $F > Cl > Br > CH_3$ agrees with it. Shimizu and Shingu[117] successfully treated the carbonyl out-of-plane deformation in a similar fashion. Because of the high electron affinity of halogens, Pauling[118] even suggested that there is some contribution from a structure of the oxonium type in phosgene and acyl halides. In spite of the fact that the structure proposed by him is not compatible with the bond dipole moments determined by Krisher and Wilson[119], there is strong experimental confirmation that induction plays a dominant role here.

The carbonyl group absorbs at a frequency higher by about 20 cm^{-1} in the vapour phase than in the liquid or solid state (see Tables 1 and 2). The

4

change of frequency with change of state probably results from inter-molecular association[108]. While there is no association in the vapour, in the condensed state the $\overset{+}{C}=\overset{-}{O}$ dipoles tend to orient themselves in such a way that the C^+ of one molecule is near the O^- of another; the polarity of the $C=O$ bond increases by induction, resulting in the observed shift to lower frequencies. Such an association would depend strongly on the dipole moment of the carbonyl group; those groups of compounds, among them acyl halides, which have a more covalent $C=O$ bond would have less tendency to self-association and would show a smaller shift than those having a more polar $C=O$ bond; actually, the shift in acetyl chloride is about 15 cm^{-1} while in dimethylformamide it is 50 cm^{-1}. This is additional evidence that in acetyl halides the $C=O$ bond is more covalent than usual.

The dependence of $\nu_{C=O}$ in acetyl chloride on the solvent is also at-tributed to local dipole–dipole associations[120]; it varies from 1810 cm^{-1} in n-hexane to 1798 cm^{-1} in methylene iodide.

2. The C—X stretching frequency (Table 3)

In contrast to the wealth of information available on the $C=O$ stretching frequency, very little systematic work was done on the C–halogen stretching vibrations of acyl halides.

As can be seen in Table 3, the usual sequence of decreasing frequencies when passing from fluoride to chloride to bromide to iodide is also found here. A comparison between the C–halogen stretching frequencies in alkyl and acyl halides (Table 4) shows that these frequencies are usually much lower in the acyl halides; the differences, however, decrease from the fluorides to the bromides and in iodides there is a reversal (the ν_{C-I} frequency being slightly higher in acetyl iodide than in methyl iodide).

The relatively low frequencies of the C–halogen stretching vibrations are directly related to the very high carbonyl stretching frequencies in the acyl halides, whether these are caused by inductive changes of polarity or by vibrational coupling between neighbouring bonds. The fact that the frequency shift relative to alkyl halides is largest for the fluorides, where $\nu_{C=O}$ and $\nu_{C-halogen}$ are the nearest, seems to support the vibrational coupling hypothesis[105].

Splitting due to rotational isomerism is sometimes found (see section II, A, 3).

3. Rotational isomerism and field effects

The splitting of vibrational absorption bands of acyl halides due to rotational isomerism was first detected by Seewann–Albert and

TABLE 3. $C-X$ stretching frequencies of some carbonyl and acyl halides

Compound	ν_{C-X} (cm^{-1})	State	Reference
HCOF	1065 vs	vapour	87
	1042 vs[a]	liquid	86
	1015 vs	solid	86
COF_2	1249 (asym)vs; 965 (sym)s	vapour	82
$COCl_2$	849 (asym)vs; 575 (sym)s	vapour	82
$COBr_2$	747 (asym)vs; 425 (sym)m	vapour	85
COFCl	1095 (asym)vs; 776 (sym)s	vapour	82
COFBr	1068 (asym)vs; 721 (sym)s	vapour	84
COClBr	806 (asym)vs; 517 (sym)m	vapour	85
CH_3COF	832 vs	vapour	105
	819 vs	liquid	105
CH_3COCl	605 vs	vapour	105
	588 vs	liquid	105
CH_3COBr	563 vs	vapour	105
	552 vs	liquid	105
CH_3COI	542 s	vapour	105
	540 s	liquid	105
CF_3COF	1339 s	vapour	104
CF_3COCl	750 s	vapour	100
	745 s[a]	liquid	100

[a] Value obtained by Raman spectroscopy.

TABLE 4. Comparison between the ν_{C-X} frequency in alkyl and acyl halides

Alkyl halide	ν_{C-X} (cm^{-1})	Acyl halide	ν_{C-X} (cm^{-1})
CH_3F	1048	CH_3COF	832
CH_3Cl	732	CH_3COCl	605
CH_3Br	611	CH_3COBr	563
CH_3I	533	CH_3COI	542

Kahovec[91, 92] and investigated by Nakagawa and co-workers[95]; it was later studied very extensively[103, 106, 121, 122]. In halogenoacetyl halides, the number of vibrational bands found, both in i.r. and in Raman spectroscopy, is much higher than can be expected from the $3n-6$ degrees of freedom of these molecules. The bands appear throughout the spectral range as doublets, in each of which the relative intensity strongly depends on the temperature (Table 5). Those bands of the doublets which decrease

TABLE 5. Carbonyl stretching vibrations in halogenoacetyl halides[95]

Substance	$\nu_{C=O}$ solid (Raman)	$\nu_{C=O}$ liquid (Raman)	$\nu_{C=O}$ vapour (i.r.)
$CH_2ClCOCl$	1799	1814, 1780*	1809, 1779*
$CH_2BrCOCl$	1797	1804, 1779*	1802, 1773*
$CH_2BrCOBr$	—	1810, 1790*	1807, 1783*

The intensity of bands marked * increases with temperature.

in intensity with decreasing temperature disappear completely in the solid state, where the $3n$–6 rule is fulfilled (e.g. in the $\nu_{C=O}$ range bromoacetyl chloride vapour has two bands, at 1802 and 1773 cm^{-1}; only the higher one is found in the solid state). This effect is almost certainly due to the presence, in the liquid and vapour states, of two rotational isomers, of which the most stable one persists at low temperatures. This hypothesis is supported by the fact that while splitting is found in all mono- and dihalogenoacetyl halides, where rotational isomerism is possible, it is never observed in unsubstituted or trihalogenated acetyl halides, for which no rotational isomers exist[121] (see Table 6). The vibrational bands

TABLE 6. Rotational isomerism in chloroacetyl chlorides[121]

Substance	$\nu_{C=O}$ in the vapour phase (cm^{-1})	$\nu_{C=O}$ in CCl$_4$ solution (cm^{-1})
CH_3COCl	1821	1806
$CH_2ClCOCl$	1835, 1798	1821, 1785
$CHCl_2COCl$	1823, 1790	1810, 1779
CCl_3COCl	1815	1803

which persist in the liquid state at low temperatures and in the solid state (e.g. the higher frequency carbonyl band of bromoacetyl chloride) are assigned to the most stable isomer, those which increase with temperature are attributed to the less stable one. From the temperature-dependence of the relative intensity of the bands of the doublets, the energy difference between the two forms could be calculated; it was found to be $1 \cdot 0 \pm 0 \cdot 1$ kcal/mole for bromoacetyl chloride, $1 \cdot 9 \pm 0 \cdot 3$ kcal/mole for bromoacetyl bromide[95].

Similar effects were found in halogeno ketones[123, 124] and it is well known that in cyclic ketones axial α-halogen substitution does not affect the carbonyl stretching frequency, while equatorial substitution raises it by about 20 cm^{-1} [125]. Such shifts cannot be due to induction which would act along the bonds of the molecule and be independent of conformation, nor can they be due to resonance which lowers the frequency. They must be due to a non-bonded force, probably a field effect occurring directly through space between the oxygen and the halogen atoms when they are close to each other. The exact nature of this force is not known, it is probably an electrostatic effect which changes the polarity of the atoms, perhaps by the lone pair–lone pair repulsion proposed by Owen and Sheppard[126]. The highest carbonyl frequency observed for substituted acetyl halides would thus correspond to that form in which the α-halogen atom is closest to the oxygen; since this is the frequency which is predominant in liquids at low temperature, and the only one persisting in the solid state, it means that this is the most stable, preferred, conformation of the molecule.

There is little doubt about the most stable conformation: microwave[172] and electron diffraction[128] studies, as well as normal coordinate calculations with a Urey–Bradley force field[103], all show this to be the conformation in which the two halogen atoms are *trans* to each other, i.e. the oxygen atom is eclipsed by the α-halogen atom, in good agreement with the field effect theory. The conformation of the other isomer is not so well established. Nakagawa and co-workers'[95] suggestion, that it is derived from the stable form by a rotation of 150°, is not easily acceptable, as it implies a six-fold barrier of rotation around the C—C bond, which would result in the existence of four rotational isomers, while only two were found. Microwave studies of acetyl fluoride[129] and acetyl chloride[130], as well as far i.r. vibrational assignments for trifluoroacetyl halides[100–102, 104], show that these compounds have a three-fold barrier, which corresponds to rotations of 120°; normal co-ordinate calculations[103] of acetyl and deuterated acetyl chloride also indicate that the less stable form is a *gauche* one, obtained from the other by rotation of 120°. However, a more recent microwave study of fluoroacetyl fluoride[127] shows the less stable conformation to be obtained from the stable one by rotation of 180°, so that the two fluorine atoms are *cis* to each other. The evidence is therefore inconclusive; it is possible that steric factors, as well as electrostatic ones, play a role, and that the conformation depends on the size of the halogen atoms; it would then be different, e.g. for fluorides and for bromides.

Rotational isomerism in acyl halides of higher molecular weight was studied by Katon and Feairheller[122] both in the liquid and solid state. In the liquid state all straight-chain acyl chlorides containing more than three carbon atoms have four strong bands in the C–halogen stretching region; in the crystalline state only one band remains, showing the origin of the splitting to be in rotational isomerism. Rotation around the α C—C bond can produce only two such isomers, therefore rotation around the β C—C bond also must play a role; thus four isomers can be obtained:

α-*trans*, β-*trans*
α-*trans*, β-*gauche*
α-*gauche*, β-*trans*
α-*gauche*, β-*gauche*

This is spectroscopic evidence for intramolecular non-bonded forces acting over considerable distances, a very astonishing result. Further confirmation is found in the fact that acetyl chloride and trimethylacetyl chloride which can have no rotational isomers show only one ν_{C-Cl} band; propionyl, isobutyryl and 3,3-dimethylbutyryl chlorides have symmetric substituents around the β C—C, but not the α C—C bond; they should exist in two isomeric forms, and in fact two ν_{C-Cl} bands are found in the liquid state, one of which disappears on crystallization.

Katon and Feairheller[122] assigned the different ν_{C-Cl} frequencies by assuming that the most stable isomers (which are found in the solid state) are of the *trans* form and that the *trans* forms have higher frequencies than the *gauche* forms. However, temperature-dependence studies must be evaluated carefully when more than two isomers are possible, as the relative intensities may then be complex functions of the temperature.

In another study of rotational isomerism, Jenkins and Ladd[106] found that in fluoroacetyl chloride and bromide the lowest ν_{C-Cl} or ν_{C-Br} frequency is almost identical with that of acetyl chloride or bromide while the ν_{C-Cl} or ν_{C-Br} frequency of the stable *trans* form is exceptionally high (Table 7). They suggest that in this conformation the increase of $\nu_{C=O}$ due to the electrostatic field effect between neighbouring O and F atoms strongly reduces the vibrational coupling between C=O and C—Cl or C—Br, thereby increasing ν_{C-Cl} or ν_{C-Br}; this, however,

TABLE 7. The ν_{C-X} splitting due to rotational isomerism[106]

Compound	ν_{C-X} (cm^{-1})	State
CH$_2$FCOCl	768 vs^a; 591 s	vapour
	765 vs^a; 590 s	liquid
	768 vs^a; 593 w	solid
CH$_2$FCOBr	751 vs^a; 544 m	vapour
	746 vs^a; 538 m	liquid
	746 vs^a; 535 m	solid

^{a} Designates the most stable isomer.

cannot be the whole explanation, as the ν_{C-Cl} or ν_{C-Br} frequencies of the *trans* conformations are even higher than those of the methyl halides.

4. Fermi splitting

The splitting of the $\nu_{C=O}$ band due to rotational isomerism must be distinguished from the splitting by Fermi resonance which appears in some acyl halides. The best-known example is that of benzoyl chloride which has two absorption bands at 1779 and 1735 cm^{-1}; this was first observed by Thompson and Norris[131] in the Raman spectrum and by Flett[132] in the i.r.

Several explanations were advanced. Lowering of the frequency by intermolecular association[90, 133] is in contradiction to the independence of the relative intensity of the two bands on the concentration[134, 135], and to cryoscopic studies[136] showing benzoyl chloride to be a monomer. The hypothesis that rotational isomerism lies at the origin of the splitting[137] must also be rejected, because in the case of benzoyl chloride no significantly different stable conformational isomers can be written; moreover, as already mentioned, rotational isomerism results in the splitting of a large number of vibration bands while only the carbonyl stretching band is split in this case. The explanation based on partial hydrolysis to benzoic acid can also be eliminated since on addition of small quantities of benzoic acid to the solution, a third carbonyl band appears[134] at 1695 cm^{-1}. There are some contradictions about the reported dependence of the relative intensities of the two bands on temperature[134, 136], and the possibility that the splitting is due to a 'hot' transition cannot be completely excluded. The most probable explanation, however, is that a Fermi resonance takes place[134, 136, 138], i.e. that the splitting is due to interaction between the C=O stretching vibration and another vibration of the same symmetry and very similar frequency. In the somewhat similar case of the $\nu_{C=O}$ splitting in cyclopentanone[139], the effect disappears on deuteration, proving beyond doubt that some intramolecular vibrational property is involved. No deuteration of benzoyl chloride was reported, but its behaviour is very similar to that of cyclopentanone, with regard to solvent polarity and temperature dependence.

A search for the vibration which could interact with the $\nu_{C=O}$ vibration shows[136]: (a) that the $\nu_{C=O}$ frequency of benzoyl chloride should appear 'normally' around 1770–1780 cm^{-1}; (b) that the strong ν_{C-Cl} vibration at 878 cm^{-1} should have its first overtone around 1745 cm^{-1} and has the same symmetry as $\nu_{C=O}$. This makes it probable that these two vibrations interact. The observation that the same $\nu_{C=O}$ splitting occurs in a large number of substituted benzoyl chlorides[90, 135], while none is found in the other benzoyl halides, strengthens the assumption that it is ν_{C-Cl} which is playing a role in this Fermi resonance. Rao and Venkataghavan[138], however, assign the 878 cm^{-1} vibration of benzoyl chloride to the Ph—C stretching vibration, and study its frequency variation on substitution: the more electron-withdrawing the substituent, the higher is the ν_{Ph-C} frequency (and that of its first overtone) and the less splitting of $\nu_{C=O}$ is observed; in 3,5-dinitrobenzoyl chloride $\nu_{Ph-C} = 988$ cm^{-1} and no Fermi splitting is seen. A similarly assigned vibration is observed in benzoyl bromide at 861 cm^{-1}; its overtone (at 1712 cm^{-1}) is too low to interact with $\nu_{C=O}$, but substitution of the phenyl by electron-withdrawing groups increases its frequency until in p-nitrobenzoyl bromide, Fermi splitting is again observed. In this connexion, it is surprising to note that in 1- and in 2-naphthoyl chloride, although a strong band appears at the same frequency (890 cm^{-1}) as in benzoyl chloride, no splitting of $\nu_{C=O}$ is observed[140].

Similar effects of splitting, found in the i.r. and Raman spectra of COF$_2$ and COFCl, have been attributed[82] to Fermi resonance between ν_2 (1942 cm^{-1}) and $2\nu_1$ (965 cm^{-1}) for COF$_2$, between ν_2 (1876 cm^{-1}) and $\nu_1 + \nu_4$ (776 cm^{-1} + 1095 cm^{-1}) for COFCl (where ν_2 represents $\nu_{C=O}$, ν_1 the in-phase ν_{C-X} and ν_4 the out-of-phase ν_{C-X}). No splitting was observed in COCl$_2$ where no band suitable for interaction with $\nu_{C=O}$ was found.

5. Oxalyl halides

A group of acyl halides that has been more thoroughly investigated is that of the oxalyl halides (FCOCOF, ClCOCOCl, BrCOCOBr and ClCOCOF), which were studied by i.r. and Raman spectroscopy[141-145] and by examination of the vibrational fine structure of u.v. absorption bands[146-152]. The complete assignment of all bands was attempted in an effort to determine the steric conformation of the oxalyl halides.

On this problem some conflicting evidence has been published. X-ray crystallography shows the oxalyl chloride to be nearly centrosymmetric, though not necessarily planar. Electron diffraction in the vapour state shows the structure to be planar and *trans*. Dipole moment measurement[153] in benzene, yielding a dipole moment of 0·92 D, would indicate some contribution from a *cis* structure, unless oxalyl chloride forms a complex with benzene. Comparison of the u.v. spectra in different solvents[154] indeed indicates some solvent–solute association in benzene, but again complex formation with benzene implies a *cis* form of the molecule.

Vibrational spectra also yield contradictory results. Saksena and Kagarise[141] found that the relative intensity of the 533 cm^{-1} Raman band of oxalyl chloride, assigned to the *cis* form, increases with temperature;

intensity variations in the vibrational fine structure of the u.v. spectrum were interpreted both by Sidman[147] and by Saksena and co-workers[146, 148] (though in different ways) as indicating the presence of the two conformations at room temperature, the *cis* form being less stable by 2·8 kcal/mole. However, Kidd and King[145] found no change in the Raman bands of oxalyl bromide on cooling and an energy difference of 2·8 kcal/mole seems too high for an appreciable amount of the *cis* form to be present at room temperature.

Complete vibrational assignments can be made on the assumption that the *trans* form only is present; the simultaneous existence of both isomers would result in many more vibrational bands than are detected experimentally. Moreover, in the *cis* form, coupling between two neighbouring carbonyl groups would result in a splitting of $\nu_{C=O}$ into symmetric and asymmetric vibrations, while no splitting can be expected in the *trans* form; experimentally such splitting has been observed, e.g. in *cis*-dimethyl oxalate[155], but never in an oxalyl halide.

It seems therefore that vibrational evidence is mainly in favour of the *trans* conformation.

6. Related Compounds

Two groups of substances related to acyl halides have been studied by i.r. spectroscopy: compounds in which the carbonyl oxygen atom was replaced by a sulphur atom, i.e. thiocarbonyl halides, and compounds in which the α-carbon atom was replaced by another element (O, N or S), i.e. chloroformates, carbamoyl chlorides or thiolchloroformates. In each group the number of substances investigated is small, and only few data are available.

a. Thiocarbonyl halides. The C=S stretching frequency of thiocarbonyl chloride[156] is 1140 cm⁻¹; in compounds like RSCSCl this frequency lies between 1060 and 1100 cm⁻¹, in compounds like RSCSF between 1075 and 1120 cm⁻¹. These frequencies are lower than those of thioketones[157] (1207–1224 cm⁻¹) and higher than those of thioamides[158] (960–980 cm⁻¹), while acyl halides have a carbonyl frequency higher than both that of ketones and that of amides. This can be attributed to the difference in polarity between the C=O and the C=S bonds[158]. The C=O bond is initially polar because O is more electronegative than C; therefore, as discussed above, inductive effects raise $\nu_{C=O}$ (as in acyl halides) while resonance effects lower it (as in amides). In C=S there is little or no initial polarity; therefore both inductive and resonative effects produce a shift of $\nu_{C=S}$ toward lower frequencies; as in the case of $\nu_{C=C}$, the highest $\nu_{C=S}$ frequency is found when the C=S bond is fully covalent, as in thioketones.

b. Chloroformates, carbamoyl halides and thiolchloroformates. All these compounds are characterized by carbonyl stretching frequencies higher than those of their unchlorinated homologues, exactly as in the case of the acyl halides themselves, and for the same reasons.

(1) *Chloroformates, ROCOCl*: One aromatic and several aliphatic chloroformates have carbonyl stretching frequencies[159–163] between 1768 and 1799 cm^{-1}, while in formates this band appears at 1722–1724 cm^{-1}.

(2) *Carbamoyl chlorides, R^{2}NCOCl*: One aromatic and two aliphatic carbamoyl chlorides[160, 162] absorb at 1739–1744 cm^{-1}, while the carbonyl frequency of dimethylformamide is found at 1687 cm^{-1}.

(3) *Thiolchloroformates, RSCOCl*: Three aliphatic and two aromatic thiolchloroformates[160, 162] have their carbonyl stretching frequency in the 1766–1775 cm^{-1} range, although Fermi resonance sometimes makes the assignment difficult.

In all these compounds, practically no data exist about the C—Cl stretching frequencies.

B. Ultraviolet Spectroscopy

The u.v. spectra of a large number of acyl halides have been determined[164]. Tables 8 and 9 give the values of λ_{max} and ε_{max} for some typical compounds. The spectra are characterized by a weak absorption band in the near u.v., generally attributed to a symmetry-forbidden $n–\pi^*$ transition. This is followed by a stronger absorption band at shorter wavelength, which in simple aliphatic acyl halides very often has not been recorded, but is only indicated by the strong increase of absorbance towards the lower limit of the wavelength range of available spectrophotometers. The second band has been variably attributed to a $\pi–\pi^*$ or to an $n–\sigma^*$ transition[165, 166]; recent studies[167, 168] seem to prove that the $\pi–\pi^*$ transition is at the origin of this absorption.

The wavelength of the $n–\pi^*$ transition in acyl halides is shorter than in the corresponding ketones and longer than in the corresponding acids (Tables 8 and 9). This was at first explained by the raising of the π^*–level of the ketones by interaction with a halogen lone pair of electrons, while the n-level remained largely unaffected[165]. More recent molecular orbital calculations[167] have, however, indicated that the blue shift is due to another mechanism. Replacement of the CH$_3$ group of acetone or acetophenone, taken as references, by a strongly σ-accepting and weakly π-donating halogen atom considerably lowers the energy of the highest filled σ-type orbital (n-electrons) and, to a lesser extent, raises the energy of the lowest empty π^*-type orbital. This results in the experimentally observed strong blue shift of the $n–\pi^*$ transition. In agreement with this

TABLE 8. Ultraviolet spectra of aliphatic acyl halides

Compound	$n \rightarrow \pi^*$ Transition		$\pi \rightarrow \pi^*$ Transition		Solvent	Reference
	λ_{max} (nm)	ε_{max}	λ_{max} (nm)	ε_{max}		
CH_3COF	205	37	<180		n-heptane	167
CH_3COCl	241	38	<180		n-heptane	167
CH_3COBr	250	94sh	<180		n-heptane	167
CF_3COF	217	60	130	2300	vapour	168
$n\text{-}C_3F_7COF$	215	66	<200		vapour	170
CF_3COCl	253	17	<200		vapour	169
$n\text{-}C_3F_7COCl$	258, 266	38	<200		vapour	170
$n\text{-}C_3F_7COBr$	273	56	217	692	cyclohexane	170
CH_3COCH_3	279	15	186	2050	n-heptane	167
CH_3COOH	203	53	<180		n-heptane	167

TABLE 9. Ultraviolet spectra of aromatic acyl halides

Compound	$n \rightarrow \pi^*$ Transition		$\pi \rightarrow \pi^*$ Transition		Solvent	Reference
	λ_{max} (nm)	ε_{max}	λ_{max} (nm)	ε_{max}		
C_6H_5COF	278	1910	230	18,400	n-heptane	167
C_6H_5COCl	282	2250	241	27,200	n-heptane	135, 167
C_6H_5COBr	282	1520	245	16,500	n-heptane	167
$C_6H_5COCH_3$	320	48	236	14,400	n-heptane	167
C_6H_5COOH	275	1040	232	13,000	n-heptane	167

theory, a progression of the $n-\pi^*$ absorption band towards shorter wavelengths is noted in the aliphatic acyl halides from bromide to fluoride, i.e. as the electronegativity of the halogen increases; in aromatic compounds this effect, while still present, is less marked. Replacement of the halogen atom by a carboxylic —OH group (i.e. in acids) further increases the hypsochromic shift. Substitution by fluorine in the α-position on the contrary results in a strong displacement of the absorption band toward longer wavelengths.

The lowest π^*-level is much less affected than the highest σ-level by introduction of the halogen atoms, and the $\pi-\pi^*$ absorption bands of acyl halides are only little shifted to the blue compared to acetone or acetophenone; in benzoyl chloride and bromide one even observes a small bathochromic effect. o-, m- or p-Substitution on the benzene ring of benzoyl halides[135] has the same effect on the $\pi-\pi^*$ absorption band as in acetophenone or in benzaldehyde. These facts considerably reduce the value of u.v. spectroscopy in the characterization of aroyl halides.

The u.v. spectra of several simple acyl halides show no clear maxima, only a slow decrease of the absorption towards longer wavelengths, which can be resolved to show vibrational and rotational fine-structure. The spectra of HCOF and DCOF in particular (which are the only stable formyl halides) have been very extensively studied in order to determine the exact geometry of the molecules in the ground and excited states, their moments of inertia, the energy barriers to rotation, etc.[171-174].

As mentioned earlier, the high-resolution u.v. spectra of another group of compounds, the oxalyl halides, have also been the subject of a number of detailed studies[146-152]. Saksena and co-workers[146, 148] and Sidman[147] suggest that oxalyl chloride, in the liquid and vapour phases, exists as two rotational isomers (the cis and trans forms) of which the trans isomer is the most stable and is predominant at room conditions, while the proportion of the cis form increases with rising temperature. Balfour and King[149-152], however, are able to explain the band structure and intensity distribution of the vapour phase spectrum of oxalyl chloride and oxalyl bromide by transitions from planar trans ground state to planar trans excited state, throwing some doubt on the presence of the cis form.

C. Electric and Magnetic Properties

I. Electric dipole moments

The dipole moments of a series of acyl chlorides were studied in the nineteen-thirties[175-179] but no systematic work on this subject has been published recently. Table 10 gives the values obtained for some of the compounds.

The moment of aliphatic derivatives increases from acetyl to propionyl chloride and then remains constant; this is due to induction along the aliphatic chain and is found in many similar homologue series. The

TABLE 10. The dipole moments of some acyl halides[179] (measured in benzene at 20°C)

Compounds	μ (D)
Acetyl chloride	2.45
Propionyl chloride	2·61
n-Butyryl chloride	2·61
n-Valeryl chloride	2·61
iso-Valeryl chloride	2·63
Phenylacetyl chloride	2·54
Chloroacetyl chloride	2·22
Trichloroacetyl chloride	1·19
Acetyl bromide	2·43
Acetyl iodide	2·22[a]
Benzoyl chloride	3·33
p-Toluoyl chloride	3·81
p-Chlorobenzoyl chloride	2·00
p-Nitrobenzoyl chloride	1·11
Benzoyl bromide	3·37
Oxalyl chloride	0·92
Malonyl chloride	2·80
Succinyl chloride	3·00
Phthaloyl chloride	5·12

[a] Acetyl iodide is unstable and this value is somewhat doubtful.

moment of phenylacetyl chloride lies between those of acetyl and propionyl chloride, as expected. The moments of acetyl bromide and iodide are a little lower than that of chloride.

From a comparison of the moments of acetyl chloride, chloroacetyl chloride and trichloroacetyl chloride, Martin and Partington[176] deduced that the moment of the aliphatic COCl group is 2·45 D and oriented at 15° to the C—C bond. Aromatic acyl halides have a much higher moment; the moment of benzoyl chloride is 3·33 D and is oriented at 12° to the C—C axis[177]; from this value, the moments of the other aromatic acyl chlorides can be deduced by vector addition, with only slight discrepancies.

The dipole moments of some dicarboxylic acid halides were determined in an attempt to study the conformation and isomerism of these compounds[178]. The measurements gave no clear-cut answer for the conformation of oxalyl chloride, which could be in the *cis* or the *trans* form, rotate freely or oscillate around a stable conformation. The value obtained for malonyl chloride corresponds to free rotation, that of succinyl chloride to almost free rotation; the liquid form of phthaloyl chloride was found to be symmetrical.

2. Nuclear quadrupole resonance

The pure quadrupole resonance of acyl halides was investigated as early as 1953 by several authors[180–182]. Only few bromides were studied: acetyl bromide shows four Br^{81} resonances at 178·082, 179·536, 180·140 and 181·350 Mc/sec; benzoyl bromide has one resonance at 191·868 Mc/sec. Chlorides were more extensively studied: the COCl group is characterized by a Cl^{35} resonance in the 28–34 Mc/sec range, at frequencies lower than the values found for chloroalkanes. This can be attributed to the vicinity of the C=O double bond and to the participation of lone-pair electrons in the bonding, giving rise to contributions of the $^-O-C=Cl^+$ structure. The ratio of 6·5 between Br^{81} and Cl^{35} resonance frequencies in similar compounds[183] was found to hold for benzoyl bromide and benzoyl chloride; no resonance, however, was found in acetyl chloride. As a rule no Cl^{35} resonance could be detected in those paraffinic acyl chlorides which contain a single COCl group; the presence of a second COCl group or at least of another chlorine atom in the molecule seems necessary. Thus resonance was found at 30·437 Mc/sec in chloroacetyl chloride, at 32·147 and 32·962 Mc/sec in dichloroacetyl chloride and at 33·721 Mc/sec in trichloroacetyl chloride, while none could be detected in unsubstituted acetyl chloride. Similarly, adipyl chloride, $ClCO(CH_2)_4COCl$, has its resonance at 28·978 Mc/sec, while no resonance is found in *n*-caproyl chloride, $CH_3(CH_2)_4COCl$.

It is interesting to note that oxalyl chloride has only one resonance frequency (at 33·621 Mc/sec)[182]. This is taken as additional proof that only one form of this compound exists in the crystalline state[184]; for, if both the *cis* and the *trans* form were present, the intramolecular chlorine–chlorine distances would be different in the two isomers, giving rise to two separate resonances.

Some carbamoyl chlorides (R^2NCOCl) and chloroformates (ROCOCl) have also been studied[182] and show resonance frequencies of the same order as acyl chlorides [$(CH_3)_2NCOCl$: 31·8 Mc/sec; $(C_2H_5)_2NCOCl$: 31·887 Mc/sec; $C_2H_5OCOCl = 33·858$ Mc/sec]. An exception is diphenyl-carbamoyl chloride[185] $(C_6H_5)_2NCOCl$, which has the very high resonance frequency of 35·2 Mc/sec.

3. Magnetic properties

a. Magnetic susceptibilities. The magnetic properties of acyl chlorides have been determined by Pascal[186-188] and by Mentzer and Pacault[189]. The susceptibility of the double-bonded oxygen ($=O$) in acyl chlorides is $+1.72 \times 10^{-6}$, the same as in ketones and aldehydes, while in carboxylic acids it is -7.95×10^{-6}. This shows the strong similarity of the $C=O$ bond in the COCl group to the $C=O$ bond in ketones and aldehydes.

b. Faraday effect. The magnetic rotation has been studied by Voigt and Gallais[190] for several aliphatic acyl chlorides. They found that the contribution of the COCl group to the molar rotations of the compounds is $\rho = 17.9 \times 10^{-5}$ radians.

III. REFERENCES

1. L. H. Greathouse, H. J. Janssen and C. H. Haydel, *Anal. Chem.*, **28**, 357 (1956).
2. C. Hoffman, *Ber.*, **22**, 2854 (1889).
3. R. E. Buckles and C. J. Thelen, *Anal. Chem.*, **22**, 676 (1950).
4. N. D. Cheronis and J. B. Entrikin, *Semimicro Qualitative Organic Analysis*, 2nd ed., Interscience, New York, 1957, pp. 228–229.
5. C. Schotten, *Ber.*, **17**, 2544 (1884).
6. E. Bauman, *Ber.*, **19**, 3218 (1886).
7. N. P. Buu-Hoï, *Bull. Soc. Chim. France*, **12**, 587 (1945).
8. S. Veibel in *Treatise on Analytical Chemistry* (Ed. I. M. Kolthoff and P. J. Elving), Interscience, New York, 1966, Part II, Vol. 13, pp. 223–300.
9. N. D. Cheronis and J. B. Entrikin, *Semimicro Qualitative Organic Analysis*, 2nd ed., Interscience, New York, 1957, pp. 342, 352–359.
10. O. C. Dermer and J. King, *J. Org. Chem.*, **8**, 168 (1943).
11. F. Feigl, *Spot Tests in Organic Analysis*, Elsevier, Amsterdam, 1960, p. 253.
12. B. E. Dixon and G. C. Hands, *Analyst*, **84**, 463 (1959).
13. E. Sawicki, D. F. Bender, T. R. Hauser, R. M. Wilson, Jr. and J. E. Meeker, *Anal. Chem.*, **35**, 1479 (1963).
14. A. M. Agree and R. L. Meeker, *Talanta*, **13**, 1151 (1966).
15. B. Alfredsson, L. Gedda and O. Samuelson, *Anal. Chim. Acta*, **27**, 63 (1962).
16. B. Larsen and A. Haug, *Acta Chem. Scand.*, **15**, 1397 (1961).
17. F. Lesquibe, *Compt. Rend.*, **251**, 2690 (1960).
18. J. Lawson and J. W. Purdie, *Mikrochim. Acta*, 415 (1961).
19. R. Raveux and J. Bové, *Bull. Soc. Chim. France*, 369 (1957).
20. J. Bové and R. Raveux, *Bull. Soc. Chim. France*, 376 (1957).
21. I. Z. Steinberg and H. A. Sheraga, *J. Am. Chem. Soc.*, **84**, 2890 (1962).
22. C. S. Marvel and R. D. Rands, Jr., *J. Am. Chem. Soc.*, **72**, 2642 (1950).
23. J. S. Fritz and N. M. Lisicki, *Anal. Chem.*, **23**, 589 (1951).
24. J. Berger, *Acta Chem. Scand.*, **8**, 427 (1954).
25. J. J. Donleavy, *J. Am. Chem. Soc.*, **58**, 1004 (1936).
26. S. Veibel and H. Lillelund, *Bull. Soc. Chim. France*, 1153 (1938).
27. S. Veibel and K. Ottung, *Bull. Soc. Chim. France*, 1434 (1939).

28. J. Berger and I. Uldall, *Acta Chem. Scand.*, **16**, 1811 (1962).
29. B. T. Dewey and R. B. Sperry, *J. Am. Chem. Soc.*, **61**, 3251 (1939).
30. B. T. Dewey and R. G. Shasky, *J. Am. Chem. Soc.*, **63**, 3526 (1941).
31. W. A. Bonner, *J. Am. Chem. Soc.*, **70**, 3508 (1948).
32. A. I. Vogel, *A Textbook of Practical Organic Chemistry*, 3rd ed., Longmans, Green, New York, London, 1962, p. 363.
33. A. Friediger and C. Pedersen, *Acta Chem. Scand.*, **9**, 1425 (1955).
34. A. Jart, *Acta Polytech. Scand.*, *Chem. Series No.* 24 (1963).
35. A. Jart, A. J. Bigler and V. Bitsch, *Anal. Chim. Acta*, **31**, 472 (1964).
36. J. B. Rather and E. E. Reid, *J. Am. Chem. Soc.*, **41**, 75 (1919).
37. W. L. Judefind and E. E. Reid, *J. Am. Chem. Soc.*, **42**, 1043 (1920).
38. R. M. Hann, E. E. Reid and G. S. Jamieson, *J. Am. Chem. Soc.*, **52**, 818 (1930).
39. T. L. Kelly and P. A. Kleff, *J. Am. Chem. Soc.*, **54**, 4444 (1932).
40. H. Lund and T. Langvad, *J. Am. Chem. Soc.*, **54**, 4107 (1932).
41. J. Berger, *Acta Chem. Scand.*, **17**, 1943 (1963).
42. J. Berger, *Acta Chem. Scand.*, **10**, 638 (1956).
43. N. L. Drake and J. Bronitsky, *J. Am. Chem. Soc.*, **52**, 3715 (1930).
44. N. L. Drake and J. P. Sweeney, *J. Am. Chem. Soc.*, **54**, 2059 (1932).
45. T. L. Kelly and E. A. Morisani, *J. Am. Chem. Soc.*, **58**, 1502 (1936).
46. E. E. Reid., *J. Am. Chem. Soc.*, **39**, 124 (1917).
47. J. A. Lyman and E. E. Reid, *J. Am. Chem. Soc.*, **39**, 701 (1917).
48. E. Lyons and E. E. Reid, *J. Am. Chem. Soc.*, **39**, 1727 (1917).
49. T. L. Kelly and M. Segura, *J. Am. Chem. Soc.*, **56**, 2497 (1934).
50. R. W. Bost and L. V. Mullen, Jr., *J. Am. Chem. Soc.*, **73**, 1967 (1951).
51. N. P. Buu-Hoi, N. D. Xuong and E. Lescot, *Bull. Soc. Chim. France*, 441 (1957).
52. N. D. Cheronis, *Mikrochim. Acta*, 925 (1956).
53. N. D. Cheronis and J. B. Entrikin, *Semimicro Qualitative Organic Analysis*, 2nd ed., Interscience, New York, 1957, p. 351.
54. J. L. Occolowitz, *J. Chromatog.*, **5**, 373 (1961).
55. S. C. Pan, A. I. Laskin and P. Principe, *J. Chromatog.*, **8**, 32 (1962).
56. G. Hammarberg and B. Wickberg, *Acta Chem. Scand.*, **14**, 882 (1960).
57. R. D. Hartley and G. J. Lawson, *J. Chromatog.*, **7**, 69 (1962).
58. C. V. Viswanathan and B. M. Bai, *J. Chromatog.*, **7**, 507 (1962).
59. A. A. Molloy and G. N. Kowkabany, *Anal. Chem.*, **34**, 491 (1962).
60. F. Drahowzal and D. Klamann, *Monatsh.*, **82**, 470 (1951).
61. J. S. Fritz and N. M. Lisicki, *Anal. Chem.*, **23**, 589 (1951).
62. M. Pesez and R. Willemart, *Bull. Soc. Chim. France*, 479 (1948).
63. R. R. Ackley and G. C. Tesoro, *Anal. Chem.*, **18**, 444 (1946).
64. S. T. Bauer, *Oil and Soap*, **23**, 1 (1946).
65. C. R. Stahl and S. Siggia, *Anal. Chem.*, **28**, 1971 (1956).
66. L. J. Lohr, *Anal. Chem.*, **32**, 1166 (1960).
67. J. Mitchell and D. M. Smith, 'Aquametry', *Chemical Analysis*, Vol. V, Interscience, New York, 1948, pp. 369–71.
68. C. Hennart and F. Vieillet, *Chim. Anal.*, **44**, 61 (1962).
69. D. Klamann, *Monatsh.*, **83**, 719 (1952).
70. K. Burger and E. Schulek, *Talanta*, **4**, 120 (1960).
71. I. Simonyi and I. Kekesy, *Z. anal. Chem.*, **215**, 187 (1965).

72. K. Stuerzer, *Z. anal. Chem.*, **216**, 409 (1966).
73. A. Patchornik and S. E. Rogozinski, *Anal. Chem.*, **31**, 985 (1959).
74. R. F. Goddu, N. F. Leblanc and C. M. Wright, *Anal. Chem.*, **27**, 1251 (1955).
75. M. M. Buzlanova, N. P. Skvortsov and L. I. Mekhryusheva, *Zh. Anal. Khim.*, **22**, 469 (1967).
76. J. G. Hanna and S. Siggia, *Anal. Chem.*, **34**, 547 (1962).
77. B. Saville, *Analyst*, **82**, 269 (1957).
78. C. Hishta and J. Bomstein, *Anal. Chem.*, **35**, 65 (1963).
79. R. A. Simonaitis and G. C. Guvernator, *J. Gas Chromatog.*, **5**, 527 (1967).
80. R. Ananthakrishnan, *Proc. Ind. Acad. Sci.*, **5A**, 285 (1937).
81. C. R. Bailey and J. B. Hale, *Phil. Mag.*, **25**, 98 (1938).
82. A. H. Nielsen, T. G. Burke, P. J. H. Woltz and E. A. Jones, *J. Chem. Phys.*, **20**, 596 (1952).
83. E. Catalano and K. S. Pitzer, *J. Am. Chem. Soc.*, **80**, 1054 (1958).
84. R. R. Patty and R. T. Lagemann, *Spectrochim. Acta*, **15**, 60 (1959).
85. J. Overend and J. C. Evans, *Trans. Faraday Soc.*, **55**, 1817 (1959).
86. H. W. Morgan, P. A. Staats and J. H. Goldstein, *J. Chem. Phys.*, **25**, 337 (1956).
87. R. F. Stratton and A. H. Nielsen, *J. Mol. Spectry*, **4**, 373 (1960).
88. E. J. Harwell, R. E. Richards and H. W. Thompson, *J. Chem. Soc.*, 1436 (1948).
89. R. S. Rasmussen and R. R. Brattain, *J. Am. Chem. Soc.*, **71**, 1073 (1949).
90. L. Kahovec and K. W. F. Kohlrausch, *Z. Phys. Chem.*, **B38**, 119 (1938).
91. H. Seewann-Albert and L. Kahovec, *Acta Phys. Austriaca*, **1**, 352 (1948).
92. H. Seewann-Albert, *Acta Phys. Austriaca*, **1**, 359 (1948).
93. R. N. Haszeldine, *Nature*, **168**, 1028 (1951).
94. R. N. Haszeldine and F. Nyman, *J. Chem. Soc.*, 1084 (1959).
95. I. Nakagawa, I. Ichishima, K. Kuratani, T. Miyazawa, T. Shimanouchi and S. Mizushima, *J. Chem. Phys.*, **20**, 1720 (1952).
96. J. C. Evans and H. J. Bernstein, *Can. J. Chem.*, **34**, 1083 (1956).
97. B. P. Susz and J. J. Wuhrmann, *Helv. Chim. Acta*, **40**, 722 (1957).
98. L. J. Bellamy and R. L. Williams, *J. Chem. Soc.*, 863 (1957).
99. L. J. Bellamy and R. L. Williams, *J. Chem. Soc.*, 4294 (1957).
100. C. V. Berney, *Spectrochim. Acta*, **20**, 1437 (1964).
101. C. V. Berney, *Spectrochim. Acta*, **21**, 1809 (1965).
102. C. V. Berney, *Spectrochim. Acta*, **25A**, 793 (1969).
103. J. Overend, R. A. Nyquist, J. C. Evans and W. J. Potts, *Spectrochim. Acta*, **17**, 1205 (1961).
104. K. R. Loos and R. C. Lord, *Spectrochim. Acta*, **21**, 119 (1965).
105. J. A. Ramsey and J. A. Ladd, *J. Chem. Soc. (B)*, 118 (1968).
106. J. E. F. Jenkins and J. A. Ladd, *J. Chem. Soc. (B)*, 1237 (1968).
107. A. W. Baker and G. H. Harris, *J. Am. Chem. Soc.*, **82**, 1923 (1960).
108. L. J. Bellamy, *Advances in Infrared Group Frequencies*, Methuen, London, 1968, pp. 123–190.
109. J. Overend and J. R. Scherer, *Spectrochim. Acta*, **16**, 773 (1960).
110. J. Overend and J. R. Scherer, *J. Chem. Phys.*, **32**, 1296 (1960).
111. J. C. Evans and J. Overend, *Spectrochim. Acta*, **19**, 701 (1963).
112. O. Burkard, *Proc. Ind. Acad. Sci.*, **8A**, 365 (1938).

113. J. O. Halford, *J. Chem. Phys.*, **24**, 830 (1956).
114. S. Bratoz and S. Besnainou, *Compt. Rend.*, **248**, 546 (1959).
115. L. J. Bellamy and R. J. Pace, *Spectrochim. Acta*, **19**, 1831 (1963).
116. R. E. Kagarise, *J. Am. Chem. Soc.*, **77**, 1377 (1955).
117. K. Shimizu and H. Shingu, *Spectrochim. Acta*, **22**, 1528 (1966).
118. L. Pauling, *The Nature of the Chemical Bond*, Cornell University Press, Ithaca, New York, 1960.
119. L. C. Krisher and E. B. Wilson, *J. Chem. Phys.*, **31**, 882 (1959).
120. L. J. Bellamy and R. L. Williams, *Trans. Faraday Soc.*, **55**, 14 (1959).
121. L. J. Bellamy and R. L. Williams, *J. Chem. Soc.*, 3465 (1958).
122. J. E. Katon and W. R. Feairheller, Jr., *J. Chem. Phys.*, **44**, 144 (1966).
123. S. Mizushima, T. Shimanouchi, T. Miyazawa, I. Ichishima, K. Kuratani, I. Nakagawa and N. Shido, *J. Chem. Phys.*, **21**, 815 (1953).
124. L. J. Bellamy, L. C. Thomas and R. L. Williams, *J. Chem. Soc.*, 3704 (1956).
125. R. N. Jones, D. A. Ramsay, F. Herling and K. Dobriner, *J. Am. Chem. Soc.*, **74**, 2828 (1952).
126. N. L. Owen and N. Sheppard, *Proc. Chem. Soc.*, 264 (1963).
127. E. Saegebarth and E. B. Wilson, Jr., *J. Chem. Phys.*, **46**, 3088 (1967).
128. Y. Morino, K. Kuchitsu and M. Sugiura, *J. Chem. Soc. Japan*, **75**, 721 (1954).
129. L. Pierce and L. C. Krisher, *J. Chem. Phys.*, **31**, 875 (1959).
130. K. M. Sinnot, *J. Chem. Phys.*, **34**, 851 (1961).
131. D. D. Thompson and J. F. Norris, *J. Am. Chem. Soc.*, **58**, 1953 (1936).
132. M. St. C. Flett, *Trans. Faraday Soc.*, **44**, 767 (1948).
133. E. Herz, L. Kahovec and K. W. F. Kohlrausch, *Monatsh.* **74**, 253 (1943).
134. R. N. Jones, C. L. Angell, T. Ito and R. J. D. Smith, *Can. J. Chem.*, **37**, 2007 (1959).
135. W. F. Forbes and J. J. J. Myron, *Can. J. Chem.*, **39**, 2452 (1961).
136. C. Garrigou-Lagrange, N. Claverie, J. M. Lebas and M. L. Josien, *J. Chim. Phys.*, **58**, 559 (1961).
137. M. L. Josien and R. Calas, *Compt. Rend.*, **240**, 1641 (1955).
138. C. N. R. Rao and R. Venkataraghavan, *Spectrochim. Acta*, **18**, 273 (1962).
139. C. L. Angell, P. J. Krueger, R. Lauzon, L. C. Leitch, K. Noack, R. J. D. Smith and R. N. Jones, *Spectrochim. Acta*, **15**, 926 (1959).
140. N. Claverie and C. Garrigou-Lagrange, *J. Chim. Phys.*, **61**, 889 (1964).
141. B. D. Saksena and R. E. Kagarise, *J. Chem. Phys.*, **19**, 987 (1951).
142. J. S. Ziomek, A. G. Meister, F. F. Cleveland and C. E. Decker, *J. Chem. Phys.*, **21**, 90 (1953).
143. J. L. Hencher and G. W. King, *J. Mol. Spectry*, **16**, 158 (1965).
144. J. L. Hencher and G. W. King, *J. Mol. Spectry*, **16**, 168 (1965).
145. K. G. Kidd and G. W. King, *J. Mol. Spectry*, **28**, 411 (1968).
146. B. D. Saksena and R. E. Kagarise, *J. Chem. Phys.*, **19**, 999 (1951).
147. J. W. Sidman, *J. Am. Chem. Soc.*, **78**, 1527 (1956).
148. B. D. Saksena and G. S. Jauhri, *J. Chem. Phys.*, **36**, 2233 (1962).
149. W. J. Balfour and G. W. King, *J. Mol. Spectry*, **25**, 130 (1968).
150. W. J. Balfour and G. W. King, *J. Mol. Spectry*, **26**, 384 (1968).
151. W. J. Balfour and G. W. King, *J. Mol. Spectry*, **27**, 432 (1968).
152. W. J. Balfour and G. W. King, *J. Mol. Spectry*, **28**, 497 (1968).
153. G. T. O. Martin and J. R. Partington, *J. Chem. Soc.*, 1178 (1936).

154. B. D. Saksena and R. E. Kagarise, *J. Chem. Phys.*, **19**, 994 (1951).
155. M. J. Schmelz, T. Miyazawa, S. Mizushima, T. J. Lane and J. V. Quagliano, *Spectrochim. Acta*, **9**, 51 (1957).
156. R. N. Haszeldine and J. M. Kidd, *J. Chem. Soc.*, 3871 (1955).
157. N. Lozach and H. Guillouzo, *Bull. Soc. Chim. France*, 1221 (1957).
158. L. J. Bellamy and P. E. Rogasch, *J. Chem. Soc.*, 2218 (1960).
159. M. A. Ory, *Spectrochim. Acta*, **16**, 1488 (1960).
160. R. A. Nyquist and W. J. Potts, *Spectrochim. Acta*, **17**, 679 (1961).
161. A. R. Katritzky, J. M. Lagowski and J. A. T. Beard, *Spectrochim. Acta*, **16**, 964 (1960).
162. A. W. Baker and G. H. Harris, *J. Am. Chem. Soc.*, **82**, 1923 (1960).
163. J. L. Hales, J. I. Jones and W. Kynaston, *J. Chem. Soc.*, 618 (1957).
164. See, e.g., *Organic Electronic Spectral Data*, Vols. I–VI, Interscience, New York, 1960–1970.
165. H. H. Jaffé and M. Orchin, *Theory and Applications of Ultraviolet Spectroscopy*, Wiley, New York, 1962, p. 178.
166. J. W. Sidman, *Chem. Rev.*, **58**, 689 (1958).
167. K. Yates, S. L. Klemenko and I. G. Csizmadia, *Spectrochim. Acta*, **25A**, 765 (1969).
168. H. Basch, M. B. Robin, N. A. Kuebler, *J. Chem. Phys.*, **49**, 5007 (1968).
169. C. V. Berney, *Spectrochim. Acta*, **20**, 1437 (1964).
170. J. F. Harris, Jr., *J. Org. Chem.*, **30**, 2182 (1965).
171. A. Foffani, I. Zanon, G. Giacometti, U. Mazzucato, G. Favaro and G. Semerano, *Nuovo Cimento*, **16**, 861 (1960).
172. L. E. Giddings, Jr. and K. K. Innes, *J. Mol. Spectry*, **6**, 528 (1961).
173. L. E. Giddings, Jr. and K. K. Innes, *J. Mol. Spectry*, **8**, 328 (1962).
174. G. Fischer, *J. Mol. Spectry*, **29**, 37 (1969).
175. C. T. Zahn, *Phys. Z.*, **33**, 686 (1932).
176. G. T. O. Martin and J. R. Partington, *J. Chem. Soc.*, 158 (1936).
177. G. T. O. Martin and J. R. Partington, *J. Chem. Soc.*, 1175 (1936).
178. G. T. O. Martin and J. R. Partington, *J. Chem. Soc.*, 1178 (1936).
179. J. R. Partington, *An Advanced Treatise on Physical Chemistry*, Vol. 5, Longmans, Green, London, 1954, pp. 513–516.
180. D. W. McCall and H. S. Gutowsky, *J. Chem. Phys.*, **21**, 1300 (1953).
181. P. J. Bray, *J. Chem. Phys.*, **22**, 1787 (1954).
182. P. J. Bray, *J. Chem. Phys.*, **23**, 703 (1955).
183. H. Zeldes and R. Livingstone, *J. Chem. Phys.*, **21**, 1418 (1953).
184. R. E. Kagarise, *J. Chem. Phys.*, **21**, 1615 (1953).
185. P. J. Bray, R. G. Barnes and S. L. Segel, unpublished results, cited by E. A. C. Lücken in *Nuclear Quadrupole Coupling Constants*, Academic Press, New York, 1969, p. 181.
186. P. Pascal, *Ann. Chim. Phys.*, **19**, 5 (1910).
187. P. Pascal, *Ann. Chim. Phys.*, **25**, 289 (1912).
188. P. Pascal, *Ann. Chim. Phys.*, **29**, 119 (1913).
189. C. Mentzer and A. Pacault, *Compt. Rend.*, **223**, 39 (1946).
190. D. Voigt and F. Gallais, *Compt. Rend.*, **247**, 1993 (1958).

CHAPTER 4

Acid–base behaviour and complex formation

D. P. N. SATCHELL
King's College, London, England
and
R. S. SATCHELL
Queen Elizabeth College, London, England

I. INTRODUCTION

As may be gathered from every chapter of this book, carbonyl halides are reactive compounds in which the C—Hal bond is easily broken heterolytically. They can behave either as Lewis bases (electron-pair donors) or as Lewis acids (electron-pair acceptors). Their interaction with an acid A can be represented, in general terms, by equation (1), where R represents any suitable substituent; that with a base B by equation (2). The factors controlling the overall basicity, or acidity, of species RCO—Hal are, in principle at least, reasonably clear. However, the relative abundances of the various possible adduct (complex) species depend upon the

particular reactants and conditions involved, and there exists at present only a rough-and-ready conception of the factors which control the positions of the competing equilibria. The systems generally are difficult to study experimentally owing to the great reactivity of the complexes. These can be troublesome to obtain pure, are usually unstable on the open bench and, for some systems, only occur (if they do occur) as transient reaction intermediates.

A. Factors Influencing Basicity

Since the basicity of RCO—Hal in equation (1) depends upon the electron availability on the oxygen and halogen atoms, in general the more electron-repelling is R the easier it will be for A to attach itself.

Which site is used preferentially appears to depend upon the nature of A, upon which halogen is involved, upon the environment, upon the extent of electron release by R, and the ability of this group to stabilize the different types of complex by resonance. The influence of particular halogen atoms on the overall basicity of RCO—Hal is complicated by the fact that these atoms possess opposing $(-I, +T)$ electronic effects. Contributions to the resonance hybrid from species **6** stabilize the free carbonyl halide, and shift some of the electron density from the halogen to the oxygen atom. Although such contributions will be most important

$$RC \underset{Hal^+}{\overset{O^-}{\diagdown}}$$

(6)

when Hal = F, acyl fluorides appear, in practice, to be more basic than the corresponding chlorides, bromides or iodides; and, moreover, they are more often attacked by acids at the halogen than at the oxygen atom, compared with the other acyl halides. The small size of fluorine is almost certainly an important factor here.

B. Factors Influencing Acidity

The acidity of RCOHal depends upon the effective positive charge on the carbonyl carbon atom, and upon its steric accessibility. In general, substituents R which withdraw electrons from this atom will increase acidity. Electron-withdrawing substituents which can also delocalize the positive charge (e.g. Ph in structure **7**), and therefore stabilize the free

$$+ \bigcirc = C \underset{Hal}{\overset{O^-}{\diagdown}}$$

(7)

carbonyl halide, will have opposing effects on acidity. The resonance effects seem particularly important. Thus acyl fluorides, for which the analogous structure **6** contributes more to the overall hybrid than do those for the other halides, appear notably less acidic than their chloro, bromo, or iodo analogues. Most of the evidence in these contexts is, however, indirect.

II. BEHAVIOUR AS BASES

A. Types of Complexes Formed with Acids

There exist various classes of acid. We shall consider in particular the proton H^+, other hydrogen acids (e.g. HCl, HF), metal ions (e.g. Hg^{2+}), molecular halogens (e.g. I_2), and covalent metal halides (e.g. $AlCl_3$, BF_3). It is found that all such acids very probably interact with carbonyl halides to give transient, or even relatively stable, species such as **1**, **2** or **3** in equation (1), but, with a single exception, up to the present time only complexes involving covalent metal halides have actually been isolated. We deal with these isolatable complexes first. The earliest reports[1] of such complexes date from 1880–1890, and structures for them like **1**, **2** and **3** were first proposed[2] in the 1920s. By the end of the 1960s a fair number of isolatable complexes had been prepared.

B. Isolatable Complexes

Writing any covalent metal halide as $(M(Hal)_n$, acid–base complex formation with species RCOHal proceeds in general terms according to equation (3). For most of the combinations studied so far $x = y = 1$, although other stoicheiometries have been reported. In the liquid state y

$$x\text{RCOHal} + y\text{M(Hal)}_n \rightleftharpoons (\text{RCOHal})_x(\text{MHal}_n)_y \qquad (3)$$

will be expected usually to be unity, and x to be determined by the normal co-ordination requirements of $M(Hal)_n$. Thus for the group III halides (e.g. $AlCl_3$) it would be expected that $x = 1$, but for group IV (e.g. $SnCl_4$) that $x = 2$. However, for complexes isolated as solids, unexpected stoicheiometries can arise owing to lattice requirements. Solid-state composition is therefore not a reliable guide to the structure in solution. Some typical complexes, together with their physical properties, are listed in Table 1. These complexes can, of course, all be considered to be derived from substituted acetyl or benzoyl halides. Apparently rather few complexes have been *isolated* which involve acetyl or benzoyl halides containing electron-withdrawing substituents. Those containing the more feebly acidic covalent halides are also uncommon. This finding is sensible, but it must be remembered that the stability of these solid complexes is determined not only by the intrinsic acidic and basic properties of the constituents, but also by the lattice energy. This energy can sometimes be expected to facilitate solid complex formation between otherwise unattractive partners. A few (very unstable) complexes have been reported between phosgene and certain powerfully acidic covalent metal halides, but there remains some disagreement about their proper compositions[13,14].

Certain of the isolated complexes have been subjected to infrared or to nuclear magnetic resonance spectroscopic examination; for a few, X-ray crystal analyses are available. Such studies have led to the conclusion that these solid complexes contain species like **1** (e.g. **8**), or like **3** (e.g. **9**).

TABLE 1. Examples of isolatable acid–base adducts with acids $M(Hal)_n$

Adduct	m.p. (°C)	Decomp. pt (°C)	Reference
$CH_3COCl-BCl_3$	$-65, -54$	—	3, 4
$CH_3COCl-BF_3$	-65	—	3
$CH_3COF-BF_3$	—	$+20$	5, 19
$C_2H_5COF-BF_3$	—	-15	6
$C_6H_5COF-BF_3$	—	-30	6
$C_6H_5COCl-AlCl_3$	$+93$	—	12
$2\text{-}CH_3C_6H_4COCl-AlCl_3$	$+55$	—	9
$2,6\text{-}(CH_3)_2C_6H_3COCl-AlCl_3$	$+82$	—	9
$CH_3COCl-AlCl_3$	$+65$	—	21
$CH_3COCl-AlBr_3$	$+90$	—	12
$CH_3COCl-GaCl_3$	$+86$	—	4
$C_6H_5COCl-GaCl_3$	$+46$	—	4
$CH_3COCl-TiCl_4$	$+20, +19$	—	7, 11
$C_2H_5COCl-TiCl_4$	$+28$	—	7
$n\text{-}C_3H_7COCl-TiCl_4$	-4	—	7
$iso\text{-}C_3H_7COCl-TiCl_4$	$+11$	—	7
$C_6H_5COCl-TiCl_4$	$+65$	—	10
$2\text{-}CH_3C_6H_4COCl-TiCl_4$	$+70$	—	9
$2,6\text{-}(CH_3)_2C_6H_3COCl-2\ TiCl_4$	$+72$	—	9
$2,4,6\text{-}(CH_3)_3C_6H_2COCl-2\ TiCl_4$	$+89$	—	8
$CH_3COCl-SbCl_5$	$+146$	—	3
$C_2H_5COCl-SbCl_5$	$+100$	—	3
$C_6H_5COCl-SbCl_5$	$+115$	—	3
$CH_3COF-SbF_5$	$+173$	—	6
$C_2H_5COF-SbF_5$	$+110$	—	6
$(CH_3)_3CCOF-SbF_5$	—	$+20$	6
$n\text{-}C_3H_7COF-SbF_5$	$+58$	—	6
$(C_2H_5)_3CCOF-SbF_5$	—	-5	6
$C_6H_5COF-SbF_5$	$+150$	—	6
$CH_3COF-PF_5$	$+30$	—	6
$C_2H_5COF-PF_5$	-5	—	6
$C_6H_5COF-PF_5$	-15	—	6
$CH_3COF-AsF_5$	$+175$	—	6
$C_2H_5COF-AsF_5$	$+110$	—	6
$C_6H_5COF-AsF_5$	$+158$	—	6

Species **1** are called oxonium complexes; species **3** are usually* termed acylium ion complexes. The spectroscopic studies suggest that species like

* Also called oxocarbonium ion complexes. Acylium ions are, of course, known to be formed in solution in a variety of ways from a host of acyl-containing precursors[15]. Complexes like **3** provide the most convenient method for obtaining such ions in the solid state.

2 do not play an important role in these systems, at least so far as isolatable complexes are concerned. This is perhaps to be expected, for in the formation of the simple coordination species 1 and 2 the oxygen atom will

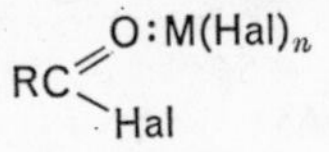

$$RC\underset{Hal}{\overset{O:M(Hal)_n}{<}}$$

(8)

$$RCO^+ \ M(Hal)^-_{n+1}$$

(9)

normally be more basic than the halogen atom. The formation of 3, however, doubtless often proceeds via the at least transient formation of species like 2.

Complex formation between $M(Hal)_n$ and RCOHal can clearly provide a route for halogen interchange between these species. Such halogen exchange has often been observed[16,17] and, because it can lead to compositional ambiguity, the most common practice is to study systems containing a single type of halogen only (e.g. all Cl). It is sometimes assumed that the existence of exchange in any system implies the presence of acylium ion complexes in that system. There is no reason, however, to suppose that the other types of complex (e.g. 2) *cannot* lead to exchange.

I. Infrared spectra of complexes*

Although an earlier study[19] of the conductivity of the $CH_3COF-BF_3$ complex in liquid sulphur dioxide had suggested that in this case an ionic structure (possibly 3) is obtained, it was not until 1957 that i.r. spectroscopic examination[5,20] of various complexes, both as mulls of the solid and in solution, showed clearly that at least two types of complex could exist; and could indeed co-exist in the same system.

I.r. spectra of systems containing either free RCOHal, or free $M(Hal)_n$, contain, among others, absorption bands attributable either to C—Hal and C=O stretching modes, or to $M(Hal)_n$ modes. When the spectrum of a solid, or dissolved, RCOHal—$M(Hal)_n$ complex is recorded either, or both, of two patterns of spectral change are observed in comparison with the spectra of the free reagents[20]:

(*i*) The free C=O absorption (usually at ca. 1800 cm^{-1}) is removed, and is replaced by a similar, shifted band at a somewhat longer wavelength. The so-called 'carbonyl shift' is usually between 60 and 250 cm^{-1}. This effect is accompanied by certain other, usually smaller, frequency shifts. Sometimes more than one shifted C=O frequency appears to be present.

---

* Acylium ions derived from benzoyl compounds have a u.v. spectrum[18] at wavelengths > 210 nm. These spectra have not yet been used in diagnostic structural studies in the present contexts.

(ii) The free C=O, C—Hal and certain $M(Hal)_n$ absorptions are removed, and new bands appear which are characteristic of $M(Hal)_{n+1}^-$, together with a strong new absorption at ca. $2200–2300$ cm^{-1}. This latter absorption usually appears at a significantly different frequency in solution in polar solvents compared with its position in mulls.

Typical results are outlined in Table 2. The usual interpretation of the two behaviour patterns is as follows. The co-ordination of species $M(Hal)_n$ to various classes of carbonyl compound (especially ketones) has been found to lead almost always to a shifting of the free C=O stretching absorption to longer wavelengths. It seems sensible therefore to interpret behaviour (i) above as reflecting the formation of oxonium species **1**.

In behaviour (ii) the appearance of the spectrum typical of the $M(Hal)_{n+1}^-$ anion, and the new band at ca. $2200–2300$ cm^{-1}, which is similar in frequency to those of the C—O bonds in carbon monoxide[25] and keten[26], strongly suggest that these spectra are those of species **3**, the acylium salt $RCO^+M(Hal)_{n+1}^-$.

2. Nuclear magnetic resonance spectra of complexes

N.m.r. (^{1}H and ^{19}F) spectra[6, 27-29] of solutions of complexes in sulphur dioxide or hydrogen fluoride support the view that more than one type of complex exists, and that the different types can co-exist in appropriate systems. The essential finding for the p.m.r. spectra is that the proton signals of aliphatic acyl halides (e.g. the signal due to the CH$_3$ group in CH$_3$COHal) are shifted to positions indicating lower shielding. Sometimes only one shifted absorption is found, sometimes two; never apparently more than two in a given solvent. The largest shifts (ca. 2 p.p.m. from the original position) are very probably due to RCO$^+$ ions, either free or as ion pairs. The smaller shifts (ca. 1 p.p.m.) are sensibly attributed to oxonium species, in view of the i.r. evidence quoted above*. This conclusion is supported by the fact that, in most systems studied, the species leading to the smaller shifts tend to be favoured, compared with RCO$^+$, in hydrogen fluoride, rather than in sulphur dioxide, solutions. This could obviously be due to species taking the form **10**, rather than **8**, in hydrogen

$$RC\overset{\diagup OH^+ \ M(Hal)_nF^-}{\diagdown Hal}$$

(10)

---

* These p.m.r. results alone would be compatible with complexes like **1** or **2**, or even with solvated acylium ion pairs. The weight of other evidence, however, argues strongly against the smaller shifts being due to ion pairs[6, 27]. The opposite impression is unfortunately left by the brief survey in reference 15.

TABLE 2. Typical i.r. spectra of RCOHal—M(Hal)$_n$ complexes

Free acyl halide		Complex	
$\bar{\nu}^a$ (cm^{-1})	Assignment	$\bar{\nu}^a$ (cm^{-1})	Assignment

$CH_3COF(gas)^{20}$ **$CH_3COF—BF_3(solid, -40°)^{20}$**

1879 vs	C=O stretch	2299 s	$-\overset{+}{C}\equiv O$ stretch
1859 vs	C=O stretch	1619 m	shifted C=O stretch?
826 s	C—F stretch	1052 vs	BF$_4^-$ asym. stretch
		527 s	BF$_4^-$ asym. bend

No absorption at 1850–1880 cm^{-1} or at 810–850 cm^{-1}

Conclusion: solid $CH_3COF—BF_3$ is predominantly $CH_3CO^+BF_4^-$

$CH_3COCl(liquid)^{24}$ **$CH_3COCl—AlCl_3$**

1807 vs	C=O stretch	(liquid)^{22}	(solid)^{21}	
594 vs	C—Cl stretch	2307 s	2305 vs	$-\overset{+}{C}\equiv O$ stretch
		2203 vs		
		1637 vs	1639 m	shifted C=O stretch?
		1567 s	1560 m	
		540 s	500 s	AlCl$_4^-$
		485 s		

No absorption at 1800–1810 cm^{-1} or at 590–600 cm^{-1}

In nitrobenzene solution absorptions at 1805 and 2200 cm^{-1} appear

Conclusions: solid $CH_3COCl—AlCl_3$ is predominantly $CH_3CO^+AlCl_4^-$, but in the liquid phase much $CH_3\overset{\displaystyle C=O:AlCl_3}{\underset{\displaystyle Cl}{}}$ is also formed. In nitrobenzene solution there exists an equilibrium between $CH_3CO^+AlCl_4^-$ and the free components

$C_6H_5COCl(liquid)^{9, 35}$ **$C_6H_5COCl—AlCl_3(solid)^{9, 35, \text{see also } 23}$**

1773 vs	C=O stretch	1665 s	shifted C=O stretch?
1732 s	C=O stretch	1605 s	shifted C=O stretch?
		1585 s	shifted C=O stretch?
		1560 s	shifted C=O stretch?

No absorption at 2200–2300 cm^{-1}. Other absorptions similar to those of C_6H_5COCl

Conclusion: solid $C_6H_5COCl—AlCl_3$ is, very probably, exclusively $C_6H_5\overset{\displaystyle CO:AlCl_3}{\underset{\displaystyle Cl}{}}$. For a fuller discussion of these, and other, i.r. spectra see reference 20.

---

^{a} Only absorptions relevant to the discussion in section II, B, 1 are included; vw = very weak, w = weak, m = medium, s = strong, vs = very strong.

fluoride. This view is consistent with the p.m.r. results for a wide range of substituted benzoyl chlorides dissolved in FSO_3H—SbF_5 mixtures[29]. Here the species formed appear to be either the protonated benzoyl halide or the acylium ion.

3. X-ray spectra of solid complexes

The X-ray work[30-32] confirms the existence of oxonium- and acylium-type complexes. It shows that, in aliphatic acylium ion complexes, the $\overset{\alpha}{C}$—C—O skeleton is linear, with a C—O bond distance similar to that in carbon monoxide, and rather insensitive to the nature of the alkyl group. The $\overset{\alpha}{C}$—C distance in CH_3CO^+ is also considerably shorter than for a normal C—C bond, shorter too than the corresponding bond in the isoelectronic compound CH_3—C≡CH. The length of the $\overset{\alpha}{C}$—C bond increases when the α-hydrogen atoms are replaced by alkyl groups. This result suggests that hyperconjugation is of some importance (as found for keten[33]). The crystal structures of the oxonium species formed between PhCOCl and $SbCl_5$, or $AlCl_3$, reveal longish M—O bond distances and an essentially undisturbed PhC—Cl system.

4. Reactant structure and complex stability

The relationship between the structure of the reactants and the nature and stability of the complexes they form (Tables 1 and 3) is not simple and is governed by a variety of factors outlined in section I, A. In practice the observed behaviour pattern is reasonably intelligible, although normally only very indirect and/or qualitative information is yet available about the stability of any complex. Only a single equilibrium constant appears to have been measured for complex formation in solution: that for oxonium complex formation between $SbCl_5$ and propionyl chloride in ethylene dichloride[36]. The fact that a solid (or liquid) complex forms readily at ordinary temperatures and has a high melting point, or decomposition temperature, certainly suggests that it is more thermodynamically stable than another which decomposes at temperatures below 0°C, but, as pointed out in section II, B, lattice energies can obscure the inherent stability of the adduct. Moreover, although the majority of complexes which readily decompose and are therefore presumably weakly bonded appear to be oxonium complexes, this does not necessarily mean that all oxonium adducts are feebly bonded, for any covalent halide may, in certain

TABLE 3. Structures of RCOHal—M(Hal)$_n$ complexes[a]

Complex	Structure[b]		Remarks
HCOF—BF$_3$		O (vs)	
CH$_3$COF—BF$_3$	A (s)	O (vw)	Table 2
C$_2$H$_5$COF—BF$_3$	A (s)	O (w)	when R in RCOF is an aliphatic group an increase in the size of R leads to an increase in the proportion of the O structure; in solution there is an equilibrium between the A and the O forms
CH$_3$COCl—BF$_3$		O ?	little interaction
CH$_3$COCl—BCl$_3$	A ?		little interaction
C$_6$H$_5$COCl—BCl$_3$	A ?		little interaction
CH$_3$COF—AlCl$_3$	A (s)	O (w)	
CH$_3$COCl—AlCl$_3$	A (s)	O (? w)	Table 2
CH$_3$COBr—AlCl$_3$	A (s)	?	
C$_6$H$_5$COCl—AlCl$_3$		O (vs)	Table 2
C$_6$H$_5$COBr—AlCl$_3$		O (s)	
2-CH$_3$C$_6$H$_4$COCl—AlCl$_3$	A (s)	O (w)	
2,6-(CH$_3$)$_2$C$_6$H$_3$COCl—AlCl$_3$	A (s)	O (vw)	
2,4,6-(CH$_3$)$_3$C$_6$H$_2$COCl—AlCl$_3$	A (vs)		
CH$_3$COCl—AlBr$_3$	A (s)	O (w)	
C$_2$H$_5$COCl—AlBr$_3$	A (s)	O (w)	
C$_6$H$_5$COCl—AlBr$_3$	A	O	in solid complexes the proportion of the A form increases as the temperature rises; mainly the O form exists below 7°C
C$_6$H$_5$COBr—AlBr$_3$	?	O (s)	
CH$_3$COCl—GaCl$_3$	A (s)	O (w)	
C$_2$H$_5$COCl—GaCl$_3$	A (s)	?	
C$_6$H$_5$COCl—GaCl$_3$	?		melt conducts electricity
CH$_3$COCl—TiCl$_4$	A (vw)	O (s)	dissociates readily in solution
C$_6$H$_5$COCl—TiCl$_4$		O (s)	
2-CH$_3$C$_6$H$_4$COCl—TiCl$_4$		O (s)	
2,6-(CH$_3$)$_2$C$_6$H$_3$COCl—2 TiCl$_4$	A (s)	O (w)	
2,4,6-(CH$_3$)$_3$C$_6$H$_2$COCl—2 TiCl$_4$	A (vs)	?	in nitrobenzene solution the A form predominates; in benzene solution the A and the O forms exist in comparable amounts
CH$_3$COF—AsF$_5$	A (s)	O (vw)	
C$_6$H$_5$COF—AsF$_5$	A (s)	O (vw)	
CH$_3$COF—SbF$_5$	A (s)	O (vw)	when R in RCOF is an

TABLE 3 (*cont.*)

Complex	Structure[b]		Remarks
$(C_6H_5)_2CHCOF—SbF_5$	A (w)	O (s)	aliphatic group, an increase in the size of R leads to an increased proportion of the O form; in solution the O and the A forms exist in equilibrium
$CF_3COF—SbF_5$	A	O	in liquid sulphur dioxide the proportion of the O form rises as the temperature rises; below $-57°C$ mainly the A form exists
$C_6H_5COF—SbF_5$	A (s)		
$CH_3COCl—SbCl_5$	A (s)	O (w)	
$C_2H_5COCl—SbCl_5$	A (w)	O (s)	
$C_6H_5COCl—SbCl_5$	A (w)	O (s)	
$CH_3COF—PF_5$	A (w)	O (s)	
$C_6H_5COF—PF_5$	A (w)	O (s)	

[a] Table compiled from results in references 8, 9, 20–24, 34, 35, 40 and 41.

[b] A = acylium complex like **9**; O = oxonium complex like **8**. Letters in parentheses (see Table 2 for key) refer to the relative importance of the given structure. The temperature is usually 25°C or below.

circumstances, conceivably prefer co-ordination to oxygen rather than to Hal. Having made these points one nevertheless finds that:

(*i*) Few adducts involving the more feebly acidic species $M(Hal)_n$ have been isolated.

(*ii*) The sequence of complexing ability with any *given* species $M(Hal)_n$ is normally $F > Cl > Br (>I)$ for Hal in RCOHal.

(*iii*) The sequence of complexing ability with any *given* species RCOHal is normally $F > Cl > Br (>I)$ for Hal in $M(Hal)_n$.

(*iv*) The rough sequence of (apparent) acidic strength is $SbF_5 \simeq AsF_5 > BF_3 \simeq GaCl_3 \simeq SbCl_5 \gtrsim AlCl_3 > PF_5 \simeq TiCl_4 (> BCl_3)$, although there are many exceptions.

(*v*) At ordinary temperatures only the most powerfully acidic covalent halides (e.g. SbF_5 and $SbCl_5$) are able to induce much acylium ion formation in unsubstituted benzoyl halides or in those containing electron-withdrawing groups. Increasing alkyl substitution, especially *ortho* substitution, leads, however, to formation of $ArCO^+$.

(vi) Most of the examined complexes of acetyl halides contain a high proportion of CH_3CO^+ regardless of the nature of $M(Hal)_n$. However, the oxonium contribution increases as the series AcF, AcCl, AcBr is traversed[37]. Alkyl substitution has the opposite effect to that found in the aromatic series and increases the proportion of oxonium adduct[27].

From consideration of these results it can be deduced that steric effects play a very important part in determining behaviour. To take a simple example: in sterically uncomplicated equilibria the sequences of acid strength $BCl_3 \gg BF_3$, and $BCl_3 > AlCl_3$ are found[38]. The inversion of these sequences in the present contexts (iii, iv above) shows that the *space* available around the central acceptor atom is of great importance. The same effect underlies the finding (ii) that acyl fluorides are better donors than the corresponding chlorides; they take up less room.

Whether the acetyl or the benzoyl derivative of a substance is the more basic (towards protonation on the carbonyl oxygen atom) is unfortunately still uncertain. Available evidence[39] implies that the difference is not very great. The behaviour of acyl halides towards acids $M(Hal)_n$ (v and vi above) suggests that the aliphatic compounds form acylium ions more readily than do the aromatic compounds, and that oxonium species are often formed preferentially by the latter. It appears, therefore, that when R is an aromatic group it is relatively better at stabilizing species 1 compared with species 3, than when it is an aliphatic group. That *ortho* substituents in the benzoyl halide lead to increased proportions of acylium complex is doubtless due to steric destabilization of the corresponding oxonium complex. This is a phenomenon well known in the context of protonation of substituted benzoyl compounds of all types[15]. In general, the formation of the acylium complex is likely to be especially favoured relative to the oxonium species by very powerful electron release, owing to the former's greater desire to disperse its more intense positive charge. The increase in the proportion of species 3 when powerful electron-releasing substituents are present in the aromatic acyl halides is therefore understandable. The, at first, curious effect (vi above) of alkyl substitution in the acetyl series (which should stabilize RCO^+ and generally increase basicity) could arise if hyperconjugation is important to the stability of CH_3CO^+ and is seriously reduced in (say) R_2CHCO^+ (see section II, B, 3 above). That $-I$ substituents lead to the expected reduction in the basicity of RCOHal is evident from the exclusive formation[13] of oxonium species by $CO(Hal)_2$. It will not be possible properly to rationalize and quantify this region of the field until many more measurements of equilibrium constants for adduct formation in solution are available. The work of rationalization seems likely to be difficult, since, at least for the solid state, there is evidence that temperature has a profound effect on the balance between oxonium and acylium complex formation[40]. Nor does the effect of temperature always appear to work in the same direction[41]. And no doubt for complex formation in solution the solvent will have an important influence on the balance between **1** and **3**.

C. Complexes as Intermediates in Acidic Systems

We have seen that virtually all the stable complexes so far isolated have been either oxonium- or acylium-type species containing, as the acid component, an entity $M(Hal)_n$ (i.e. **8** or **9**). As is well known, such species occur as intermediates in liquid-phase acylation of substances by RCOHal, under catalysis by $M(Hal)_n$ (equation 4). Depending upon the particular reactants (RCOHal and $M(Hal)_n$), the solvent, and the temperature involved, either **8** or **9**, or the free acylium ion, will predominate in the reaction mixture. However, since all these, and possibly certain other

$$RCOHal + M(Hal)_n \rightleftharpoons RC{\overset{O:M(Hal)_n}{\underset{Hal}{\diagdown}}} \rightleftharpoons RCO^+ \, M(Hal)^-_{n+1} \rightleftharpoons RCO^+ + M(Hal)^-_{n+1}$$

$$(8) \qquad\qquad\qquad (9)$$

$$SH \Big|? \qquad\qquad SH\Big|? \qquad\qquad SH\Big|? \qquad\qquad (4)$$

$$RCOS + HHal + M(Hal)_n$$

$$(SH = ArH \text{ or } R'OH, \text{ etc.})$$

(e.g. **2**), potential acylating agents are present in rapid equilibrium with each other, an important mechanistic question is to decide which species is the actual acylating agent, or if more than one species is active in this respect[42]. No doubt different answers apply for different systems. At present the answer is not known definitely for any system, although it would certainly seem likely, on intuitive grounds, that for those systems in which acylium-type species greatly predominate over oxonium species, the former provide most of any acylation occurring. This matter is discussed in detail in an earlier volume of the series[42, 43].

Although acids other than $M(Hal)_n$ rarely appear to lead to isolatable adducts with carbonyl halides, nevertheless they will certainly be expected to interact in appropriate circumstances, and so to provide labile complexes, or even semi-stable intermediates. We deal with a selection of known examples below.

5

I. Hydrogen acids (HX)

The dissolution of suitable (e.g. mesitoyl) carbonyl halides in 100% H_2SO_4 leads to complete ionization[44] as in equation (5). This can be con-

$$RCOCl + 3 H_2SO_4 \rightleftharpoons RCO^+ + HSO_3Cl + H_3O^+ + 2 HSO_4^- \qquad (5)$$

sidered to arise in two stages, equations (6) and (7). Clearly a proton

$$RCOCl + H_2SO_4 \rightleftharpoons HCl + HSO_4^- + RCO^+ \qquad (6)$$

$$HCl + 2 H_2SO_4 \rightleftharpoons HSO_3Cl + H_3O^+ + HSO_4^- \qquad (7)$$

attaches itself to the departing halide ion; perhaps ionization is initiated by hydrogen-bonded species like **11**. Other carbonyl halides will undergo reaction (5) in sulphuric acid to various extents[45] and other similar strong acids like disulphuric ($H_2S_2O_7$) and fluorosulphuric (FSO_3H) can also be used to effect analogous ionizations[29, 46]. Disulphuric acid seems par-

$$RC\underset{Cl\cdots H-O}{\overset{O}{<}}\overset{O\underset{O}{\overset{O}{\underset{\|}{S}}}}{}OH$$

(11)

ticularly effective and will completely convert acetyl chloride and bromide to acetylium ions[45]. Sulphuric and halogeno-sulphuric acids will apparently[29, 45, 47] only produce oxonium complexes with acetyl and with unsubstituted benzoyl halides, as in equation (8), acylium species being

$$RCOHal + H_2SO_4 \rightleftharpoons \underset{Hal}{\overset{|}{RCO^+H}} + HSO_4^- \rightleftharpoons \underset{OSO_3H}{\overset{OH}{\overset{|}{RC-Hal}}} \qquad (8)$$

(12)

undetectable spectroscopically*. Increasing numbers of alkyl substituents, especially *ortho* substituents, in the benzoyl halide lead, however, to increasing proportions of acylium ions[29]. The pattern of behaviour is very similar to that found for complexes with $M(Hal)_n$ (section II, B). The structure of the carbonyl addition species **(12)** is perhaps uncertain, but crystalline solids of this stoicheiometric composition can be isolated[43] when R = Ph.

Although spectroscopically undetectable, a small amount of acylium, or oxonium, complex formation, probably nearly always induced by hydrogen bonding from solvent species, doubtless occurs for all carbonyl halides in hydrogen-bonding solvents of high dielectric constant. Thus a significant

---

* Phosgene apparently leads to little complex formation of any type, even in disulphuric acid[46].

contribution from an acylium ion route can be detected[48,49] in the spontaneous hydrolysis of benzoyl halides in neutral aqueous media, equations (9) and (10). Even the so-called bimolecular route, in which

$$PhCOCl + nH_2O \xrightleftharpoons{fast} PhCOCl...HOH(H_2O)_{n-1}$$

$$\xrightleftharpoons{slow} PhCO^+ + Cl^-(H_2O)_n \quad (9)$$

$$PhCO^+ + H_2O \xrightarrow{fast} PhCO_2H_2^+ \xrightarrow{fast} PhCO_2H + H_3O^+ \quad (10)$$

H_2O attacks the carbonyl carbon atom in a slow step, probably involves assistance of leaving-group departure by hydrogen bonding from the solvent (equation 11). An alternative representation of the bimolecular

$$ 3H_2O + RC\overset{O}{\underset{Hal}{\diagdown}} \; \xrightleftharpoons{slow} \; \left[\begin{array}{c} \text{T.S.} \end{array} \right] \; \rightleftharpoons \; RCO_2H + H_3O^+ \\ + \quad (11) \\ Hal^-(H_2O) $$

route involves carbonyl addition (equations 12 and 13). In such a route, first the carbonyl oxygen atom, and then the halogen-leaving group, will

$$ R-C\overset{O}{\underset{Hal}{\diagdown}} + 4H_2O \xrightleftharpoons{slow} \left[\begin{array}{c} \text{T.S.} \end{array} \right] \rightleftharpoons \overset{OH}{\underset{OH}{RC-Hal}} + 3H_2O \quad (12) $$

$$ \overset{OH}{\underset{OH}{RC-Hal}} + 2H_2O \xrightleftharpoons{slow} \left[\begin{array}{c} \text{T.S.} \end{array} \right] $$

$$ \Big\Updownarrow $$

$$ RCO_2H + H_3O^+ + Hal^-(H_2O) \quad (13) $$

act as basic sites for solvent hydrogen bonding. The various hydrogen-bonding water molecules may all themselves be linked by hydrogen bonds, so leading to cyclic transition states, but the importance of complexes like 1 and 2 to the progress of these acylations is very evident. Moreover, the importance in acylation generally of the assistance of leaving-group departure by acid–base co-ordination effects cannot be overemphasized[50, 51]. However, of all classes of acylating agent RCOX, those, the carbonyl halides (in which X = Hal) make the least use of such assistance. This is essentially because halide ions are relatively weak bases, and have little tendency compared with (say) RCO_2^-, OR^- or OH^- ions, to co-ordinate with protons or other acid species. A related consequence is that C—Hal bonds tend to cleave heterolytically relatively easily; C—Hal bond-breaking is therefore usually well advanced in transition states of acylations involving RCOHal. Assistance of leaving-group departure is thus energetically of somewhat minor importance here, and any provided by a hydrogen-bonding solvent is often entirely adequate, the accelerating effect of any deliberately added acid catalyst being negligible. Since the basicity of the halide anions is in the sequence $I^- < Br^- < Cl^- < F^-$ one will expect acid catalysis of leaving-group departure to be most important for species RCOF and least for RCOI. This is indeed found for hydrogen acid catalysis. Thus, for bimolecular hydrolysis in aqueous solution, only acyl fluorides are subject[52] to catalysis by added hydrogen ions* (equations 14–15). For hydrolysis via a route like (9)–(10), the energetics of whose transition state is more dependent on the full cleavage of the C—Hal bond, it will be expected that H_3O^+ catalysis will be relatively more advantageous than in a route like (14)–(15) based on (11). This expectation is realized in that such catalysis is detectable for appropriate acyl *chlorides* such as mesitoyl chloride[54] (equation 16).

Studies of hydrolyses and alcoholyses of acyl halides in non-hydroxylic media have revealed features all of which are interpretable in terms of the necessity for intermediates like **1** or **2**. Thus, in aromatic or ethereal solvents, various acyl halides react not with monomeric, but preferentially with polymeric, water or alcohol species[55]. (This very probably also happens in hydroxylic media (e.g. equation 11) but there goes undetected.) As well as providing base catalysis without the necessity of a termolecular collision, such polymeric species can also provide cyclic transition states involving proton transfer to oxygen or halogen, e.g. **13**. More significant perhaps is the important catalysis exhibited by carboxylic acids in reactions of acyl chlorides with amines in non-hydroxylic media[51]. Clearly $RCOHal—RCO_2H$ complexes of some kind are involved here (nothing is to be gained by protonating the amine). The most obviously relevant complexes would be those providing hydrogen bonding to the halogen atom.

* For a fuller discussion of acid-catalysed acylation generally see references 50 and 53.

$$RCOF + H_3O^+ \rightleftharpoons RC\overset{\displaystyle O}{\overset{\|}{}}\!-\!\overset{+}{F}H + H_2O \quad \text{fast} \tag{14}$$

$$R\overset{+}{C}OFH + 2H_2O \longrightarrow \left[\begin{array}{c} O \\ \| \\ RC-FH \\ \uparrow \quad + \\ O \\ H \diagup \; \diagdown H \quad \ddot{O}H_2 \end{array}\right]$$

T.S.

$$\downarrow$$

$$RCO_2H + HF + H_3O^+ \quad \text{slow} \tag{15}$$

$$ArCOCl + H_3O^+ \rightleftharpoons Ar\overset{+}{C}OClH + H_2O \rightleftharpoons ArCO^+ + H_3O^+ + Cl^-$$

$$\xrightarrow{\;H_2O\;} ArCO_2H + H_3O^+ \tag{16}$$

In acetonitrile or nitromethane solutions anhydrous hydrogen chloride is a strong enough acid to induce acylium ion formation from acetyl and

$$\begin{array}{c} O \\ \| \\ RC-Hal \\ \diagdown \\ R^1\!\diagup^{O}\diagdown_{H}\cdots O\diagdown_{R^1}^{H} \end{array}$$

(13)

other straight-chain acyl chlorides. The acylation of a small concentration of a phenol in such a system follows[56] the general scheme (17). Acetyl

$$CH_3COCl + HCl \rightleftharpoons CH_3CO^+HCl_2^- \rightleftharpoons CH_3CO^+ + HCl_2^- \quad \text{fast}$$

↓ phenol	↓ phenol	↓ phenol	slow
products	products	products	(17)

bromide and hydrogen bromide show similar behaviour[57]. In view of the results in section II, B it will not be surprising to find that benzoyl chloride shows less extensive ionization under these conditions, but mesitoyl chloride should certainly exhibit a similar effect.

It will be evident from the foregoing discussion that hydrogen acids can form a variety of complexes with carbonyl halides, and that such complexes play important roles in hydrogen acid-catalysed acylation, especially in non-hydroxylic media. There appears to be no reason why species like $RCO^+HCl_2^-$ should not eventually be isolated.

2. Molecular halogen acids (Hal_2)

Molecular halogen acids are appreciably weaker acids than species $M(Hal)_n$, but are known to give adducts with bases in many contexts[58]. Ions like I_3^- and Br_3^- are relatively stable. It would therefore be expected that molecular halogens can catalyse reactions of acyl halides, under appropriate conditions, via complex formation. The most favourable conditions for detecting this effect are likely to be those obtaining in non-acidic, aprotic solvents where all other sources of assistance for leaving-group departure are absent. As with acids $M(Hal)_n$, the situation is potentially complicated by the fact that systems containing RCOHal and Hal_2 undergo halogen exchange reactions (equation 18). Where a single

$$RCOHal^* + Hal_2 \rightleftharpoons RCOHal + Hal{-}Hal^* \qquad (18)$$

type of halogen is involved this process can still be detected and followed using radio-isotopes. Kinetic studies[59] of the systems RCOI, I_2 and RCOBr, Br_2, using the solvents hexane and carbon tetrachloride, have shown that exchange in the iodine system is the faster, and that the rate equation takes the general form (19). Various considerations suggest that

$$Rate = k_2[RCOHal][Hal_2] + k_3[RCOHal][Hal_2]^2 \qquad (19)$$

the transition state corresponding to the bimolecular term approximates to $RCO^+ Hal_3^-$ (cf. 3); that corresponding to the termolecular term may well resemble 14 (cf. 2). In these processes an essentially electrophilic

$$
\begin{array}{c}
\overset{\displaystyle O}{\overset{\displaystyle \|}{RC}}\diagdown \overset{\delta-}{Hal} \\
\delta- \diagup \overset{\delta+}{Hal} \quad \overset{\delta+}{Hal} \\
Hal \quad Hal \\
\delta+ \quad \delta-
\end{array}
$$

(14)

substrate is catalysing its own acylation via acid–base complex formation at the leaving group.

3. Metal and similar ions

The catalytic effect of metal ions on the reactions of acyl chlorides has been demonstrated in two cases. Thus the mercuric ion catalyses the hydrolysis of benzoyl chloride in water[60], the reaction very probably proceeding via acylium ion formation (equations 20 and 21), although the results do not exclude the interaction between acyl halide and Hg^{2+} being

$$PhCOCl + Hg^{2+} \rightleftharpoons PhCO^+ + HgCl^+ \tag{20}$$

$$PhCO^+ + 2 H_2O \longrightarrow PhCO_2H + H_3O^+ \tag{21}$$

at the carbonyl group. Curiously, n-butyl chloroformate (BuOCOCl) exhibits no catalysis under similar conditions[60]. Even stranger is the finding[61] that while silver ion catalyses the nitration of isopropyl chloroformate in acetonitrile (equation 22 and 23), the methyl derivative is not

$$i\text{-}PrOC{\overset{O}{\underset{Cl}{\diagup}}} + Ag^+ \rightleftharpoons i\text{-}Pr\overset{+}{O}C{=}O + AgCl \tag{22}$$

$$i\text{-}Pr\overset{+}{O}\overset{\cdot}{C}O + NO_3^- \longrightarrow i\text{-}PrOC{\overset{O}{\underset{ONO_2}{\diagup}}}$$

$$\downarrow$$

$$i\text{-}PrONO_2 + CO_2 \tag{23}$$

subject to such catalysis. The effect of Hg^{2+} and Ag^+ in these reactions is doubtless due to their great affinity for halogen.

III. BEHAVIOUR AS ACIDS

A. Types of Complexes Formed with Bases

The adducts formed between carbonyl halides and acids (discussed above), although very reactive towards attack by any base present, are not, in isolation, inherently unstable with respect to further products; i.e. species **1**, **2** and **3** do not spontaneously decompose, save to their components. The possible complexes formed between carbonyl halides and bases, **4** and **5**, are in contrast often, indeed usually, unstable towards rapid dehydrohalogenation. This means that although carbonyl halides are particularly susceptible to attack by bases, nevertheless rather few of the resulting adducts are long enough lived to be isolatable. Typical decomposition equations for **4** and **5**, in which BH represents the base, would be (24) and (25). Appropriate choice of BH, however, can provide relatively stable adducts. These cases are discussed below.

$$RCOHal + BH \rightleftharpoons \underset{\underset{(4)}{\overset{\displaystyle +BH}{|}}}{\overset{\overset{\displaystyle O^-}{|}}{RC}}-Hal \rightleftharpoons \underset{\underset{}{\overset{\displaystyle B}{|}}}{\overset{\overset{\displaystyle OH}{|}}{RC}}-Hal \rightleftharpoons RCOB + HHal \quad (24)$$

$$\left[\underset{\underset{\mathbf{(5)}}{\overset{\displaystyle BH^+}{|}}}{\overset{\overset{\displaystyle O}{\|}}{RC}} \right] Hal^- \underset{\displaystyle \rightleftharpoons}{\overset{\displaystyle \rightleftharpoons}{}} \begin{array}{l} RCOBH^+ + Hal^- \\[2em] \Big\updownarrow \\[1em] RCOB + HHal \quad (25) \end{array}$$

B. Isolatable Complexes

If BH in equations (24) and (25) has a structure such that proton is not readily abstracted from it (e.g. if $BH = NR_3^2$) then a stable complex can result (equation 26). Equilibria like (26) appear to lie well towards species

$$R^1COHal + R_3^2N \underset{\displaystyle \rightleftharpoons}{\overset{\displaystyle \rightleftharpoons}{}} \begin{array}{l} \underset{\underset{+}{\overset{\displaystyle NR_3^2}{|}}}{\overset{\overset{\displaystyle O^-}{|}}{R^1C}}-Hal \quad \mathbf{(16)} \\[3em] \Big\updownarrow \\[1em] \overset{\overset{\displaystyle O}{\|}}{[R^1C}-NR_3^2]^+ \, [Hal]^- \quad (26) \\[0.5em] \mathbf{(15)} \end{array}$$

15 under most conditions. Salt-like complexes can be prepared by mixing the pure liquids R^1COHal and R_3^2N, or by mixing these reagents in solution in an inert solvent such as benzene, petroleum ether or dioxan. The preparation is best carried out at temperatures below 0°C. The complex usually separates as a white, crystalline material, easily decomposed by moisture. The complexes dissolve to some extent in dipolar, aprotic solvents (such as excess of R^1COCl) when they probably dissociate to give conducting solutions[62]. Table 4 includes examples of known complexes. They are, in general, poorly characterized. Rather fewer adducts of this character have been described in the literature than for complex formation between RCOHal and acids. This is somewhat surprising, since the first report was at least[66] as early as 1892. It is surprising too that these complexes have been little examined spectroscopically. The two available i.r. studies[67, 68] are not in good agreement, although it appears very probable that all the results are compatible with co-ordination of the tertiary

nitrogen atom to the carbonyl carbon atom, with concommitant salt formation essentially as indicated in **15**. An X-ray structural analysis of these complexes will be welcome.

TABLE 4. Examples of isolatable adducts with tertiary nitrogen bases

Adduct[a]	m.p. (°C)	Reference
CH_3COCl — pyridine	98	62, 63
$(CH_3COCl)_3$ — (α-picoline)$_2$	61	62, 65
CH_3COCl — β-picoline	78	62, 65
CH_3COCl — γ-picoline	125	62, 65
Furoyl chloride — pyridine	60	64
$PhCOCl$ — Et_3N	238	63
$PhCOCl$ — $(C_5H_{11})_3N$	117	63
$PhCOCl$ — $Et_2NC_6H_5$	115	63

[a] All the adducts very probably have structures like **15** (section III, B); when pure they are usually colourless and crystalline.

Reactions between acyl halides and tertiary amines do not invariably lead to stable complexes. Thus the phosgene complexes decompose[69] at room temperature as in equation (27).

$$[Cl-\overset{\overset{\text{O}}{\|}}{C}-NEt_3]^+Cl^- \longrightarrow EtCl + Et_2NCOCl \tag{27}$$

Under some conditions, especially it seems at room temperature and above, an α-hydrogen atom is eliminated from the acyl group, if this group is of suitable structure, and a keten results[70] (equation 28). The factors controlling the balance between processes (26) and (28) are not well

$$R^3CH_2COHal + NR_3^2 \longrightarrow R^3CH{=}C{=}O + R_3^2NH^+Cl^- \tag{28}$$

understood. It may be that keten formation does not proceed via species **15** or **16**, but by a direct attack of the amine on the α-hydrogen atom.

Isolatable crystalline, hygroscopic complexes are also formed from acyl bromides and suitable tertiary amides like N,N-dimethylformamide[71]. Similar complexes probably form with acyl chlorides, but have not been isolated. In an excess of liquid amide, partial dissociation of the complexes provides conducting solutions. Two types of interaction have been suggested[71], equations (29) and (30), the latter being preferred in the light

$$RCOBr + Me_2NC\overset{O}{\underset{H}{\diagdown}} \rightleftharpoons \left[Me_2\overset{+}{N}\overset{COR}{\underset{\underset{H}{C=O}}{\diagup}} \right] Br^- \qquad (29)$$

$$\textbf{(17)}$$

$$Me_2\overset{+}{N}\overset{COR}{\underset{\underset{H}{C=O}}{\diagup}} + Br^-$$

$$RCOBr + Me_2NC\overset{O}{\underset{H}{\diagdown}} \rightleftharpoons \left[Me_2\overset{+}{N}=C\overset{OCOR}{\underset{H}{\diagdown}} \right] Br^- \qquad (30)$$

$$\textbf{(18)}$$

$$Me_2\overset{+}{N}=\underset{\underset{H}{|}}{C}OCOR + Br^-$$

of the products obtained on reacting the complexes with water and amines. This interpretation is perhaps supported by the nature of the complexes formed by the related *N,N*-dimethyl-*N'*-phenylformamidines[72] (equation 31).

$$RCOCl + Me_2NCH=NPh \rightleftharpoons \left[Me_2N-C=N\overset{Ph}{\underset{COR}{\diagdown}} \right]^+ Cl^- \qquad (31)$$

Too little is known about complexes **15** and **18** to justify the drawing of generalizations about their stability as a function of structure. It is evident, however, that, at least in **15**, bulky bases will encounter steric difficulties.

C. Complexes as Intermediates in Basic Systems

Complexes like **15** are probably intermediates in the tertiary amine-catalysed preparation of anhydrides from acyl halides and carboxylic acids[73], or from acyl halides and limited amounts of water. Certainly if a deficit of water is added to the preformed complex in a suitable solvent reactions (32) occur[64]. Even if all the reagents are added simultaneously,

$$R^1CO\overset{+}{N}R_3^2Hal^- + H_2O \longrightarrow R^1CO_2H + R_3^2\overset{+}{N}HHal^-$$

$$\xrightarrow{R^1CO\overset{+}{N}R_3^2Hal^-} (R^1CO)_2O + 2 R_3^2\overset{+}{N}HHal^- \qquad (32)$$

provided that a relatively large concentration of tertiary amine is present, scheme (32) will probably obtain.

As noted in section III, A, most bases which react with acyl halides contain abstractable hydrogen atoms, and in undergoing acylation lead to rapid production of hydrogen halide without any complexes being isolatable. The first question to arise therefore is whether such bases form intermediate complexes like **4** or **5**, or whether the elimination of hydrogen halide is a purely synchronous process, e.g. equation (33). A second

$$
\underset{\substack{\|\\O}}{RC}-Hal\ +\ 2\,BH \;\rightleftharpoons\; \left[\underset{\substack{B\\\diagdown\\H\cdots B}}{\overset{\substack{O\\\|\\ }}{RC}}\!\!-\!Hal\!\!\underset{H}{\nearrow}\right]\ \longrightarrow\ RCOB\ +\ BH\ +\ HCl\ (33)
$$

T.S.

question, if they *do* form complexes, is which type of complex is preferred under different circumstances? The discussion in section II, C concerning the acylation of substrates (bases*) *under acid catalysis*, shows that synchronous processes may be prevalent under such conditions (e.g. equation (15) and species **13** and **14**). Equation (33) is, of course, just another example of this sort. Otherwise, under acid catalysis, acylium ion routes prevail, so that transient complexes like **5** are involved when the base attacks. In short, for acid-catalysed interaction between RCOHal and bases, there appears little evidence that addition of the base leads to complexes like **4** involving carbonyl addition. One of the best tests for the occurrence of carbonyl addition complexes in aqueous systems involving species RCOX (where $X = NH_2$, OH, OR, OCOR, Hal, etc.), is the detection of oxygen exchange occurring simultaneously with any

$$
O\ +\ \underset{\substack{\|\\^{18}O}}{RC}-X \;\rightleftharpoons\; \underset{\substack{|\\OH_2^+}}{\overset{^{18}O^-}{RC}}-X \;\rightleftharpoons\; \underset{\substack{|\\OH}}{\overset{^{18}OH}{RC}}-X \;\rightleftharpoons\; \underset{\substack{|\\O^-}}{\overset{^{18}OH_2^+}{RC}}-X \;\rightleftharpoons\; \underset{\substack{\|\\O}}{RC}-X\ +\ H_2{}^{18}O
$$

(19)   (20)   (21)   (34)

$$
RCO^{18}OH_2^+\ +\ X^-\ \rightarrow\ RCO^{18}OH\ +\ HX
$$

hydrolysis[74]. If exchange occurs, the carbonyl group must be disturbed, and it seems likely that complexes **19**, **20** and **21**, similar to **4**, will be involved (equation 34). The tenor of our discussion so far suggests that such exchange during hydrolysis may not be very evident when X = Hal

---

* In acylation by RCOX the carbonyl carbon atom is always an electrophile, so that the substrate always acts as a nucleophile, i.e. as a base.

for other routes to hydrolysis, not involving carbonyl addition, are available to, and apparently preferred by, acyl halides (e.g. equation 11). The existing results[74] concerning oxygen exchange conform to these expectations: relatively little exchange is detected even in neutral solution. That *some* exchange is found suggests, however, that carbonyl halides do form (transient) species like **4**, even when the base contains replaceable hydrogen atoms. Whether such species become more prominent in acylation in non-hydroxylic media which do not facilitate acylium, or synchronous, routes remains something of an open question[56, 75]. It is certainly a mistake, however, to assume automatically, as is often done, that in acylation by RCOHal, if intermediate species like **5** are not involved, then species like **4** are; in all cases intermediates (or transition states) like **13** require consideration.

IV. BEHAVIOUR AS SOLVENTS

Pure, liquid carbonyl halides have been relatively little used as solvents. This is probably due to their great reactivity (especially towards water) which makes reliable experimentation difficult and limits their use as 'inert' media. Studies from a few laboratories are, however, available. Mostly these studies have been concerned with various acid–base equilibria occurring in carbonyl halides, and are therefore relevant to this chapter. Indeed many of the phenomena observed can be predicted from what we have already learnt in sections I to III.

The carbonyl halides which have been most thoroughly examined as solvents are phosgene[76], acetyl chloride[62], benzoyl chloride[62] and benzoyl fluoride[77], and our attention will be confined to these. Their physical characteristics are given in Table 5. Phosgene proves of limited usefulness,

TABLE 5. Physical constants of carbonyl halide solvents

Carbonyl halide	Mol. wt.	m.p.	b.p.	D	Specific conductance $(ohm^{-1}\ cm^{-1})$	Dipole moment (Debye)
CH_3COCl	78·5	−113	52	15·9[b]	4×10^{-7a}	2·40
$PhCOCl$	140·4	−0·6	197	22·9[a]	9×10^{-8b}	3·33
$PhCOF$	124·1	−28·5	155·6	22·7[b]	1×10^{-8b}	—
$COCl_2$	99	−128	8	4·7[d]	7×10^{-9c}	—

[a] At 25°C.
[b] At 20°C.
[c] At 10°C.
[d] At 0°C.

but has been used as one early model for the solvent-systems definitions[78] of acid–base behaviour. Since this approach is unfortunately still used by many inorganic chemists, in spite of its being quite unsatisfactory as a general representation of acid–base behaviour, it is appropriate at this stage to discuss the proper formulation of acid–base effects, and its application to the systems being considered*. The only completely satisfactory definitions of acids and bases are those of Lewis[80]. Both the widely used alternatives to Lewis's, i.e. the definitions usually attributed to Brønsted and Lowry[81], and the solvent-systems definitions[78], are obsolescent. The unsuitability of the Brønsted view, in which all acids must contain donatable protons, is evident from the very existence of this chapter: one cannot discuss the acid–base properties of carbonyl halides on the Brønsted basis. The solvent-systems approach limits acid–base phenomena to those occurring in self-ionizing solvents. No one, on reflexion, can believe that this is realistic. Moreover, even for suitable solvents, this definition can be very misleading, since it overemphasizes ionization phenomena. Thus an acid is defined as a species which, when added to the solvent, increases the concentration of the typical solvent cation. Similarly a base is any species which increases the concentration of solvent anion. Simple co-ordination is not regarded as an acid–base phenomenon.

The restrictions of both the Brønsted–Lowry and the solvent-systems definitions can be overcome by using Lewis's, which, in fact, contain both the other approaches as special cases†.

Lewis's essential acid–base reaction is the forward step of (35), its

$$A + :B \rightleftharpoons A:B \tag{35}$$
$$\text{acid} \quad \text{base} \qquad \text{adduct (complex)}$$

inverse is bond heterolysis. Related heterolytic processes are (36), (37) and (38). Thus, the formation of complexes **1** (or **2**) in equation (1) is an example

$$A^1 + A^2:B \rightleftharpoons A^1:B + A^2 \tag{36}$$
$$\text{acid} \quad \text{adduct} \qquad \text{adduct} \quad \text{acid}$$

$$A:B^1 + B^2 \rightleftharpoons A:B^2 + B^1 \tag{37}$$

$$A^1:B^1 + A^2:B^2 \rightleftharpoons A^1:B^2 + A^2:B^1 \tag{38}$$

of equation (35). The formation of **3** from **2** is another example of equation (35), its formation from RCOHal and A an example of equation (36). Reactions of acyl halides with bases which lead eventually to acylation of the base (e.g. equation 24) are examples of equation (38).

* See reference 79 for a full discussion of acid–base definitions.

† We have, of course, used Lewis's definitions throughout this chapter (see Introduction).

In the solvent-systems treatment of carbonyl halides as solvents they are considered to self-ionize to some (undisclosed) extent as in equation (39).

$$RCOHal \rightleftharpoons RCO^+Hal^- \rightleftharpoons RCO^+ + Hal^- \tag{39}$$

A solvent-systems acid will therefore be any species [e.g. $RCO^+M(Hal)_{n+1}^-$ or simply $M(Hal)_n$] which will increase the amount of RCO^+ present. A typical solvent-systems base (e.g. $Et_4\overset{+}{N}Hal^-$ or C_5H_5N) will increase the concentration of Hal^-. Thus solvent-systems neutralization reactions can be written as equations (40) and (41). In (40) the essential neutralization

$$RCO^+M(Hal)_{n+1}^- + Et_4N^+Hal^- \rightleftharpoons RCOHal + Et_4N^+M(Hal)_{n+1}^- \tag{40}$$

$$\tag{41}$$

reaction is (42), so that both (40) and (41) are simply examples of equation

$$RCO^+ + Hal^- \rightleftharpoons RCOHal \tag{42}$$

(35). Thus the Lewis formulation includes the solvent-systems approach, the solvent (here RCOHal) *always* behaving as an adduct. However, on Lewis's basis, RCOHal can *also* act either as a base (e.g. in formation of **1** or **2**) or as an acid (e.g. in formation of **4**). In the solvent-systems description these complexes are all termed solvates[62].

A. Solution of Species M(Hal)$_n$ in Carbonyl Halides

As for the other solvents RCOHal which have been studied (Table 5), the specific conductance of purified phosgene is very low. The residual conductance may be attributable to some very small amount of self-ionization like (43), but is more probably essentially due to traces of water

$$COCl_2 \rightleftharpoons COCl^+ + Cl^- \tag{43}$$

impurity which lead to reaction (44). Most metal chlorides are poorly

$$COCl_2 + 3 H_2O \longrightarrow CO_2 + 2 H_3O^+ + 2 Cl^- \tag{44}$$

soluble in phosgene, but $AlCl_3$ has been shown to dissolve readily. Some increase in conductivity results. On the solvent-systems theory this would normally be explained along the lines of equation (45). However, no

$$COCl_2 + AlCl_3 \rightleftharpoons [COCl]^+[AlCl_4]^- \rightleftharpoons COCl^+ + AlCl_4^- \tag{45}$$

halogen exchange occurs between $COCl_2$ and $AlCl_3$ under such conditions (cf. section II, B). It follows that equation (45) cannot represent the behaviour of $AlCl_3$ dissolved in $COCl_2$. The increase in conductivity can be accounted for by assuming $COCl_2$ to behave as a Lewis oxygen base, its class **1** adduct with $AlCl_3$ being capable of two different heterolyses (equation 46). In this context therefore, on the solvent-systems basis, $AlCl_3$

$$COCl_2 + AlCl_3 \rightleftharpoons Cl_2C{=}O{:}AlCl_3 \rightleftharpoons Cl_2CO{:}AlCl_2^+ + Cl^- \quad (46)$$

should be regarded as a base.

Phosgene is no doubt exceptional, and although, as we have seen (section II, B), benzoyl chloride also interacts with MCl_n mainly as a Lewis oxygen base, the increase in conductivity produced in this solvent more probably arises from the formation of small quantities of free acylium and MCl_{n+1}^- ions (equation 47). For acetyl chloride acylium

$$PhCOCl + MCl_n \rightleftharpoons \underset{\underset{\textstyle Cl}{|}}{PhC}{=}OMCl_n \rightleftharpoons [PhCO]^+[MCl_{n+1}]^-$$

$$\rightleftharpoons PhCO^+ + MCl_{n+1}^- \quad (47)$$

ion formation is certainly likely on the addition of MCl_n (see section II, B).

When species RCOHal and $M(Hal)_n$ are mixed and crystalline solids precipitated, or isolated, the composition of the solid may not directly reflect the acid–base interactions involved in forming the complex (section II, B). Thus phosgene and $AlCl_3$ have been reported[13, 14] to lead to distinct crystals of composition $AlCl_3$: $COCl_2$ 2 : 1, 1 : 1, 1 : 2 and 1 : 3 respectively. Since the essential acid–base adduct in this case is **22**, the extra molecules of acid or base in the other crystals (if real) are simply molecules

$$Cl_2C{=}O : AlCl_3$$
$$\textbf{(22)}$$

of crystallization. A further disadvantage of the solvent-systems approach is that it leads to indiscriminate use of the term solvate for all these species, regardless of how the solvent is held. The structures of different RCOHal—$M(Hal)_n$ complexes have been discussed in section II.

The range of solubility of different MCl_n species in phosgene, acetyl chloride, benzoyl chloride and benzoyl fluoride is illustrated in Table 6. It will be apparent from the foregoing discussion that the past interpretation, by some workers, of the behaviour of compounds $M(Hal)_n$ in carbonyl halide solvents as *exclusively* either dissociation (equation 48) or acylium ion induction (equation 49) is oversimple, and indeed fundamentally incorrect. Fortunately, this attitude is slowly changing.

$$M(Hal)_n \rightleftharpoons M(Hal)_{n-1}^+ + Hal^- \quad (48)$$

$$M(Hal)_n + RCOHal \rightleftharpoons [RCO]^+[M(Hal)_{n+1}]^- \rightleftharpoons RCO^+ + M(Hal)_{n+1}^- $$
$$(49)$$

TABLE 6. Solubility of typical species $M(Hal)_n$ in carbonyl halide solvents[62, 76, 77]

$M(Hal)_n$	Solvent[a]			
	$COCl_2$	CH_3COCl	$PhCOCl$	$PhCOF$
NaCl	−	−	−	−
KCl	−	−	−	−
$CaCl_2$	−	−	−	
$BaCl_2$	−	−	−	
$HgCl_2$		+	+	
BF_3				+
BCl_3	+	+	+	
$AlCl_3$	+	+	+	
$SnCl_4$		+	+	+
$TiCl_4$		+	+	+
$FeCl_3$		+	+	+
$SbCl_3$		+	+	
$SbCl_5$		+	+	+
PCl_5		+	+	
$NiCl_2$		−		
ICl			+	+

[a] + reasonably soluble or completely miscible; − very slightly soluble.

A final point about solutions of species $M(Hal)_n$ is that they are reported to be often coloured[62]. This is the case for solutions in acetyl chloride and is surprising since CH_3CO^+ is colourless. The observed colours may arise from impurities.

B. Solutions of Tertiary Nitrogen Bases in Carbonyl Halides

The details of section III suggest that bases like pyridine and triethylamine, dissolved in benzoyl or acetyl chloride, will undergo with the solvent reactions like (50). Some increase in conductivity will be expected owing to the formation of small quantities of the free ions. It is by no means certain, as appears often to be assumed, that, in these solutions, the equilibrium position in (50) will always lie well towards complex

$$R^1COCl + R_3^2N \rightleftharpoons \left[\begin{array}{c} R^1CO \\ | \\ NR_3^2 \end{array} \right]^+ [Cl]^- \rightleftharpoons R^1CO\overset{+}{N}R_3^2 + Cl^- \quad (50)$$

formation; much free amine may exist in many cases, especially with sterically hindered amines.

C. Reactions Between Solute Acids and Bases in Carbonyl Halides

Much of the work with carbonyl halide solvents has concerned equilibria between dissolved acids and bases[62, 77]. Two main types of study have been: (i) the titration of strong bases, such as pyridine and quinoline, against powerful acids like $SbCl_5$, $TiCl_4$ and $SnCl_4$; (ii) the measurement[82] of equilibrium constants for acceptance, by a variety of acids MCl_n, of Cl^- from species such as Ph_3CCl, or $Et_4N^+Cl^-$.

(i) The titrations can be followed either by observing the concomitant changes in conductivity or, less satisfactorily, by the use of visual indicators (e.g. crystal violet or bromophenol blue). The titrations with $SbCl_5$ can be formulated (ignoring for the present any co-ordinated solvent) as in equation (51). The resulting acid–base adduct, if isolated, sometimes

$$SbCl_5 + \underset{}{\text{(pyridine)}} N \longrightarrow \underset{}{\text{(pyridine)}} N \rightarrow SbCl_5 \qquad (51)$$

contains solvent of crystallization, but, in our opinion, there is little evidence to suggest[62, 83] that the principal reaction should really be represented as in equation (52). Any RCOCl crystallizing with the adduct

$$RCO^+SbCl_6^- + [RCONC_5H_5]^+Cl^- \longrightarrow RCOCl + [RCONC_5H_5]^+SbCl_6^-$$
$$\qquad\qquad\qquad\qquad\qquad \textbf{(23)} \qquad\qquad\qquad (52)$$

is readily released and can therefore hardly be bound as in species **23**. In most systems it will of course be more correct, in writing equation (51), to include the co-ordinated solvent. For $SbCl_5$ in many acyl halides the principal sequence is probably (53), when the base is pyridine. Small quantities of species like **23** may also be formed and contribute to observed changes in conductivity. These changes are difficult to explain convincingly, and it is evident that there is still much to be learnt about these titrations.

$$RCO^+SbCl_6^- \ (\text{or } RC\overset{Cl}{=}O\!:\!SbCl_5) \ + \ [C_5H_5NCOR]^+Cl^-$$

$$\Big\updownarrow \qquad\qquad\qquad\qquad /\!/ \qquad\qquad\qquad (53)$$

$$RCOCl + SbCl_5 \ + \ RCOCl + C_5H_5N \longrightarrow 2RCOCl + C_5H_5N\!:\!SbCl_5$$

When a dibasic acid like $TiCl_4$ is used, the evidence suggests that both 1:1, and 2:1, base:acid complexes are formed as acid is added to base. The 2:1 complexes, which form first owing to the presence of excess of base, tend to precipitate, but redissolve as more acid is added, to give the

$$2\,C_5H_5N + TiCl_4 \longrightarrow (C_5H_5N)_2TiCl_4 \qquad (54)$$

$$(C_5H_5N)_2TiCl_4 + TiCl_4 \longrightarrow 2\,C_5H_5NTiCl_4 \qquad (55)$$

1:1 complex. Here again, isolated complexes sometimes contain solvent of crystallization, but, as before, there is little justification for writing the principal neutralization product as $[C_5H_5NCOR]_2^+TiCl_6^{2-}$.

(ii) The carbonium ion Ph_3C^+ has a spectrum which permits equilibria like (56) to be easily followed spectrophotometrically. Typical results for the association constant for reaction (56) are in Table 7. One interesting

$$Ph_3CCl + MCl_n \; \xrightleftharpoons \; [Ph_3C]^+MCl_{n+1}^- \tag{56}$$

and curious point which emerges from this Table, and also from results[84] of similar equilibria in various other dipolar aprotic solvents, is that the

TABLE 7. Association constants (K) for reaction (56) in benzoyl chloride (temperature $\approx 20°C$?)[82]

M(Hal)$_n$	K (l/mole)
FeCl$_3$	$> 10^4$
SbCl$_5$	$> 10^4$
GaCl$_3$	$\approx 10^4$
SnCl$_4$	300
BCl$_3$	70
ZnCl$_2$	60
TiCl$_4$	58
SbCl$_3$	1·8
AlCl$_3$	1·5
PCl$_5$	0·2

sequence of acid strengths of species MCl_n towards Cl^-, although mainly in agreement with that found for oxygen and nitrogen bases[38], is exceptional in the low positions occupied by BCl_3 and $AlCl_3$. We believe this result arises at least partly from a steric effect. The chloride ion is appreciably bulkier than either an oxygen or a nitrogen atom, and there is likely therefore to be relatively more steric strain involved in its adducts with MCl_n. Where M is a large atom itself, there will normally be sufficient space easily to accommodate the chloride ion, but for $AlCl_3$ and BCl_3 important steric repulsions may result, relative to those incurred by oxygen and nitrogen bases. This sort of explanation is applicable also to the difficulty which has been found in forming BF_3Cl^- and BCl_3Br^-. The unusually low acidity of BCl_3 and of $AlCl_3$ toward Cl^- has, of course, important implications for the choice of catalyst in MCl_n catalysed Friedel–Crafts acylation by acyl halides and similar reactions: BCl_3 and $AlCl_3$ will be expected to be poorer catalysts than might have been anticipated from their behaviour towards oxygen and nitrogen bases. This seems to be so[85].

V. CONCLUSION

The behaviour of carbonyl halides as Lewis acids or bases is understood qualitatively and in outline, but there still remains some doubt, in several contexts, as to the principal type of adduct (complex) which they form. Isolatable complexes have so far only been obtained for a very limited number of the possible acid–base interactions. The quantitative study of the acid–base properties has scarcely begun, and the determination of a reasonable number of equilibrium constants for acid–base complex formation in solution is an essential pre-requisite of meaningful, further theoretical discussion in this field.

VI. REFERENCES

1. G. Perrier, *Compt. Rend.*, **116**, 1298 (1893); M. Nencki, *Ber.*, **32**, 2414 (1899).
2. J. Boeseken, *Rec. Trav. Chim.*, **39**, 622 (1920); K. H. Klipstein, *Ind. Eng. Chem.*, **18**, 1328 (1926); P. Pfeiffer, *Organische Molekülverbindungen*, 2nd ed., F. Enke, Stuttgart, 1927.
3. H. Meerwein and H. Maier-Hüser, *J. Prakt. Chem.*, **134**, 51 (1932).
4. N. N. Greenwood and K. Wade, *J. Chem. Soc.*, 1527 (1956).
5. G. A. Olah and S. J. Kuhn, *Chem. Ber.*, **89**, 866 (1956).
6. G. A. Olah, S. J. Kuhn, W. S. Tolgyesi and E. B. Baker, *J. Am. Chem. Soc.*, **84**, 2733 (1962).
7. N. M. Cullinane, S. J. Chard and D. M. Leyshon, *J. Chem. Soc.*, 4106 (1952).
8. D. Cassimatis, P. Gagnaux and B. P. Susz, *Helv. Chim. Acta*, **43**, 424 (1960).
9. B. P. Susz and D. Cassimatis, *Helv. Chim. Acta*, **44**, 395 (1961).
10. A. Bertrand, *Bull. Soc. Chim. France*, (2) **34**, 631 (1881).
11. A. Bertrand, *Bull. Soc. Chim. France*, (2) **33**, 403 (1880).
12. B. Menshutkin, *Chem. Zentr.* **I**, 481 (1911).
13. D. E. H. Jones and J. L. Wood, *J. Chem. Soc. (A)*, 1140 (1967); K. O. Cristie, *Inorg. Chem.*, **6**, 1706 (1967).
14. Z. Iqbal and T. C. Waddington, *J. Chem. Soc. (A)*, 1745 (1968).
15. D. Bethell and V. Gold, *Carbonium Ions: an Introduction*, Academic, London, 1967.
16. F. Fairbrother, *J. Chem. Soc.*, 503 (1937).
17. G. Oulevey and B. P. Susz, *Helv. Chim. Acta*, **44**, 1425 (1961).
18. *Carbonium Ions*, Vol. 1 (Ed. G. A. Olah and P. von R. Schleyer), Interscience, New York, 1968.

19. F. Seel, *Z. Anorg. Allgem. Chem.*, **250**, 331 (1943).
20. B. P. Susz and J. J. Wuhrmann, *Helv. Chim. Acta*, **40**, 722 (1957); D. Cook, 'Spectroscopic Investigations', in *Friedel–Crafts and Related Reactions*, Vol. I (Ed. G. A. Olah), Interscience, New York, 1963.
21. B. P. Susz and J. J. Wuhrmann, *Helv. Chim. Acta*, **40**, 971 (1957).
22. D. Cook, *Can. J. Chem.*, **37**, 48 (1959).
23. G. A. Olah, S. J. Kuhn and S. H. Flood, *J. Am. Chem. Soc.*, **84**, 1688 (1962); G. A. Olah, W. S. Tolgyesi, S. J. Kuhn, M. E. Moffatt, I. J. Bastien and E. B. Baker, *J. Am. Chem. Soc.*, **85**, 1328 (1963).
24. J. Overend, R. A. Nyquist, J. C. Evans and W. J. Potts, *Spectrochim. Acta*, **17**, 1205 (1961).
25. E. Amaldi, *Z. Physik.*, **79**, 492 (1932).
26. W. F. Arendale and W. H. Fletcher, *J. Chem. Phys.*, **26**, 793 (1957).
27. G. A. Olah, S. J. Kuhn, S. H. Flood and B. A. Hardie, *J. Am. Chem. Soc.*, **86**, 2203 (1964).
28. G. A. Olah and M. B. Comisarow, *J. Am. Chem. Soc.*, **88**, 4442 (1966).
29. D. A. Tomalia, *J. Org. Chem.* **34**, 2583 (1969).
30. J. M. Le Carpentier, B. Chevrier and R. Weiss, *Bull. Soc. Fr. Mineral. Cristalog.*, **91** (6), 544 (1968).
31. S. V. Rasmussen and N. C. Broch, *Acta Chem. Scand.*, **20**, 1351 (1966).
32. F. P. Boer, *J. Am. Chem. Soc.*, **88**, 1572 (1966).
33. P. J. Lillford and D. P. N. Satchell, *J. Chem. Soc.* (*B*), 1016 (1970).
34. J. C. Jaccard and B. P. Susz, *Helv. Chim. Acta*, **50**, 97 (1967).
35. I. Cooke, B. P. Susz and C. Herschmann, *Helv. Chim. Acta*, **37**, 1280 (1954).
36. G. Olofsson, *Acta Chem. Scand.*, **21**, 1114 (1967).
37. G. A. Olah, 'Intermediate Complexes', in *Friedel–Crafts and Related Reactions* Vol. I, (Ed. G. A. Olah), Interscience, New York, 1963.
38. D. P. N. Satchell and R. S. Satchell, *Chem. Rev.*, **69**, 251 (1969).
39. D. D. Perrin, *Dissociation Constants of Organic Bases in Aqueous Solution*, Butterworths, London, 1965.
40. H. H. Perkampus and W. Weiss, *Angew. Chem. Intern. Ed. Engl.*, **7**, 70 (1968).
41. E. Lindner and H. Kranz, *Chem. Ber.*, **99**, 3800 (1966).
42. D. P. N. Satchell and R. S. Satchell, 'Formation of Aldehydes and Ketones by Acylation, Formylation and some Related Processes', in *The Chemistry of the Carbonyl Group* (Ed. S. Patai), Wiley, London, 1966.
43. See also R. Corriu, C. Coste and G. Dubosi, *Compt. Rend.*, **261**, 3632 (1965); R. Corriu and C. Coste, *Bull. Soc. Chim. France*, 3276 (1969); H. Ruotsalainen, L. A. Kumpulainen and P. O. I. Virtanen, *Suom. Kemistilehti*, (*B*), **43**, 91 (1970).
44. R. J. Gillespie and E. A. Robinson, *J. Am. Chem. Soc.*, **86**, 5676 (1964); R. J. Gillespie and E. A. Robinson, *J. Am. Chem. Soc.*, **87**, 2428 (1965).
45. F. Carré and R. Corriu, *Bull. Soc. Chim. France*, 2898 (1967); A. Casadevall, G. Canquil and R. Corriu, *Bull. Soc. Chim. France*, 204 (1964).
46. R. C. Paul, V. P. Kapila and K. C. Malhotra, *Chem. Commun.*, 644 (1968).
47. F. Carré, R. Corriu and G. Dabosi, *Bull. Soc. Chim. France*, 2905 (1967).
48. V. Gold, J. Hilton and E. G. Jefferson, *J. Chem. Soc.*, 2756 (1954).
49. D. A. Brown and R. F. Hudson, *J. Chem. Soc.*, 3352 (1953); B. L. Archer and R. F. Hudson, *J. Chem. Soc.*, 3259 (1950).

50. D. P. N. Satchell and R. S. Satchell, 'Substitution in the Groups COOH and COOR', in *The Chemistry of Carboxylic Acids and Esters* (Ed. S. Patai), Wiley, London, 1969; P. J. Lillford and D. P. N. Satchell, *J. Chem. Soc. (B)*, 897 (1968); I. I. Secemski and D. P. N. Satchell, *J. Chem. Soc. (B)*, 1013 (1970).
51. L. M. Litvinenko, D. M. Aleksandrova and N. I. Pilyuk, *Ukrain. Khim. Zhur.*, **25**, 81 (1959).
52. D. P. N. Satchell, *J. Chem. Soc.*, 555 (1963).
53. D. P. N. Satchell, *Quart. Rev. (London)*, **17**, 160 (1963).
54. M. L. Bender and M. C. Chen, *J. Am. Chem. Soc.*, **85**, 30 (1963).
55. R. F. Hudson and I. Stelzer, *Trans. Faraday Soc.*, **54**, 213 (1958).
56. J. M. Briody and D. P. N. Satchell, *J. Chem. Soc.*, 168 (1965); D. P. N. Satchell, *J. Chem. Soc.*, 558 (1963).
57. J. M. Briody and D. P. N. Satchell, *J. Chem. Soc.*, 3724 (1964).
58. L. J. Andrews and R. M. Keefer, *Molecular Complexes in Organic Chemistry*, Holden Day, San Francisco, 1964.
59. A. Goldman and R. M. Noyes, *J. Am. Chem. Soc.*, **79**, 5370 (1957).
60. H. K. Hall and C. H. Lueck, *J. Org. Chem.*, **28**, 2818 (1967).
61. D. N. Kevill and G. H. Johnson, *Chem. Commun.*, 235 (1966).
62. R. C. Paul and S. S. Sandhu, in *The Chemistry of Non-aqueous Solvents* (Ed. J. J. Lagowski), Vol. 3, Academic, New York, 1970.
63. K. Freudenberg and D. Peters, *Ber.*, **52**, 1463 (1919); W. M. Dehn, *J. Am. Chem. Soc.*, **34**, 1399 (1912); W. M. Dehn and A. A. Ball, *J. Am. Chem. Soc.*, **36**, 2091 (1914).
64. H. Adkins and Q. E. Thompson, *J. Am. Chem. Soc.*, **71**, 2242 (1949).
65. R. C. Paul, D. Singh and S. S. Sandhu, *J. Chem. Soc.*, 315 (1959).
66. G. Minnunni, *Gazz. Chim. Ital.*, **22**, 213 (1892).
67. R. C. Paul and S. L. Chadha, *Spectrochim. Acta*, **22**, 615 (1966).
68. J. Holecek, I. Pavlik and J. Klikorka, *Sb. Ved. Praci. Vysoka Skola Chem. Technol. Pardubice*, **2**, 23 (1964); *Chem. Abstr.*, **64**, 5815g.
69. Y. A. Strepikheev, T. G. Perlova and L. A. Zhivechkova, *Zhur. Org. Khim.* **4**, 1891 (1968).
70. J. B. Hanford and A. B. Sauer, *Org. Reactions*, **3**, 124 (1946).
71. H. K. Hall, *J. Am. Chem. Soc.*, **78**, 2717 (1956).
72. F. Falk, *J. Prakt. Chem.*, **15**, 228 (1962).
73. R. Kuhn and I. Löw, *Ber.*, **77**, 202 (1944).
74. M. L. Bender, *J. Am. Chem. Soc.*, **76**, 1626 (1951); D. Samuel and B. L. Silver, *Adv. Phys. Org. Chem.*, **3**, 123 (1965).
75. D. N. Kevill and F. D. Foss, *J. Am. Chem. Soc.*, **91**, 5054 (1969).
76. V. Gutmann, *Halogen Chemistry*, Vol. 2, Academic, London, 1967.
77. G. Jander and L. Schweik, *Z. Anorg. Allgem. Chem.*, **310**, 1, 12 (1961).
78. A. F. O. Germann, *J. Am. Chem. Soc.*, **47**, 2461 (1925).
79. D. P. N. Satchell and R. S. Satchell, *Quart. Rev. (London)*, **25**, 171 (1971).
80. G. N. Lewis, *Valence and the Structure of Atoms and Molecules*, Chemical Catalogue Co., New York, 1923.
81. J. N. Brønsted, *Rec. Trav. Chim.*, **42**, 718 (1923); T. M. Lowry, *Chem. Ind.*, **42**, 43 (1923).
82. V. Gutmann and G. Hampel, *Monatsh.*, **92**, 1048 (1961).

83. R. C. Paul, M. L. Lakhaupal, P. S. Gill and J. Singh, *Indian J. Chem.*, **2**, 262 (1964); K. Goyal, R. C. Paul and S. S. Sandhu, *J. Chem. Soc.*, **322** (1959).
84. M. Baaz, V. Gutmann and J. R. Masaguer, *Monatsh.*, **92**, 582 (1961).
85. R. M. Evans and R. S. Satchell, *J. Chem. Soc. (B)*, 1667 (1970).

CHAPTER **5**

Directing and activating effects of COX groups

P. H. GORE

Brunel University, Uxbridge, Middlesex, England

I. INTRODUCTION

This chapter describes the effect of substituent carbonyl halide groups on the qualitative and quantitative behaviour of alkyl or aryl groups to which they are attached. These substituents have not received as much attention as some other electron-attracting substituents (e.g. $-NO_2$, $-COOH$), and results in the literature are mostly sporadic. The available data relate in the main to the $-COCl$ substituent, with some data on $-COBr$ and $-COF$ groups; no data on the substituent effects of a $-COI$ group appear to have been published. In fact, the only detailed studies in the whole field covered by the title to this chapter have been on free-radical halogenations of alkane carbonyl halides. In consequence, general structure–reactivity correlations could not be attempted. It is clear that this field is still wide open for the experimental chemist.

A. The Polar Effects of Carbonyl Halide Groups

The polar character of a carbonyl halide group is a vital factor in determining its effect upon aliphatic or aromatic groups to which it may be attached. The COX group (**1**, X = F, Cl, Br, I) is essentially a carbonyl group, but modified in its overall polar characteristics by the halogen

$$-C\overset{O}{\underset{X}{\diagup}}\qquad\qquad -C\overset{O}{\underset{X}{\diagup}}\longleftrightarrow -C\overset{O^-}{\underset{X^+}{\diagup}}$$

(1)   (1a)

atom. A halogeno-group may attract electrons by the inductive electronic mode $(-I)$, whose sequence of magnitudes[1a] is given in sequence (1). It may also release electrons by the conjugative (mesomeric) electronic mode[1b] $(+M)$, in the same order (sequence 2). It has been suggested that

Sequence of $-I$ effects: $-F > -Cl > -Br > -I$ (1)

Sequence of $+M$ effects: $-F > -Cl > -Br > -I$ (2)

the latter effect actually involves a contribution by the π-inductive (I_π) effect[2-4]. Evidence for these electronic mechanisms comes mainly from physical properties of ground states of halogenated molecules (e.g. proton magnetic resonance spectra of CH_3X molecules[5]). The carbonyl oxygen atom powerfully attracts electrons both by the inductive $(-I)$ and the mesomeric $(-M)$ effects[1a, c]. Electron distributions in **1** will therefore be affected by the electronic effects originating at X, as in **1a**. In a series of carbonyl derivatives (RCOX), allowing X to become any of a succession of groups having progressively reduced $+M$ effects, sequences are formed along which the internal mesomeric electron displacement will diminish[1b, c], *viz.* (3) and (4).

$+M$ effects within RCOX: $X = -O^- > -NR_2 > -OR > -Cl > -H$ (3)

$+M$ effects within RCOX: $X = -F > -Cl > -Br > -I$ (4)

Where a COX group is linked to a saturated grouping the overall effect of this mesomerism is a transfer of electronic charge from the halogen atom to the oxygen atom. The extent of electron-deficiency of the carbon atom is not thereby materially altered. The overall electron-attracting properties $(-I)$ of a COX substituent, which are exerted on the saturated group to which it is attached, are therefore, in addition to the strong electron-attracting nature of the carbonyl group, graded by the superimposed $-I$ effects of the halogen atoms. Support for this reactivity sequence (5) is

$-I$ effects of $-COX$ substituents: $-COF > -COCl > -COBr > -COI$ (5)

lacking, of the kind available for other substituents, e.g. carboxyl[6] or amino groups[7]. The overall inductive effect of a COX group will be a combination[8] of the σ-inductive effect (I_σ) and the field effect (F). The former describes the transmission of charge by successive and diminishing polarization of a series of adjacent $C-C$ σ bonds, as in **(2)**. Factors such

$$\underset{C}{\overset{\delta\delta\delta\delta+}{}} \longrightarrow \underset{C}{\overset{\delta\delta\delta+}{}} \longrightarrow \underset{C}{\overset{\delta\delta+}{}} \longrightarrow \underset{C}{\overset{\delta+}{}} \overset{O}{\underset{X}{\diagdown}}$$

(2)

as $\frac{1}{3}$ have been suggested[9, 10] for the proportion of charge disturbance relayed from one carbon to its neighbour. The I_σ effect is considered by many workers[11-16] to be of minor importance compared to the F effect, an electrostatic effect operating directly through space. The precise separation of the I_σ and F effects is still subject to argument[17, 18], and for the present purposes they are considered together, as 'inductive effects' (I)[19].

When a carbonyl halide group is bound through its carbon atom to an unsaturated residue **(3)** or an aryl residue **(4)**, the group as a whole must withdraw electrons[1d], since the $C=O$ group is conjugated with the unsaturated (or aryl) group, while the group X is not thus conjugated. Group

(3)   (4).

X competes with the $-M$ effect of the whole group by providing an alternative source of electrons for the electronegative oxygen atom. The relative $-M$ effects of the complete COX groups will be given (sequence 6) by inverting the series representing the relative mesomeric displacements

Sequence of $-M$ effects: $-COI > -COBr > -COCl > -COF$ (6)

within the group. Evidence of conjugation of type 3 or 4 has been obtained, for example, from exaltations of molecular refraction[1e, 20], and from electronic spectra[21]. Thus, the u.v. light absorption of benzoyl chloride has been shown to be very similar to that of benzaldehyde or acetophenone (Table 1). However, the apparent magnitude of the $-M$

TABLE 1. U.v. absorption bands of compounds C_6H_5COY in cyclohexane solution[a]

Compound	B-band		C-band	
	λ_{max} (nm)	ε_{max}	λ_{max} (nm)	ε_{max}
C_6H_5COCl	242	14,500	282	1300
	(ca. 250)	12,000	(ca. 291)	900
C_6H_5CHO	241	14,500	277–278	1200
	(247)	12,500	287	1000
$C_6H_5COCH_3$	237–238	12,500	277	900

[a] Data from reference 21. Data in parentheses refer to inflexions.

effects follow the sequence of substituents (7). The B-band is associated

Sequence of $-M$ effects: $-NO_2 > -COCl > -CHO > -COCH_3 > -CO_2Et$
(7)

with resonance of type 5, which schematically represents interactions between the carbonyl group and the benzene ring. To this mesomerism

(5)

the chlorine atom does not appreciably contribute. This type of conjugation is reduced in the presence of bulky *ortho* substituents (steric inhibition of resonance). This is shown, for example, by a reduction of molar extinction (ε_{max}) of the *B*-band, in the order of effective size (8),

Sequence of interference radii: $-NO_2 > -COCH_3 > -COCl > -CHO$ (8)

i.e. in the order of the effective interference radii of these substituents[21]. In these cases of *ortho*-substituted derivatives, the carbonyl halide (or other) substituent cannot exert its full electronic effect on the aromatic nucleus.

The significance of conjugation of type **4** becomes evident in aromatic substitution reactions. Inductive withdrawal of electrons from the aromatic nucleus ($-I$) is enhanced by that of the conjugative mechanism ($-M$), which effects specifically the *ortho*- and *para*-positions. The presence of electron-deficient sites in these nuclear positions should activate them towards attack by nucleophilic reagents (*Nu*), e.g. should promote the formation of a transition state of type **6**. However, nucleophiles appear

(6)

to react invariably at the carbonyl C-atom of the side chain, and not at the aromatic nucleus. A discussion of these side-chain reactions is outside the scope of the present chapter.

In benzoyl halides the electronic effects ($-I, -M$) combine to withdraw electrons from the aromatic ring; whilst the $-M$ effect extracts electrons specifically from *para*- and *ortho*-positions. Therefore, in electrophilic substitutions attack will occur predominantly at the *meta*-positions of a deactivated substrate.

II. DIRECTING AND ACTIVATING EFFECTS OF COX GROUPS IN ALIPHATIC CARBONYL HALIDES

A. Substitution Reactions

I. Fluorinations

Vapour-phase fluorination of acetyl fluoride has been achieved[22] by passing a mixture of the halide with fluorine, diluted with nitrogen, over heated copper. Fluorinated materials accounted for about half of the products by weight (reaction 9). An indirect method of analysis showed

$$CH_3COF \longrightarrow FCH_2COF \text{ (ca. 35\%)} + F_2CHCOF \text{ (ca. 5\%)} + F_3CCOF \text{ (trace)}$$
(9)

that substantial amounts of the acetyl fluoride had remained unchanged; this result agreed with an earlier report[23] of the low α-reactivity in the halogenations of acyl halides. It is significant that polysubstitution occurred to an appreciable extent, pointing to a reactivity for monofluoroacetyl fluoride not much lower than for acetyl fluoride.

The fluorination of butyryl chloride has been reported by the use of an *in situ* electrolytic method of generation of fluorine[23]. The products were mainly 4-fluoro-derivatives, with smaller amounts of the mono-3-fluoro-derivative (reaction 10). Traces of 4,4,4-trifluorobutyryl chloride,

$$CH_3CH_2CH_2COCl \longrightarrow CH_2FCH_2CH_2COCl + CHF_2CH_2CH_2COCl$$
$$\text{(main product)}$$
$$+ CH_3CHFCH_2COCl \qquad (10)$$

$CF_3CH_2CH_2COCl$, must have been formed, since succinic acid was detected among the hydrolysis products. Products of chlorination were also detected in this reaction. These were thought to be due to an initial halogen exchange reaction (11), which was followed by chlorination in

$$CH_3CH_2CH_2COCl \xrightarrow{\text{F}_2} CH_3CH_2CH_2COF \qquad (11)$$

the carbon chain of the butyryl fluoride formed.

An electrolytic method for the perfluorination of propionyl chloride and of butyryl chloride has been reported[24]. Low temperature fluorination of *n*-propylcarbamoyl chloride, $CH_3CH_2CH_2NHCOCl$, gave mainly the *N*-fluoro-derivative, but about 8% of C-fluoro-compounds were also formed[25].

2. Chlorinations

Side-chain chlorinations have been achieved with carbonyl halides by two main methods, which differ according to the type of attacking reagent. In the presence of certain catalysts ionic chlorinations proceed, through attack (mainly at the 2-position) by electrophilic chlorine. In the absence of such a catalyst, and in the presence of light or radical promoters, the chlorination will proceed by a free-radical mechanism. By this process usually more than one C-atom along the chain is liable to be substituted.

a. Ionic chlorinations. In the presence of 'halogen-carriers' aliphatic carbonyl halides can undergo ionic chlorination α- to the carbonyl group[26-29]. Thus, Markownikoff[30] found in 1869 that chlorination of

$$(CH_3)_2CHCOCl \xrightarrow[\text{(I}_2)]{\text{Cl}_2} (CH_3)_2CClCOCl \qquad (12)$$

isobutyryl chloride was catalysed by iodine (12). These catalysed chlorinations proceed readily in the presence of a variety of catalysts (13); in the

$$CH_3CH_2COCl \xrightarrow[\text{(PCl}_3)]{\text{Cl}_2} CH_3CHClCOCl + CH_3CCl_2COCl \qquad (13)$$

presence of light free-radical chlorination will compete. The catalysts include halides and oxyhalides of sulphur[27, 29] and of phosphorus[26, 27, 29], metallic halides[29] and iodine[26, 28, 30, 31]. α-Chlorination of carbonyl chlorides can be achieved by means of sulphuryl chloride, with iodine as catalyst[32].

By the prolonged action of phosphorus pentachloride on acetyl chloride (or chloroacetyl chloride), chlorination took place with formation of trichloroacetyl chloride[33]. No mechanism was suggested for this reaction (14).

$$CH_3COCl \xrightarrow[\text{28 days}]{PCl_5} CCl_3COCl \tag{14}$$

Simultaneous vicinal dichlorination has been reported[34], when butyryl chloride is boiled (in the dark) with excess of iodine trichloride (15); much

$$CH_3CH_2CH_2COCl \xrightarrow{ICl_3} CH_3CHClCHClCOCl \ (16\%) \tag{15}$$

butyryl chloride remained unaltered. The action of the halide reagent was believed to involve the dissociation[35] (16).

$$ICl_3 \rightleftharpoons ICl + Cl_2 \tag{16}$$

A liquid-phase chlorination of hexanoyl chloride appeared to give a mixture of isomers[36]. Neither light nor a catalyst was specifically excluded from the reaction, the mechanism of which thus remains uncertain.

Two kinetic studies of the chlorination of carbonyl chlorides have appeared in the literature. Watson and Roberts[31] found that iodine powerfully catalysed the chlorination of acetyl chloride, in carbon tetra-chloride solution (k_1 at $25° = 0.0113$ min^{-1}). The reaction proceeded only as far as the first stage (17).

$$CH_3COCl + Cl_2 \longrightarrow ClCH_2COCl + HCl \tag{17}$$

Hertel, Becker and Clever[37] showed that chlorination in the dark was a pseudo first-order reaction. The mechanism was believed to involve an initial enolization of the carbonyl chloride as the rate-determining step (18).

$$CH_3-C{\overset{O}{\underset{Cl}{\lessgtr}}} \xrightarrow{slow} CH_2=C{\overset{OH}{\underset{Cl}{\lessgtr}}} \xrightarrow[(+Cl_2)]{rapid} CH_2Cl-\overset{OH}{\underset{Cl}{C}}-Cl \xrightarrow[(-HCl)]{rapid} CH_2Cl-C{\overset{O}{\underset{Cl}{\lessgtr}}} \tag{18}$$

The rates for these halogenations were found to follow the sequence: acetyl chloride > butyryl chloride > propionyl chloride. Activation energies for these three halides were, approximately, 21, 24 and 27 kcal/mole, respectively.

b. Free-radical chlorinations. Two main methods have been used for the introduction of chlorine atoms into the paraffinic chain of carbonyl halides by a free-radical mechanism. One procedure involves the use of sulphuryl chloride, with benzoyl peroxide as radical promoter[32, 38]. Chlorine atoms are thereby readily introduced at C-atoms which are some distance removed from the carbonyl chloride group. Acetyl chloride does not react.

A second method involves the action of chlorine on carbonyl halides with irradiation by a u.v. source ('photochlorination'). Acetyl chloride, preferably in the absence of a solvent, may be converted in this way to chloroacetyl chloride, but only in low yield[39, 40] (reaction 19).

$$CH_3COCl \xrightarrow{Cl_2, h\nu} CH_2ClCOCl (2\%) + HCl \qquad (19)$$

A third method of radical chlorination, which seems to have been used only once[41], involves the irradiation (at 335 nm) of acetyl chloride in carbon tetrachloride solution. There is a short induction period, during which there is a build-up of chlorine atoms (20). The subsequent chlorination of acetyl chloride is said to proceed quantitatively.

$$2 CCl_4 \longrightarrow C_2Cl_6 + 2 Cl^\bullet \qquad (20)$$

As expected from the low reactivity of acetyl chloride, photochlorination of monochloroacetyl chloride[31, 41] or of dichloroacetyl chloride[42] does not take place.

An abnormal reaction was found to occur[43] when acetyl chloride, chloroacetyl chloride or dichloroacetyl chloride, in solution or in the gas phase at 100–200°C, are allowed to interact with chlorine under irradiation with energy-rich light. In each case tetrachlorosuccinyl dichloride is formed in good yield (reaction 21), presumably by coupling of radical fragments.

$$ClCH_2COCl \xrightarrow[\text{boiling-point, 40h}]{Cl_2,\ h\nu} \begin{array}{c} CCl_2COCl \\ | \\ CCl_2COCl \end{array} (75\%) \qquad (21)$$

Free-radical chlorination of the higher carbonyl chlorides takes a more complex course. An attempt has been made[44] to explain the orientations essentially as a modifying influence of the substituents ('influence indices') on the statistical isomeric distribution.

The free-radical chlorinations which have been reported for propionyl chloride are summarized in Table 2. Typically, both 2- and 3-chloro-substituted products are obtained, with the latter predominating. The one exception to this rule is the observation by Smit and den Hertog[45, 46, 47]

TABLE 2. Free-radical chlorinations of propionyl chloride

Conditions	Product composition		References
	2- (%)	3- (%)	
SO$_2$Cl$_2$, benzoyl peroxide	40	60	32
Cl$_2$, hv, 0–5°	11	86	49
Cl$_2$, hv	30	70	39, 40
Cl$_2$, hv, 20°C	40	60	50
Cl$_2$, hv, 20°C	26	74	45, 46, 47
Cl$_2$, hv, −70°C	17	83	45, 46
Cl$_2$, C$_6$H$_6$, hv, 20°C	42	58	45
Cl$_2$, CS$_2$, hv, 20°C	43	57	45, 46
Cl$_2$, CS$_2$, hv, −70°C	61	39	45, 46

that in carbon disulphide solution at a low temperature photochlorination, in contrast to that conducted at room temperature, gives mainly 2- chloro-propionyl chloride.

The reported free-radical chlorinations of *n*-butyryl chloride are given in Table 3. All the results are in agreement with a sequence of reactivities

TABLE 3. Free-radical chlorinations of *n*-butyryl chloride

Conditions	Product composition			References
	2- (%)	3- (%)	4- (%)	
SO$_2$Cl$_2$, benzoyl peroxide, CCl$_4$, 85°C[a]	3	49	48	38
Cl$_2$, hv, 20°C[b]	20	60	20	51
Cl$_2$, hv, 20°C	2·5	49·5	48	52, 53
Cl$_2$, hv, 20°C	3	64	33	45, 47
Cl$_2$, hv, 101°C	5	65	30	39, 40
Cl$_2$, hv, 101°C	2·5	55	42·5	52
Cl$_2$, hv, 101·5°C[c]	6·2	49	44·8	48
Cl$_2$, hv, −70°C	1·5	67	31	45
Cl$_2$, C$_6$H$_6$, hv, 20°C	2	77	21	47

[a] Cf. chlorination, under the same conditions, of 1,1,1-trichlorobutane: 8% 2-, 42% 3-, 50% 4-derivatives.

[b] A 4% yield of dichlorobutyryl chloride (presumably 2,4-) was also formed.

[c] Cf. results obtained similarly for butyric acid (164°C): 7·8% 2-, 52·4% 3-, 39·8% 4-derivatives; for butyronitrile (117·6°C): 51·1% 3-, 48·9% 4-derivatives; methyl butyrate: (59° or 74°C): 57·8% 3-, 42·2% 4-derivatives.

3- > 4- > 2- of the alkyl chain. The earlier report[32] of the formation of 15% 2-chlorobutyryl chloride was later amended[38] to a lower value. Unless very pure starting materials were used, some ionic chlorination (giving the 2-derivative) was believed to compete with the predominant radical reaction[38]. Brown and Ash[38] suggested that a —COCl group is very similar to a —CCl₃ group in its effect of directing further substitution in the aliphatic chain (Table 2). Magritte and Bruylants[48] similarly showed that the —CO₂H, —CO₂Me and —CN derivatives also react to give closely similar results. The low 2- reactivity in chlorinations of *n*-butyryl chloride (or indeed of propionyl chloride and acetyl chloride) was ascribed to the inductive effect ($-I$) of the —COCl group[38].

Table 4 lists free-radical chlorinations of branched-chain carbonyl chlorides. Isobutyryl chloride gives predominantly the 3-derivative, and

TABLE 4. Free-radical chlorinations of branched-chain carbonyl chlorides

Conditions	Monochlorination products			References
	2- (%)	3- (%)	4- (%)	
(a) Isobutyryl chloride (CH₃)₂CHCOCl				
SO₂Cl₂, benzoyl peroxide	20	80		32
Cl₂, *hv*, 0–5°C	25	65		50, 72
Cl₂, *hv*	31	69		39, 40
Cl₂, *hv*, 20°	60–70	30–40		51
Cl₂, *hv*, 90°	28·5	71·5		49
(b) Isovaleryl chloride (CH₃)₂CHCH₂COCl				
Cl₂, *hv*[a]	32	60	8	39
Cl₂, *hv*, 20°C[b]	—	50	—	51
(c) Pivaloyl chloride (CH₃)₃CCOCl				
SO₂Cl₂, benzoyl peroxide[c]	—	100	—	32

[a] Under similar conditions isovaleric acid gave: 26% 2-, 69% 3-, 5% 4-derivatives; and methyl isovalerate gave: 49% 2-, 49% 3-, 2% 4-derivatives; in the latter case some dichloro-derivatives were also formed.

[b] The 3-chloro-derivative was the main product; the 2- and 4-compounds were also formed.

[c] The reaction was very rapid.

this is expected from reactivity as well as statistical considerations. Dichlorination may also take place[49]. Wautier and Bruylants[49] believed that chlorine attack occurs preferentially at the electron-rich 3-carbon atom, which results from a combination of hyperconjugation and inductive effects. The results from the photochlorination of isovaleryl chloride are very close to those obtained (under similar conditions) with isovaleric acid and methyl isovalerate (see Table 4).

The photochlorinations of three carbonyl fluorides and four carbonyl chlorides with longer chain lengths are recorded in Table 5. The gas-phase

TABLE 5. Free-radical chlorinations of pentanoyl, hexanoyl, heptanoyl and octanoyl halides

Conditions	Monochlorination product composition (%)							References
	2-	3-	4-	5-	6-	7-	8-	
(a) Pentanoyl fluoride								
Cl₂, hv, 65–160°C	1·5	23	60	15				54, 55, 56
(b) Pentanoyl chloride								
Cl₂, hv	5	65	30	0				39
Cl₂, hv, 20°C	<1.5	15–20	50–55	30–35				52, 53
Cl₂, hv, 20°C	0·7	19	58	23				47, 57
Cl₂, C₆H₆, hv, 20°C	0·4	14	76	10				47
Cl₂, CCl₄, hv, 52°C	4·3	16·9	46·0	32·8				55
Cl₂, CH₃CN, hv, 50°C	2·7	18·8	47·3	31·2				55, 56
(c) Hexanoyl fluoride								
Cl₂, hv, 50°C	1·4	12·6	34·8	38·8	12·9			55
Cl₂, hv, 105°C	2·5	13·6	32·0	37·7	14·2			55
(d) Hexanoyl chloride								
Cl₂, hv, 20°C	<1	5–10	25–30	40–45	20–25			53
Cl₂, hv, 20°C	0	10	30	44	16			45, 47, 57
Cl₂, C₆H₆, hv	0	5·3	35	53	6·6			45, 47, 57
Cl₂, CCl₄, hv, 52°C	2·6	10·0	26·9	34·4	25·8			55
Cl₂, CH₃CN, hv, 0°C	0·6	8·2	28·9	40·3	22·0			55
Cl₂, CH₃CN, hv, 52°C	2·7	8·6	27·2	36·6	25·0			55
(e) Heptanoyl fluoride								
Cl₂, hv, 60°C	1·3	10·8	25·4	25·3	27·5	9·2		55, 56
Cl₂, hv, 105°C	1·4	10·2	25·3	25·4	27·6	10		55
(f) Heptanoyl chloride								
Cl₂, CH₃CN, hv, 52°C	1·1	7·7	16·5	20·9	25·9	27·9		55, 56
(g) Octanoyl chloride								
Cl₂, hv, 20°C	0	5·0	14	20	25	26	10	45, 47
Cl₂, hv, −48°C	0	4·3	13	21	27	25	8·7	45, 46
Cl₂, C₆H₆, hv, 20°C	0	2·0	12·5	22	32	28	3·4	45, 46, 47
Cl₂, C₆H₆, hv, 73°C	0·9	3·4	13	22	28	26	5·8	45

chlorinations of pentanoyl, hexanoyl and heptanoyl fluorides were studied by Singh and Tedder[54, 55] at several temperatures. A comparison of their reactions at 60°[54, 55] is made in Table 6, where the 4-carbon atom is given

TABLE 6. Gas-phase photochlorinations of pentanoyl, hexanoyl and heptanoyl fluorides at 60°

Formula and results expressed as relative reactivities

FOC	—CH$_2$—	—CH$_2$—	—CH$_2$—	—CH$_3$		
	0·02	0·37	1	0·24		
FOC	—CH$_2$—	—CH$_2$—	—CH$_2$—	—CH$_2$—	—CH$_3$	
	0·04	0·37	1	1·1	0·25	
FOC	—CH$_2$—	—CH$_2$—	—CH$_2$—	—CH$_2$—	—CH$_2$—	—CH$_3$
	0·05	0·40	1	1·0	1·1	0·25

arbitrary unit reactivity. Clearly, the COF group exerts no appreciable effect beyond the 3-position of the alkyl chain. Chlorinations of each of these fluorides at other temperatures gave essentially the same relative proportions of isomers[54, 55]. Results suggested that differences in reactivity were due principally to differences in activation energies[55].

Chlorinations of the corresponding carbonyl chlorides in solution, however, gave different results[55, 56] (Table 7). Here, the rate of attack at

TABLE 7. Photochlorinations of pentanoyl, hexanoyl and heptanoyl chlorides in acetonitrile solution at 52°

Formula and results expressed as relative reactivities

ClOC	—CH$_2$—	—CH$_2$—	—CH$_2$—	—CH$_3$		
	0·06	0·40	1	0·44		
ClOC	—CH$_2$—	—CH$_2$—	—CH$_2$—	—CH$_2$—	—CH$_3$	
	0·09	0·32	1	1·4	0·62	
ClOC	—CH$_2$—	—CH$_2$—	—CH$_2$—	—CH$_2$—	—CH$_2$—	—CH$_3$
	0·06	0·50	1	1·3	1·5	1·1

the terminal C-atom increases with increasing chain length. Therefore, the substituent is affecting chlorination at least five atoms down the chain[55]. Unfortunately the reactivities of these halides were not studied relatively to each other*.

* In Tables 6–11 and 14, unit reactivities for the chosen position along the chain are not necessarily the same for the different compounds.

An interesting comparison of reactivities for the photochlorinations of various derivatives of n-butane and pentanoic acid has been given by Singh and Tedder[54] (Table 8). It can be seen that in gas-phase chlorinations

TABLE 8. Gas-phase photochlorinations of derivatives of n-butane at 75°C

X	Relative reactivities			
	X——CH_2^α——CH_2^β——CH_2^γ——CH_3^δ			
H	1	3·6	3·6	1
F	0·9	1·7	3·7	1
Cl	0·8	2·1	3·7	1
CF_3	0·04	1·2	4·3	1
COF	0·08	1·6	4·2	1
COCl	0·2	2·1	3·9	1
CO_2CH_3	0·4	2·4	3·6	1

none of the substituents has an appreciable effect at the γ-position. Deactivation at the β-position by a carbonyl halide is of the same order as that induced by a halogen atom. At the α-position the carbonyl group is more deactivating than a halogen group. A similar analysis of photobrominations is given in section II, A, 3, b.

A comparison of results obtained for liquid-phase photochlorinations of carbonyl chlorides from propionyl chloride to octanoyl chloride[47] is instructive (Table 9). Data are given for reactions conducted in benzene

TABLE 9. Relative reactivities in photochlorinations of carbonyl halides in benzene solution at 20°C

Compound	Relative reactivities						
	2-	3-	4-	5-	6-	7-	8-
Propionyl chloride	0·72	1·0					
Butyryl chloride	0·1	3·6	1·0				
Pentanoyl chloride	0·04	1·5	7·6	1·0			
Hexanoyl chloride	0	0·8	5·2	8·0	1·0		
Octanoyl chloride	0	0·6	3·7	6·5	9·4	8·4	1·0

solution, since selectivity of chlorination is increased in this solvent. The reactivities of the 2-positions are lower than the ω-positions in each halide. The reactivities increase with increasing distance from the deactivating

COCl group up to the $(\omega-1)$ or $(\omega-2)$ C-atoms. The inductive effect of the COCl group is still felt at the 5-position in octanoyl chloride. This confirms the results given for the photochlorinations of carbonyl chlorides given above (Table 8). In Table 10 a comparison is made between the chlorinations of octanoyl chloride and *n*-heptane under the same conditions[45]. It is clear that the deactivation 2->3->4->5- must be due to the progressively weaker inductive effect $(-I)$ of the COCl grouping. A similar comparison has been made (Table 11) for the photochlorinations of octanoyl chloride and heptane-1,7-dicarbonyl dichloride (azelayl chloride)[45]. The 3- and 4-positions of azelayl chloride appear to be affected only by the near (not the distant) COCl group; the 5-position, equidistant from both COCl groups, is somewhat reduced in reactivity relative to octanoyl chloride.

It has generally been assumed[58] that in the photochlorination reaction chlorine atoms are first formed[45, 46] (reaction 22). These atoms are strongly electronegative, and abstract H-atoms in a rapid, exothermic reaction; the transition state involves only a slight breaking of the C—H bond[59] (reaction 23). In solution the transition state in hydrogen abstractions by chlorine atoms is very much more polar than in the gas phase[55].

$$Cl_2 \xrightarrow{h\nu} 2\,Cl^\bullet \qquad (22)$$

$$R{:}H + Cl^\bullet \longrightarrow [R{:}H^\bullet Cl \longleftrightarrow R^\bullet + H{:}Cl^- \longleftrightarrow R^\bullet H{:}Cl]$$

$$\longrightarrow R^\bullet + H{:}Cl \quad (23)$$

The chlorine atoms predominantly attack C—H bonds with the highest electron-availability, while the strength of the C—H bond is not very important. In spite of the fact that H-abstraction at the 2-position of alkane carbonyl chlorides produces a free-radical somewhat stabilized by resonance (7), the 2-chloro compound is the one formed in smallest

(7)

amount, owing to the dominance of the inductive effect of the COCl substituent[45].

3. Brominations

a. Ionic brominations. In the presence of halogen-carriers carbonyl halides may be brominated α- to the carbonyl function[30]. Thus, in the presence of iron powder and phosphoryl chloride, bromine and anhydro-camphoronyl chloride (8) gave a 75% yield of the 2-bromo-derivative[60].

TABLE 10. Photochlorination of octanoyl chloride and n-heptane in benzene solution at 20°C

Compound	Relative reactivities						

Octanoyl chloride

CH$_3$—CH$_2$—CH$_2$—CH$_2$—CH$_2$—CH$_2$—CH$_2$—COCl
1·0   8·4   9·4   6·5   3·7   0·6   0

n-Heptane

CH$_3$—CH$_2$—CH$_2$—CH$_2$—CH$_2$—CH$_2$—CH$_3$
1·0   8·1   9·6   9·1   9·6   8·1   1·0

TABLE 11. Photochlorination of azelayl chloride and octanoyl chloride at 20°C in the absence of a solvent

Compound	Relative reactivities						

Azelayl chloride

ClOC—CH$_2$—CH$_2$—CH$_2$—CH$_2$—CH$_2$—CH$_2$—CH$_2$—COCl
     1·0   2·6   3·1   2·6   2·6   5·2   1·0

Octanoyl chloride

CH$_3$—CH$_2$—CH$_2$—CH$_2$—CH$_2$—CH$_2$—CH$_2$—COCl
2·0   5·2   5·0   4·0   2·9   1·0   1·0

$$\underset{O}{\overset{CH_3}{\underset{\diagdown}{\underset{CO-C(CH_3)_2}{\overset{CO-C-CH_2COCl}{\diagup}}}}}$$

(8)

Isovaleryl chloride on bromination at 100°C gave a 72% yield of 2-bromo-derivatives[61] (reaction 24). Bromination in the carbon chain preceded the

$$(CH_3)_2CHCH_2COCl \xrightarrow[\text{PCl}_3 \text{ (trace)}]{\text{Br}_2} (CH_3)_2CHCHBrCOBr + (CH_3)_2CHCHBrCOCl \quad (24)$$

$$(58\%) \qquad\qquad\qquad (14\%)$$

replacement of the acyl halogen[61]. Acetyl bromide could not be brominated in the presence (or absence) of acetic acid in the temperature range 10–118°C[62].

The kinetics of bromination of acetyl chloride and acetyl bromide were first investigated by Watson[63]. The reactions probably proceeded by an ionic mechanism, but the reaction vessel was not specifically shielded from light. First-order rate coefficients for the acetyl bromide reaction showed a progressive decrease, but a constant value was obtained by binding bromine in the equilibrium (reaction 25). Iodine monobromide

$$Br_2 + HBr \rightleftharpoons HBr_3 \quad (25)$$

proved to be effective as a catalyst, but ferric chloride or sulphuric acid were not (Table 12). The bromination of acetyl chloride was found to

TABLE 12. Rate constants for the bromination of acetyl chloride and acetyl bromide at 25°C in the absence of a solvent

Substrate	Added component	k_1 (min^{-1})
Acetyl chloride	—	0·00087
Acetyl bromide	—	0·0023
Acetyl bromide	FeCl$_3$ (0·035 M)	0·0022
Acetyl bromide	H$_2$SO$_4$ (0·035 M)	0·0022
Acetyl bromide	IBr (0·035 M)	0·0033

proceed at about one-third the rate for acetyl bromide. The initial rate with acetyl chloride increased somewhat with time, due to slow formation of the more reactive acetyl bromide (reaction 26). A rapid initial enolization

$$CH_3COCl + HBr \rightleftharpoons CH_3COBr + HCl \quad (26)$$

was proposed, followed by a comparatively slow addition of bromine. The kinetic results are collected in Table 12.

The kinetics of bromination of acetyl bromide have also been studied by Cicero and Matthews[64]. Good second-order rate constants were obtained in nitrobenzene solution. A mechanism was suggested in which bromine addition took place to the enol form (reaction 27). The kinetic data are summarized in Table 13.

$$\left[CH_3COBr \rightleftharpoons CH_2=C{\overset{Br}{\underset{OH}{\Big\langle}}} \right] \longrightarrow \quad \longrightarrow$$

$$+ \quad HBr \qquad (27)$$

TABLE 13. Kinetics of bromination of acetyl bromide in nitrobenzene solution

Temperature (°C)	$10^4 k_2$ (l/mole sec)	Derived data
50·9	1·21 ± 0·14	E_{act} = 20·33 kcal/mole
60·6	3·00 ± 0·32	A = 5·84 × 10⁹
70·3	6·90 ± 0·13	ΔH^* (60·6°C) = 19·67 kcal/mole
		ΔS^* = −16 e.u.

b. Free-radical brominations. n-Butyryl chloride, when reacted[50, 65] with bromine in direct sunlight and in the absence of a solvent, gives mainly the 3-bromo-derivative (reaction 28). In a semi-quantitative study,

$$CH_3CH_2CH_2COCl \xrightarrow[h\nu]{Br_2} \begin{array}{l} CH_3CHBrCH_2COCl \\ + CH_3CH_2CHBrCOCl \\ + CH_3CH_2CHBrCOBr \end{array} \qquad (28)$$

Kharasch and Hobbs[66] found that (*i*) brominations of alkane carbonyl chlorides were accelerated by light, (*ii*) reactions carried out in the dark were not materially affected by the presence of oxygen, (*iii*) the photobromination of acetyl chloride was substantially unaffected by oxygen, but those of propionyl chloride and butyryl chloride were appreciably inhibited. The reactivities of the halides in the photobromination reaction increased in the sequence acetyl chloride < propionyl chloride < butyryl chloride.

The photobrominations of n-valeryl fluoride and of n-valeryl chloride have been studied by Singh and Tedder[54] in the gas phase at 150–160°C.

In each case the 4-position proved the most reactive position. By comparing the data with other derivatives of *n*-butane (Table 14), it was concluded

TABLE 14. Photobrominations of various derivatives of
n-butane at 160°C

X	Relative reactivities[a]			
	$X \text{——} CH_2^\alpha \text{——} CH_2^\beta \text{——} CH_2^\gamma \text{——} CH_3^\delta$			
H	1	80	80	1
F	9	7	90	1
Cl	34	32	(80)	1
CF$_3$	1	7	90	1
COF	34	26	(80)	1
COCl	29	32	77	1
CO$_2$CH$_3$	41	35	77	1

[a] Values in parentheses are estimated.

that, as with chlorination (section II, A, 2, *b*), the substituent exerted no effect beyond the β-carbon atom. This represents the limit of the inductive effect $(-I)$ down the chain in this system, and contrasts with that found in halogenations in solution (Table 9). For selectivity in brominations, polar factors are more important, and relative bond-strengths are less important, than for chlorinations[54].

4. Chlorocarbonylations

Kharasch, Eberly and Kleiman[67] allowed hexahydrobenzoyl chloride to react with trichloromethyl chloroformate (diphosgene) in a sealed tube for 10 h at 225°C, when a second chlorocarbonyl group was introduced α- to the first group. This reaction (29) gave best yields when substitution

occurred at a tertiary group, and lowest yields at a primary group. The method is useful for the preparation of dialkylmalonic acids; yields reported were: $(C_2H_5)_2C(COCl)_2$ 90%, $(CH_3)_2C(COCl)_2$ 70%, $n\text{-}C_4H_9C(C_2H_5)(COCl)_2 > 30\%$, $CH_3CH(COCl)_2$ 15% and $C_6H_5CH(COCl)_2$ 2%.

A similar chlorocarbonylation method, using oxalyl chloride in place of diphosgene, was used by Runge and Koch[68] for phenyl-substituted carbonyl chlorides (reaction 30). Diethyl benzylmalonate and diethyl diphenylmalonate were obtained similarly from $C_6H_5CH_2CH_2COCl$ and $(C_6H_5)_2CHCOCl$, respectively.

$$C_6H_5CH_2COCl \xrightarrow[100-120^\circ C]{(COCl)_2, 30h} [C_6H_5CH(COCl)_2]$$

(30)

$$\xrightarrow{C_2H_5OH} C_6H_5CH(COOC_2H_5)_2 \quad (27\%)$$

5. Sulphoxidation and sulphochlorination

Sulphoxidation[69] is a process by which a sulphonic acid residue can be introduced photochemically into an alkyl chain, by means of a mixture of sulphur dioxide and oxygen (reaction 31). With propionyl chloride and butyryl chloride the major products were 3-sulphonic acids (Table 15),

$$RH + SO_2 + \tfrac{1}{2}O_2 \xrightarrow{h\nu} RSO_3H$$

(31)

but the relative reactivities (2-/3-) of the two carbonyl chlorides vary considerably[71].

TABLE 15. Sulphoxidation and sulphochlorination of carbonyl chlorides

Reagents	Substrate	Products of substitution		
		2- (%)	3- (%)	4- (%)
SO_2, O_2, $h\nu$	Propionyl chloride	47	53	—
SO_2, O_2, $h\nu$	Butyryl chloride	10	58	32
SO_2, Cl_2, $h\nu$	Propionyl chloride	0	100	—
SO_2, Cl_2, $h\nu$	Butyryl chloride	0	28	72

Sulphochlorination (reaction 32) is usually a less selective reaction[70] than sulphoxidation, but with the two carbonyl chlorides investigated

$$RH + SO_2 + Cl_2 \xrightarrow{h\nu} RSO_2Cl + HCl$$

(32)

(Table 15) it is highly selective[71]. A surprising feature is that sulphochlorination does not take place at all at the 2-positions. The 3-position is unexpectedly also not favoured in the reaction with butyryl chloride. The overall rates are about the same for both substrates, however[71].

B. Elimination Reactions

β-Chloroisobutyryl chloride can be dehydrohalogenated by heating in the presence of certain catalysts, e.g. phosphoric acid, phosphorus pentoxide, copper phosphate or peat charcoal[72]. It is not certain whether the carbonyl chloride substituent enhances the rate of the reaction (33).

$$
\underset{CH_3}{\overset{ClCH_2}{>}}CH{-}COCl \longrightarrow \underset{CH_3}{\overset{CH_2}{>}}C{-}COCl \tag{33}
$$

Such an enhancement is indeed likely at least for an $E2$-type of elimination mechanism[1f].

Pyrolysis of 2,3-dichloropropionyl chloride causes loss of two molecules of hydrogen chloride with formation of propiolyl chloride[73] (reaction 34).

$$
ClCH_2CHClCOCl \xrightarrow{170-210^\circ} HC{\equiv}CCOCl \tag{34}
$$

It was not established if either olefinic carbonyl chloride, $CH_2{=}CClCOCl$ or $ClCH{=}CHCOCl$, were intermediates in this elimination reaction.

C. Solvolyses of Dicarbonyl Dihalides

The qualitative effect of a COX group on the reactions of another COX group within the same molecule may be assessed by considerations of kinetic data. Pseudo first-order rate coefficients, apparent activation energies (E_{act}) and pre-exponential factors (A) for the heterogeneous alkaline hydrolysis of adipyl chloride ($ClCO(CH_2)_4COCl$) and sebacoyl chloride ($ClCO(CH_2)_8COCl$) have been measured for the water–benzene system[74] and the water–chlorobenzene system[75]. These reactions appear to occur at the water–organic interface. The data (Table 16) clearly show

TABLE 16. Heterogeneous alkaline hydrolysis of adipyl chloride and sebacoyl chloride

Compound	System: KOH—H_2O–benzene[a]			System: NaOH—H_2O–chlorobenzene[b]	
	k_1 (min^{-1}) (25°C)	E_{act} (kcal/mole)	A	k_1[c]	E_{act} (kcal/mole)
Adipyl chloride	$8 \cdot 28 \times 10^{-2}$	8·98	$2 \cdot 65 \times 10^4$	416	11·50
Sebacoyl chloride	$8 \cdot 86 \times 10^{-4}$	7·62	$5 \cdot 61 \times 10^4$	140	10·58

[a] Reference 74.
[b] Reference 75.
[c] Arbitrary first-order rate units.

that hydrolysis occurs more rapidly with adipyl chloride. The differences in rate constant are too large to be due to an inductive effect ($-I$) originating at the distant carbonyl chloride group of adipyl chloride. The differences in rates were considered to be due to differences in water-solubility of the halides[75] or, alternatively, to differences of rates of diffusion across the interface[74]. The possibility of ring formation in an intermediate stage in reactions of adipyl chloride may also be a factor.

III. DIRECTING AND ACTIVATING EFFECTS OF COX GROUPS IN AROMATIC CARBONYL HALIDES

A. Addition Reactions

The addition of chlorine to benzoyl chloride was first investigated by Bornwater and Holleman[76]. Liquid chlorine (3 mole) and benzoyl chloride (1 mole) were brought together in a sealed tube and exposed to sunlight. The main product was γ-benzoyl chloride hexachloride, $C_6H_5Cl_6COCl$, together with an impure benzene hexachloride and traces of chloro-substitution products. It was also found that this addition occurred only very slowly in the dark.

In a more detailed study van der Linden[77] exposed benzoyl chloride and an excess of chlorine to sunlight for several days. The formation of substantial amounts of hydrogen chloride gave evidence of nuclear chloro-substitution products. Fractional crystallization afforded three main products (1) a molecular compound, m.p. 153°C, of δ-1,2,3,4,5,6-hexa-chloro-1-cyclohexanecarbonyl chloride (m.p. 159°C; acid, m.p. 247°C) and β-1,2,3,4,4',5,6-heptachloro-1-cyclohexanecarbonyl chloride (m.p. 144°C; acid, m.p. 277°C), (2) γ-1,2,3,4,5,6-hexachloro-1-cyclohexane-carbonyl chloride, m.p. 112–113°C, which was believed to be identical with the product obtained by Bornwater and Holleman[76], and (3) α-1,2,3,4,4',5,6-heptachloro-1-cyclohexanecarbonyl chloride (m.p. 162°C; acid, m.p. 291°C). Evidence was obtained for the formation in this reaction of at least four different isomers of benzoyl chloride hexachloride, and probably at least four different isomers of 4-chlorobenzoyl chloride hexachloride. The formation of isomers having undergone chloro-substitution only at the *para*-position suggested to van der Linden[77] that in this reaction chlorine-addition to the benzene ring preceded chloro-substitution. The reaction scheme is summarized in (35). The stereochemistry of these products remains unknown.

(4 isomers)   (4 isomers)

B. Substitution Reactions

1. Nitrations

By the action of nitric acid (or nitric anhydride) in acetic anhydride on benzoyl chloride at about 60°, Karslake and Huston[78] observed the formation of *ortho*-acetylbenzoic acid and *meta*-nitrobenzoic acid. The claim[79] that aromatic substitution was here preceded by hydrolysis of the benzoyl chloride to benzoic acid has not been definitely established. Cooper and Ingold[79] examined the nitration of benzoyl chloride in an anhydrous medium (N_2O_5, P_2O_5 in CCl_4) at $-12.5°C$. *meta*-Nitrobenzoyl chloride was formed in 90·3% yield. In a recent summary Ingold[1g] reported the isomer proportions as 90·3% *meta*-, 8% *ortho*- and < 2% *para*-. Thus *meta*-orienting powers of substituents followed the sequence (36)[79].

$$-COCl\,(90\%) > -CO_2H\,(82\%) > -CONH_2\,(69\%) > -COCH_3\,(55\%) \qquad (36)$$

2. Chlorinations

a. Benzoyl chloride. Chlorine was found to react exothermically with benzoyl chloride, in the presence of catalytic amounts of ferric chloride[80]. When approximately molar equivalents of reactants were used the product comprised 13·5% recovered benzoyl chloride, 76% monochloro-derivatives and 5% dichloro-derivatives[80]. The isomer proportions reported were: 79–84% *meta*-, 14–20% *ortho*- and 0·7–2% *para*-chlorobenzoyl chlorides[80, 81]; these values have been confirmed in a more recent study[82]. When chlorination was continued until the weight increase corresponded to 1·6 times the theoretical for the introduction of one chlorine, Hope and Riley[81] found 54% mono- and 37% dichlorination to occur. The dichlorination products comprised 70% 2,5-, 23% 2,3- and 7% 3,4-dichlorobenzoyl chlorides[81]. Traces of 3,5-dichlorobenzoyl chloride have also been detected[82]. The scheme of the catalysed chlorination of benzoyl chloride can therefore be summarized as in (37).

Chlorination of methyl benzoate at 160° with an iron catalyst has been reported to give *m*-chlorobenzoyl chloride[83]. In this reaction formation of benzoyl chloride was considered to precede nuclear halogenation. Chlorination of liquid benzoyl chloride has also been effected with gaseous chlorine at 30–160°C in the presence of a mixture of iron(III) chloride and antimony(III) chloride[84], promoted by small amounts of water. The identity of the products was not given. The formation of dichlorobenzoyl chlorides was reported from chlorinations of benzoyl chloride at 10–180°C in the presence of small amounts of antimony sulphide or other catalysts[85].

The chlorination of benzoyl chloride has also been studied in the gas phase. A mixture of chlorine and benzoyl chloride, with nitrogen as carrier

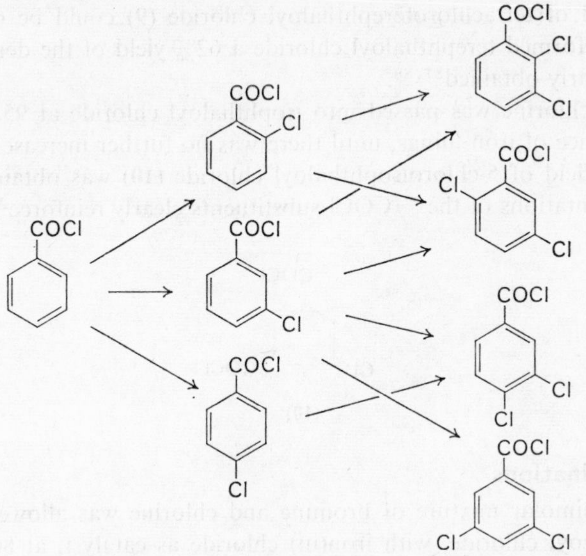

COCl

COCl
Cl

COCl
Cl

COCl
Cl
Cl

COCl
Cl

COCl
Cl

COCl
Cl
Cl

COCl
Cl

COCl
Cl
Cl

(37)

gas, was heated at 375°C for 45 s, when a 17% molar conversion took place to give mixed monochlorobenzoyl chlorides[86]. As gauged by the percentage of conversion of chlorine, the overall substrate reactivities were found to increase as in (reaction 38)[86].

$$C_6H_5CN < C_6H_5COCl < C_6H_5CCl_3 < C_6H_5Cl < C_6H_6 \qquad (38)$$

b. Isophthaloyl chloride and terephthaloyl chloride. Dimethyl terephthalate, on treatment with chlorine in the presence of iron filings at 220°C, has been quantitatively converted to monochloroterephthaloyl chloride (reaction 39). The nuclear substitution occurs after conversion of the ester functions to carbonyl chloride groups[83]. With excess of chlorine a

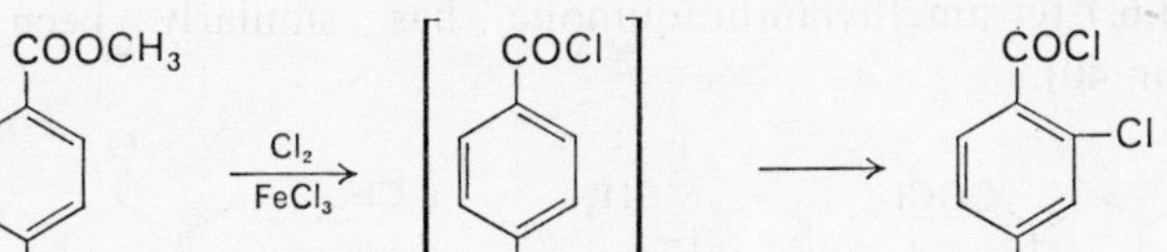

COOCH₃ ——Cl₂/FeCl₃——→ [COCl / COCl] ——→ COCl, Cl, COCl

COOCH₃
COCl

(39)

Cl, COCl, Cl / Cl, Cl / COCl

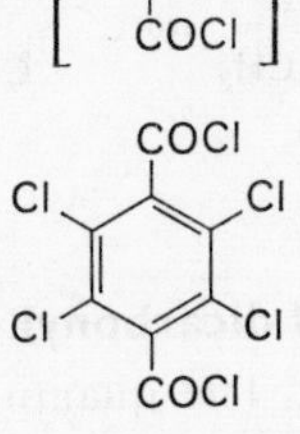

(9)

54% yield of tetrachloroterephthaloyl chloride (9) could be obtained[83]. From preformed terephthaloyl chloride a 62% yield of the derivative (9) was similarly obtained[87, 88].

When chlorine was passed into isophthaloyl chloride at 95–100°C, in the presence of iron filings, until there was no further increase in weight, an 81% yield of 5-chloroisophthaloyl chloride (10) was obtained[87]. The *meta*-orientations of the —COCl substituents clearly reinforce each other here.

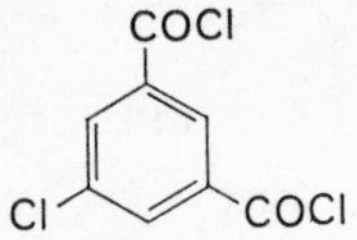

(10)

3. Brominations

An equimolar mixture of bromine and chlorine was allowed to react with benzoyl chloride, with iron(III) chloride as catalyst, at 80–85°C. A quantitative yield of *meta*-bromobenzoyl chloride was reported[89]. No nuclear chlorination took place under these conditions, nor was there replacement of chlorine by bromine in the carbonyl halide side chain.

4. Friedel–Crafts self-condensations

Benzoyl chloride can act as the substrate, as well as the reagent, in a self-condensation brought about by heating with aluminium chloride at 140°C[90]. A low yield of anthraquinone was reported by this method. The expected predominant *meta*-condensation leads to polymeric products; only in so far as initial substitution occurs at the *ortho*-position can anthraquinone be formed. From 3,4-dimethylbenzoyl chloride a 2% yield of 2,3,6,7-tetramethylanthraquinone has similarly been reported[91] (reaction 40).

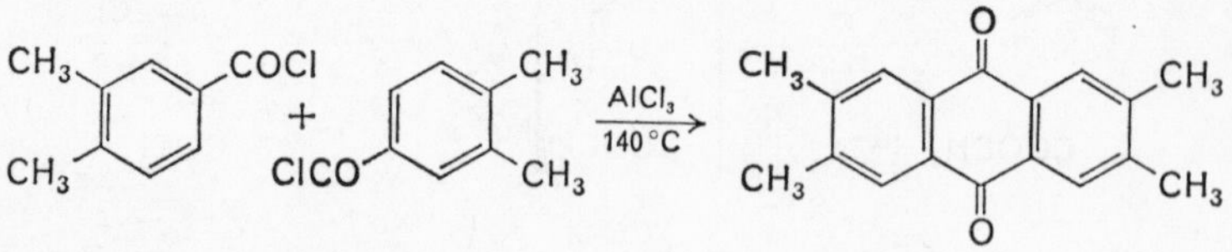

(40)

C. Side-chain Reactions

I. Nucleophilic reactions of dicarbonyl dihalides

a. Friedel–Crafts acylations. The quantitative substituent effect of a *meta*- or *para*-carbonyl chloride group can be gauged from work published recently by Hoornaert and Slootmaekers[92]. The kinetics of the Friedel–Crafts acylations of toluene (reaction 40a), catalysed by aluminium chloride, were determined in *ortho*-dichlorobenzene solution, for a range of derivatives of benzoyl chloride (Table 17). The percentage formation

TABLE 17. Activation parameters and relative reactivities for the aluminium-chloride-catalysed acylation of toluene with substituted benzoyl chlorides in *ortho*-dichlorobenzene solution

X in XC$_6$H$_4$COCl	$\Delta H^{\ddagger}$ (kcal/mole)	$\Delta S^{\ddagger}$ (cal/deg mole)	$10^4 k_2$ (25°C) (l/mole s)	k/k_0 (25°C)
—H	12·9	−30·4	4·97	1
m-OCH$_3$	12·7	−30·6	5·93	1·19
m-Cl	12·4	−28·7	25·8	5·20
m-NO$_2$	12·3	−28·9	27·9	5·61
p-NO$_2$	11·7	−30·5	34·9	7·03
m-SO$_2$Cl	12·3	−27·7	47·5	9·57
p-SO$_2$Cl	12·0	−28·6	48·6	9·77
m-COCl	12·1	−27·7	71·5	14·40
p-COCl	11·0	−31·3	79·8	16·05

of isomers was not determined specifically for iso- or terephthaloyl chlorides, but the composition of the acylated products in these systems

$$\text{(40a)}$$

was usually 94 (±2)% *para*-, 0·9 (±0·4)% *meta*- and 4·6 (±1·1)% *ortho*-. Kinetic data are also presented for some other electron-attracting substituent groups. The influence of these groups, substituted *meta*- or *para*- in benzoyl chloride, is similar to that of a carbonyl chloride group. There is generally a sharp increase in rate coefficient on substitution by an electronegative group, and this effect is more pronounced for substitution at the *para*-position. This observation suggests that rate enhancement occurs by a combined resonance ($-M$) and inductive ($-I$) mechanism. In the case of *meta*-substitution the $-I$ effect operates alone.

The increases in rate constants are due mainly to decreases in enthalpies of activation ($\Delta H^{\ddagger}$). For the *meta*-substituents the increases in reactivities follow the sequence (41); $-OCH_3$ and $-Cl$ substituents in the *para*-position proved to be deactivating[92]. The substituent action of carbonyl

$$-H < -OCH_3 < -Cl < -NO_2 < -SO_2Cl < -COCl \qquad (41)$$

chloride groups, therefore, is stronger than that of any other group here investigated. The data obtained[92] are not complicated by possible competitive or statistical factors, since by using only one molar equivalent of the catalyst, one only of the two (equivalent) carbonyl chloride groups was allowed to react. In the case of isophthaloyl chloride, however, the reaction product was shown to contain some diketone, formed presumably as the result of a secondary reaction. All *meta*-substituents, including the *meta*-COCl substituent, were reasonably well correlated in a linear plot of log k/k_0 (see Table 17) versus the Hammett σ-constants, following the procedure of Charton[93]. *para*-Substituents, on the other hand, were not well correlated. Values of the σ-constants were estimated as *meta*-ClCO = 0·53, and *para*-ClCO = 0·62[92].

b. Solvolytic reactions. The reactivities of phthaloyl, isophthaloyl and terephthaloyl chlorides have been compared with each other, and sometimes with benzoyl chloride in several systems (Table 18). The kinetics of the homogeneous hydrolyses in the presence of alkali of phthaloyl and terephthaloyl chlorides have been investigated in aqueous dioxan solution. The first-order rate constants for the *para*-compound were greater than for the *ortho*-compound[94]. In the absence of added alkali a solvolysis occurred (in water–dioxan) for which second-order rate constants[95] suggested a greater reactivity for the *ortho*-isomer. It should be realized, however, that in the experiments reported for phthaloyl chloride no account was taken of the possibility of the involvement of its tautomerism (reaction 42)[96, 97]. The significance of these results is therefore uncertain.

$$\qquad (42)$$

Most of the available data relate to heterogeneous reactions, however. The heterogeneous solvolysis of benzoyl and phthaloyl chlorides has been determined using a mixture of chlorobenzene and water; reaction here occurs at the interface[98]. The low temperature coefficients were thought to indicate here the predominance of the diffusion process over the chemical process.

The pseudo first-order rate constants determined for the hydrolysis with alkali in a water–benzene system of terephthaloyl chloride and isophthaloyl chloride were of the same order of magnitude[74]. The reaction of terephthaloyl chloride here involves both a much lower activation energy and a much lower pre-exponential factor than that of isophthaloyl chloride.

A series of polycondensation reactions has been investigated by Korshak and co-workers[99]. These reactions are of significance to the study of the formation of polyesters and polyamides, which are very important as fibre-forming polymers. The hydrolysis of the carbonyl chloride function in these compounds is an important stage in the chain-termination process. The rate constants for the reactions of isophthaloyl chloride and terephthaloyl chloride with several diols in the range 110–150°C were of comparable magnitude. There are considerable differences, however, in their activation parameters. The authors[99] considered the polycondensation process to be sterically more favourable for the *para*-compound.

Several condensation reactions of these carbonyl chlorides with amines, such as piperazine and piperidine, have also been studied. These reactions have been described as 'superfast', often with half-times of under 1 min. They follow complex kinetics, although only one acyl group (when two are present) actually reacts with the amine. In one system[100] (Table 18) the reaction (43) consists of two stages. The first stage proceeds by a combination of two mechanisms: a first-order reaction, k_1^I, and as econd-order reaction, k_2^I; the second stage is a first-order reaction, k_1^{II}. Each of

$$A + B \xrightarrow{\ k^I\ } AB \xrightarrow{\ k^{II}\ } C \tag{43}$$

these component rate coefficients follows the sequence of magnitudes benzoyl chloride > terephthaloyl chloride > phthaloyl chloride. The carbonyl chloride substituent clearly exerts a more powerful deactivating effect from the *ortho*-position than from the *para*-position.

2. Reductions of substituted nitrobenzenes

The kinetics of the catalytic hydrogenation over a rhodium-on-carbon catalyst of a set of substituted nitrobenzenes, presumably to the corresponding anilines, have been studied in ethanolic medium at atmospheric pressure[103] (Table 19). The substituent effect on this reduction of a *para*-carbonyl chloride group is substantially greater than that of the corresponding *meta*-group. The carbonyl chloride function was not itself reduced by this procedure. A linear relation was obtained between the activation energies and the 'solvatochromic shifts' of the conjugation

TABLE 18. Kinetics of various reactions of benzoyl, phthaloyl, isophthaloyl and terephthaloyl chlorides

Reaction	Data	Benzoyl chloride	Phthaloyl chloride	Isophthaloyl chloride	Terephthaloyl chloride	References
Hydrolysis, KOH, H_2O/dioxan[a]	$10^2 k_1$ (35°C) (s^{-1})	—	2·31	—	2·64	94
Hydrolysis, H_2O/dioxan[b]	$10^4 k_2$ (35°C) (l/mole s)	—	4·12	—	2·45	95
Hydrolysis, KOH, H_2O/benzene[c]	$10^4 k_1$ (25°C) (min^{-1})	—	—	2·65	2·13	74
	E_{act} (kcal/mole)	—	—	13·2	9·6	
	A	—	—	$1·26 \times 10^6$	$2·52 \times 10^3$	
Hydrolysis,[d] H_2O/chlorobenzene	$10^3 k_1$ (30°C) (min^{-1})	16·6	4·7	—	—	98
	E_{act} (kcal/mole)	8·1	11·0	—	—	
	A	$1·17 \times 10^2$	$1·67 \times 10^3$	—	—	
Polycondensation with 1,1,1-trimethylolethane[e]	$10^4 k_2$ (150°C) (l/mole s)	—	—	27·6	31·5	99
	E_{act} (kcal/mole)	—	—	10·8	15·4	
	A	—	—	$1·05 \times 10^3$	$3·83 \times 10^5$	

Table 18 (cont.)

Reaction	Parameter				Ref.
Polycondensation with 1,1,1-trimethylolpropane[e]	$10^4 k_2$ (150°C) (l/mole s)	—	27·1	30·4	99
	E_{act} (kcal/mole)	—	14·0	17·1	
	A	—	$5·14 \times 10^4$	$3·16 \times 10^6$	
Polycondensation with 2,2-bis(4-hydroxyphenyl)propane	$10^5 k_2$ (150°C) (l/mole s)	—	5·72	2·99	99, 101
	E_{act} (kcal/mole)	—	19·7	22·7	
	A	—	$6·46 \times 10^5$	$1·35 \times 10^7$	
Condensation with piperazine, H_2O/heptane[f]	$10^5 k_2$ (20°C) (l/mole s)	0·87	—	2	102
Condensation with piperazine, heptane	$10^3 k_2^{I}$ (20°C) (s⁻¹)	9·2	990	67[g]	100
	$10^{-2} k_2^{I}$ (20°C) (l/mole s)	3·2	560	33[g]	
	$10^4 k_2^{II}$ (20°C) (s⁻¹)	1·1	35	6·8[g]	

[a] Initial [KOH] = $1·38 \times 10^{-4}$ M.

[b] [H_2O] in dioxan = 4·35 M. Rates increase with increasing polarity of the medium.

[c] Heterogeneous reaction; other water/organic systems were also studied.

[d] Heterogeneous reaction; the rates are given for M solutions. At lower initial concentrations rates are somewhat lower. E_{act} and A values were computed from the published rate constants (reference 98).

[e] Solvent was 'dinyl' viz., the azeotrope (b.p. 258°) of diphenyl (26·5%) and diphenyl ether (73·5%).

[f] Heterogeneous reaction.

[g] Data at 18·8°C.

TABLE 19. Kinetics of catalytic reduction of substituted nitrobenzenes

Substituent	$10^2 k_1$ (min^{-1}) at 30°C	E_{act} (kcal/mole)[a]
p-OH	2·7	11·3
p-CH$_3$	3·5	9·9
H	4·6	8·9
m-CH$_3$	4·8	8·7
m-COCl	6·8	7·3
p-Cl	7·7	6·6
m-Cl	8·1	6·3
o-Cl	8·5	6·0
p-COCl	10·2	5·5

[a] Values read off from graph (reference 102).

bands in electronic spectra. The precise significance of this relation was not made clear.

IV. EFFECTS OF COX SUBSTITUENTS ON THE ARYL GROUP IN ω-ARYLALKANE CARBONYL HALIDES

A. Substitutions

Phenylacetyl chloride undergoes rapid sulphonation in 100% sulphuric acid, even at 10°C[103a]. It was assumed that predominant *para*-substitution occurred. In view of the results from nitration experiments[104] of a series of derivatives (11) (where X is an electron-attracting group, such as —NO$_2$, —CN or —CO$_2$C$_2$H$_5$), it is probable that substantial amounts of *ortho*- and *meta*-isomers could have been formed in the sulphonation of phenylacetyl chloride.

(11)

B. Self-condensations

When heated in carbon disulphide suspension with aluminium chloride, phenylacetyl chloride can undergo a Friedel–Crafts self-condensation. This condensation involves predominantly the *para*-position, and to a lesser extent the *meta*-position, and gives dimeric, trimeric and even polymeric products[105-109] (reaction 44).

$$n \quad \langle \bigcirc \rangle\text{-CH}_2\text{COCl} \xrightarrow{\text{AlCl}_3} \langle \bigcirc \rangle\text{-CH}_2\text{CO}\left[\text{-}\langle \bigcirc \rangle\text{-CH}_2\text{CO}\right]_{n-2}\langle \bigcirc \rangle\text{-CH}_2\text{COCl} \quad (44)$$

ω-Phenylalkane carbonyl chlorides with longer alkyl chains tend to undergo condensations of another kind[110,111]. When treated with aluminium chloride in a large excess of solvent, either a normal Friedel–Crafts acylation to give an open-chain ketone (12), or cyclization into the *ortho*-position of the phenyl ring to give (13), occurs. In the latter case the

$$\begin{array}{ccc} \text{CH}_2\text{CH}_2\text{COCl} & \text{CH}_2\text{CH}_2\text{COAr} & \text{CH}_2\text{—CH}_2 \\ \bigcirc & \xrightarrow[\text{ArH}]{\text{AlCl}_3} \quad \bigcirc & + \quad \bigcirc\text{CO} \end{array} \quad (45)$$

	(12)	(13)
ArH = Benzene	0%	90%
Ethylbenzene	40%	50%
Anisole	90%	0%

reactivity of the *ortho*-position must be appreciably higher than that of the solvent species, present in large excess (reaction 45). The extent of cyclization is thus a measure of the *ortho*-reactivity of the aromatic ring in the carbonyl chloride. The data for the cyclization reactions suggest the approximate sequence of substrate reactivities (46).

$$\text{Cl-}\langle\bigcirc\rangle\text{-(CH}_2)_2\text{COCl} < \langle\bigcirc\rangle\text{-(CH}_2)_2\text{COCl} < \text{CH}_3\text{-}\langle\bigcirc\rangle\text{-(CH}_2)_2\text{COCl} <$$

$$\langle\bigcirc\rangle\text{-(CH}_2)_3\text{COCl} < \langle\bigcirc\bigcirc\rangle\text{-(CH}_2)_2\text{COCl} \quad (46)$$

Carbonyl chlorides of the same general type but with at least six C-atoms in the side chain have also been reacted with aluminium chloride, but in the absence of a reactive solvent[112]. With 7-phenylheptanoyl chloride a 5% yield of a dimeric ketone (14, $n = 6$) (a dioxocyclophane) was obtained (47), together with a 0·4% yield of a trimeric ketone (15). 10-Phenyldecanoyl chloride gave a 0·7% yield of the diketone (14, $n = 9$). The effect

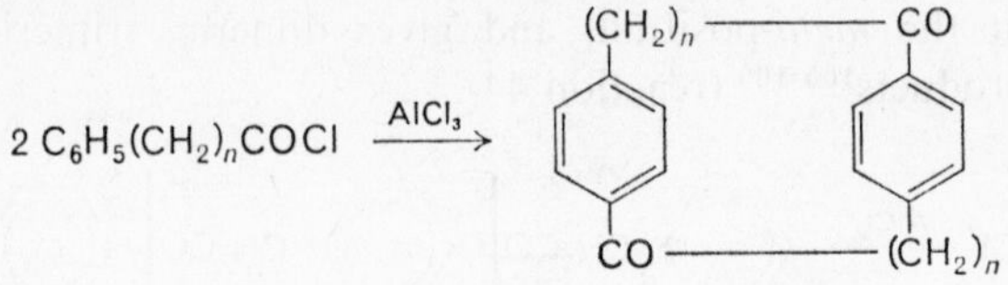

$$2\ C_6H_5(CH_2)_nCOCl \xrightarrow{AlCl_3}$$

(14) (47)

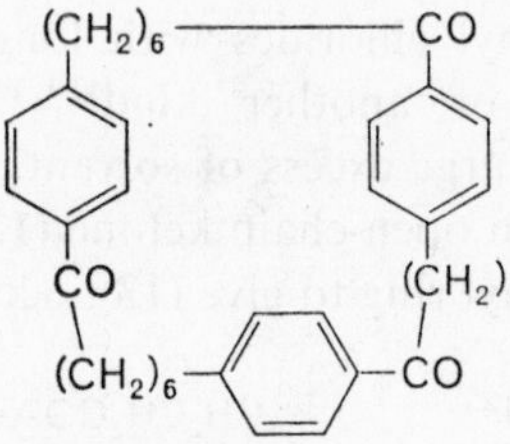

(15)

on the phenyl nucleus in these substrates of a $-(CH_2)_nCOCl$ side chain is thus essentially that of a simple alkyl chain.

The length of the alkyl chain, in a related series of condensations of carbonyl chlorides containing an ω-thienyl substituent, determines the nature of the acylation product[113] (reaction 48). Whilst the role of the carbonyl chloride group is clearly that of a reagent, the proximity of the group to reactive positions of the thiophene nucleus effectively determines the course of the reaction. Yields of the diketone (**16**) were 8% ($n = 5$), 4% ($n = 8$) and 15% ($n = 9$)[113].

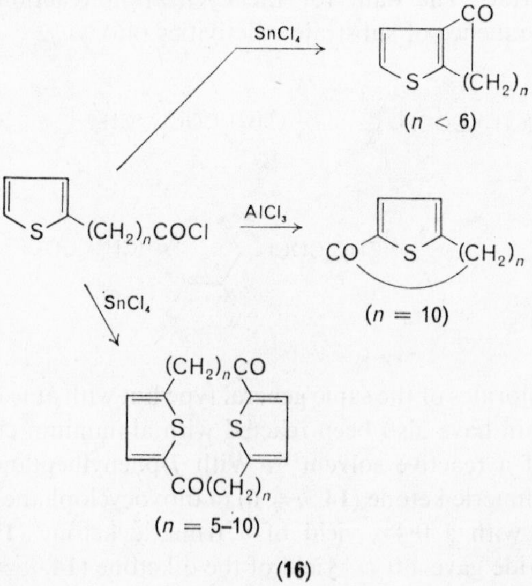

$(n < 6)$

$(CH_2)_n COCl$ — via $AlCl_3$

$(n = 10)$ (48)

$(n = 5{-}10)$

(16)

V. SUBSTITUENT EFFECTS OF COX GROUPS IN NUCLEAR MAGNETIC RESONANCE SPECTROSCOPY

Nuclear magnetic resonance spectra can throw light on the effect of substituents, e.g. —COCl, on the chemical shifts of other groups (H or F) within the same molecule. These chemical shifts are known to be affected by the environment of the atom involved in the magnetic resonance and thereby can give a guide, qualitative and quantitative, to electronic properties of substituents. Thus, it is well known that *para*-proton chemical shifts show good correlations with ^{13}C chemical shifts of the *para* carbon-atom, ^{19}F chemical shifts of *para*-substituted fluorobenzenes, Hammett σ_p-constants, the calculated π-electron density on the *para*-carbon atom, etc[114].

A. Olefinic Protons

The ^{1}H-chemical shifts of olefinic protons have been correlated with the incremental contributions (Z_i) of chemical shift for a number of substituents, according to their stereochemical position relative to the proton[115]. Table 20 gives the semi-empirical rule which has been developed,

TABLE 20. Semi-empirical rules for chemical shift increments of substituents in mono-olefins[a]

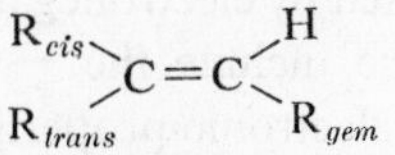

$$\tau \text{ (p.p.m.)} = 4.72 + \sum_i Z_i$$

R_{cis} \ / H
   >C=C<
R_{trans} / \ R_{gem}

Z_i (p.p.m.) for

Substituent	R_{gem}	R_{cis}	R_{trans}
COOH	−1·00	−1·35	−0·74
COOR	−0·84	−1·15	−0·56
CHO	−1·03	−0·97	−1·21
C(=O)N	−1·37	−0·93	−0·35
C(=O)Cl	−1·10	−1·41	−0·99
CO (unconjugated)	−1·10	−1·13	−0·81
CO (conjugated)	−1·06	−1·01	−0·95

[a] In carbon tetrachloride solution, with tetramethylsilane as internal standard.

and the Z_i values for *gem-*, *cis-* and *trans-*substituents, from which chemical shifts may be obtained. The substituents listed comprise various groups containing a carbonyl function and include the —COCl group.

B. Aromatic Protons

Proton chemical shifts in monosubstituted benzenes have been much studied[114,116,117]. In Table 21 are collected chemical shifts, relative to

TABLE 21. Proton chemical shifts and coupling constants in monosubstituted benzenes

	Substituent shifts relative to benzene ($\tau = 2.73$)[a]			Coupling constants (Hz)					
	ortho-	meta-	para-	J_{23}	J_{34}	J_{26}	J_{35}	J_{24}	J_{25}
H[b]	0	0	0	7.56	7.56	1.38	1.38	1.38	0.68
COCl[c]	−0.84	−0.22	−0.36	7.97	7.49	1.96	1.40	1.27	0.58
COBr	−0.80	−0.21	−0.37	8.00	7.47	1.95	1.43	1.27	0.52
CO_2CH_3	−0.71	−0.11	−0.21	7.88	7.52	1.85	1.33	1.31	0.61
$COCH_3$	−0.62	−0.14	−0.21	7.82	7.38	1.90	1.34	1.28	0.58
CCl_3	−0.64	−0.13	−0.10	8.13	7.45	2.39	1.44	1.14	0.53
CN	−0.36	−0.18	−0.28	7.79	7.68	1.76	1.30	1.28	0.63
SO_2Cl	−0.77	−0.35	−0.45	8.09	7.53	2.20	1.34	1.16	0.54
NO_2	−0.95	−0.26	−0.38	8.35	7.46	2.46	1.47	1.17	0.48

[a] At infinite solution in carbon tetrachloride.
[b] Data from reference 118.
[c] Data for substituted benzenes from references 114, 116 and 117.

benzene, of benzene derivatives monosubstituted by electronegative groups at the ortho-, meta- or para-positions. These include the −COCl and −COBr group, and other groups similar in electronic properties (mostly −I, −M). A good linear correlation was found[114] between the para-^{1}H chemical shifts and Taft's[119] ^{19}F chemical shifts measured in carbon tetrachloride solution. Table 21 also includes the various coupling constants obtained from a detailed analysis[117], and these are linearly correlated with the electronegativities of the substituents. From the magnitudes of the chemical shifts it can be seen that generally the effect of substituents decreases in the sequence ortho->para->meta-, i.e. that both electronegativity ($-I$) and resonance ($-M$) effects here operate. A sequence (49) of increasing $-I$ effects can be gauged from the magnitudes of the chemical shifts of meta-protons.

$$H < -CO_2CH_3 < -CCl_3 < -COCH_3 < -CN < -COBr < -COCl < -NO_2$$
$$< -SO_2Cl \quad (49)$$

Proton chemical shift correlations in ortho-disubstituted benzenes appear to be unreliable. For meta-disubstituted benzenes the chemical shift of a particular proton may be obtained[120] by applying one of four relationships, which combine the contributions of the individual substituents (Table 22). The negative sign indicates a shift towards lower

TABLE 22. Proton chemical shifts in *meta*-disubstituted benzenes[a]

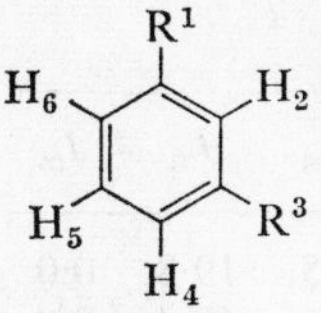

Chemical shifts[b] given by:

$H_2 = R^1$ (*ortho-*) $+ R^3$ (*ortho-*) $H_5 = R^1$ (*meta-*) $+ R^3$ (*meta-*)

$H_4 = R^1$ (*para-*) $+ R^3$ (*ortho-*) $H_6 = R^1$ (*ortho-*) $+ R^3$ (*para-*)

Chemical shift increments

Substituent R	*ortho-*	*meta-*	*para-*
CHO	−0·54	−0·20	−0·2
COCH$_3$	−0·64	−0·09	−0·1
COCl	−0·83	−0·16	−0·2
CN	−0·27	−0·10	−0·1

[a] Relative to benzene, $\tau = 2\cdot73$, in carbon tetrachloride solution.
[b] In τ (p.p.m.).

field, as expected for electronegative groups. Chemical shifts appear to be simply additive for *para*-disubstituted benzenes[120, 121].

C. Aromatic Fluorine Nuclei

Chemical shifts have been investigated for a series of *para*-substituted fluorobenzenes[122] (17). The ^{19}F chemical shifts (in p.p.m. from fluorobenzene in carbon tetrachloride solution) gave a value for *para*-COCl =

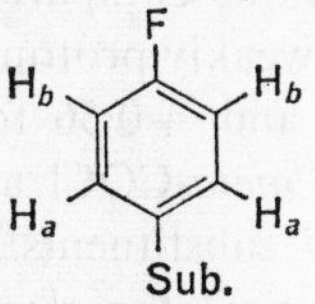

(17)

−11·32, comparable to other strongly electron-attracting substituents, viz. *para*-NO$_2$ = −9·50, *para*-NO = −11·06 and *para*-SO$_2$Cl = −12·30. A linear relationship was found between δ_F (p.p.m.) and Hammett σ^- substituent[123] constants; this gave a value $\sigma^- = 1\cdot24$ for the *para*-COCl group, a high value for the series. Coupling-constants were found to be: $J_{FH_b} = 8\cdot1$, $J_{FH_a} = 5\cdot1$ and $J_{H_aH_b} = 9\cdot1$ Hz. The coupling-constants did not show much variation with the nature of the *para*-substituent in **(17)**.

A study of substituted pentafluorobenzenes[124] has provided values of chemical shifts and coupling-constants; Table 23 lists values obtained for

TABLE 23. ^{19}F Chemical shifts and coupling constants in pentafluorobenzenes[a]

Substituent	Chemical shifts (p.p.m.)			Coupling constants (Hz)					
	ortho-	*meta-*	*para-*	J_{23}	J_{24}	J_{25}	J_{26}	J_{34}	J_{35}
—COCl	59·2	80·5	67·3	−21·0	6·1	+8·3	7·5	19·8	0·0
—CO$_2$H	58	82	68	−21·1	4·1	+8·2	5·1	19·4	0·0
—CF$_3$	60·5	80·7	68·3	−20·5	5·9	+8·3	7·9	19·1	0·0

[a] Shifts (p.p.m.) to high field of CF_3CO_2H (external) for 5 mole % solutions in carbon tetrachloride solution.

three substituents, —COCl, —CO$_2$H and —CF$_3$. The data could be correlated by the use of two equations (50, 51), where Φ_p, Φ_m are chemical shifts for *para-* and *meta*-fluorine nuclei, and σ_R^0 and σ_I are Taft's

$$\Phi_p = -39\cdot7\ \sigma_R^0 \ -12\cdot9\ \sigma_I \ +154\cdot4 \tag{50}$$

$$\Phi_m = -7\cdot2\ \sigma_R^0 \ -5\cdot3\ \sigma_I \ +162\cdot9 \tag{51}$$

resonance-effect and inductive-effect polar substituent constants, respectively[125]. Conversion from the C_6H_5F to the CF_3CO_2H scale, using the mean difference of 79·6 p.p.m., gave calculated Taft parameters for the —COCl group: $\sigma_R^0 = +0\cdot03$ and $\sigma_I = +0\cdot49$. Values for other substituents were mostly in good agreement with parameters determined previously[119, 126]. The latter value compares with values of σ_I (from ^{19}F shielding) $= +0\cdot42 \pm 0\cdot01$ for weakly protonic solvents, $+0\cdot39$ for 'normal solvents', $+0\cdot34$ for dioxan and $+0\cdot56$ for trifluoroacetic acid[126]. The substituent effects found for *meta*-COCl and *meta*-COF substituents[126], or *para*-COCl and *para*-COF substituents[119], were found throughout to be extremely similar, and typical for electron-attracting groups of the carbonyl type.

VI. REFERENCES

1. C. K. Ingold, *Structure and Mechanism in Organic Chemistry*, 2nd ed., G. Bell, London 1969: (a) p. 83; (b) p. 88; (c) p. 90; (d) p. 91; (e) p. 151; (f) p. 651; (g) p. 313.
2. D. T. Clark, J. N. Murrell and J. M. Tedder, *J. Chem. Soc.*, 1250 (1963).

3. D. A. Brown and M. J. S. Dewar, *J. Chem. Soc.*, 2406 (1953).
4. H. H. Jaffé, *J. Am. Chem. Soc.*, **77**, 274 (1955).
5. G. V. D. Tiers, *J. Phys. Chem.*, **62**, 1151 (1958).
6. G. Kohnstam and D. L. H. Williams, in *The Chemistry of the CO_2H and CO_2R Groups* (Ed. S. Patai), Interscience, 1970, p. 768.
7. G. Chuchani, in *The Chemistry of the Amino Group* (Ed. S. Patai), Interscience, 1968, p. 205.
8. A. R. Katritzky and R. D. Topsom, *Angew. Chem. Intern. Ed.*, **9**, 88 (1970).
9. S. Ehrenson, in *Progress in Physical Organic Chemistry* Vol. 2 (Ed. S. G. Cohen, A. Streitwieser, Jr. and R. W. Taft), Interscience, 1964, p. 206 and references therein.
10. I. Mochida and Y. Yoneda, *Bull. Chem. Soc. Japan*, **41**, 1479 (1968).
11. M. J. S. Dewar and P. J. Grisdale, *J. Am. Chem. Soc.*, **84**, 3539, 3548 (1962), and references therein.
12. F. W. Baker, R. C. Parish and L. M. Stock, *J. Am. Chem. Soc.*, **89**, 5677 (1967).
13. C. F. Wilcox and C. Leung, *J. Am. Chem. Soc.*, **90**, 336 (1968).
14. P. E. Peterson, R. J. Bopp, D. M. Chevli, E. L. Curran, D. E. Dillard and R. J. Kamat, *J. Am. Chem. Soc.*, **89**, 5902 (1967).
15. N. Bodor, *Rev. Roumaine Chim.*, **13**, 555 (1968).
16. K. Bowden and D. C. Parkin, *Can. J. Chem.*, **47**, 185 (1969).
17. V. Palm, *Reakts. Sposobnost. Org. Soedin. Tartusk. Gos. Univ.*, **5**, 583 (1968).
18. J. A. Pople and M. Gordon, *J. Am. Chem. Soc.*, **89**, 4253 (1967).
19. K. Bowden, *Can. J. Chem.*, **41**, 2781 (1963).
20. F. Eisenlohr, *Spektrochemie organischer Verbindungen*, Enke, Stuttgart, 1912.
21. W. F. Forbes and J. J. J. Myron, *Can. J. Chem.*, **39**, 2452 (1961).
22. W. T. Miller, Jr. and M. Prober, *J. Am. Chem. Soc.*, **70**, 2602 (1948).
23. W. Bockemüller, *Ann.*, **506**, 20 (1933).
24. Chen-Ya Hu and Chieh Ch'eb, *Chung Kuo K'o Hsueh Yuan Ying Yung Hua Hsueh Yen Chin So Chi K'an*, 73 (1965); *Chem. Abstr.*, **64**, 11077e (1966).
25. R. F. Merritt, *J. Org. Chem.*, 32, 1633 (1967).
26. A. B. Ash and H. C. Brown, *Rec. Chem. Progr.*, **9**, 81 (1948).
27. H. H. Guest and C. M. Goddard, Jr., *J. Am. Chem. Soc.*, **66**, 2074 (1944).
28. R. Wolffenstein and J. Rolle, *Ber.*, **41**, 733 (1908).
29. S. L. Bass, *U.S. Pat.*, 2,010,685 (1935).
30. W. Markownikoff, *Ann.*, **153**, 228 (1870).
31. H. B. Watson and E. H. Roberts, *J. Chem. Soc.*, 2779 (1928).
32. M. S. Kharasch and H. C. Brown, *J. Am. Chem. Soc.*, **62**, 925 (1940).
33. J. de Pascual Teresa and F. G. Espinosa, *Anales Real. Soc. Españ. Fis. Quim.*, **52B**, 447 (1956).
34. E. Campaigne and W. Thompson, *J. Am. Chem. Soc.*, **72**, 629 (1950).
35. C. K. Ingold and E. H. Ingold, *J. Chem. Soc.*, 1314 (1926).
36. H. H. Guest, *J. Am. Chem. Soc.*, **69**, 300 (1947).
37. E. Hertel, G. Becker and A. Clever, *Z. Physik. Chem.*, **B27**, 303 (1934).
38. H. C. Brown and A. B. Ash, *J. Am. Chem. Soc.*, **77**, 4019 (1955).

39. A. Bruylants, M. Tits, C. Dieu and R. Gauthier, *Bull. Soc. Chim. Belg.*, **61**, 366 (1952).

40. A. Bruylants, M. Tits and R. Dauby, *Bull. Soc. Chim. Belg.*, **58**, 310 (1949).

41. A. Benrath and E. Hertel, *Z. Wiss. Phot.*, **23**, 30 (1924).

42. I. F. Spasskaya, V. S. Etlis and G. A. Razuvaev, *Zh. Obshch. Khim.*, **28**, 3004 (1958).

43. Farbwerke Hoechst A.-G., *Ger. Pat.*, 1,116,647 (1959).

44. H. Magritte and A. Bruylants, *Ind. Chim. Belge*, **22**, 547 (1957).

45. P. Smit, *Doctoral Thesis*, University of Amsterdam, 1968.

46. P. Smit and H. J. den Hertog, *Tetrahedron Letters*, 595 (1971).

47. P. Smit and H. J. den Hertog, *Rec. Trav. Chim.*, **83**, 891 (1964).

48. H. Magritte and A. Bruylants, *Bull. Soc. Chim. Belg.*, **66**, 367 (1957).

49. J. Wautier and A. Bruylants, *Bull. Soc. Chim. Belg.*, **72**, 222 (1963).

50. W. Schmidt and F. Schloffer, *Ger. Pat.*, 738,398 (1943).

51. A. Michael and W. W. Garner, *Ber.*, **34**, 4034 (1901).

52. H. J. den Hertog, B. de Vries and J. van Bragt, *Rec. Trav. Chim.*, **74**, 1561 (1955).

53. P. Smit and H. J. den Hertog, *Rec. Trav. Chim.*, **77**, 73 (1958).

54. H. Singh and J. M. Tedder, *J. Chem. Soc.*, 4737 (1964).

55. H. Singh and J. M. Tedder, *J. Chem. Soc. (B)*, 605 (1966).

56. H. Singh and J. M. Tedder, *Chem. Commun.*, 5 (1965).

57. H. J. den Hertog and P. Smit, *Proc. Chem. Soc.*, 132 (1959).

58. W. A. Waters, *The Chemistry of Free Radicals*, Oxford University Press, London, 1948, p. 174.

59. G. S. Hammond, *J. Am. Chem. Soc.*, **77**, 334 (1955).

60. W. F. Goebel, *J. Am. Chem. Soc.*, **45**, 2770 (1923).

61. Nippon New Drug Co., *Japan Pat.*, 156,782 (1943).

62. L. Sajus, *Bull. Soc. Chim. France*, 2263 (1964).

63. H. B. Watson, *J. Chem. Soc.*, 1137 (1928).

64. C. Cicero and D. Matthews, *J. Phys. Chem.*, **68**, 469 (1964).

65. A. Michael and E. Scharf, *Ber.*, **46**, 135 (1913).

66. M. S. Kharasch and L. M. Hobbs, *J. Org. Chem.*, **6**, 705 (1941).

67. M. S. Kharasch, K. Eberly and M. Kleiman, *J. Am. Chem. Soc.*, **64**, 2975 (1942).

68. F. Runge and U. Koch, *Chem. Ber.*, **91**, 1217 (1958).

69. F. Asinger, B. Fell and A. Commichau, *Chem. Ber.*, **98**, 2154 (1965).

70. F. Asinger, *Chemie und Technologie der Paraffinkohlenwasserstoffe*, Akademie-Verlag, Berlin, 1956, p. 591.

71. F. Asinger, B. Fell and A. Commichau, *Tetrahedron Letters*, 3095 (1966).

72. W. Schmidt, *U.S. Pat.*, 2,396,609 (1946).

73. F. C. Schaefer, *U.S. Pat.*, 2,388,660 (1945).

74. V. V. Korshak, T. M. Frunze, S. V. Vinogradova, V. V. Kurashev and A. S. Lebedeva, *Izv. Akad. Nauk S.S.S.R., Otd. Khim. Nauk*, 1807 (1962).

75. T. I. Shein, G. I. Kudryavtsev and L. N. Vlasova, *Khim. Volokna*, No. 5, 13 (1960).

76. J. Th. Bornwater and A. F. Holleman, *Rec. Trav. Chim.*, **31**, 221 (1912).

77. T. van der Linden, *Rec. Trav. Chim.*, **53**, 703 (1934).

78. W. J. Karslake and R. C. Huston, *J. Am. Chem. Soc.*, **31**, 479 (1909).

79. K. E. Cooper and C. K. Ingold, *J. Chem. Soc.*, 836 (1927).

80. E. Hope and G. C. Riley, *J. Chem. Soc.*, **121**, 2510 (1922).
81. E. Hope and G. C. Riley, *J. Chem. Soc.*, **123**, 2470 (1923).
82. D. Brox, *Monatsber. Deut. Akad. Wiss. Berlin*, **6**, 661 (1964).
83. Chemische Werke Written, G.m.b.H., *Ger. Pat.*, 1,097,973 (1961).
84. Ya. P. Skarinskii, E. V. Sergeev, B. F. Filimonov and L. I. Grigoruk, *U.S.S.R. Pat.*, 194,803 (1967).
85. Ya. P. Skarinskii, E. V. Sergeev, A. A. Ryzhkov, S. A. Antipova and L. M. Pavlovich, *U.S.S.R. Pat.*, 239,311 (1969).
86. J. W. Engelsma and E. C. Kooyman, *Rec. Trav. Chim.*, **80**, 537 (1961).
87. N. Rabjohn, *J. Am. Chem. Soc.*, **70**, 3518 (1948).
88. E. Profft and D. Timm, *Arch. Pharm.*, **299**, 577 (1966).
89. E. C. Britton and R. M. Tree, Jr., *U.S. Pat.*, 2,607,802 (1952).
90. C. Seer, *Monatsh.*, **32**, 143 (1911); **33**, 33 (1912).
91. G. T. Morgan and E. A. Coulson, *J. Chem. Soc.*, 2323 (1931).
92. G. Hoornaert and P. J. Slootmaekers, *Bull. Soc. Chim. Belg.*, **77**, 295 (1968).
93. M. Charton, *J. Org. Chem.*, **28**, 3121 (1963).
94. R. P. Tiger, E. Ya. Nevel'skii, I. V. Epel'baum and S. G. Entelis, *Izv. Akad. Nauk S.S.S.R., Ser. Khim. Nauk*, **(11)**, 1969 (1964).
95. S. G. Entelis, R. P. Tiger, E. Ya. Nevel'skii and I. V. Epel'baum, *Izv. Akad. Nauk S.S.S.R., Ser. Khim. Nauk*, 245 (1963).
96. W. Csányi, *Monatsh.*, **40**, 81 (1919).
97. E. Ott, *Ann.*, **392**, 245 (1912).
98. D. Karve and K. K. Dole, *J. Univ. Bombay*, **7**, 108 (1938).
99. S. V. Vinogradova, V. V. Korshak, P. M. Valetskii and Yu. V. Mironov, *Izv. Akad. Nauk S.S.S.R., Ser. Khim. Nauk*, 70 (1966).
100. S. G. Entelis and O. V. Nesterov, *Dokl. Akad. Nauk S.S.S.R.*, **148**, 1323 (1963).
101. S. V. Vinogradova and V. V. Korshak, *Dokl. Akad. Nauk S.S.S.R.*, **123**, 849 (1958).
102. S. G. Entelis, E. Yu. Bekhli and O. V. Nesterov, *Kinetika i Kataliz.*, **6**, 331 (1965).
103. A. V. Finkel'shtein and Z. M. Kuz'mina, *Reakts. Sposobnost Org. Soedin. Tartusk. Gos. Univ.*, **3**, 72 (1966).
103a. M. Liler, *J. Chem. Soc.* (*B*), 205 (1966).
104. J. R. Knowles and R. O. C. Norman, *J. Chem. Soc.*, 2938, 3888 (1961).
105. J. Schmitt, M. Suquet, J. Boitard and P. Comoy, *Compt. Rend.*, **240**, 2538 (1955).
106. J. Schmitt and J. Boitard, *Bull. Soc. Chim. France*, [5] **22**, 1033 (1955).
107. J. Schmitt, M. Suquet and P. Comoy, *Bull. Soc. Chim. France*, [5] **22**, 1055 (1955).
108. J. Schmitt, J. Boitard, M. Suquet and P. Comoy, *Compt. Rend.*, **242**, 649 (1956).
109. J. Schmitt, P. Comy, J. Boitard and M. Suquet, *Bull. Soc. Chim. France*, 636 (1956).
110. N. P. Buu-Hoï, Ng. Hoan and Ng. D. Xuong, *J. Chem. Soc.*, 3499 (1951).
111. P. H. Gore, in *Friedel–Crafts and Related Reactions*, Vol. III (Ed. G. A. Olah), Interscience, 1964, Part 1, p. 63.

112. W. M. Schubert, W. A. Sweeney and H. K. Latourette, *J. Am. Chem. Soc.*, **76**, 5462 (1954).
113. Y. L. Gol'dfarb, S. Z. Taits and C. I. Belen'kii, *Zh. Obshch. Khim.*, **29**, 3564 (1959).
114. K. Hayamizu and O. Yamamoto, *J. Mol. Spectry.*, **25**, 89 (1968).
115. C. Pascual, J. Meier and W. Simon, *Helv. Chim. Acta*, **49**, 164 (1966).
116. J. L. Garnett, L. J. Henderson, W. A. Sellich and G. Van Dyke Tiers, *Tetrahedron Letters*, 516 (1961).
117. K. Hayamizu and O. Yamamoto, *J. Mol. Spectry.*, **25**, 422 (1968).
118. J. M. Read, Jr., R. E. Mayo and J. H. Goldstein, *J. Mol. Spectry.*, **21**, 235 (1966).
119. R. W. Taft, E. Price, I. R. Fox, I. C. Lewis, K. K. Anderson and G. T. Davis, *J. Am. Chem. Soc.*, **85**, 3146 (1963).
120. J. S. Martin and B. P. Dailey, *J. Chem. Phys.*, **39**, 1722 (1963).
121. G. W. Smith, *J. Mol. Spectry.*, **12**, 146 (1964).
122. H. Suhr, *Ber. Bunsenges. Phys. Chem.*, **68**, 169 (1964).
123. D. M. McDaniel and H. C. Brown, *J. Org. Chem.*, **23**, 420 (1958).
124. R. Fields, J. Lee and D. J. Mowthorpe, *J. Chem. Soc. (B)*, 308 (1968).
125. R. W. Taft, S. Ehrenson, I. C. Lewis and R. E. Glick, *J. Am. Chem. Soc.*, **81**, 5352 (1959).
126. R. W. Taft, E. Price, I. R. Fox, I. C. Lewis, K. K. Anderson and G. T. Davis, *J. Am. Chem. Soc.*, **85**, 709 (1963).

CHAPTER **6**

Mechanisms of substitution at the COX group

ANTTI KIVINEN

University of Helsinki, Finland

I. INTRODUCTION

Acyl halides, especially acyl chlorides, are the most extensively employed reactants in organic syntheses[1]. Owing to their great reactivity, however, the clarification of the mechanisms of reactions of acyl halides has progressed less than that of the mechanisms of reactions of many other carbonyl compounds.

Substitution reactions of acyl halides can be represented by the simple equation (1). The most thoroughly investigated reactions of acyl halides

$$RCO-X + Y = RCO-Y + X \tag{1}$$

from the mechanistic viewpoint are (a) the reactions with hydroxyl compounds (especially water and alcohols) (b) reactions with amines (especially anilines) and (c) Friedel–Crafts reactions.

The Friedel–Crafts reactions have been thoroughly documented by Olah[2], and will not be discussed here. Moreover, many of these reactions are heterogeneous and we shall limit the following discussion mainly to homogeneous reactions.

The main problem in the reaction of an acyl halide with a hydroxylic compound is whether the bimolecular reaction takes place by a synchronous or an addition–elimination mechanism. These mechanisms may be represented, when water is the reactant, by equations (2) and (3).

$$RCOX + H_2O \longrightarrow RCOOH + HX \tag{2}$$

$$RCOX + H_2O \longrightarrow RC(OH)_2X \longrightarrow RCOOH + HX \tag{3}$$

The situation is quite analogous when the nucleophile is an amine. Before discussing the mechanisms of the reactions we will first briefly survey in each section the available kinetic data.

Solvolytic reactions that are catalysed by tertiary amines and related reactions will be discussed separately (section IV). Due to limitations of space, we cannot discuss all reactions of acyl halides about whose mechanisms information is available. Technologically important polymerization reactions will not be treated either since studies dealing with them provide only very little information that has a bearing on reaction mechanisms.

Reactions of the esters of chloroformic acid will only be touched upon here as these reactions are examined in detail in another chapter of this volume and have been previously reviewed by Matzner, Kurkjy and Cotter[3].

The reactions of aliphatic acyl chlorides have been extensively reviewed by Sonntag[1], who has also discussed mechanisms of the reactions to some extent. A general survey of acylation reactions has been published by Praill[4] and of acylation at carbon by House[5].

Summaries on the mechanisms of acylation reactions have been written by Hudson[6] and Satchell[7]. Reactions of acyl halides have been discussed in some detail also in general reviews on the reactivities of carbonyl compounds and on the mechanisms of their reactions[8-10].

II. REACTIONS WITH WATER, ALCOHOLS AND PHENOLS

A. Acyl Chlorides

I. General

From the mechanistic standpoint, the solvolytic reactions of acyl halides can be divided into two main types, unimolecular and bimolecular. The terms unimolecular mechanism, ionization mechanism and S_N1 or 'S_N1-like' mechanism are generally used as synonyms. As mentioned above, two quite different points of view prevail in the case of bimolecular mechanisms. The mechanism is either described as an S_N2 or 'S_N2-like' mechanism or as an addition–elimination (AE) mechanism. In the latter, also called the carbonyl addition mechanism, the reaction is believed to proceed through a tetrahedral intermediate.

In its simplest form, the unimolecular mechanism can be represented by equations (4). The rate-determining step is the ionization of the carbonyl

$$\text{(a)} \quad R^1COX \xrightarrow{\ \text{slow}\ } R^1CO^+ + X^-$$

$$\text{(b)} \quad R^1CO^+ + R^2OH \xrightarrow{\ \text{fast}\ } R^1COOR^2 + H^+ \tag{4}$$

halide to an acylium ion R^1CO^+ and a halide ion. The acylium ion then reacts rapidly with a hydroxylic solvent.

The bimolecular S_N2 reaction is a one-step conversion of the reactants to the products, i.e. a direct synchronous displacement of the halogen by an alkoxy group (equation 5).

$$R^1COX + R^2OH \longrightarrow R^1COOR^2 + HX \qquad (5)$$

Since oxygen exchange is observed in solvolytic reactions, it has been proposed that the bimolecular reaction takes place not as a synchronous substitution, but by way of a tetrahedral intermediate:

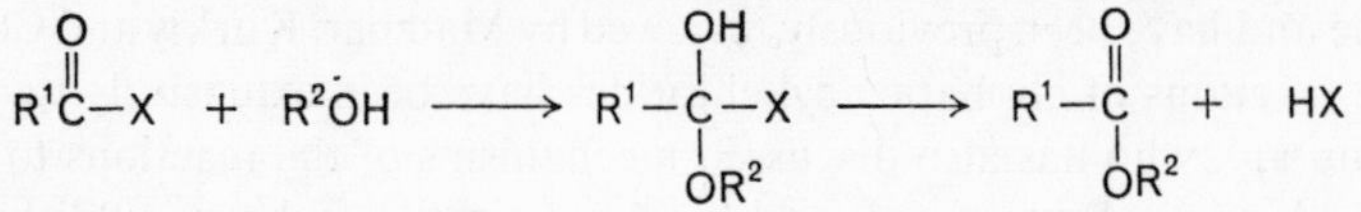

The formation of a tetrahedral intermediate is considered very probable in the reactions of several carbonyl compounds[8-11]. It should be noted, however, that it is only recently that reliable evidence has been presented which shows that this intermediate lies on the reaction path (in the ethanolysis of ethyl trifluoroacetate[12]). The various mechanisms will be discussed in more detail in section II, D.

If the view is accepted that the reactions of acyl halides may be of the S_N type, one should discuss many aspects that they have in common with reactions where halogen displacement occurs at a saturated carbon atom. These aspects have, however, been thoroughly documented[13-17] and will therefore be considered only occasionally.

Reviews on solvolytic reactions of acyl chlorides have been written by Kivinen[18] and Minato[19] and the reactions of alkyl chloroformates have been discussed by Kivinen[20] and Queen[21]. The most extensive survey of the solvolytic reactions of acyl halides has been presented by Hudson and co-workers in a series of papers[22-47].

2. Early work

The earliest studies on the solvolytic reactions of acyl chlorides dealt primarily with the effect of structure on the rate, although the influence of solvent was investigated to some extent. In a study of the hydrolysis of mono- and disubstituted benzoyl chlorides in 1 : 1 acetone–water mixtures, Olivier[48-50] found that the rate of hydrolysis of 2-bromobenzoyl chloride was slightly faster, whereas that of 2,6-dibromobenzoyl chloride was a hundred-fold slower than that of the unsubstituted compound. The low rate of the 2,6-substituted compound is evidently due to steric factors. Olivier concluded that the mechanisms of the reactions are in accordance with equation (2). Ashdown[51] studied the reactions of p-nitrobenzoyl chloride with 29 alcohols in diethyl ether and found the rates to be second-order in the alcohols. Some of the results obtained by Branch and Nixon[52] in a study of the ethanolysis of monosubstituted benzoyl chlorides and chloroacetyl chlorides in 60 : 40 (v/v) ether–ethanol mixtures are presented in Table 1. Norris and co-workers[53-55] studied in detail the alcoholysis of aromatic acyl chlorides in ethanol (Table 1) and the influence of solvent

TABLE 1. Kinetic data and relative rates of ethanolysis of acyl chlorides in 60% ether–ethanol[52] and in ethanol[53] at 25°

| Acyl chloride | 60% ether–ethanol | | | Relative rates in | |
	10^4k (s^{-1})	E (kcal/mole)	$\log A$	Ether–ethanol	Ethanol
CH_3COCl	164	12·50	7·4	171	—
p-$NO_2C_6H_4COCl$	20·4	11·10	5·5	19·0	21·6
p-BrC_6H_4COCl	2·23	13·45	6·2	2·1	2·1
p-ClC_6H_4COCl	2·05	13·85	6·5	1·9	1·9
p-IC_6H_4COCl	2·10	13·45	6·2	1·9	1·9
p-FC_6H_4COCl	1·22	14·65	7·8	1·1	—
C_6H_5COCl	1·08	14·40	6·6	1·0	1·0
p-$CH_3C_6H_4COCl$	0·64	15·90	7·5	0·60	0·78
p-$CH_3OC_6H_4COCl$	0·53	18·65	9·4	0·49	0·81

on the rate of the reaction between benzoyl chloride and ethanol[56]. These workers discussed two possible reaction mechanisms which are essentially those presented in equations (4) and (6).

Leimu[57] reported rates of alcoholysis for a large number of aliphatic acyl chlorides and alkyl chloroformates. Some of his data are given in Table 2. An interesting observation is the very low rate of alcoholysis of pentachloropropionyl chloride.

TABLE 2. Relative rates (CH_3COCl = 1) of alcoholysis of aliphatic acyl chlorides RCOCl at at 25°. Solvent: 2 molar solution of 2-chloroethanol in dioxan[57]

R	k_{rel}	R	k_{rel}
CH_3	1	$CH_2ClCHCl$	1·62
CH_2Cl	1·48	$CHCl_2CH_2$	0·23
$CHCl_2$	4·46	CCl_3CCl_2	0·047
CCl_3	32·67	$CH_2Cl(CH_2)_2$	0·64
CH_3CHCl	2·09	CH_3CH_2O	0·00011
CH_2ClCH_2	0·29	$COCl_2$ (at 15°)	11·99
CH_3CCl_2	1·98		

3. Solvent effects

a. Hydrolysis. The only acyl chlorides in addition to *N,N*-dialkylcar-
bamoyl chlorides whose rates of neutral hydrolysis in water and water-rich
solvent mixtures can be easily measured are the alkyl chloroformates. The
reason for the low rates of these compounds compared to those of other
acyl chlorides is that their ground states are stabilized by resonance (7).

$$R-O-\underset{\underset{\|}{O}}{C}-Cl \rightleftharpoons R-\overset{+}{O}=\underset{\underset{\|}{O^-}}{C}-Cl \qquad (7)$$

Most aromatic acyl chlorides are much more reactive and the investigation of their reactions with water is limited also by their low solubility
and/or low rate of solution in water-rich solvents. For these reasons,
information is available only on their rates in solvent mixtures of high
inert solvent content.

The rates of hydrolysis of aliphatic acyl chlorides are so high that their
rates in water-rich solvents can be studied only by special techniques.
Using the 'stopped flow' technique, Hudson and Moss[30] found that the
first-order rate constant of the hydrolysis of acetyl chloride in a 24·4%
(v/v) dioxan–water mixture at 27°C is 292 s^{-1}, which corresponds to a
half-life of $2\cdot4 \times 10^{-3}$ s.

In many hydrolytic reactions the measured rate parameters vary
regularly in the order: esters and anhydrides of carboxylic acids—acyl
chlorides— alkyl and aryl halides. Rates of neutral hydrolysis of carboxylic
esters can only be measured for esters of exceptional structure[58] because
the neutral hydrolysis is often masked by a rapid acid- or base-catalysed
reaction. The rates of hydrolysis of carboxylic anhydrides can be measured
over a wider pH region near neutrality[11, 12]. The hydrolyses of acyl chlorides
are catalysed by acids only in special cases (section II, 11).

In most studies of solvent effects in the hydrolysis of acyl chlorides, the
solvents have been dioxan–water or acetone–water mixtures[23, 26, 30, 59-70].
The rates decrease as the proportion of the organic component in the
binary mixture increases, but this decrease is much greater when the
reaction takes place by a unimolecular mechanism than when it takes
place by a bimolecular mechanism.

The effect of solvent is intimately connected with solvent structure and
solvent–solute interactions. A discussion of these lies outside the scope of
this paper, but it may be noted that our knowledge of the structures of
binary solvent mixtures in which at least one component is a hydroxylic
compound is still very limited.

Solvolytic reactions in water have been reviewed by Robertson[71]. A
paper of Kohnstam[72] deals primarily with the effect of solvent on activation

parameters. The structure of water is discussed in references 73 and 74 and a general review of hydrogen-bonded solvent systems is given in reference 75. The solvent properties of ordinary and heavy water have been compared by Arnett and McKelvey[76]. Parker[77] and Ritchie[78] have discussed the influence of dipolar aprotic solvents on the rates of hydrolytic reactions and Schneider[79] the selective solvation of ions in mixed solvents.

Only recently have direct thermal measurements been made to determine to what extent the variations in the heat and entropy of activation of a reaction, when the solvent is changed, are due to changes in the solvation of the initial state and of the transition state. It has been proposed that changes in the former are important in the solvolysis of, for example, t-butyl chloride in alcohol–water mixtures[80], but it has also been stated that a change in the solvation of the transition state is more important in other reactions[81, 82]. General discussions of these questions are found in references 76–78. These effects have been discussed in the case of the solvolysis of acyl chlorides by Hudson[38], who concluded that the heat of solvation of the transition state in the bimolecular reactions of acetyl chloride and ethanol in carbon tetrachloride–ethanol mixtures is almost independent of solvent composition. Direct experimental measurements for the clarification of this question in the case of acyl halides have not been carried out, however.

The tendency of substituted benzoyl chlorides to react by a bimolecular mechanism decreases in the order: 2,4,6-trinitro $\simeq$ 4-nitro > H > 4-methoxy $\gg$ 2,4,6-trimethyl[26]. This is also the expected order of the reaction rates if the reactions are bimolecular; for unimolecular reactions the reverse order is expected. A change in mechanism from bimolecular to unimolecular with increasing water content is also clearly evident from the data of Brown and Hudson[26]. It can be concluded that 4-nitrobenzoyl chloride still reacts exclusively by the bimolecular mechanism in a 50% acetone–water mixture.

The so-called solvent parameters provide a valuable means of classifying solvents[83, 84]. The values of these were calculated[85–88] for the solvolytic reactions of a number of acyl chlorides.

The equation (8) of Grunwald–Winstein[89, 90] has been used to decide

$$\log(k/k_0) = m \cdot Y \tag{8}$$

whether the solvolysis of an acyl chloride takes place by a unimolecular or a bimolecular mechanism. Values of m close to one are generally considered to signify S_N1 reactions, while lower values signify increasing S_N2 character.

The values of m for the hydrolysis of acetyl chloride in dioxan–water mixtures[30] and for the hydrolysis of benzoyl chloride[23, 26] suggest that hydrolysis of the former takes place primarily by the unimolecular mechanism, whereas the mechanism of hydrolysis of the latter changes from bimolecular to unimolecular with increasing water content. Frölich and co-workers[70] measured the n.m.r. shifts of the water protons in aqueous (5–30% v/v) acetone, tetrahydrofuran and 1,4-dioxan mixtures and found linear relationships between the shifts and the Gibbs energies of activation for the hydrolysis of benzoyl chloride in these solvents.

Instead of using equation (8), similar information can often be gained by determining the apparent order of the rate of a solvolytic reaction in respect of water[91, 92]. If the rate law is formally written

$$dx/dt = k_1(a-x) = k'[H_2O]^n(a-x) \qquad (9)$$

then

$$\log k_1 = \log k' + n \log[H_2O] \qquad (10)$$

where k_1 is the experimentally determined first-order rate constant. It is assumed that k' is independent of solvent composition.

The values of n for the hydrolysis of an alkyl or aryl halide in dioxan–water and acetone–water mixtures are about 6–7 if the reaction proceeds by the S_N1 reaction, but about 2 if the reaction proceeds by the S_N2 mechanism[91–93]. Accordingly, the value of n can be considered a mechanistic criterion for S_N reactions in certain solvent mixtures. This seems to apply also to the solvolytic reactions of acyl chlorides in dioxan–water and acetone–water mixtures[18, 20]. It may be noted that in the above cases, the correlation between the logarithm of the rate constant and water concentration is at least as good as the correlation between the logarithm of the rate constant and water activity.

The physical significance of the apparent order n has been extensively discussed. It has been proposed that it represents the number of water molecules bound in the transition state (or the difference between the numbers of water molecules bound in the transition state and the initial state). It has been concluded that if 6 or 7 water molecules are 'frozen' in the transition state, the temperature dependence (dE/dT) of the activation energy should be of the same order of magnitude as the experimentally determined values[92], but this view is not shared by all investigators.

Zimmerman and Yuan[60] found the rate of hydrolysis of acetyl chloride to be proportional to the 1·92th power of the water concentration in acetone–water mixtures ($[H_2O] = 0·547–2·83M$) and Koskikallio[63] to the 1·9th power in dioxan–water mixtures ($[H_2O] = 0·480–2·22M$). From the data of Hudson and Moss[30] for the hydrolysis of acetyl chloride in dioxan–water mixtures, it may be estimated that the value of n is about 5 at water contents from 19·6 to 75·6%. It can thus be concluded that the hydrolysis of acetyl chloride proceeds by the bimolecular mechanism in mixtures of low water content and that a shift to the unimolecular mechanism occurs as the water concentration increases.

Böhme and Schürhoff[59] determined the values of n for the hydrolysis of benzoyl chloride and ethyl chloroformate in aqueous dioxan, tetra-hydrofuran, 1,2-dimethoxyethane and bis[methoxyethyl] ether. The values varied from 1·2 to 2·5 at 25°, indicating bimolecular mechanisms.

Entelis and co-workers[67-69] investigated the hydrolysis of phthaloyl and terephthaloyl dichlorides in dioxan–water mixtures. Their data suggest that the rates refer to the first stage of hydrolysis. The values of n are low, varying from 1 to 2 for both dichlorides. Kelly and Watson[61] found that the value of n is about 4 for the hydrolysis of benzoyl chloride in acetone–water mixtures containing from 40 to 50% water, but increases to over 7 in mixtures of higher water content. This implies a shift to the unimolecular mechanism.

It has been proposed that dioxan and acetone may function as nucleophiles[94-97]. In dioxan–water mixtures, for instance, the reaction of RX may be represented by scheme (11), according to which the substitution of the halogen by a hydroxyl group may take place either directly

or after the formation of an oxonium ion. The scheme (which has also been criticized[98]) includes the case where azide ions are present in the reaction mixture in a 'trapping' experiment. This scheme may apply also to the solvolysis of acyl chlorides (see section II, D), making the interpretation of the kinetic data still more difficult.

Many questions relating to solvent effects in solvolytic reactions still remain unanswered. For instance, it is not possible to say what forms of water, monomers, dimers or higher aggregates participate in the reactions. In addition to these, complexes of varying composition may exist in dioxan–water mixtures[99]. It has also been proposed that the non-polar dioxan acts essentially as a diluent, merely changing the concentration of the polar species involved in the equilibrium process[100].

Addition of dimethyl sulphoxide (DMSO) to water has been found to increase the rates of hydrolysis of alkyl chloroformates that react by the bimolecular mechanism and to lower rates of hydrolysis of alkyl chloroformates that react by the unimolecular mechanism[101]. DMSO greatly retards the neutral hydrolysis of ethylene oxide[102] and diethyl dicarbonate[103]. The rates of hydrolysis of substituted benzyl chlorides which hydrolyse by the unimolecular mechanism are lowered by DMSO[92, 104], as are also the rates of hydrolyses of t-butyl halides[107]. On the other hand, the rates of hydrolysis of dinitrofluorobenzene[105] and methyl iodide[106] increase almost as much as the rate of hydrolysis of ethyl chloroformate when DMSO is added to water. The effect of added DMSO on the bimolecular hydrolysis of alkyl chloroformates is thus similar to its effect on reactions that proceed by S_N2-like mechanisms. (For the solvolysis of alkyl chloroformates, see references 20, 21, 108–110, and for the solvation of halide ions in DMSO–water mixtures the paper of Langford and Stengle[111].)

In contrast to the solvolytic reactions of carboxylic esters and alkyl and aryl halides (and also alkyl chloroformates), the hydrolytic reactions of acyl chlorides generally satisfy the Arrhenius equation, i.e. the activation energies do not vary with temperature. This may suggest that the transition state is less polar in the reactions of the acyl chlorides. It should be noted, however, that the rates of solvolysis of acyl chlorides can only rarely be measured with an accuracy that would reveal any deviations from the Arrhenius equation.

Both negative and positive values of the differential dE/dT ($E=$ energy of activation) have been reported for the reactions of some substituted benzoyl chlorides in acetone–water mixtures. These can at least partly be connected with changes in reaction mechanism with increasing temperatures (see section II, A, 6).

Different values of the entropy of activation are obtained for solvolytic reactions according to whether these values are based on first-order (s^{-1}) or on second-order (l/mole s) rate constants.

We shall denote by A_1 values of the frequency factor based on first- and by A_2 values based on second-order rate constants. The relationship between A_1 and A_2 is

$$\log A_2 = \log A_1 - \log[\text{ROH}] \qquad (12)$$

where ROH is the solvolytic reagent. Similarly, we shall use the symbols ΔS_1^* and ΔS_2^* for entropies of activation.

Activation energy values varying from 13 to 18 kcal/mole have been generally reported for solvolytic reactions of acyl chlorides. The values of log A_1 lie mostly between 3 and 10.

b. Alcoholysis. The lower rates of alcoholysis of acyl chlorides make it possible to study solvent effects over wider ranges of solvent composition than is the case with hydrolysis. Since the lower alcohols are miscible with most other organic solvents, alcoholysis may be studied under widely varying conditions.

A general feature of alcoholysis reactions is that their rate constants are monotonous functions of alcohol concentration. Similarly, the activation parameters vary fairly regularly with solvent composition and do not pass through minima or maxima like the parameters of various hydrolysis reactions.

Hudson and co-workers[40] have studied the ethanolysis of *p*-nitrobenzoyl chloride in ether, dioxan, acetone, acetonitrile and nitromethane containing relatively low proportions of alcohol. In all cases the rate of the reaction has been found to be approximately proportional to the second power of the ethanol concentration. The same is true for the ethanolysis of acetyl chloride in alcohol–ether mixtures[36] ([ROH] = 0·690–6·90M); at lower alcohol concentrations the value of n (equation 10) decreases to about 1·52.

On the other hand, the value of n for the ethanolysis of *p*-nitrobenzoyl chloride is about 2 at low ethanol concentrations in carbon tetrachloride but decreases to 1·1 as the ethanol concentration increases.

By comparing kinetic and spectral data, Hudson and Loveday[39] concluded that the rate of alcoholysis of an acyl chloride is proportional to the concentration of alcohol aggregates, the most probably reactive being the trimer (see section II, D).

Pure ethanol seems to be the only solvent where the rates of alcoholysis of aliphatic and aromatic acyl chlorides and ethyl chloroformate have been measured under the same conditions. Data for reactions that probably take place by the bimolecular mechanism are shown in Table 3. The only exception is the reaction of *p*-methoxybenzoyl chloride whose mechanism is closer to the unimolecular than to the bimolecular mechanism[18] as reflected in the higher energy of activation and value of log *A*. The values of ΔS_2^* vary from -28 to -34 cal/deg mole, with the exception of *p*-methoxybenzoyl chloride[18] (-16 cal/deg mole). The half-life of the ethanolysis of acetyl chloride estimated from the first-order rate constant is about half a minute at 25°C.

Kivinen[18] studied the ethanolysis of benzoyl chlorides in binary mixtures of ethanol with ether, acetone and benzene. Relative rate constants of alcoholysis are presented in Table 4. Cocivera[114] concluded from these

TABLE 3. Kinetic data for the ethanolysis of acyl chlorides in ethanol at 25°

Acyl chloride	10^3k (s^{-1})	E (kcal/mole)	log A	Reference
CH$_3$COCl	147	13·23	8·87	113
Me$_3$CCOCl	18·1	13·15	7·90	112
p-NO$_2$C$_6$H$_4$COCl	11·4	12·75	7·40	18
m-NO$_2$C$_6$H$_4$COCl	10·6	12·56	7·23	18
m-BrC$_6$H$_4$COCl	2·66	13·68	7·45	18
m-CH$_3$OC$_6$H$_4$COCl	0·798	15·12	7·98	18
C$_6$H$_5$COCl	0·776	15·25	8·05	18
m-CH$_3$C$_6$H$_4$COCl	0·719	15·66	8·33	18
p-CH$_3$OC$_6$H$_4$COCl	0·949	19·10	10·98	18
Ethyl chloroformate	0·0211	16·04	7·08	20

TABLE 4. Relative rates k_R/k_H of ethanolysis of substituted benzoyl chlorides RC$_6$H$_4$COCl in diethyl ether–ethanol and benzene–ethanol mixtures at 25° [18]

R	Diethyl ether (wt. %)			
	0	20	60	90
p-NO$_2$	14·7	—	24·3	32·7
m-NO$_2$	13·7	16·1	20·7	28·9
m-Br	3·43	3·97	4·38	5·19
m-CH$_3$O	1·03	1·09	1·03	1·22
m-CH$_3$	0·927	0·916	0·875	0·993
p-CH$_3$O	1·22	0·893	0·516	0·425

R	Benzene (wt. %)			
	0	20	60	90
p-NO$_2$	15·3	14·5	14·4	12·1
m-NO$_2$	13·5	12·4	11·1	9·93
m-Br	3·53	3·39	3·01	2·77
m-CH$_3$O	1·07	1·04	1·04	1·05
m-CH$_3$	0·945	0·927	0·906	0·918
p-CH$_3$O	1·28	1·21	0·971	0·714

results that the mechanisms are best described as addition–elimination. Fagley and co-workers[115-122] studied the alcoholysis of monosubstituted benzoyl chlorides in benzene–ethanol mixtures and presented a thermodynamic interpretation of the solvent effect.

The methanolysis of p-nitrobenzoyl chloride[123] in methanol–acetonitrile mixtures is discussed in section II, 10. The solvolysis (and decomposition) of t-butyl-peroxy chloroformate was studied by Bartlett and Minato[124], who concluded that simple methanolysis occurs in methanol with very little loss of peroxidic oxygen. Ross[125] measured the rates of ethanolysis of benzoyl chloride and mononitrobenzoyl chlorides in acetone and chloroform, and discussed the various possible forms of hydrogen-bonded transition states. Cason and Kraus[126] have shown that the solvolysis of acyl chlorides is highly sensitive to steric effects.

It has been proposed that, as in aqueous mixtures, the value of n is of the order of 6 for unimolecular and of the order of 2 for bimolecular reactions[20] in ether–ethanol mixtures. Some values of n for reactions in ether–ethanol and benzene–ethanol mixtures are shown in Table 5. It

TABLE 5. Apparent order n of ethanol in the ethanolysis of substituted benzoyl chlorides RC_6H_4COCl in benzene–ethanol and diethyl ether–ethanol mixtures at 25° [18]

R	Benzene–ethanol mixtures	Diethyl ether (wt. %) in diethyl ether–ethanol mixtures		
		0–20	20–60	60–90
p-NO$_2$	1·4	1·9	1·9	1·9
m-NO$_2$	1·5	2·0	2·0	2·0
m-Br	1·4	2·2	2·0	2·0
m-CH$_3$O	1·3	2·5	2·1	2·0
H	1·3	2·3	2·2	2·0
m-CH$_3$	1·3	2·5	2·4	2·0
p-CH$_3$O	1·5	4·1	2·7	2·4

has been estimated that the value of n for p-methoxybenzoyl chloride rises to about 5 in pure ethanol. No variation in the value of n has been observed for mononitrobenzoyl chlorides which react by the bimolecular mechanism over the whole range of solvent composition. The data in Table 5 show also that ether–ethanol mixtures are differentiating solvents, whereas benzene–ethanol mixtures are levelling solvents, which means that the value of n depends only slightly on whether the reaction is uni- or bimolecular.

From the data of Biordi[127] for the methanolysis of benzoyl chloride in dioxan–methanol mixtures, the value of n can be estimated to be about

1·1 when the methanol is between 52 and 66% and about 1·65 when it is between 81 and 100%.

Some of the kinetic data reported by Geuskens and co-workers[128] for the methanolysis of polycyclic aromatic acyl chlorides in a 1 : 1 acetone–methanol mixture are listed in Table 6. They estimated the isokinetic temperature to be 220°K.

TABLE 6. Kinetic data for the methanolysis of some aromatic acyl chlorides in 1 : 1 acetone–methanol mixtures at 25° [128]

Acyl chloride	$10^3 k_1$ (s^{-1})	E (kcal/mole)	$\log A$
Benzoyl	1·45	13·64	7·16
Naphthoyl-1	2·26	13·76	7·45
Naphthoyl-2	1·48	13·96	7·41
Phenanthrene-carbonyl-9	2·81	14·90	8·38
Chrysene-carbonyl-2	6·31	14·77	8·63
Pyrene-carbonyl-3	12·7	16·67	10·33
Pyrene-carbonyl-4	2·07	13·12	6·94

Murculescu and Demetrescu[129] investigated the alcoholysis of benzoyl chloride in five lower alcohols and found that the rate of methanolysis was about twice that of the others.

Akiyama and Tokura[130, 131] studied the acylation of *l*-menthol with *para*-substituted benzoyl chlorides and substituted acetyl chlorides (Table 7) in chloroform, benzene, liquid sulphur dioxide and tetrahydrofuran. They concluded that all the reactions are of the S_N2 type. The

TABLE 7. Kinetic data for the acylation of *l*-menthol by acyl chlorides in tetrahydrofuran at 25° [130]

Acyl chloride	$10^5 k_2$ (l/mole s)	E (kcal/mole)	ΔS^* (cal/deg mole)
Propionyl	5·96	10·6	− 43·8
Monochloroacetyl	99·3	9·6	—
Dichloroacetyl	312	6·9	− 48·5
Trichloroacetyl	776	3·2	− 59·4

unexpected order of the rates, propionyl > monochloroacetyl > dichloro-acetyl (and *p*-methoxybenzoyl > *p*-methylbenzoyl > benzoyl > *p*-chloro-benzoyl), may be due to the fact that the reactions were run at temperatures above the isokinetic temperature.

Addition of DMSO to ethanol accelerates the ethanolysis of benzoyl chloride[132], but the kinetics are complex, evidently owing to the concurrent reaction of the acyl chloride with DMSO[133].

The parameters of the Arrhenius equation for alcoholysis of acyl chlorides vary regularly with solvent composition. Thus, for most *para*- and *meta*-substituted benzoyl chlorides in ether–ethanol mixtures, when the mole fraction of alcohol decreases from 1 to about 0·1–0·05, the activation energies decrease linearly by about 2 kcal/mole[18]. The same happens in benzene–ethanol mixtures, but, at mole fractions lower than 0·2, the E_A values decrease abruptly by several kilocalories per mole[18].

A similar abrupt decrease in the enthalpy of activation at low alcohol concentrations has been reported for the alcoholysis of *m*-fluorobenzoyl chloride, *m*-trifluoromethylbenzoyl chloride and *p*-methoxybenzoyl chloride in benzene–ethanol mixtures[119] and for the alcoholysis of *p*-methoxybenzoyl chloride in benzene-2-propanol mixtures[121]. After correcting the values of the activation enthalpy by adding the values of the partial molar enthalpy of the alcohol to them, Fagley and co-workers[122] found that the corrected values of the enthalpy of activation (Table 8) are

TABLE 8. Experimental and corrected activation enthalpies (kcal/mole) for the ethanolysis of *m*-trifluoromethylbenzoyl and *p*-methoxybenzoyl chloride in benzene–ethanol mixtures[122]

m-Trifluoromethylbenzoyl chloride			*p*-Methoxybenzoyl chloride		
x_{EtOH}	ΔH^*	ΔH^*_{corr}	x_{EtOH}	ΔH^*	ΔH^*_{corr}
0·05	7·81	11·16	0·10	11·11	20·72
0·10	9·29	11·26	0·30	15·95	19·72
0·20	10·18	11·14	0·40	16·45	19·13
0·30	10·63	11·16	0·50	16·82	18·72
0·40	10·89	11·18	0·60	17·86	19·17
0·50	11·01	11·18	0·70	18·67	19·52
			0·80	20·05	20·55

independent of solvent composition. (Similar corrections can be applied to entropies of activation.) Similar data for reactions in other binary solvent mixtures would be of great interest.

4. Substituent effects

a. Aliphatic acyl chlorides. Systematic studies of the effects of substituents on the solvolytic reactions of aliphatic chlorides are few[57]. The

Taft[134, 135] or other related free-energy relationships have not been tested. Ugi and Beck[66] measured the rates of hydrolysis of a large number of acyl chlorides in acetone–water mixtures: some of their data are given in Table 9.

TABLE 9. Rate constants k_1 (s^{-1}) for the hydrolysis of aliphatic and aromatic acyl chlorides in acetone containing 10·9 vol. % water at 25° [66]

Aliphatic acyl chloride		Substituted benzoyl chloride	
R in RCOCl	$10^4 k_1$	Substituent	$10^4 k_1$
CH$_3$	10·9	H	0·0417
(CH$_3$)$_3$C	0·74	2-Methyl	0·0644
CH$_3$(CH$_2$)$_8$	4·40	2-Nitro	0·212
CH$_2$=CHCH$_2$	4·44	4-Nitro	2·33
C$_6$H$_5$CH$_2$	3·60	4-Methoxy	0·0146
CH$_3$CH=CH	0·37	2,6-Dimethyl	1·31
ClCH$_2$	203	2,6-Dimethoxy	9·70
Cl$_2$CH	31000^a	3,5-Dinitro	60·4
Cl$_3$C	> 100000	2,4,6-Trimethyl	20·0
HO$_2$C	> 100000	4-Methoxy-2,6-dimethyl	1110
C$_5$H$_{10}$N	0·0021^a	4-Nitro-2,6-dimethyl	0·0012^a
[(CH$_3$)$_2$CH]$_2$N	0·0839	2,4,6-Trinitro	0·071
CH$_3$O	0·0012	3,5-Dinitro-2,4,6-trimethyl	0·0016^a
Cl	10200		

^{a} Extrapolated value.

The effect of structure on the rates of hydrolysis of several alkyl chloroformates has been investigated by Queen[21] and Hudson[28, 29].

b. Aromatic acyl chlorides. Shorter[136, 137] has written a summary on various linear free-energy relationships. Swain and Lupton[138] have recently proposed a general structure–reactivity equation and Charton[139, 140] has proposed an equation for the effects of *ortho* substituents.

The relationship between structure and reactivity in the solvolysis of acyl chlorides has been studied by Swain and Dittmer[141] and Brownstein[88].

In most studies of the solvolysis of aromatic acyl chlorides, the range of substituents has been small and most investigators have employed the Hammett equation (13)[142]. In general, this is satisfactory except when compounds with different substituents react by different mechanisms. The investigators have generally employed the original σ values[142, 143].

$$\log k_X/k_H = \rho\sigma \tag{13}$$

Electron-attracting substituents in benzoyl chlorides increase the positive charge on the carbon atom of the chlorocarbonyl group and thus strengthen the carbon–chlorine bond. Consequently, benzoyl chlorides with such substituents have a tendency to react by a bimolecular mechanism. Electron-releasing groups promote reaction by a unimolecular mechanism by reducing the strength of the carbon–halogen bond. Accordingly, the reaction constant ρ is positive for reactions in less polar solvents in which the bimolecular mechanism is favoured over the unimolecular one.

Hudson and co-workers have studied the influence of *para*-substituents on the rates of solvolysis of benzoyl chlorides in acetone–water mixtures containing 50 vol. %[22] and 5 vol. %[24] water and in several formic acid–water mixtures[27]. Some of their results are shown in Table 10; these results yield for ρ a value of about $+ 2.0$.

TABLE 10. Kinetic data for the hydrolysis of *p*-substituted benzoyl chlorides in 95 vol. % acetone–water at 25° [24]

Substituent	$10^5 k_1$ (s^{-1})	E (kcal/mole)	log A
CH$_3$O	2·92	14·7	5·89
CH$_3$	2·69	12·7	4·75
H	5·40	11·3	3·61
Br	15·5	11·0	3·27
NO$_2$	169	6·67	2·12

Kivinen[18] has found that the data for the alcoholysis of *meta*- and *para*-substituted benzoyl chlorides except *p*-methoxybenzoyl chloride in ethanolic ether, acetone and benzene mixtures obey the Hammett equation satisfactorily. The discrepancy noted with *p*-methoxybenzoyl chloride is due to a shift from a bimolecular to a unimolecular mechanism as the alcohol content of the solvent increases; all the other acyl chlorides react by the bimolecular mechanism. The value of ρ increases from $+ 1.57$ for the reactions in neat ethanol to $+ 1.92$ in ether containing 2·5 wt. % ethanol and to $+ 1.71$ in acetone containing 2·5 wt. % ethanol, but decreases to $+ 1.30$ in benzene containing 2·5 wt. % ethanol. This suggests that addition of benzene to ethanol does not cause as great a shift toward the bimolecular mechanism as the addition of ether or acetone.

Most of the rate constants reported by Heidbuchel[144] for the ethanolysis of *ortho*-, *meta*- and *para*-substituted phenylacetyl chlorides in a 0·553M solution of ethanol in benzene (4 vol. % ethanol) are in agreement with the Hammett equation. The deviation of the rate constant for p-t-butyl-phenylacetyl chloride from the plot is probably due to a shift toward the unimolecular mechanism. The value of ρ is $+0\cdot160$, i.e. much smaller than that found $(+1\cdot4)$ for the ethanolysis of benzoyl chlorides[18]. Heidbuchel attributed the difference to a strong inhibition of the meso-meric interaction of the substituted phenyl group with the reaction centre by the intermediate methylene group.

Bender and Chen[145] studied the hydrolysis of para-substituted 2,6-dimethylbenzoyl chlorides in 99% acetonitrile–water and obtained the ρ values $-3\cdot85$, $-3\cdot73$ and $+1\cdot20$ for the neutral, acid-catalysed and base-catalysed reactions, respectively. The large negative values of ρ imply that the neutral and the acid-catalysed hydrolysis proceed by unimolecular mechanisms.

5. Salt effects

Very few published papers deal with salt effects in the hydrolysis of acyl chlorides[25, 31]. Mercury(II) perchlorate generally accelerates the hydrolysis of acyl chlorides by the unimolecular mechanism[109], but has very little effect on the reactions that proceed by the bimolecular mechanism. Similar effects of mercury(II) chloride have been observed in the unimolecular solvolysis of acyl chlorides in ethanol[146].

Lithium bromide greatly accelerates the hydrolysis of benzoyl chloride, especially in acetone–water mixtures of low water content[25]. The reason is probably that the exchange reaction (14) occurs alongside the hydrolysis

$$\text{RCOCl} + \text{Br}^- \rightleftharpoons \text{RCOBr} + \text{Cl}^- \tag{14}$$

and that benzoyl bromide hydrolyses more rapidly than benzoyl chloride[31, 146, 147].

6. Heat capacities of activation

The heat capacity of activation[72] is defined by equation (15) and can be

$$\Delta C_p^* = d(\Delta H^*)/dT = dE/dT - R \tag{15}$$

evaluated after fitting experimental rate constants at different temperatures to the three-parameter equation (16). The review of Robertson[71, 148] dealing with heat capacities of activation refers primarily to reactions in water, and that of Kohnstam[72] to reactions in mixed solvents. Deviations

$$\log K = A/T + B \log T + C \tag{16}$$

from the Arrhenius equation have also been discussed by Hulett[149]. A negative value of ΔC_p^* implies a markedly polar transition state[149].

Deviations from the Arrhenius equation have been reported for solvolytic reactions of only a few acyl chlorides. Archer and Hudson[23] reported the values -21, $+53$ and $+61$ cal/deg mole for dE/dT in the hydrolysis of benzoyl chloride in $61 \cdot 05\%$, $70 \cdot 24\%$ and $81 \cdot 68\%$ (w/w) acetone–water mixtures respectively. They attributed the positive temperature coefficients to a change from a bimolecular to a unimolecular mechanism which has the higher activation energy and gains importance at higher temperatures.

Values of the temperature coefficient between 60 and 120 cal/deg mole have been reported for the ethanolysis of benzoyl chloride in $0 \cdot 16$ and $0 \cdot 08$ molar solutions of ethanol in benzene[18]. One possible reason for the positive values is the variation of the self-association of the alcohol with temperature.

Extreme precision is required of the rate data before significant values of the heat capacity can be obtained (see section II, A, 3). The study of Queen[21] of the hydrolysis of alkyl chloroformates seems to be the only one in which accurate values of heat capacities of activation have been determined.

7. Effect of pressure

Valuable information about reaction mechanisms can be gained by studying the effect of pressure on the rates of reactions in solution. According to Evans and Polanyi[150], the effect of pressure on the rate constant at constant temperature can be expressed by equation (17) where ΔV^*, the activation volume, is the difference between the partial molar

$$(\partial \ln k/\partial p)_T = - \Delta V^*/RT \tag{17}$$

volumes of the transition and initial states. The kinetic effects of pressure have been recently reviewed by Kohnstam[151] and Weale[152]. Because the activation volume usually varies with pressure, the activation volume at a pressure of 1 atm is evaluated either by graphical methods or from the best fit of the experimental points to the equation (18)[151]. Heydtmann and co-workers measured the rates of hydrolysis of para-substituted benzoyl

$$\ln k = a + bp + cp^2 \tag{18}$$

chlorides in tetrahydrofuran–water mixtures[153] and the rates of hydrolysis of benzoyl chloride in dioxan–water mixtures[154] at pressures from 1 to 1000 atm. Some of their data are presented in Table 11. A change in the mechanism of hydrolysis of p-methoxybenzoyl chloride from a bimolecular to a unimolecular one occurs with increasing water content in tetrahydrofuran–water mixtures: this is revealed by the dependence of the rate constant on water concentration and by the ρ value which is $+ 2.4$ in the solvents containing 2.2 and 6.1% water, but -1.4 in the solvent containing 29.7% water. The difference between the activation volumes in

TABLE 11. Activation volumes ΔV_0^* (at 1 atm) for the solvolysis of acyl chlorides (a = acetone, d = dioxan, tf = tetrahydrofuran, w = water)

Acyl chloride	Solvent (wt. %)	T (°C)	$-\Delta V_0^*$ (cm³/mole)	Reference
C_6H_5COCl	95·7% d–w	25	20·8	154
C_6H_5COCl	93% d–w	25	23·0	154
C_6H_5COCl	90·2% d–w	25	27·0	154
C_6H_5COCl	85·5% d–w	25	27·4	154
C_6H_5COCl	81·2% d–w	25	24·7	154
C_6H_5COCl	77·2% d–w	25	24·8	154
C_6H_5COCl	72·3% d–w	25	28·0	154
C_6H_5COCl	70·7% d–w	25	31·4	154
C_6H_5COCl	97·8% tf–w	20	33·1	153
C_6H_5COCl	93·9% tf–w	20	31·0	153
C_6H_5COCl	84·2% tf–w	20	37·4	153
C_6H_5COCl	80·6% tf–w	20	38·4	153
C_6H_5COCl	70·3% tf–w	20	41·0	153
$p\text{-}CH_3OC_6H_4COCl$	97·8% tf–w	20	27·5	153
$p\text{-}CH_3OC_6H_4COCl$	94·3% tf–w	20	25·3	153
$p\text{-}CH_3OC_6H_4COCl$	84·2% tf–w	20	23·6	153
$p\text{-}CH_3OC_6H_4COCl$	80·6% tf–w	20	28·1	153
$p\text{-}CH_3OC_6H_4COCl$	70·3% tf–w	20	31·2	153
$p\text{-}BrC_6H_4COCl$	97·8% tf–w	20	33·8	153
$p\text{-}BrC_6H_4COCl$	93·9% tf–w	20	28·3	153
$p\text{-}NO_2C_6H_4COCl$	97·8% tf–w	20	43·0	153
Ethyl chloroformate	20% a–w[a]	0	12·5	156
Ethyl chloroformate	water	0	11·7	155
C_6H_5COCl	ethanol	0	29·1	155
$p\text{-}CH_3OC_6H_4COCl$	ethanol	0	20·1	155
Ethyl chloroformate	ethanol	25	23·8	156

[a] Vol. %.

the reactions of unsubstituted benzoyl chloride and p-methoxybenzoyl chloride increases from about $5.6 \text{ cm}^3/\text{mole}$ in the solvent containing 2.2% water to about $10.2 \text{ cm}^3/\text{mole}$ in the solvent containing 29.7% water. Heydtmann and co-workers[153, 154] concluded that hydrogen bonding to the carbonyl oxygen in the hydrolysis of a benzoyl chloride polarizes the carbon–oxygen bond and probably also changes the solvent structure in the surroundings of the substrate molecule. This is also a possible explanation for the different effects of pressure on the solvolytic reactions of acyl chlorides and on S_N-type hydrolysis reactions where substitution occurs at a saturated carbon atom.

In a study of the effect of pressure on the hydrolysis and ethanolysis of acyl chlorides, Kivinen and Viitala[155, 156] concluded that the difference, about 9 cm^3/mole, between the activation volumes in the ethanolysis of benzoyl and p-methoxybenzoyl chlorides (Table 11) is due to the fact that the former compound reacts by a bimolecular mechanism and the latter by a unimolecular mechanism. The activation volumes are less negative for the hydrolysis of ethyl chloroformate than for the neutral hydrolysis of carboxylic anhydrides in pure water[157]. Too little information is available on rates at high pressures to decide whether the solvolytic reactions of acyl halides take place by a synchronous or an addition–elimination mechanism.

8. Oxygen exchange

Oxygen exchange reactions of organic and organometallic compounds have been reviewed by Samuel and Silver[158]. Carbonyl oxygen exchange with water often occurs in the hydrolysis of carbonyl compounds. This has been taken as strong evidence for the formation of a tetracovalent addition intermediate and was initially the main justification of the proposed addition–elimination mechanism in the hydrolysis of acyl halides. In simplified form, the carbonyl oxygen exchange can be represented by equation (19). However, only the work of Johnson[159] on the solvolysis of

$$
\begin{array}{c}
\overset{O}{\overset{\|}{RC}}-Cl \; + \; {}^{18}H_2O \; \underset{k_2}{\overset{k_1}{\rightleftharpoons}} \\[2em]
\overset{{}^{18}O}{\overset{\|}{RC}}-Cl \; + \; H_2O \; \underset{k_2}{\overset{k_1}{\rightleftharpoons}}
\end{array}
\quad
\begin{array}{c}
\overset{OH}{\underset{{}^{18}OH}{\overset{|}{RC}-Cl}} \; \overset{k_3}{\longrightarrow} \; \overset{{}^{18}O}{\overset{\|}{RC}}-OH \; + \; HCl \quad (19)
\end{array}
$$

ethyl trifluoroacetate provides the first convincing evidence that the tetrahedral intermediate actually lies on the reaction path of the solvolysis. The criticism of Burwell and Pearson[160] does not seem to invalidate the conclusion.

It follows from equation (19) that the ratio of k_h (hydrolysis) to k_{ex} (isotope exchange) is equal to $2k_3/k_2$. The experimental rate constant k_{obsd} is hence given by equation (20)[9]. Values of k_2/k_3 smaller than one imply that the addition step (k_1) is rate-determining and this seems to be the case in the hydrolysis of acyl chlorides.

$$k_{obsd} = k_1 k_3/(k_2 + k_3) = k_1/(k_2/k_3 + 1) \tag{20}$$

The few studies on oxygen exchange in the hydrolysis of acyl chlorides are not fully comparable because of different experimental conditions (Table 12). The marked dependence of the ratio k_h/k_{ex} on solvent composition for the reactions in dioxan–water mixtures may indicate that

TABLE 12. Oxygen exchange during the hydrolysis of benzoyl chlorides. k_h = rate constant of hydrolysis, k_{ex} = rate constant of oxygen exchange (w = water, d = dioxan, an = acetonitrile)

Acyl chloride	Solvent (vol. %)	T (°C)	k_h/k_{ex}	Reference
C_6H_5COCl	95% d–w	25	large	161
C_6H_5COCl	95% d–w	25	large	62
C_6H_5COCl	75% d–w	25	25	161
C_6H_5COCl	67% d–w	25	18	161
C_6H_5COCl	67% d–w	25	14	62
p-$CH_3C_6H_4COCl$	95% d–w	25	large	161
p-$CH_3C_6H_4COCl$	67% d–w	25	51	161
$2,4,6(NO_2)_3C_6H_2COCl$	99% an–w	25	large	145
$2,4,6(NO_2)_3C_6H_2COCl$	95% d–w	25	31	161

proton transfer between oxygen and chlorine affects the partitioning of the tetracovalent intermediate[158, 162, 163]. It has also been suggested that water molecules associate in a mixed solvent of low water content in such a way that proton transfer occurs along an associated chain and prevents oxygen exchange[62].

The oxygen exchange studies show that the synchronous mechanism cannot be the only one by which acyl chlorides hydrolyse, but it is possible that it may be the dominating one in solvolytic reactions.

9. Isotope effects

a. Kinetic solvent isotope effects. Values of the kinetic deuterium isotope effect (k_H/k_D) determined for solvolytic reactions of a number of acyl chlorides[21, 65, 164–168] are shown in Table 13. It is probable that the reactions of all the listed compounds except the carbamoyl chlorides and, perhaps, i-propyl chloroformate are bimolecular.

Solvent isotope effects[169] and the solvent properties of water and deuterium oxide[170, 171] have been reviewed.

Large differences are not usually observed in the values of k_H/k_D for S_N1 and S_N2 solvolyses of alkyl and aryl halides. The values vary from 1·2 to 1·5[169]. On the other hand, values between 2 and 5 have been reported

TABLE 13. Values of the kinetic solvent deuterium isotope effects k_H/k_D for the solvolysis of acyl chlorides (d = dioxan, w = water)

Acyl chloride	Solvent	T (°C)	k_H/k_D	Reference
C_6H_5COCl	85% d–w	25·1	1·6	167
C_6H_5COCl	80% d–w	25	1·5	166
C_6H_5COCl	67% d–w	25	1·9	166
C_6H_5COCl	60% d–w	25	1·7	164
C_6H_5COCl	60% d–w	25·1	1·7	167
p-CH_3CH_4COCl	91% d–w	25	1·67	65
p-$CH_3C_6H_4COCl$	73% d–w	25	1·41	65
p-$CH_3C_6H_4COCl$	50% d–w	25	1·41	65
p-$NO_2C_6H_4COCl$	91% d–w	25	1·6	65
p-$NO_2C_6H_4COCl$	73% d–w	25	1·74	65
p-$NO_2C_6H_4COCl$	50% d–w	25	1·80	65
$2,4,6(NO_2)_3C_6H_2COCl$	95% d–w	25	1·65	164
$2,4,6(NO_2)_3C_6H_2COCl$	95% d–w	25·1	1·7	167
$2,4,6(NO_2)_3C_6H_2COCl$	93% d–w	25	1·5	166
Diphenylcarbamoyl chloride	water	25	1·1	9
Methyl chloroformate	water	25·0	1·89	21
Ethyl chloroformate	water	25	1·95	168
Ethyl chloroformate	water	24·5	1·82	21
i-Propyl chloroformate	water	24·5	1·35	21
Phenyl chloroformate	water	7·5	1·79	21

for the neutral hydrolysis of carboxylic esters and anhydrides[9]. The values of k_H/k_D for solvolytic reactions of acyl chlorides fall in the intermediate range, and relatively high values (> 1·5) of the ratio have been considered to be at variance with an S_N-type mechanism[21]. Bunton and co-workers[167] have calculated values of the solvent isotope effect for various assumed models of the transition state of the hydrolysis of acyl chlorides. For S_N2-like reactions, their estimate of k_H/k_D is 1·5, which is of the same order as the experimentally determined values.

The values 1·568 and 1·564 have been reported for the ratio k_H/k_D in the hydrolysis of methanesulphonyl chloride and benzenesulphonyl chloride, respectively, in water at 20°C[172]. It seems that these are best described by a slightly modified S_N-type mechanism[172, 173]. Relatively large values of k_H/k_D in the hydrolysis of acyl chlorides do not therefore seem to be conclusive evidence that the bimolecular reactions proceed by an addition–elimination mechanism.

b. Secondary deuterium isotope effects. Secondary isotope effects have been reviewed[174, 175]. Until quite recently, the only report on β-secondary deuterium effects in the solvolysis of acyl chlorides was that by Bender and Feng[176]. The available data are collected in Table 14. Papaioannou[177]

TABLE 14. Values of the secondary deuterium isotope effects k_H/k_D for the hydrolysis of acyl chlorides (a = acetone, tf = tetrahydrofuran, w = water)

Acyl chloride	Solvent (vol. %)	T (°C)	k_H/k_D	Reference
CH₃COCl	95% a–w	−22·6	1·008	179
CH₃COCl	90% a–w	−22	1·51	176
CH₃COCl	90% a–w	−22·0	1·059	179
CH₃COCl	90% a–w	−28·9	1·044	179
CH₃COCl	85% a–w	−21·2	1·101	179
CH₃COCl	80% a–w	−22	1·62	176
CH₃COCl	80% a–w	−22·0	1·106	179
CH₃COCl	80% a–w	−31·0	1·122	179
CH₃COCl	75% a–w	−30·2	1·133	179
CH₃COCl	95% tf–w	25	1·01	176
8-Methyl-1-naphthoyl chloride	95% a–w[a]	24·4	1·03	177
8-Methyl-1-naphthoyl chloride	75% a–w[a]	−20·5	1·13	177

[a] Wt. %.

postulated that whereas 8-methyl-d_3-1-naphthoyl chloride solvolyses by a limiting S_N1 mechanism, 1-naphthoyl chloride solvolyses by a mixed unimolecular–bimolecular mechanism. Bartell's method[182] has been used[177–181] to estimate the relative contributions of hyperconjugation and non-bonded interactions to secondary isotope effects in the reactions listed in Table 14. In most cases the calculated and experimental values are in agreement, but reactions where hyperconjugation is possible[175, 181] are exceptions.

The results of Bender and Feng[176] differ significantly from the results of other investigators. The differences are due to differences in the rates of solvolysis of protium compounds[179]. The β-isotope effect in the hydrolysis of acetyl chloride increases with the polarity of the solvent. Evans[179] has concluded that the overall rate constant of the reaction over the studied range of solvent composition is the sum of the rate constants of two reactions (equation 21), where k_{lim} is the rate constant of the S_N1 reaction

$$k_{obsd} = k_{lim} + k_{nucl} \tag{21}$$

and k_{nucl} the rate constant of the bimolecular reaction that proceeds by way of a tetrahedral intermediate. The increase of the isotope effect with temperature in the hydrolysis of acetyl chloride in acetone–water mixtures containing 90 and 95% acetone may be taken to indicate that the contribution of the unimolecular reaction to the overall rate also increases[179, 183].

Swain[184] obtained the value 0·911 for the oxygen isotope effect $[k(^{16}O)/k(^{18}O)]$ in the methanolysis of benzoyl chloride in methanol at 25°.

10. Reactions in special solvents

a. *Acetonitrile and nitromethane*. A number of interesting studies have been made to clarify the mechanisms of solvolysis of acyl chlorides by measuring rates of solvolysis in nitromethane and, especially, acetonitrile containing a relatively low concentration of a hydroxylic compound. (Ionic reactions in acetonitrile have been reviewed by Coetzee[185].) Bender and Chen's data on the hydrolysis of 4-substituted 2,6-dimethylbenzoyl chlorides in 99% acetonitrile–water[145] will be discussed below. Akiyama and Tokura[130, 131, 186] studied the reactions of l-menthol with acyl chlorides in both acetonitrile and nitromethane and Satchell and co-workers[187–189] the acylation of phenols in the same solvents.

Kevill and co-workers[123, 190, 191] studied the methanolysis of p-nitro-benzoyl chloride (PNBC) at several temperatures in acetonitrile–methanol mixtures from 0·01 to 1·60 molar in methanol. The rates were represented by equation (22). Added phenol does not catalyse the reaction, but

$$d[HCl]/dt = k_2[PNBC][CH_3OH] + [RNBC][CH_3OH]^2 \tag{22}$$

chloride ion, which is a strong base in acetonitrile, effects a pronounced acceleration. Kevill and Foss[123] proposed that a tetrahedral intermediate is first formed and that this is deprotonated by either an acetonitrile molecule (second-order kinetics) or a second methanol molecule (third-order kinetics) to give a new tetrahedral intermediate which collapses to the products (see section II, D). They also disputed the conclusion of Briody and Satchell[189] that reactions between chloroacyl chlorides and phenols proceed by a synchronous S_N2-type mechanism and proposed an alternative explanation in terms of ionization and tetrahedral intermediate mechanisms.

A few remarks may be made concerning the work of Kevill and Foss[123]. It is known that hetero-association of methanol and phenol takes place at much lower concentrations of these reactants than the self-association of methanol in inert solvents[192, 193]. In this association methanol functions as a proton acceptor and phenol as a proton donor. This can be expected to alter the basicity of the methanol oxygen at least to some extent. In addition, acyl chlorides are not very strong proton acceptors. Phenol, a stronger acid, can replace methanol as a hydrogen donor to the acyl oxygen and thereby alter to some extent the charge distribution in the acyl group. These remarks do not necessarily invalidate the arguments of Kevill and Foss, but serve to show that the interpretation of kinetic data for reactions in mixed hydroxylic solvents may be very difficult.

Jones and Foster[194] concluded from a study of chloride exchange in the solvolysis of ethyl chloroformate in acetone–water mixtures containing from 40 to 85% water that the chloride ion is a more reactive nucleophile toward the carbonyl carbon atom than is water.

b. Liquid sulphur dioxide. Tokura and co-workers studied the acylation of *l*-menthol[130, 131, 186] and the benzoylation of methanol in liquid sulphur dioxide. The overall rates can be expressed by equation (23) where

$$\text{Rate} = k_1^0[\text{RCOCl}][\text{ROH}]_0 + k_2[\text{RCOCl}][\text{ROH}] \qquad (23)$$

$[\text{ROH}]_0$ is the initial concentration of the alcohol, $[\text{ROH}]$ is the concentration of the alcohol and $[\text{RCOCl}]$ of the acyl chloride at time t, k_1^0 a first-order and k_2 a second-order rate constant. The rate law apparently changes gradually from one of second-order to one of first-order as the reaction proceeds. The reactions are faster in sulphur dioxide than in other solvents.

Nagai and co-workers[195] compared kinetic data for the benzoylation of methanol with n.m.r. data for the self-association of methanol in sulphur dioxide–carbon tetrachloride mixtures of varying composition. They concluded that the reaction more likely involves methanol monomers than methanol associates. This conclusion is the opposite of that of Hudson and Loveday[39] according to whom acylations of an alcohol involve associated forms of the alcohol in carbon tetrachloride.

c. Formic and acetic acids. Howald has studied the effects of salts on the solvolysis of acetyl chloride in glacial acetic acid[196] and Crunden and Hudson the solvolysis of substituted benzoyl chlorides in formic acid containing water[27]. Most acyl chlorides react by the ionization mechanism in these strongly ionizing solvents. The reaction of *p*-nitrobenzoyl chloride is still bimolecular in these solvents; the rate is approximately proportional to the water concentration and is increased by formate ions[27].

Satchell[197, 198] found that the reactions of 2-naphthol with various acyl halides in acetic acid are catalysed by hydrogen halides and salts. He derived approximate values for the equilibrium constants for various halogen exchange reactions from the measured rates of the reactions.

II. Acid catalysis

In contrast to acyl fluorides (see section II, B), the solvolytic reactions of acyl chlorides are not catalysed by acids[22, 48, 199]. The ethanolysis of 2,4,6-trimethylbenzoyl chloride in dilute solutions of ethanol in carbon tetrachloride is, however, strongly catalysed by the hydrogen chloride formed[37]. The reaction is practically irreversible and the autocatalysis becomes less pronounced as the alcohol concentration increases. The only case where the acid catalysis has been extensively examined seems to be a study of the hydrolysis of 4-substituted 2,6-dimethylbenzoyl chlorides in 99% acetonitrile–water[145]. Perchloric acid is a stronger catalyst than hydrochloric acid, and the common ion effect counteracts the catalysis by the latter acid. No acid catalysis was detected in the hydrolysis of unsubstituted benzoyl chloride under the same conditions.

12. Reactions with hydroxide, alkoxide and phenoxide ions

The effect of added alkali on the rates of solvolysis of acyl chlorides has been studied in a few cases. A pronounced acceleration of the rate by hydroxide (or alkoxide) ion is generally considered to point to a bimolecular mechanism. The work of Hudson and co-workers[22, 26, 28], for example, shows that the effect of hydroxide ion is generally in agreement with other mechanistic criteria although the effect is not always of the expected magnitude[26].

Hydroxide ion (1M OH⁻) effects a hundred-fold increase in the rate of hydrolysis of 2,4,6-trimethylbenzoyl chloride in 95% dioxan–water[200], but does not accelerate the reaction in 50% acetone–water[22]. Obviously the reaction is unimolecular in the latter solvent. The rate of hydrolysis of dimethylcarbamoyl chloride in water does not vary with hydroxide ion concentration, but ethyl chloroformate is very reactive toward the hydroxide ion, in agreement with the fact that the former reaction is unimolecular and the latter bimolecular[108].

Bender and Chen[145] found that tetramethylammonium hydroxide greatly accelerates the hydrolysis of 4-substituted 2,6-dimethylbenzoyl chlorides in 99% acetonitrile–water. The value of ρ for the base-catalysed reaction is + 1·2. The authors concluded that the neutral (and acid-catalysed) and the base-catalysed solvolysis proceed by quite different mechanisms. If this is true, base catalysis may not be a reliable mechanistic criterion.

Peeling[147] found that ethoxide ion increased the rates of alcoholysis of some substituted benzoyl chlorides in an 80% acetone–ethanol mixture, but some of his conclusions have been criticized[31, 146]. In a study of the reactions of ethyl chloroformate with p-substituted phenoxide ions in an 85% acetone–water mixture, Hudson and Loveday[34] obtained the value 0·78 for the constant α of the Brønsted relation.

13. Reactions with phenols

Satchell and co-workers[188, 189, 201] found that the acylations of 2-naphthol and p-methoxyphenol in acetonitrile have significant acid-catalysed components. The reactions can be represented by the scheme (24). When the acylation reagent is chloroacetyl chloride, the reactions exhibit several special features[189].

$$ROCl + HCl \rightleftharpoons RCO^+HCl_2^- \rightleftharpoons RCO^+ + HCl_2^- \quad (fast)$$

$$\downarrow ArOH \qquad\qquad \downarrow ArOH \qquad\qquad \downarrow ArOH \qquad\qquad (slow) \quad (24)$$

Products · Products Products

B. Acyl Fluorides, Bromides and Iodides

Solvolytic reactions of other acyl halides than chlorides have been studied to only a limited extent. Some information is available on the reactions of fluorides[202], but very little on the reactions of acyl bromides and iodides. The usual order of reactivity is $RCOF < RCOCl < RCOBr < RCOI$. The electronegativities of halogens decrease from fluorine to iodine and correspondingly the positive partial charge on the carbonyl carbon atom in acyl halides decreases in the same order. Swain and Scott[203] ascribed the low reactivity of acyl fluorides to the greater strength of the carbon–fluorine bond.

The solvolytic reactions of acyl fluorides differ from the reactions of other acyl halides in that they are catalysed by acids[43, 87, 197, 203, 204, 205]. One explanation[204] is that the leaving groups in the acyl chlorides, bromides and iodides may not form sufficiently stable complexes with protons in aqueous mixtures.

Bunton and Fendler[205] studied the hydrolysis of acetyl fluoride in water. The activation energy was 12·5 kcal/mole and the activation entropy -27 cal/deg mole. Over the pH range 2·80–6·80 the rate of hydrolysis was constant (about 3×10^{-3} s^{-1} at 0·4°C). A similar plateau (neutral hydrolysis) has been observed with carboxylic anhydrides and several aliphatic esters[58].

The similarities observed between acetyl fluoride and acetic anhydride (including also solvent effects) may be explained by assuming similar, i.e. addition–elimination mechanisms. According to Bunton and Fendler, the structure **2** of the transition state is highly polar and stabilized by hydrogen bonding with a water solvent.

$$
\begin{array}{ccc}
\overset{\displaystyle O}{\underset{\displaystyle \delta+ \;\; \|\;\; \delta-}{H_2O\cdots C\cdots X}}
& \quad
\overset{\displaystyle \delta-}{\underset{\displaystyle \delta+ \;\; |\vdots}{\underset{H_2O\cdots C-X}{O}}}
& \quad
\overset{\displaystyle \delta-}{\underset{\displaystyle \delta+ \;\; |\vdots\;\; \delta-}{\underset{H_2O\cdots C\cdots X}{O}}} \\
\textbf{(1)} & \textbf{(2)} & \textbf{(3)}
\end{array}
$$

If the bimolecular hydrolysis of an acyl halide takes place by an S_N-type mechanism, the transition state can be taken to have a structure similar to **1**. The transition state in the addition–elimination mechanism is probably one similar to **2**. In structure **1** the carbon–halogen bond is considerably extended and one would expect that the ratio of the rates of solvolysis of an acyl fluoride and acyl chloride (in the following called 'the fluorine–chlorine ratio') is much less than one[202]. In structure **2** the stretching of the carbon–halogen bond is slight or nil. The fluorine–chlorine ratio may be expected to be much greater than one when the structure is **2** because, owing to the greater positive charge on the carbonyl carbon atom in an acyl fluoride than in an acyl chloride, the repulsion energy when the transition state is formed from the attacking reagent and the acyl halide is smaller in the case of the fluoride than in the case of the chloride. The same applies also to other halogen–halogen ratios.

The majority of the reported halogen–halogen ratios are collected in Table 15. Most values are much smaller than unity, and according to the above discussion these may be interpreted as indicating an S_N-type mechanism. The situation is, however, more complicated if there is an intermediate on the reaction path. The various proposed explanations for the observed fluorine–chlorine ratios have been examined by Parker[202], who has suggested the possibility that a generalized transition state (**3**), involving varying degrees of carbon–halogen bond stretching, may exist.

In the hydroxide ion-catalysed hydrolyses of acyl halides, the fluorine–chlorine ratio is slightly greater than one. This is in agreement with the opinion that the mechanisms of reactions in which strong nucleophiles participate may differ from the mechanism of neutral hydrolysis[145]. The reaction with hydroxide ion as nucleophile would hence take place by an addition–elimination mechanism.

It is probable that a rapid protonation of the fluorine atom (rather than the oxygen atom) takes place in the acid-catalysed reactions of acyl

TABLE 15. Fluorine–chlorine, chlorine–bromine and bromine–iodine replacement ratios for the reactions of acyl halides (a = acetone, d = dioxan, e = ethanol, w = water)

Acyl halide	Reagent	Solvent	T (°C)		Ratio	Reference
CH_3COCl	H_2O	75% a–w	25	F/Cl	$1·3 \times 10^{-4}$	203
CH_3COCl	$2-C_{10}H_7OH$	AcOH	40	F/Cl	$1·4 \times 10^{-4}$	197
C_6H_5COCl	H_2O	75% a–w	25	F/Cl	$1·1 \times 10^{-2}$	203
C_6H_5COCl	H_2O	50% a–w	0	F/Cl	$2·6 \times 10^{-2}$	203
C_6H_5COCl	OH^-	50% a–w	0	F/Cl	$1·4$	203
C_6H_5COCl	H_2O	80% e–w	25	F/Cl	$2·4 \times 10^{-2}$	87
$p-NO_2C_6H_4COCl$	H_2O	80% e–w	25	F/Cl	$2·0 \times 10^{-2}$	87
$p-CH_3C_6H_4COCl$	H_2O	80% e–w	25	F/Cl	$6·2 \times 10^{-2}$	87
CH_3COCl	H_2O	89% a–w	-20	Cl/Br	$3·1 \times 10^{-3}$	66
CH_3COCl	H_2O	AcOH	40	Cl/Br	$3·7 \times 10^{-2}$	197
CH_3COCl	$2-C_{10}H_7OH$	AcOH	40	Cl/Br	$6·7 \times 10^{-2}$	197
CH_3COCl	H_2O	89% a–w	-20	Br/I	5×10^{-2}	66

fluorides and that this is followed by a rate-limiting attack of water upon the conjugate acid (equations 25 and 26)[202, 204, 205]. The fact that hydrogen

$$RC{-}F + H_3O^+ \rightleftharpoons RCFH^+ + H_2O \quad \text{(fast)} \qquad (25)$$

$$RCFH^+ + H_2O \longrightarrow \text{Products} \quad \text{(slow)} \qquad (26)$$

fluoride produced in the hydrolysis of acetyl fluoride does not catalyse the reaction suggests that the reaction is not catalysed by general acids[205].

C. Phosgene and Related Compounds

Phosgene is the only carbonyl halide of the type COX_2 or COXY, where X and Y are halogen atoms, whose rates of solvolysis have been measured (see Tables 2 and 9)[57, 66]. It seems that the hydrolysis of phosgene does not proceed in two clearly distinguishable stages[66].

D. Reaction Mechanisms

The mechanisms of acyl halide solvolysis have been discussed from various aspects by Ugi and Beck[66], Minato[19], Kivinen[18, 20], Queen[21],

Johnson[9], Heydtmann and Stieger[153], Evans[179] and Kevill and Foss[123]. The question from the mechanistic standpoint is whether the reactions are best described by a synchronous S_N-type mechanism or an addition–elimination mechanism (see section I). Since the carbonyl carbon in acyl halides has an appreciable partial positive charge, nucleophilic attack is easier here than at a saturated carbon. In addition, a trigonal carbon atom is sterically less hindered than a tetrahedral saturated carbon. Hence, even when substitution reactions of acyl halides take place by a synchronous mechanism, these can be expected to differ from reactions of alkyl halides.

I. The unimolecular mechanism

The unimolecular hydrolysis (or, more generally, solvolysis) of an acyl halide may be assumed to take place either with the initial formation of an acylium ion in accordance with equation (27) (cf. equation (4) in

$$\underset{\underset{\displaystyle X}{|}}{\overset{\overset{\displaystyle O}{\|}}{RC}} \xrightarrow{\text{slow}} \; RC\overset{+}{=}O \; \xrightarrow[\text{H}_2\text{O}]{\text{fast}} \; \text{Products} \qquad (27)$$

section II, A, 1) or by way of a hydrated tetrahedral intermediate (equation 28). Scheme (28) does not seem plausible, as it presupposes that

$$\overset{\overset{\displaystyle O}{\|}}{RC}{-}X \; + \; H_2O \; \underset{}{\overset{\text{fast}}{\rightleftharpoons}} \; \underset{\underset{\displaystyle OH}{|}}{\overset{\overset{\displaystyle OH}{|}}{RC}}{-}X \; \xrightarrow{\text{slow}} \; \underset{\underset{\displaystyle OH}{|}}{\overset{\overset{\displaystyle OH}{|}}{RC^+}} \; + \; X^- \; \xrightarrow{\text{fast}} \; \text{Products} \; (28)$$

the geminal diol formed is relatively stable, and this is not very likely[21]. The isotope effects observed in the solvolysis of acetyl chloride are also best explained by scheme (27). In addition, the fact that the oxygen exchange in aqueous mixtures is very slow[164] does not fit scheme (28) (see also section II, A, 8).

Minato[19] has proposed a refined addition–ionization mechanism for the solvolysis of acyl chlorides. In this mechanism (29) the rate constants k_1

$$R^1C{-}Cl \; \underset{\underset{\displaystyle \text{fast}}{k_{-1}}}{\overset{\overset{\displaystyle \text{slow}}{k_1,\,R^2\text{OH}}}{\rightleftharpoons}} \; \left[\underset{\underset{\displaystyle O^-}{|}}{\overset{\overset{\displaystyle H\diagdown O \diagup R^2}{|}}{R^1C}}{-}Cl \right] \; \underset{\underset{\displaystyle k_{-2},\,R^2\text{OH}}{}}{\overset{\overset{\displaystyle k_2,\,R^2\text{OH}}{}}{\rightleftharpoons}} \; \underset{\underset{\displaystyle OH}{|}}{\overset{\overset{\displaystyle OR^2}{|}}{R^1C}}{-}Cl \; \xrightarrow{k_3} \; \underset{\underset{\displaystyle OH}{|}}{\overset{\overset{\displaystyle OR^2}{|}}{R^1C^+}} \; + \; Cl^- \; (29)$$

and k_2 are small, and k_{-1} is large. Although the steps to which k_2 and k_{-2} refer could be intramolecular proton transfers, a more plausible process would be the formation of a six-membered transition state incorporating a solvent molecule. In proposing his scheme, Minato[19] greatly stressed the influence of the dielectric constant of the solvent. There are, however, many solvolytic reactions of acyl halides and other substrates whose rates do not vary in proportion to the dielectric constant of the solvent[21]. Minato's mechanism has been criticized because it is difficult to reconcile it with observed substituent[20] and isotope effects[179] and with the pressure dependence of the rates[153]. The best description of the unimolecular mechanism seems to be the simple scheme (27) involving the formation of an acylium ion.

2. The bimolecular mechanism

The bimolecular mechanism of solvolysis of acyl halides may be either a synchronous $S_N 2$-type mechanism (30) or an addition–elimination

$$R^1C(=O)-X \; + \; R^2OH \; \longrightarrow \; \overset{R^2}{\underset{H}{>}}O\overset{\delta+}{\cdots}\overset{\overset{O}{\|}}{\underset{R^1}{C}}\overset{\delta-}{\cdots}X \; \longrightarrow \; \text{Products} \qquad (30)$$

mechanism. Although considerable evidence has been collected in favour of the addition–elimination mechanism, especially for the solvolysis of alkyl chloroformates[21], the kinetic data for the solvolysis of other acyl halides can in most cases equally well be explained by equation (30). It may be that the hydroxylic solvent forms a hydrogen bond with the halogen atom in an acyl halide and thus facilitates the release of the halogen.

The addition–elimination mechanism has been formulated in different ways. The most usual schemes are (31)–(33).

$$R^1C(=O)-X \; + \; R^2OH \; \rightleftharpoons \; \underset{R^2\overset{+}{O}H}{\overset{O^-}{\underset{|}{R^1C}-X}} \; \longrightarrow \; \text{Products} \qquad (31)$$

$$R^1C(=O)-X \; + \; R^2OH \; \rightleftharpoons \; \underset{\underset{(4)}{R^2\overset{+}{O}H}}{\overset{O^-}{\underset{|}{R^1C}-X}} \; \rightleftharpoons \; \underset{\underset{(5)}{R^2O}}{\overset{O^-}{\underset{|}{R^1C}-X}} \; \longrightarrow \; \text{Products} \qquad (32)$$

$$R^1C(=O)-X \; + \; R^2OH \; \rightleftharpoons \; \underset{\underset{(6)}{R^2\overset{+}{O}H}}{\overset{O^-}{\underset{|}{R^1C}-X}} \; \rightleftharpoons \; \underset{R^2O}{\overset{OH}{\underset{|}{R^1C}-X}} \; \longrightarrow \; \text{Products} \qquad (33)$$

The observed rates of carbonyl oxygen exchange in acyl halides in water (see section II, A, 8) fit the formation of a tetrahedral intermediate (6), but there is no unequivocal evidence for its presence.

Kevill and Foss[123] concluded that the methanolysis of *p*-nitrobenzoyl chloride in acetonitrile proceeds by mechanism (32). The intermediate (4) first formed normally reverts to the reactants unless deprotonation leads to a new tetrahedral intermediate (5) which may yield the products. The deprotonation of 5 may be effected by the weakly basic acetonitrile (solvent) molecules or by methanol.

Ross[125] studied the ethanolysis of benzoyl and *p*-nitrobenzoyl chlorides in acetone and chloroform to clarify the significance of hydrogen bonding in the formation of the transition state and concluded that every rate-determining transition state must also contain an acceptor for hydrogen bonding. Ross's explanation for the large increase in the rate of methanolysis of *p*-nitrobenzoyl chloride caused by added chloride ions differs somewhat from that of Kevill and Foss[123].

The kinetic data of Heydtmann and co-workers[153, 154] and Fagley and co-workers[119, 122] for reactions of acyl halides in solvents of relatively higher hydroxylic component content can be explained by a mechanism similar to (30), which does not involve the formation of a tetrahedral intermediate.

Other reported results are in accordance with a synchronous mechanism. Clark[206] compared kinetic data for decarboxylation reactions with those obtained for the ethanolysis of substituted benzoyl chlorides and concluded that the rate-determining steps are probably similar for decarboxylation and solvolysis[18], the mechanism being of the S_N-type for both reactions. The effect of dimethyl sulphoxide on the rates of hydrolysis of alkyl chloroformates is similar to its effect on reactions that are known to proceed by a synchronous mechanism (section II, A, 3).

Kinetic data are not sufficient to decide whether the mechanism of the neutral hydrolysis of acyl halides is a synchronous or an addition–elimination mechanism. The mechanisms may also differ depending on the structure of the acyl halide or the solvent and the reaction may proceed concurrently by both mechanisms.

When an acyl halide reacts with a hydroxide ion or another strong nucleophile (section II, A, 12), the addition–elimination mechanism (34) is more probable.

$$\underset{\substack{\|\\O}}{RC}-X \; + \; OH^- \; \rightleftharpoons \; \underset{\substack{|\\OH}}{\overset{\substack{O^-\\|}}{RC}}-X \; \longrightarrow \; \text{Products} \qquad (34)$$

Structure **7** has been proposed for the transition state in the neutral solvolysis of acyl halides. This would exclude oxygen exchange[62] in solvents of low polarity and would also be in agreement with the opinion[39] that acyl halides react with associated forms of alcohols.

(7)   (8)   (9)

Akiyama and Tokura[131] suggested that the bimolecular solvolysis of acyl chlorides proceeds by an S_N2 mechanism in which the transition state has a cyclic four-centred structure (**8**). In the phenolysis of 1-phenyl-ethyl chloride where displacement occurs at a saturated carbon atom, a transition state similar to **8** has been proposed[207]. If the carbon–chlorine bond in acyl halides is at least partly converted to an ion pair, structure **8** may also fit the ion-pair mechanism of nucleophilic substitution proposed by Sneen and co-workers[95-97]. This hypothesis, when applied[208] to the data of Gold, Hilton and Jefferson[209] for the reactions of benzoyl chloride with water and o-nitroaniline in acetone–water mixtures, was found to give excellent agreement between theory and experiment. The cyclic transition state may also have structure **9** in which the hydrogen bond is formed with the halogen atom instead of the carbonyl carbon atom.

We conclude that more work must be done before a detailed mechanism of solvolysis of acyl halides can be accepted.

III. REACTIONS WITH AMINES

A. Acyl Chlorides

I. General

Most amines are powerful nucleophiles and are more readily acylated than alcohols or water. Sometimes the reactions are violent. The normal order of carbonyl reactivity is observed, acyl halides being more reactive than anhydrides and esters. Substitution reactions at an amino nitrogen have been reviewed by Challis and Butler[210].

Tertiary amines are not usually very reactive towards acyl halides. The formation of acyl quaternary amine salts and their solvolysis will be discussed in section IV. The dehydrochlorination of acyl halides (possessing an α-hydrogen) by tertiary amines to a ketene[211] will not be discussed in this review. In the following, the reactions of acyl halides with primary and secondary amines are examined.

The reactions between acyl halides and amines proceed in general quantitatively in accordance with equations (35) and (36). The formed

$$R^1NH_2 + R^2COCl \longrightarrow R^1NHCOR^2 + HCl \tag{35}$$

$$R^1NH_2 + HCl \longrightarrow R^1NH_3^+Cl^- \tag{36}$$

amine hydrochloride usually precipitates from the solution. In a few cases the hydrogen halide formed does not react with a second amine molecule and the overall reaction is then (35)[212]. With primary amines the second hydrogen atom on the nitrogen may also be displaced by an acyl group.

The conditions where reactions (35) have been studied kinetically have been such that the mechanism can be assumed to be bimolecular, i.e. the existence of an acylium ion in some stage of the reaction is improbable. Acyl halides usually ionize in acid media where amines, in salt form, are poorly reactive or in solvent mixtures containing a hydroxylic component in which solvolysis renders interpretation of the results difficult.

The kinetics of acylation of amines has been studied by Hinshelwood and Venkataraman and co-workers[213-224], Litvinenko and co-workers[225-246] and Vorob'ev and Kuritsyn[247-249]. The majority of the studies have dealt with reactions of substituted anilines with aromatic acyl chlorides and substituent effects, solvent effects and energy–entropy relationships have been examined.

In only a few studies have attempts been made to determine the detailed reaction mechanisms. Whether the reactions (35) take place by an S_N2-like or by the addition–elimination mechanism will be discussed later.

When the kinetics of the benzoylation of anilines is studied, the customary method is to remove the precipitated anilinium hydrochloride by filtration and to titrate the chloride ion in the filtrate by Volhard's method or potentiometrically. The validity of this procedure has been questioned[225, 226]. The reported rate constants for the benzoylation of aniline in benzene at 25°C have varied from 6.00×10^{-2} to 7.48×10^{-2} l/mole s and the values of the activation energy from 6.91 to 7.60 kcal/mole[214, 217, 226, 247, 250-254].

2. Solvent effects

The rate constants of the benzoylation of anilines in individual runs in different solvents usually exhibit a good constancy, but exceptional solvents in this respect are hexane[213], heptane[253], octane[248], decane[248], cyclohexane[253] and carbon tetrachloride[213]. The reason is evidently that a heterogeneous reaction accompanies the homogeneous reaction[213]. The

8

anilinium hydrochloride particles accelerate the reaction, but as they fall
to the bottom of the reaction vessel, the reaction slows down. Vorob'ev[248]
has found that the homogeneous reaction predominates above a certain
critical temperature for each solvent and irregularities are no longer
observed.

The benzoylation of anilines has been studied in numerous solvents[213-220, 247, 249, 250, 253]. Some of the data reported are shown in
Table 16.

TABLE 16. Kinetic data for the reaction of benzoyl chloride
with aniline in different solvents at 25°

Solvent	$10^2 k$ (l/mole s)	E (kcal/mole)	$\log A$	Reference
Carbon tetrachloride	0·80	7·10	3·11	247
p-Xylene	3·99	7·35	4·08	247
Diethyl ether	16·3	6·10	3·7	253
Diethyl ether	16·6	6·40	3·9	249
Anisole	29·6	4·30	2·62	247
Nitrobenzene	103	6·30	4·63	247
Ethyl acetate	196	5·90	4·8	249
Acetone	427	5·75	4·85	247
Dioxan	661	9·40	7·7	249
Acetonitrile	871	6·90	6·0	249

The low activation energy values are attributable to the weak repulsion
between the nitrogen atom of the amino group and the carbonyl carbon
atom of the acyl group[253]. The entropies of activation for the benzoylations
of anilines are usually of the order of -40 to -55 cal/deg mole.

The benzoylation of anilines has been studied also in various mixed
solvents, especially in binary mixtures composed of benzene and a polar
solvent (see Table 17)[220, 253, 254]. A common feature is that the activation
energy decreases 1–3 kcal/mole when 10–20 per cent of the polar solvent is
added to benzene, but further additions of the polar solvent alter the
activation energy only slightly.

The data for the acylation of amines show that, as in the solvolysis of
acyl halides, dielectric effects are not of prime importance. Thus, for
example, the rates of acylation of anilines are higher in phenyl cyanide–
benzene mixtures than in nitrobenzene–benzene mixtures, although the
dielectric constants of the former mixtures are lower[220].

Kondo and Tokura[255, 256] derived an equation for the variation of the
rate constant in binary mixed solvents. An expression for the activation

energy is obtained by differentiation of this equation. The agreement between calculated and experimental[220, 257] data for the benzoylation of p-nitroaniline is relatively good.

TABLE 17. Kinetic data for the reaction of benzoyl chloride with m-nitroaniline in benzene–diethyl ether mixtures at 25° [253]

Ether (wt. %)	$10^4 k$ (l/mole s)	E (kcal/mole)	log A	ΔS^* (cal/deg mole)
0	3·78	10·20	4·061	− 41·9
8·2	8·66	8·06	2·837	− 47·5
16·6	11·0	8·00	2·911	− 47·2
34·7	13·5	8·08	3·056	− 46·5
54·4	13·0	8·59	3·410	− 44·9
75·9	11·6	8·90	3·587	− 44·1
100	10·1	9·21	3·756	− 43·3

3. Substituent effects

Most of the investigations on the acylation of amines have dealt with substituent effects. Since they clarify the mechanism to a limited extent only, they will be discussed briefly.

In reactions between benzoyl chlorides and anilines, the effect of a substituent in the benzene ring of either reactant is exerted almost entirely on the activation energy, whereas the entropy of activation remains almost constant. Electron-repelling substituents in the acyl chloride increase and electron-attracting substituents decrease the activation energy. These effects are reversed when the substituents are in the aniline[219, 253].

In order to compare the effects of various substituents, Stubbs and Hinshelwood[217] used the determined frequency factor (A) for aniline to calculate corrected activation energies E' for the various substituted derivatives. Generally the experimentally determined values of the activation energy E and the corrected values E' are quite similar. A measure of the effect of a substituent on the reaction is the difference $\Delta E' = E'_X - E'_H$, where E'_X and E'_H are the values for the substituted aniline and aniline, respectively.

The additivity of substituent effects in reactions between benzoyl chlorides and anilines has been studied[217, 218, 223, 224, 253] by comparing the values of E' for disubstituted compounds with the sums of the E' values for monosubstituted compounds. The additivity is good for disubstituted anilines, the differences between the experimental and calculated activation energy values being generally less than ± 200 cal/mole. The differences are slightly greater for disubstituted benzoyl chlorides.

The substituent effects depend slightly on the solvent. The additivity is good for reactions in benzene, but larger deviations are observed for reactions in nitrobenzene, for example. The latter deviations are possibly partly due to experimental inaccuracy (see section III, A, 2).

The acylation of aromatic amines other than anilines has also been studied, mainly in benzene[227-233, 258]. Although many of these amines have had complex structures, the kinetic parameters have values closely similar to those for the acylation of anilines. The values of log A vary from 2 to 6, but are in most cases about 4; the activation energies vary from 4·5 to 11·5 kcal/mole.

Hall[259] measured the rates of reaction of dimethylcarbamoyl chloride with amines of various types in ethanol. The data satisfy the Swain–Scott equation[86] with s values close to 2·0, which implies a strong dependence on the nucleophilicity of the amine.

The data in Table 18 show that the kinetic parameters for other acylating agents are similar in magnitude to those for the reaction of benzoyl chloride.

TABLE 18. Kinetic data for the acylation of
m-nitroaniline in benzene at 25° [219]

Acyl chloride	10^2k (l/mole s)	E (kcal/mole)	log A
Benzoyl	0·0324[a]	10·55	4·378
Hexanoyl	0·687	8·00	3·784
Acetyl	1·231	7·70	3·816
Phenylacetyl	2·079	8·80	4·860
Trichloroacetyl	27·69	5·70	3·682
Chloroacetyl	41·64	6·50	4·445

[a] Calculated from the Arrhenius equation.

One may summarize the data by stating that the values of log A, and hence those of the entropy of activation, are only slightly influenced by the structures of the amine and acyl chloride. This may be taken to indicate that the reaction mechanism is the same in all cases.

4. Isotope effects

Very little information is available on isotope effects in the acylation of amines. Elliott and Mason[252] obtained the values 0·06, 0·07, 0·065 and

0.074 l/mole s for the rate constants of the benzoylation of aniline, N,N-dideuteroaniline, 2,4,6-trideuteroaniline and N,N-2,4,6-pentadeutero-aniline, respectively, in benzene at 25°. Hence, breaking of a nitrogen–hydrogen bond cannot be rate-determining; it this were so, the N-deuter-ated anilines would react more slowly than the protium analogues.

5. Acid catalysis

Litvinenko studied the catalysis of the benzoylation of amines in benzene by carboxylic acids[235, 236, 239-245, 260] and amides[261] and also bifunctional catalysis[262]. Some of his data are shown in Table 19.

TABLE 19. Second-order rate constants for the uncatalysed (k_0) and acid-catalysed (k_{cat}) reactions between benzoyl chloride and substituted anilines in benzene at 25°

Substituent	Acid	$10^2 k_0$ (l/mole s)	$10^2 k_{cat}$ (l/mole s)	Reference
H	acetic	6·99	1350	240
p-Cl	acetic	1·23	352	240
m-Cl	acetic	0·46	101	240
p-NO$_2$	acetic	0·0042	1·2	240
H	benzoic	7·0	2230	235
m-Cl	benzoic	—	388	243

The overall rate k_{obs} of a reaction catalysed by a carboxylic acid can be represented by equation (37) where k_0 and k_{cat} are the rate constants

$$k_{obs} = k_0 + k_{cat} m \alpha \qquad (37)$$

of the uncatalysed and catalysed reactions, respectively, m is the concentration of the catalysing acid and α its degree of dissociation. Only the monomeric form of a carboxylic acid is catalytically active[235, 260]. The rate constants of the catalysed reactions obey the Arrhenius equation.

According to Litvinenko[243-245], the catalysis involves the formation of a cyclic transition state (**10** or **11**).

(10)   (11)

The decrease observed in the rate of the reaction at high acid concentration is evidently due to the formation of a salt by the catalyst acid and the amine. The catalytic effects of carboxylic acids increase with increasing acid strength in the order acetic < benzoic < monochloroacetic < trichloroacetic acid, although this order applies to trichloroacetic acid only at low concentrations[262].

Values of the activation parameters have been reported for only a few catalysed reactions. The values of the activation energy for the uncatalysed benzoylation of p-nitroaniline and the reaction catalysed by acetic acid are 11·8 and 5·7 kcal/mole and the values of log A 4·29 and 2·25, respectively. The low activation energy of the catalysed reaction can be attributed to the formation of a cyclic transition state[240].

6. Effect of basicity of amines

Bose and Hinshelwood[218] found that for the reactions of benzoyl chlorides with various amines in benzene, nitrobenzene and chlorobenzene the relationship between the logarithm of the catalytic rate constant k_B and the basicity constant K_B (in water) of the amine is approximately linear. The value of β in the Brønsted relation[263, 264] (38) derived from their data

$$\log k_B = -\beta \log K_B + \text{constant} \tag{38}$$

is one. This implies that the basicity of the amine determines the rate of the reaction.

B. Acyl Fluorides, Bromides and Iodides

Kinetic studies have been conducted on only a few acylations of amines by acyl halides other than chlorides. The rate increases with the strength of the carbon–halogen bond in the order F < Cl < Br < I. It seems that the size of the halogen has a greater effect in the reactions of acyl halides with aromatic amines than in their solvolytic reactions. The data in Table 20 show that the halogen–halogen rate ratios are of the same magnitude in the reactions of acyl halides with amines as in their solvolytic reactions (section II, B, Table 15). Bender and Jones[265] explained the great difference between the rates of reaction of benzoyl fluoride and benzoyl chloride with morpholine in cyclohexane by assuming that the electrophilic catalysis of the removal of the fluoride ion which occurs in aqueous solution cannot occur in cyclohexane. Tables 15 and 20 show, however, that the fluorine–chlorine reactivity ratio is almost equal in the hydrolysis and in the reaction with morpholine. An alternative explanation for the low fluorine–chlorine ratio is that proposed for the solvolytic reactions in section II, B.

TABLE 20. Fluorine–chlorine, chlorine–bromine and bromine–iodine replacement ratios for the reactions of acyl halides with anilines and morpholine at 25°

Acyl halide	Amine	Solvent	Ratio		Reference
Benzoyl	morpholine	C_6H_{12}	F/Cl	$4\cdot0 \times 10^{-4}$	265
Acetyl	p-$NO_2C_6H_4NH_2$	C_6H_6	Cl/Br	$1\cdot3 \times 10^{-2}$	219
i-Butyryl	m-$ClC_6H_4NH_2$	C_6H_6	Cl/Br	$3\cdot0 \times 10^{-2}$	212
Benzoyl	p-$NO_2C_6H_4NH_2$	C_6H_6	Cl/Br	$7\cdot4 \times 10^{-3}$	215, 217, 219
Benzoyl	m-$NO_2C_6H_4NH_2$	C_6H_6	Cl/Br	$1\cdot3 \times 10^{-2}$	215, 217, 219
Benzoyl	o-$NO_2C_6H_4NH_2$	C_6H_6	Cl/Br	$6\cdot7 \times 10^{-3}$	215, 217, 219
Benzoyl	o-$ClC_6H_4NH_2$	C_6H_6	Cl/Br	$7\cdot0 \times 10^{-3}$	217, 219
Benzoyl	morpholine	C_6H_{12}	Cl/Br	$4\cdot0 \times 10^{-2}$	265
Benzoyl	morpholine	C_6H_{12}	Br/I	$2\cdot8 \times 10^{-1}$	265

In the acylation of substituted anilines, the effect of the substituent on the activation energy is almost the same for acyl chlorides as for acyl bromides (Table 21).

TABLE 21. Values of the differences in activation energy in the acylation of aniline (E_H) and substituted anilines (E_X) in benzene[219]

Substituent in aniline	$\Delta E = E_X - E_H$ (kcal/mole)			
	CH_3COCl	CH_3COBr	C_6H_5COCl	C_6H_5COBr
m-NO_2	2·94	—	3·04	3·35
p-NO_2	4·40	4·80	4·43	4·30
o-NO_2	—	5·85	6·00	5·95
o-Cl	—	—	2·98	2·90

Litvinenko and co-workers[260] studied the catalysis of the reactions of benzoyl halides and substituted anilines by carboxylic acids. For the ratio k_{cat}/k_0 in the acylations of p-methoxyaniline by benzoyl fluoride, benzoyl chloride and benzoyl bromide, they reported the values 294,000, 141 and 31, respectively. These values suggest that the transition state involves hydrogen-bond formation between the catalysing carboxylic acid and the halogen of the acyl halide.

C. Reaction Mechanisms

As for the solvolytic reactions, the reactions with amines can be described by a synchronous or by an addition–elimination mechanism. Most investigators prefer the former in which the transition state (12) is that proposed by Venkataraman and Hinshelwood[219]. The constancy of the activation entropy especially fits structure 12 since when the highly polar

(12)   (13)   (14)

reaction centre is the same, the substituents, be they in the aniline or in the benzoyl chloride, are far from it.

Structure 13 was proposed by Elliott and Mason[252] on the basis of isotope effects (section III, A, 4).

Mather and Shorter[251] presented the schematic structure 14 for the transition state in the reactions of mono- and disubstituted benzoyl chlorides with aniline in benzene. The two benzene rings lie parallel in this structure and have little freedom of rotation.

Bender and Jones[265], on the other hand, have concluded that amines are acylated by the addition–elimination mechanism (39) and have explained

the low fluorine–chlorine ratio in the reactions of benzoyl halides with morpholine in cyclohexane (Table 20) by assuming that the ratio k_2/k_3 is high and variable and that there are differences in the resonance stabilization of the acylating reactants.

Tokura and co-workers have successfully applied their theory[255] of reaction rates in binary mixed solvents to the kinetic data for the benzoylation of m-nitroaniline in benzene–nitrobenzene mixtures[220] and their electrostriction theory[266] to the data for the reaction between m-nitroaniline and benzoyl chloride in various solvents[216]. These data are in better accord with the synchronous than with the addition–elimination mechanism.

Although a mechanism similar to (39) seems more plausible for reactions between strongly nucleophilic amines and acyl halides, most experimental data can be more simply explained by an S_N2-type mechanism in which the transition state has one of the structures 12–14. More experimental work is required before a final decision can be made between these two alternative mechanisms.

IV. SOLVOLYTIC REACTIONS CATALYSED BY TERTIARY AMINES

A. General

Tertiary amines strongly catalyse solvolytic reactions of acyl halides[8]. Pyridine, for example, increases the rates of hydrolysis of acetyl and benzoyl chlorides in 89·1% acetone–water more than 10^5-fold[66]. The reaction catalysed by an amine may be thought[267] to occur so that the acyl halide reacts with a hydrogen-bonded alcohol–amine or water–amine complex (equation 40) or forms first an acyl quaternary amine salt which is decomposed by the solvolytic agent (equation 41).

$$R^1COX + R^2OH + R^3_3N \longrightarrow R^1COX + R^2OH \cdots R^3_3N \longrightarrow \text{Products}$$
(40)

$$R^1COX + R^2OH + R^3_3N \longrightarrow R^1CON\overset{+}{R^3_3}, X^- + R^2OH \longrightarrow \text{Products}$$
(41)

Owing to the very high rates of the amine-catalysed reactions, very little information is available about the mechanisms. It is, for instance, difficult to estimate the relative contributions of reactions (40) and (41)[267]. The acyl quaternary amine halide that is an intermediate in reaction (41) can be prepared by allowing the acyl halide and the tertiary amine to react with each other directly or in an inert solvent.

B. Quaternarization of Acyl Halides

The reactions of acid chlorides with tertiary amines have been reviewed by Leduc and Chabrier[268]. The quaternary acyl halide adducts are mostly unstable and hygroscopic. The N,N-dialkylcarbamoylpyridinium chlorides are, however, less reactive and their solvolytic behaviour has been studied to some extent[269-271]. Spectral evidence has been collected for the existence of the ion $RC \equiv O^+$ in adducts formed by acetyl and benzoyl chlorides with pyridine, 3-picoline and 4-picoline[272]. On the other hand, there is a strong carbonyl absorption band in the spectra of dialkylcarbamoyl chlorides[269] and acetylpyridinium chloride[273]. It is also possible that the ionic character varies from one adduct to another.

1-(N,N-Dimethylcarbamoyl)pyridinium chloride decomposes in non-hydroxylic solvents according to equation (42). The equilibrium constant

$$(CH_3)_2\overset{+}{N}COPyr, Cl^- \rightleftharpoons (CH_3)_2NCOCl + Pyr$$
(42)

of this reaction varies greatly with the initial concentration of dimethyl-carbamoyl chloride[269]. In hydroxylic solvents, the adduct is stabilized by hydrogen bonding[269]. The ease with which dimethylcarbamoyl chloride forms complexes with amines diminishes in the order[274]: pyridine >

N-methylpiperidine $\gg$ triethylamine $> N,N$-dimethylaniline. The catalytic effects of tertiary amines on the hydrolysis of acyl chlorides in 89·1% acetone–water decrease in the order: pyridine $>$ trimethylamine $>$ triethylamine $>$ collidine[66]. 2,6-Lutidine, 2,4,6-collidine and 2,6-dimethoxypyridine do not react with dimethylcarbamoyl chloride when equimolar amounts of amine and the acyl chloride are mixed[269]. No information is available on the corresponding reactions of carboxylic acid chlorides and tertiary amines.

Amide–hydrogen halide adducts can be prepared from acyl halides and secondary amines under carefully controlled conditions[275].

The activation energies and values of log A are usually low, and hence the activation entropies strongly negative, when quaternary ammonium salts are formed from alkyl halides[276]. The kinetic parameters are thus of the same magnitude as those for reactions of acyl halides with primary and secondary amines (see section III, A). The quaternarization of acyl halides does not seem to have been studied by kinetic methods. It is possible that the quaternary salt formation proceeds by way of a tetrahedral intermediate[277, 278].

C. Aliphatic Amines as Catalysts

The available information on mechanistic details of solvolytic reactions of acyl halides catalysed by tertiary aliphatic amines is meagre. An unexpected shift of a double bond from the α,β-position to the β,γ-position has been observed when crotonoyl chloride undergoes alcoholysis in the presence of a tertiary amine[279]. It is possible that an acyl quaternary ammonium salt is formed as an intermediate[279, 280].

Ketenes are prepared by allowing acyl halides containing an α-hydrogen atom to react with tertiary amines[211]. Truce and Bailey[277, 278] have presented evidence which shows that the alcoholysis of acyl halides proceeds by two routes when a tertiary amine is present. One route is an elimination–addition process where a ketene is an intermediate and the other route is a substitution reaction where an acyl quaternary ammonium salt is first formed. The reactions can be represented by equations (43).

$$R_2CHCOX + Et_3N \begin{array}{c} \xrightarrow{B} R_2C{=}C{=}O + Et_3NHX \xrightarrow{MeOD} R_2CDCO_2Me \\ \Big\updownarrow \\ \xrightarrow{A} R_2CHCO\overset{+}{N}Et_3, X^- \xrightarrow{MeOD} R_2CHCO_2Me \end{array} \quad (43)$$

The existence of the ketene as intermediate has been confirmed by carrying out the alcoholysis in the presence of methanol-d. It is interesting to note that the intermediates are interconvertible. The contribution of the substitution reaction increases in the order: acetyl chloride < acetyl bromide < acetyl iodide.

D. Aromatic Amines as Catalysts

The formation of acyl quaternary amine halides as intermediates in the solvolytic reactions of acyl halides in the presence of amines is very probable. Recently an acetylpyridinium ion has been established as an intermediate in a pyridine-catalysed acyl transfer, i.e. in the hydrolysis of acetic anhydride catalysed by pyridine[281].

Johnson and Rumon[270] have studied the nucleophilic reactions of 1-(N,N-dimethylcarbamoyl)pyridinium chloride in water. The neutral solvolytic reaction is a direct nucleophilic reaction, as shown by the moderately large kinetic solvent isotope effect (2·26) and the low value of the activation entropy (−27 cal/deg mole). These values differ greatly from the values for the neutral hydrolysis of dimethylcarbamoyl chloride which is a unimolecular reaction[108].

V. COMPARISON OF REACTIVITIES OF COF, COCl, COBr AND COI GROUPS

Very little can yet be said about any essential differences in reaction mechanisms when the halogen atom in an acyl halide changes from fluorine through chlorine and bromine to iodine. In the case of all reactions discussed above the reactivities increase regularly in the sequence F < Cl < Br < I (see sections II, B and III, B).

The reaction mechanisms are probably the same when the halogen is chlorine, bromine or iodine. The high electronegativity and small size of fluorine compared to the other halogens gives cause to assume that if there are differences in reaction mechanism, they should become evident when reactions of acyl fluorides are compared with reactions of other acyl halides. In fact, the solvolytic reactions of acyl fluorides exhibit a number of special features (section II, B), but kinetic evidence is still too meagre to permit one to decide whether only differences in degree are in question or whether acyl fluorides react by different mechanisms than the other acyl halides.

VI. ACKNOWLEDGEMENT

The author wishes to express his deep gratitude to Miss Ritva Takala, Ph.M., for her valuable assistance in the preparation of this article.

VII. REFERENCES

1. N. O. V. Sonntag, *Chem. Rev.*, **52**, 237 (1953).
2. *Friedel–Crafts and Related Reactions*, Vol. 3 (Ed. G. A. Olah), Interscience, London, 1964.
3. M. Matzner, R. P. Kurkjy and R. J. Cotter, *Chem. Rev.*, **64**, 645 (1964).
4. P. F. G. Praill, *Acylation Reactions*, Pergamon, London, 1963.
5. H. O. House, *Modern Synthetic Reactions*, Benjamin, New York, 1965, Chap. 9.
6. R. F. Hudson, *Chimia*, **15**, 394 (1961).
7. D. P. N. Satchell, *Quart. Rev. Chem. Soc.*, **17**, 160 (1963).
8. M. L. Bender, *Chem. Rev.*, **60**, 53 (1960).
9. S. L. Johnson in *Advances in Physical Organic Chemistry*, Vol. 5 (Ed. V. Gold), Academic, London, 1967, p. 237.
10. D. P. N. Satchell and R. S. Satchell in *The Chemistry of Carboxylic Acids and Esters* (Ed. S. Patai), Interscience, London, 1969, Chap. 9.
11. T. C. Bruice and S. Benkovic, *Bioorganic Mechanisms*, Vol. 1, Benjamin, New York, 1966.
12. S. L. Johnson, *J. Am. Chem. Soc.*, **86**, 3819 (1964).
13. A. Streitwieser, Jr., *Solvolytic Displacement Reactions*, McGraw–Hill, New York, 1962.
14. C. A. Bunton, *Nucleophilic Substitution at a Saturated Carbon Atom*, Elsevier, Amsterdam, 1963.
15. E. R. Thornton, *Solvolysis Mechanisms*, The Ronald Press, New York, 1964.
16. E. S. Amis, *Solvent Effects on Reaction Rates and Mechanisms*, Academic, New York, 1966.
17. C. K. Ingold, *Structure and Mechanism in Organic Chemistry*, 2nd ed., Cornell University Press, Ithaca, 1969.
18. A. Kivinen, *Ann. Acad. Sci. Fenn. Ser. A2*, No. 108 (1961).
19. H. Minato, *Bull. Chem. Soc. Japan*, **37**, 316 (1964).
20. A. Kivinen, *Acta Chem. Scand.*, **19**, 845 (1965).
21. A. Queen, *Can. J. Chem.*, **45**, 1619 (1967).
22. R. F. Hudson and J. E. Wardill, *J. Chem. Soc.*, 1729 (1950).
23. B. L. Archer and R. F. Hudson, *J. Chem. Soc.*, 3259 (1950).
24. D. A. Brown and R. F. Hudson, *J. Chem. Soc.*, 883 (1953).
25. B. L. Archer, R. F. Hudson and J. E. Wardill, *J. Chem. Soc.*, 888 (1953).
26. D. A. Brown and R. F. Hudson, *J. Chem. Soc.*, 3352 (1953).
27. E. W. Crunden and R. F. Hudson, *J. Chem. Soc.*, 501 (1956).
28. E. W. Crunden and R. F. Hudson, *J. Chem. Soc.*, 3748 (1961).
29. M. Green and R. F. Hudson, *J. Chem. Soc.*, 1076 (1962).
30. R. F. Hudson and G. E. Moss, *J. Chem. Soc.*, 5157 (1962).
31. R. F. Hudson and G. E. Moss, *J. Chem. Soc.*, 2982 (1964).
32. R. F. Hudson and M. Green, *J. Chem. Soc.*, 1055 (1962).
33. R. F. Hudson and G. Klopman, *J. Chem. Soc.*, 1062 (1962).
34. R. F. Hudson and G. Loveday, *J. Chem. Soc.*, 1068 (1962).
35. R. F. Hudson and B. Saville, *J. Chem. Soc.*, 4114 (1955).
36. R. F. Hudson and B. Saville, *J. Chem. Soc.*, 4121 (1955).
37. R. F. Hudson and B. Saville, *J. Chem. Soc.*, 4130 (1955).

38. R. F. Hudson, *J. Chem. Soc. B*, 761 (1966).
39. R. F. Hudson and G. W. Loveday, *J. Chem. Soc. B*, 766 (1966).
40. R. F. Hudson, G. W. Loveday, S. Fliszar and G. Salvadori, *J. Chem. Soc. B*, 769 (1966).
41. R. F. Hudson and I. Stelzer, *J. Chem. Soc. B*, 775 (1966).
42. D. Brown and R. F. Hudson, *Nature*, **167**, 819 (1951).
43. C. W. Bevan and R. F. Hudson, *J. Chem. Soc.*, 2187 (1953).
44. R. F. Hudson and I. Stelzer, *Trans Faraday Soc.*, **54**, 213 (1958).
45. M. Green and R. F. Hudson, *Proc. Chem. Soc. London*, 149 (1959).
46. R. F. Hudson, *Chimia*, **16**, 173 (1962).
47. R. F. Hudson, *Ber. Bunsenges. Phys. Chem.*, **68**, 215 (1964).
48. G. Berger and S. C. J. Olivier, *Rec. Trav. Chim. Pays-Bas*, **46**, 516 (1929).
49. S. C. J. Olivier, *Rec. Trav. Chim. Pays-Bas*, **48**, 227 (1929).
50. S. C. J. Olivier, *Chem. Weekbl.*, **26**, 521 (1929).
51. A. A. Ashdown, *J. Am. Chem. Soc.*, **52**, 268 (1930).
52. G. E. K. Branch and A. C. Nixon, *J. Am. Chem. Soc.*, **58**, 2499 (1936).
53. J. F. Norris, E. V. Fasce and C. J. Staud, *J. Am. Chem. Soc.*, **57**, 1415 (1935).
54. J. F. Norris and H. H. Young, Jr., *J. Am. Chem. Soc.*, **57**, 1420 (1935).
55. J. F. Norris and V. W. Ware, *J. Am. Chem. Soc.*, **61**, 1418 (1939).
56. J. F. Norris and E. C. Haines, *J. Am. Chem. Soc.*, **57**, 1425 (1935).
57. R. Leimu, *Ann. Univ. Turku. Ser. A*4, No. 3 (1935); *Ber.*, **70**, 1040 (1937).
58. E. K. Euranto in *The Chemistry of Carboxylic Acids and Esters* (Ed. S. Patai), Interscience, London, 1969, Chap. 11.
59. H. Böhme and W. Schürhoff, *Chem. Ber.*, **84**, 28 (1951).
60. G. Zimmerman and C. Yuan, *J. Am. Chem. Soc.*, **77**, 332 (1955).
61. M. J. Kelly and G. M. Watson, *J. Phys. Chem.*, **62**, 260 (1960).
62. M. L. Bender and R. D. Ginger, *Suom. Kemistilehti B*, **33**, 25 (1960).
63. J. Koskikallio, *Suom. Kemistilehti B*, **33**, 107 (1960).
64. E. J. Cairns and J. M. Prausnitz, *J. Chem. Phys.*, **32**, 169 (1960).
65. S. L. Johnson, quoted in reference 9, p. 315.
66. I. Ugi and F. Beck, *Chem. Ber.*, **94**, 1839 (1961).
67. S. G. Entelis, R. P. Tiger, E. Ya. Nevelskii and I. V. Epelbaum, *Izv. Akad. Nauk SSSR, Otd. Khim. Nauk*, 245 (1963); *Chem. Abstr.*, **58**, 13746a (1963).
68. S. G. Entelis, R. P. Tiger, E. Ya. Nevelskii and I. V. Epelbaum, *Izv. Akad. Nauk SSSR, Otd. Khim. Nauk*, 429 (1963); *Chem. Abstr.*, **59**, 5830e (1963).
69. R. P. Tiger, E. Ya. Nevelskii, I. V. Epelbaum and S. G. Entelis, *Izv. Akad. Nauk SSSR, Ser. Khim.*, 1969 (1964); *Chem. Abstr.*, **62**, 11650b (1965).
70. P. Frölich, W. Köhler and R. Radeglia, *Z. Chem.*, **8**, 467 (1968).
71. R. E. Robertson in *Progress in Physical Organic Chemistry*, Vol. 4 (Ed. A. Streitwieser, Jr. and R. W. Taft), Interscience, New York, 1967, p. 213.
72. G. Kohnstam in *Advances in Physical Organic Chemistry*, Vol. 5 (Ed. V. Gold), Academic, London, 1967, p. 121.
73. D. Eisenberg and W. Kauzmann, *The Structure and Properties of Water*, Oxford University Press, Oxford, 1969.
74. P. A. H. Wyatt, *Annu. Rep. Chem. Soc. A*, **66**, 93 (1969), and the references cited therein.
75. A. K. Covington and P. Jones, *Hydrogen-bonded Solvent Systems*, Taylor and Francis, London, 1968.

224 Antti Kivinen

76. E. M. Arnett and D. R. McKelvey in *Solute–Solvent Interactions* (Ed. J. F. Coetzee and C. D. Ritchie), Marcel Dekker, New York, 1969, p. 344.
77. A. J. Parker, *Chem. Rev.*, **69**, 1 (1969).
78. C. D. Ritchie in *Solute–Solvent Interactions* (Ed. J. F. Coetzee and C. D. Ritchie), Marcel Dekker, New York, 1969, p. 219.
79. H. Schneider in *Solute–Solvent Interactions* (Ed. J. F. Coetzee and C. D. Ritchie), Marcel Dekker, New York, 1969, p. 301.
80. E. M. Arnett, *J. Am. Chem. Soc.*, **87**, 151 (1965).
81. P. Haberfield, A. Nudelman, A. Bloom, H. Guizberg and P. Steinhertz, *Chem. Commun.*, 194 (1968).
82. P. Haberfield, L. Clayman and J. S. Cooper, *J. Am. Chem. Soc.*, **91**, 787 (1969).
83. C. Reichardt, *Lösungsmittel-Efekte in der organischen Chemie*, Verlag Chemie, Weinham, 1969; C. Reichardt, *Angew. Chem. Int. Ed. Engl.*, **4**, 29 (1965).
84. M. R. J. Dach, *Chem. Brit.*, **6**, 347 (1970).
85. C. G. Swain, *J. Am. Chem. Soc.*, **70**, 1119 (1948).
86. C. G. Swain and C. B. Scott, *J. Am. Chem. Soc.*, **75**, 141 (1935).
87. C. G. Swain, R. B. Mosely and D. E. Brown, *J. Am. Chem. Soc.*, **77**, 3731 (1955).
88. S. Brownstein, *Can. J. Chem.*, **38**, 1590 (1960).
89. E. Grunwald and S. Winstein, *J. Am. Chem. Soc.*, **70**, 846 (1948).
90. A. H. Fainberg and S. Winstein, *J. Am. Chem. Soc.*, **78**, 2770 (1956), and earlier papers.
91. E. Tommila, E. Paakkala, U. K. Virtanen, A. Erva and S. Varila, *Ann. Acad. Sci. Fenn.*, Ser. *A*2, No. 91 (1959).
92. E. Tommila, *Acta Chem. Scand.*, **20**, 923 (1966), and references cited therein.
93. K. J. Laidler and R. Martin, *Intern. J. Chem. Kinetics*, **1**, 113 (1969).
94. H. Weiner and R. A. Sneen, *J. Am. Chem. Soc.*, **87**, 287 (1965).
95. H. Weiner and R. A. Sneen, *J. Am. Chem. Soc.*, **87**, 292 (1965).
96. R. A. Sneen and J. W. Larsen, *J. Am. Chem. Soc.*, **91**, 362 (1969).
97. R. A. Sneen and F. R. Rolle, *J. Am. Chem. Soc.*, **91**, 2140 (1969).
98. E. P. Grimsrud and J. W. Taylor, *J. Am. Chem. Soc.*, **92**, 739 (1970).
99. G. G. Hammes and W. Knoche, *J. Chem. Phys.*, **45**, 4041 (1966).
100. L. A. Dunn and W. L. Marshall, *J. Phys. Chem.*, **73**, 2619 (1969); W. L. Marshall, *J. Phys. Chem.*, **74**, 346 (1970).
101. A. Kivinen and S. Koskela, unpublished results.
102. P. O. I. Virtanen, *Suom. Kemistilehti B*, **38**, 135 (1965).
103. A. Kivinen, unpublished results.
104. E. Tommila, *Ann. Acad. Sci. Fenn.*, Ser. *A*2, No. 139 (1967).
105. J. Murto and A. M. Hiiro, *Suom. Kemistilehti B*, **37**, 177 (1964).
106. J. Murto, *Suom. Kemistilehti B*, **34**, 92 (1961).
107. K. Heinonen, *State Inst. Techn. Res. Finland Publ.*, No. 116 (1967); *Chem. Abstr.*, **67**, 76677c (1967).
108. H. K. Hall, Jr., *J. Am. Chem. Soc.*, **77**, 5993 (1955).
109. H. K. Hall, Jr. and C. H. Lueck, *J. Org. Chem.*, **28**, 2818 (1963).
110. A. Queen, T. A. Nour, M. N. Paddon-Row and K. Preston, *Can. J. Chem.*, **48**, 522 (1970).

111. C. H. Langford and T. R. Stengle, *J. Am. Chem. Soc.*, **91**, 4014 (1969), and references cited therein.
112. A. Kivinen, *Suom. Kemistilehti B*, **38**, 209 (1965).
113. E. K. Euranto and R. S. Leimu, *Acta Chem. Scand.*, **20**, 2028 (1966).
114. M. Cocivera, Ph.D. Thesis, University of California; *Diss. Abstr.*, **24**, 3110 (1964).
115. J. W. Sims, Ph.D. Thesis, Tulane University; *Diss. Abstr.*, **25**, 135 (1964).
116. G. L. Bertrand, Ph.D. Thesis, Tulane University; *Diss. Abstr.*, **26**, 5579 (1965).
117. J. J. Rathmell, Ph.D. Thesis, Tulane University; *Diss. Abstr.*, **26**, 5590 (1965).
118. G. A. Von Bodungen, Ph.D. Thesis, Tulane University; *Diss. Abstr. B*, **27**, 1448 (1966).
119. T. F. Fagley, G. A. Von Bodungen, J. J. Rathmell and J. D. Hutchinson, *J. Phys. Chem.*, **71**, 1374 (1967).
120. J. D. Hutchinson, Ph.D. Thesis, Tulane University; *Diss. Abstr.*, *B*, **27**, 3482 (1967).
121. J. S. Bullock, Ph.D. Thesis, Tulane University; *Diss. Abstr. B*, **30**, 2628 (1969).
122. T. F. Fagley, J. S. Bullock and D. W. Dycus, *J. Phys. Chem.*, **74**, 1840 (1970).
123. D. N. Kevill and F. D. Foss, *J. Am. Chem. Soc.*, **91**, 5054 (1969).
124. P. D. Bartlett and H. Minato, *J. Am. Chem. Soc.*, **85**, 1858 (1963).
125. S. D. Ross, *J. Am. Chem. Soc.*, **92**, 5998 (1970).
126. J. Cason and K. W. Kraus, *J. Org. Chem.*, **26**, 2624 (1961).
127. J. C. Biordi, *J. Chem. Eng. Data*, **15**, 166 (1970).
128. G. Geuskens, M. Planchon, J. Nasielski and R. H. Martin, *Helv. Chim. Acta*, **42**, 522 (1959).
129. I. G. Murculescu and I. Demetrescu, *Rev. Roum. Chim.*, **14**, 149 (1969).
130. F. Akiyama and N. Tokura, *Bull. Chem. Soc. Japan*, **39**, 131 (1966).
131. F. Akiyama and N. Tokura, *Bull. Chem. Soc. Japan*, **41**, 2690 (1968).
132. A. Kivinen, unpublished results.
133. R. Michelot and B. Tchoubar, *Bull. Soc. Chim. France*, 3039 (1966).
134. R. W. Taft, Jr. in *Steric Effects in Organic Chemistry* (Ed. M. S. Newman), Wiley, London, 1956, Chap. 13.
135. R. W. Taft, Jr., *J. Phys. Chem.*, **64**, 1805 (1960), and earlier papers.
136. J. Shorter, *Quart. Rev. Chem. Soc.*, **24**, 433 (1970).
137. J. Shorter, *Chem. Brit.*, **5**, 269 (1969).
138. C. G. Swain and E. C. Lupton, Jr., *J. Am. Chem. Soc.*, **90**, 4328 (1968).
139. M. Charton, *J. Am. Chem. Soc.*, **91**, 6649 (1969).
140. M. Charton in *Progress in Physical Organic Chemistry*, Vol. 8 (Ed. A. Streitwieser, Jr. and R. W. Taft), Interscience, New York, 1970, p. 235.
141. C. G. Swain and D. C. Dittmer, *J. Am. Chem. Soc.*, **75**, 4627 (1953).
142. L. P. Hammett, *Physical Organic Chemistry*, 2nd ed., McGraw–Hill, New York, 1970, Chap. 11.
143. C. Laurence and B. Wojtkowiak, *Ann. Chim. Paris*, **5**, 163; 191 (1970).
144. P. W. Heidbuchel, *Bull. Soc. Chim. Belg.*, **77**, 149 (1968).
145. M. L. Bender and M. C. Chen, *J. Am. Chem. Soc.*, **85**, 30 (1963).
146. A. Kivinen, *Suom. Kemistilehti B*, **36**, 163 (1963).

226 Antti Kivinen

147. E. R. A. Peeling, *J. Chem. Soc.*, 2307 (1959).
148. H. S. Golinkin, D. M. Parbhoo and R. E. Robertson, *Can. J. Chem.*, **48**, 1296 (1970), and earlier papers.
149. J. R. Hulett, *Quart. Rev. Chem. Soc.*, **18**, 227 (1964).
150. M. G. Evans and M. Polanyi, *Trans. Faraday Soc.*, **31**, 875 (1935).
151. G. Kohnstam in *Progress in Reaction Kinetics*, Vol. 5 (Ed. G. Porter), Pergamon, London, 1970, p. 335.
152. K. E. Weale, *Chemical Reaction Rates at High Pressures*, E. and F. N. Spon, London, 1967.
153. H. Heydtmann and H. Stieger, *Ber. Bunsenges. Phys. Chem.*, **70**, 1095 (1966).
154. D. Büttner and H. Heydtmann, *Ber. Bunsenges. Phys. Chem.*, **73**, 640 (1969).
155. A. Kivinen and A. Viitala, *Suom. Kemistilehti B*, **40**, 19 (1967).
156. A. Kivinen and A. Viitala, unpublished results.
157. A. Kivinen and A. Viitala, *Suom. Kemistilehti B*, **41**, 372 (1968).
158. D. Samuel and B. L. Silver in *Advances in Physical Organic Chemistry* Vol. 3 (Ed. V. Gold), Academic, London, 1965, p. 123.
159. S. L. Johnson, *Tetrahedron Letters*, 1481 (1964).
160. R. L. Burwell, Jr. and R. G. Pearson, *J. Phys. Chem.*, **70**, 300 (1966).
161. C. A. Bunton, T. A. Lewis and D. A. Llewellyn, *Chem. Ind. London*, 1154 (1954).
162. A. Moffat and H. J. Hunt, *J. Am. Chem. Soc.*, **81**, 2082 (1959).
163. M. L. Bender and H. d'A. Heck, quoted in reference 9, p. 269.
164. C. A. Bunton, N. Fuller, S. G. Perry and V. J. Shiner, Jr., *Chem. Ind. London*, 1130 (1960).
165. A. R. Butler and V. Gold, *Chem. Ind. London*, 1218 (1960).
166. A. R. Butler and V. Gold, *J. Chem. Soc.*, 2212 (1962).
167. C. A. Bunton, N. A. Fuller, S. G. Perry and V. J. Shiner, *J. Chem. Soc.*, 2918 (1963).
168. A. Kivinen, *Suom. Kemistilehti B*, **38**, 205 (1965).
169. P. M. Laughton and R. E. Robertson in *Solute–Solvent Interactions* (Ed. J. F. Coetzee and C. D. Ritchie), Marcel Dekker, New York, 1969, p. 319.
170. E. M. Arnett and D. R. McKelvey in *Solute–Solvent Interactions* (Ed. J. F. Coetzee and C. D. Ritchie), Marcel Dekker, New York, 1969, p. 343.
171. V. Gold in *Advances in Physical Organic Chemistry*, Vol. 7 (Ed. V. Gold), Academic, London, 1969, p. 259.
172. R. E. Robertson, B. Rossali, S. E. Sugamori and L. Treindl, *Can. J. Chem.*, **47**, 4199 (1969).
173. O. Rogne, *J. Chem. Soc. B*, 1056 (1970), and references cited therein.
174. E. A. Halevi in *Progress in Physical Organic Chemistry*, Vol. 1 (Ed. S. G. Cohen, A. Streitwieser, Jr. and R. W. Taft), Interscience, New York, 1963, p. 109.
175. E. R. Thornton, *Ann. Rev. Phys. Chem.*, **17**, 349 (1966).
176. M. L. Bender and M. S. Feng, *J. Am. Chem. Soc.*, **82**, 6318 (1960).
177. C. G. Papaioannou, Ph.D. Thesis, Michigan State University; *Diss. Abstr. B*, **28**, 1859 (1967).
178. G. C. Sonnichsen, Ph.D. Thesis, Michigan State University; *Diss. Abstr. B*, **28**, 4953 (1968).

179. T. A. Evans, Ph.D. Thesis, Michigan State University; *Diss. Abstr. B*, **30**, 120 (1969).
180. G. J. Karabatsos, G. C. Sonnichsen, C. G. Papaioannou, S. E. Scheppele and R. L. Shone, *J. Am. Chem. Soc.*, **89**, 463 (1967).
181. G. J. Karabatsos and C. G. Papaioannou, *Tetrahedron Letters*, 2629 (1968).
182. L. S. Bartell, *J. Am. Chem. Soc.*, **83**, 3567 (1961).
183. L. Hakka, A. Queen and R. E. Robertson, *J. Am. Chem. Soc.*, **87**, 161 (1965).
184. C. G. Swain, quoted in reference 158, p. 167.
185. J. F. Coetzee in *Progress in Physical Organic Chemistry*, Vol. 4 (Ed. A. Streitwieser, Jr. and R. W. Taft), Interscience, New York, 1967, p. 45.
186. N. Tokura and F. Akiyama, *Bull. Chem. Soc. Japan*, **37**, 1723 (1964).
187. J. M. Briody and D. P. N. Satchell, *Proc. Chem. Soc. London*, 268 (1964).
188. D. P. N. Satchell, *J. Chem. Soc.*, 558 (1963).
189. J. M. Briody and D. P. N. Satchell, *J. Chem. Soc.* 168 (1965).
190. D. N. Kevill and F. D. Foss, *Tetrahedron Letters*, 2837 (1967).
191. D. N. Kevill and W. F. K. Wang, *Chem. Commun.*, 1179 (1967).
192. L. J. Bellamy and R. J. Pace, *Spectrochim. Acta*, **22**, 525 (1966).
193. L. J. Bellamy, *Advances in Infrared Group Frequencies*, Methuen, London, 1968, Chap. 8, and references quoted therein.
194. L. B. Jones and J. P. Foster, *J. Org. Chem.*, **32**, 2900 (1967).
195. T. Nagai, H. Abe and N. Tokura, *Bull. Chem. Soc. Japan*, **42**, 1705 (1969).
196. R. A. Howald, *J. Org. Chem.*, **27**, 2043 (1962).
197. D. P. N. Satchell, *J. Chem. Soc.*, 1752 (1960).
198. D. P. N. Satchell, *J. Chem. Soc.*, 564 (1963).
199. V. Gold and J. Hilton, *J. Chem. Soc.*, 838 (1955).
200. C. A. Bunton and T. A. Lewis, *Chem. Ind. London*, 180 (1956).
201. J. M. Briody and D. P. N. Satchell, *J. Chem. Soc.*, 3724 (1964).
202. R. E. Parker in *Advances in Fluorine Chemistry*, Vol. 3 (Ed. M. Stacey J. C. Tatlow and A. G. Sharpe), Butterworths, London, 1963, p. 63.
203. C. G. Swain and C. B. Scott, *J. Am. Chem. Soc.*, **75**, 246 (1953).
204. D. P. N. Satchell, *J. Chem. Soc.*, 555 (1963).
205. C. A. Bunton and J. H. Fendler, *J. Org. Chem.*, **31**, 2307 (1966).
206. L. W. Clark in *The Chemistry of Carboxylic Acids and Esters* (Ed. S. Patai), Interscience, London, 1969, Chap. 12, p. 589.
207. K. Okamoto, T. Kinoshita and H. Shingu, *Bull. Chem. Soc. Japan*, **43**, 1545 (1970), and references cited therein.
208. R. A. Sneen and J. W. Larsen, *J. Am. Chem. Soc.*, **91**, 6031 (1969).
209. V. Gold, J. Hilton and E. G. Jefferson, *J. Chem. Soc.*, 2756 (1954).
210. B. C. Challis and A. R. Butler in *The Chemistry of the Amino Group* (Ed. S. Patai), Interscience, London, 1968, p. 277.
211. R. N. Lacey in *The Chemistry of Alkenes* (Ed. S. Patai), Interscience, London, 1964, Chap. 14.
212. P. J. Lillford and D. P. N. Satchell, *J. Chem. Soc. B*, 360 (1967).
213. G. H. Grant and C. N. Hinshelwood, *J. Chem. Soc.*, 1351 (1933).
214. E. G. Williams and C. N. Hinshelwood, *J. Chem. Soc.*, 1079 (1934).
215. W. B. S. Newling, L. A. K. Staveley and C. N. Hinshelwood, *Trans. Faraday Soc.*, **30**, 597 (1934).
216. N. J. T. Pickels and C. N. Hinshelwood, *J. Chem. Soc.*, 1353 (1936).

217. F. J. Stubbs and C. N. Hinshelwood, *J. Chem. Soc. Suppl.*, **1**, 71 (1949).
218. A. N. Bose and C. N. Hinshelwood, *J. Chem. Soc.*, 4085 (1958).
219. H. S. Venkataraman and C. N. Hinshelwood, *J. Chem. Soc.*, 4977 (1960).
220. H. S. Venkataraman and C. N. Hinshelwood, *J. Chem. Soc.*, 4986 (1960).
221. A. E. Shilov and H. S. Venkataraman, *J. Chem. Soc.*, 4992 (1960).
222. H. S. Venkataraman and J. Madhusudana Rao, *Indian J. Chem.*, **3**, 363 (1965).
223. H. S. Venkataraman and J. Madhusudana Rao, *Indian J. Chem.*, **3**, 413 (1965).
224. H. S. Venkataraman, J. Madhusudana Rao and L. Iyenger, *Indian J. Chem.*, **5**, 255 (1967).
225. L. M. Litvinenko, *Ukr. Khim. Zh.*, **28**, 131 (1962); *Chem. Abstr.*, **57**, 6661d (1962).
226. L. M. Litvinenko and A. P. Grekov, *Ukr. Khim. Zh.*, **21**, 66 (1955); *Chem. Abstr.*, **49**, 9532b (1955).
227. L. M. Litvinenko, S. V. Tsukerman and A. P. Grekov, *Dokl. Akad. Nauk SSSR*, **101**, 265 (1955); *Chem. Abstr.*, **50**, 3333b (1956).
228. L. M. Litvinenko and A. P. Grekov, *Zh. Obshch. Khim.*, **26**, 3391 (1956); *Chem. Abstr.*, **51**, 9539b (1957).
229. L. M. Litvinenko and A. P. Grekov, *Zh. Obshch. Khim.*, **27**, 234 (1957); *Chem. Abstr.*, **51**, 12866c (1957).
230. L. M. Litvinenko, S. V. Tsukerman, A. P. Grekov and E .A. Slobodkina, *Ukr. Khim. Zh.*, **23**, 223 (1957); *Chem. Abstr.*, **51**, 14651i (1957).
231. L. M. Litvinenko and A. P. Grekov, *Ukr. Khim. Zh.*, **23**, 228 (1957); *Chem. Abstr.*, **51**, 14652c (1957).
232. L. M. Litvinenko, R. S. Chesko and A. D. Gofman, *Zh. Obshch. Khim.*, **27**, 758 (1957); *Chem. Abstr.*, **51**, 14651e (1957).
233. L. M. Litvinenko, R. S. Chesko and S. V. Tsukerman, *Dokl. Akad. Nauk SSSR*, **118**, 946 (1958); *Chem. Abstr.*, **52**, 12810h (1958).
234. L. M. Litvinenko, A. P. Grekov and S. V. Tsukerman, *Ukr. Khim. Zh.*, **21**, 510 (1955); *Chem. Abstr.*, **50**, 5590h (1956).
235. L. M. Litvinenko and D. M. Aleksandrova, *Dokl. Akad. Nauk SSSR*, **118**, 321 (1958); *Chem. Abstr.*, **52**, 19364e (1958).
236. L. M. Litvinenko and D. M. Aleksandrova, *Ukr. Khim. Zh.*, **26**, 621 (1960); *Chem. Abstr.*, **55**, 15381b (1961).
237. L. M. Litvinenko and D. M. Aleksandrova, *Ukr. Khim. Zh.*, **27**, 212 (1961); *Chem. Abstr.*, **55**, 23389h (1961).
238. L. M. Litvinenko, D. M. Aleksandrova and A. A. Zhilinskaya, *Ukr. Khim. Zh.*, **26**, 476 (1960); *Chem. Abstr.*, **55**, 10022h (1961).
239. L. M. Litvinenko and D. M. Aleksandrova, *Ukr. Khim. Zh.*, **27**, 336 (1961); *Chem. Abstr.*, **56**, 3381i (1962).
240. L. M. Litvinenko and D. M. Aleksandrova, *Ukr. Khim. Zh.*, **27**, 487 (1961); *Chem. Abstr.*, **56**, 12798g (1962).
241. L. M. Litvinenko, D. M. Aleksandrova and S. F. Prokopovich, *Ukr. Khim. Zh.*, **27**, 494 (1961); *Chem. Abstr.*, **56**, 12799a (1962).
242. L. M. Litvinenko and D. M. Aleksandrova, *Ukr. Khim. Zh.*, **27**, 634 (1961); *Chem. Abstr.*, **56**, 12799d (1962).
243. L. M. Litvinenko, N. M. Oleinik and G. D. Titskii, *Dokl. Akad. Nauk SSSR*, **157**. 1153 (1964): *Chem. Abstr.*, **62**, 2682h (1965).

244. L. M. Litvinenko and N. M. Oleinik, *Reakts. Sposobnost Org. Soedin.*, **2**, 57 (1965); *Chem. Abstr.*, **63**, 17860h (1965).
245. L. M. Litvinenko and N. M. Oleinik, *Ukr. Khim. Zh.*, **32**, 174 (1966); *Chem. Abstr.*, **64**, 15690g (1966).
246. G. V. Semenyuk, N. M. Oleinik and L. M. Litvinenko, *Reakts. Sposobnost Org. Soedin.*, **4**, 760 (1967); *Chem. Abstr.*, **70**, 2931y (1969).
247. N. K. Vorob'ev and L. V. Kuritsyn, *Izv. Vyssh. Ucheb. Zaved. Khim. Khim. Technol.*, **6**, 591 (1963); *Chem. Abstr.*, **60**, 3529h (1964).
248. N. K. Vorob'ev and L. V. Kuritsyn, *Izv. Vyssh. Ucheb. Zaved. Khim. Khim. Technol.*, **7**, 34 (1964); *Chem. Abstr.*, **61**, 5476d (1964).
249. N. K. Vorob'ev and L. V. Kuritsyn, *Izv. Vyssh. Ucheb. Zaved. Khim. Khim. Technol.*, **7**, 930 (1964); *Chem. Abstr.*, **62**, 16008e (1965).
250. R. A. Benkeser, C. E. DeBoer, R. E. Robinson and D. M. Sauve, *J. Am. Chem. Soc.*, **78**, 682 (1956).
251. J. G. Mather and J. Shorter, *J. Chem. Soc.*, 4744 (1961).
252. J. J. Elliott and S. F. Mason, *Chem. Ind. London*, 488 (1959).
253. E. Tommila and T. Vihavainen, *Acta Chem. Scand.*, **22**, 3224 (1968).
254. E. Tommila and T. Vihavainen, *Suom. Kemistilehti B*, **42**, 89 (1969).
255. Y. Kondo and N. Tokura, *Bull. Chem. Soc. Japan*, **40**, 1433, 1438 (1967).
256. Y. Kondo, Y. Hondo and N. Tokura, *Bull. Chem. Soc. Japan*, **41**, 987 (1968).
257. J. B. Rossel, *J. Chem. Soc.*, 5183 (1963).
258. M. Simonetta and S. Carra, *Atti Acad. Naz. Lincei Rend. Cl. Sci. Fis. Mat. Natur.*, **22**, 176 (1957).
259. H. K. Hall, Jr., *J. Org. Chem.*, **29**, 3539 (1964).
260. L. M. Litvinenko, N. M. Oleinik and G. V. Semenuyk, *Ukr. Khim. Zh.*, **35**, 278 (1969); *Chem. Abstr.*, **71**, 2694u (1969).
261. L. M. Litvinenko, G. D. Titskii and V. A. Tarasov, *Reakts. Sposobnost Org. Soedin.*, **5**, 325 (1968), and references cited therein; *Chem. Abstr.*, **70**, 19388h (1969).
262. G. V. Semenuyk, N. M. Oleinik and L. M. Litvinenko, *Zh. Obshch. Khim.*, **38**, 2009 (1968); *Chem. Abstr.*, **70**, 19475a (1969).
263. J. N. Brønsted and K. Pedersen, *Z. Phys. Chem.*, *Leipzig*, **108**, 185 (1924).
264. J. N. Brønsted, *Chem. Rev.*, **5**, 231 (1928).
265. M. L. Bender and J. M. Jones, *J. Org. Chem.*, **27**, 3771 (1962).
266. Y. Kondo and N. Tokura, *Bull. Chem. Soc. Japan*, **42**, 1660 (1969).
267. V. V. Korshak, S. V. Vinogradova and V. A. Vasnev, *Dokl. Akad. Nauk SSSR*, **191**, 614 (1970).
268. P. Leduc and P. Chabrier, *Bull. Soc. Chim. France*, 2271 (1963).
269. S. L. Johnson and K. A. Rumon, *J. Phys. Chem.*, **68**, 3149 (1964).
270. S. L. Johnson and K. A. Rumon, *J. Am. Chem. Soc.*, **87**, 4782 (1965).
271. S. L. Johnson and K. A. Rumon, *Tetrahedron Letters*, 1721 (1966).
272. R. C. Paul and S. L. Chadha, *Spectrochim. Acta*, **22**, 615 (1966).
273. D. Cook, *Can. J. Chem.*, **40**, 2362 (1962).
274. D. L. Goldhamer, M. Onyszkewycz and A. Wilson, *Tetrahedron Letters*, 4077 (1968).
275. W.-W. Tso, C. H. Snyder and H. B. Powell, *J. Org. Chem.*, **35**, 849 (1970).
276. E. Tommila and E. Hämäläinen, *Acta Chem. Scand.*, **17**, 1985 (1963), and references cited therein.

277. P. S. Bailey, Jr., Ph.D. Thesis, Purdue University; *Diss. Abstr.*, **30B**, 4965 (1970).
278. W. E. Truce and P. S. Bailey, Jr., *J. Org. Chem.*, **34**, 1341 (1969).
279. T. Ozeki and M. Kusaka, *Bull. Chem. Soc. Japan*, **39**, 1995 (1966).
280. P. W. Hickmott, *J. Chem. Soc.*, 883 (1964).
281. A. R. Fersht and W. P. Jencks, *J. Am. Chem. Soc.*, **91**, 2125 (1969).

CHAPTER 7

Reduction

OWEN H. WHEELER

*Caribbean Research Laboratories, Omni Research Inc., Mayaguez,
Puerto Rico*

I. INTRODUCTION

Strong reducing reagents, such as lithium aluminium hydride, reduce acid
chlorides to carbinols (equation 1) (see section III). However, modified

$$-COX + 4 H \longrightarrow -CH_2OH + HX \qquad (1)$$

metal hydrides reduce acid chlorides to aldehydes (equation 2) and this

$$-COX + 2 H \longrightarrow -CHO + HX \qquad (2)$$

can also be achieved catalytically (Rosenmund reduction; see section II).
Bimolecular reduction to acyloins (equation 3) and glycols (equation 4)

$$2 COX + 2 H \longrightarrow -CO-CHOH- + 2 HX \qquad (3)$$

$$2 COX + 4 H \longrightarrow -CHOH-CHOH- + 2 HX \qquad (4)$$

can be performed with magnesium–magnesium iodide (see section IV).
Reduction to a methyl group would involve hydrogenolysis of the carbon-
oxygen bond, and has also been reported. Reduction to an alkyl halide
($-CH_2X$) is not possible, since hydrogenolysis of the carbon–halide bond
would occur.

Acyl bromides are more easily reduced than acyl chlorides, but no data
have been published on the reduction of acyl fluorides or iodides.

231

II. CATALYTIC REDUCTION

Acid chlorides can be hydrogenated selectively to aldehydes (equation 5) by using a suitable catalyst[1, 2], and the reaction is known as the Rosenmund

$$RCOCl + H_2 \longrightarrow RCHO + HCl \qquad (5)$$

reduction[3]. The catalyst is usually 5% palladium on barium sulphate, although platinum, osmium and nickel catalysts on kieselguhr, calcium carbonate or activated carbon have also been used[1]. To prevent further reduction of the aldehyde to a carbinol, the catalyst is poisoned[4] with a mixture of quinoline and sulphur[5]. Thiourea has also been employed as a poison[6]. The catalyst is suspended in a heated solution of the acid chloride in xylene or toluene and a stream of hydrogen passed through the solution to remove the hydrogen chloride. Benzene, tetralin and decalin have also been used as solvents[1], as have ethers[7]. The acid chloride is prepared using thionyl chloride, but phosphorus halides cannot be used, since phosphorus compounds deactivate the catalyst[8, 9].

Aliphatic, aromatic and heterocyclic acid chlorides are converted into the corresponding aldehydes in 50–80% yields[1]. The main by-products in general are carbinol (equation 6) and hydrocarbon (equation 7) formed

$$RCHO + H_2 \longrightarrow RCH_2OH \qquad (6)$$

$$RCH_2OH + H_2 \longrightarrow RCH_3 + H_2O \qquad (7)$$

by further reduction[10, 11], and ester formed from the carbinol (equation 8).

$$RCH_2OH + RCOCl \longrightarrow RCH_2OCOR + HCl \qquad (8)$$

In the presence of traces of oxygen or water, acid anhydrides are important by-products (equation 9)[12]. Palmitoyl chloride in xylene gave 86%

$$RCOCl + RCO_2H \longrightarrow (RCO)_2O + HCl \qquad (9)$$

hexadecanal[13]. The yields can be improved (to 80–90%) in the reduction of butyryl, nonanoyl, palmitoyl, oleoyl and stearoyl chloride by using a palladium–barium sulphate catalyst regulated with dimethylaniline[14]. The nitro group in p-nitrobenzoyl chloride was not reduced, and o-chlorobenzoyl chloride afforded o-chlorobenzaldehyde[4].

The Rosenmund reduction of some heterocyclic acid chlorides leads to decarbonylation (equation 10). The carbonyl group in p-anisoyl chloride[10]

$$RCOCl + H_2 \longrightarrow RH + CO + HCl \qquad (10)$$

and 3,4,5-trimethoxybenzoyl chloride[15] was also removed, whereas triphenylacetyl chloride afforded triphenylmethane[16]. The acid chlorides of dibasic acids give only low yields of aldehyde. Succinyl chloride (1) formed butyrolactone (2)[17], and adipyl chloride (3) gave formyl-valeric acid (**4**), together with cyclopentene-1-carboxylic acid (**5**)[17]. *o*-Phthalyl

$$\begin{matrix} CH_2COCl \\ | \\ CH_2COCl \end{matrix} \qquad \begin{matrix} CH_2CO \\ | \qquad \rangle O \\ CH_2CH_2 \end{matrix}$$

(1)   (2)

$$\begin{matrix} CH_2CH_2COCl \\ | \\ CH_2CH_2COCl \end{matrix} \qquad \begin{matrix} CH_2CH_2CHO \\ | \\ CH_2CH_2CO_2H \end{matrix} \qquad \begin{matrix} CH_2-CH_2 \\ | \qquad\qquad \rangle C-CO_2H \\ CH_2-CH \end{matrix}$$

(3)   (4)   (5)

chloride (**6**) afforded phthalide (**7**)[18], although the *meta* and *para* isomers formed the dialdehydes in 80% yield[12]. However, 4-carbomethoxybutyroyl

(6)   (7)   (8)

$$CH_3O_2C(CH_2)_3COX$$

chloride (**8**, X = Cl) was reduced to methyl 4-formyl butyrate (**8**, X = H)[19], and 9-carbethoxypelargonyl, 11-carbethoxyundecanoyl and 15-carbethoxypentadecanoyl chloride were all reduced to the corresponding ester aldehydes in yields of 65%, 89% and 95%, respectively[20]. Glutarimide-β-acetyl chloride (**9**, X = Cl) has also been reduced to the corresponding aldehyde (**9**, X = H) in dioxan and xylene at 120–130°, in 99·5% yield[21].

(9)

The acid chloride was formed from the acid using thionyl chloride and dimethylformamide as catalyst.

Trifluoracetyl chloride was reduced under Rosenmund conditions to trifluoracetaldehyde (64%)[22], although another publication reports the formation of the corresponding carbinol[23]. Trichloracetyl chloride, however, afforded dichloracetaldehyde (50–60%) but no chloral, on reduction in a high boiling petroleum fraction. Polymerization occurred in xylene solution[24].

Phenothiazine-1-carboxaldehyde was prepared in good yield by Rosenmund reduction of the corresponding acid chloride[25]. Salicyl aldehydes were synthesized by reduction of the acid chlorides at 80° but without the quinoline–sulphur poison[26]. Stearoyl chloride was reduced in 96% yield to the aldehyde by adding to pre-reduced 5% palladium on barium sulphate in acetone and using dimethylacetamide to neutralize the hydrogen chloride[27]. The Rosenmund preparation of β-naphthaldehyde is recorded in *Organic Synthesis*[5].

Deuterium-labelled benzaldehyde (PhCDO) has been prepared by Rosenmund reduction of benzoyl chloride using deuterium. Labelled *p*-phenylbenzaldehyde was similarly prepared from the corresponding acid chloride[28]. 1-^{14}C-Dodecanoic acid has been reduced in the same fashion via its acid chloride, to the corresponding labelled aldehyde in 42% overall yield[29].

The effect of sulphur compounds as poisons for a palladium–barium sulphate catalyst in the reduction of benzoyl chloride to benzaldehyde in toluene solution has been studied[30]. Tetramethylthiourea was very effective, since the addition of 4×10^{-6} moles resulted in an 80% yield of aldehyde. Thiourea was less effective, while thiophene had a smaller effect and dibenzothiophene had little effect. Palladium sulphide (5%) on barium sulphate produced 40% benzaldehyde, which was increased to 80% on the addition of tetramethylthiourea. The remaining product was benzyl alcohol. This was formed by subsequent reduction of the aldehyde on the catalyst, since when ^{14}C-labelled benzaldehyde (Ph[14]CHO) was added, the

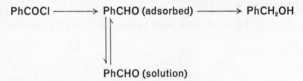

activity of the benzyl alcohol showed that 57% of the labelled aldehyde had been reduced. The labelled benzaldehyde was prepared by Rosenmund reduction of ^{14}C-labelled benzoyl chloride. Removal of the benzaldehyde formed in the stream of hydrogen gave a 58% yield, whereas only 28% was formed under the same conditions when the aldehyde was not removed. The reduction of benzaldehyde to benzyl alcohol probably requires two active sites for hydrogen and some of these adjacent sites will be occupied by molecules of the poison. However, in the reduction of benzoyl chloride, the chlorine can be replaced by active hydrogen already adsorbed on the catalyst surface, and a poisoned catalyst will still be active in such a reduction[30].

Aromatic and higher aliphatic acid chlorides have been hydrogenated at reduced pressure with a nickel, nickel chloride and platinum catalyst.

Octanoyl chloride gave the highest yield (50%) of aldehyde[31]. However, there seems no advantage in using low pressures. The catalytic reduction of n-butyroyl chloride at low pressure using a palladium–asbestos catalyst produced n-butyraldehyde and 3-methylalheptene. Isovaleroyl chloride gave a nearly quantitative yield of the corresponding aldehyde, and succinoyl chloride formed γ-butyrolactone (54%). Benzoyl chloride gave benzaldehyde (89·5%), but phenacetyl chloride afforded ethylbenzene and β-phenylethanol, whereas o-anisoyl chloride formed 2-methylanisole. Sulphur and phosphorus compounds poisoned the catalyst, and Raney nickel was not an effective catalyst[32].

Vapour phase hydrogenation at 190° over 20% palladium charcoal transformed pivaloyl chloride and 2,2-dimethylbutyroyl chloride into the corresponding aldehydes[33]. Similar reduction of cyclopentane carboxylic acid chloride produced 85% cyclopentane carboxaldehyde, together with cyclopentane and cyclopentene. Cyclohexane carboxylic acid chloride and 1-methylcyclopentane and 1-methylcyclohexane carboxylic acid chlorides were also reduced in the same manner[34].

Benzoyl chloride and hydrogen over a nickel catalyst gave benzaldehyde[35]. However, at 300° under pressure diphenyl, toluene and benzoic acid with a trace of benzaldehyde resulted, although no benzyl alcohol was formed. At reduced hydrogen pressure (50–300 mm) and 300°, 60% benzaldehyde and some benzyl alcohol were formed. A platinum catalyst at 250° reduced benzoyl chloride to toluene and benzyl alcohol and a small amount of benzaldehyde. Phenacyl chloride at 200° gave some phenacetaldehyde and β-phenylethanol, with large amounts of 1,4-diphenyl-2-butanol (10). n-Valeroyl and n-hexanoyl chlorides afforded low yields of

$$PhCH_2CH(OH)(CH_2)_2Ph$$
(10)

aldehyde under these conditions[36]. At 400° isovaleroyl chloride formed isobutylene and propylene, and propionyl chloride gave ethylene, methane, carbon monoxide and hydrogen chloride. However, acetyl chloride decomposed violently under the same conditions[37].

The reduction of benzoyl chloride to benzaldehyde by hydrogen transfer from cyclohexene in the presence of a platinum-black catalyst only resulted in 10% reduction in 16 hours[38].

III. METAL HYDRIDE REDUCTION

Lithium aluminium hydride reduces acid chlorides to carbinols and the yields are better than those obtained from the acids and esters[39,40]. Benzoyl chloride gave 72% benzyl alcohol, and o-phthaloyl chloride formed 95% phthalyl alcohol; trimethylacetyl chloride, sorboyl chloride, isocaproyl chloride and palmitoyl chloride all gave 86–98% of the corresponding carbinols[41]. Mono-, di- and trichloracetyl chloride formed 62–64% of the chloroethanols[42] and trifluoracetyl chloride afforded 85% trifluorethanol[43]. *O*-Methylpodocaproyl chloride was reduced (93%) to the corresponding carbinol[44]. However, triphenylacetyl chloride formed triphenylacetaldehyde[39]. The reduction of caproyl chloride and benzoyl chloride was complete in $\frac{1}{2}$ h in tetrahydrofuran at 0°[45].

Sodium borohydride in dioxan or diethylcarbitol reduces acid chlorides to carbinols. Thus, *n*-butyroyl, benzoyl and palmitoyl chlorides gave the corresponding alcohols in 76–87% yield, and monoethylsuccinate acid chloride afforded butyrolactone (40%). However, *o*-phthaloyl chloride gave phthalide (49%) and phthalyl alcohol (15%), whereas cinnamoyl chloride formed hydrocinnamyl alcohol (12%)[46]. However, sodium borohydride in diglyme in the presence of triethylamine reduced benzoyl chloride to benzaldehyde (65%). *o*-Chlorobenzaldehyde and *p*-nitrobenzaldehyde were also formed from the corresponding acid chlorides in yields of 79% and 65% respectively[47]. Sodium borohydride plus aluminium chloride reduced benzoyl chloride to benzyl alcohol, and the same reagent also reduced esters and carboxylic acids to carbinols[48, 49].

The following reductions of acid chlorides to carbinols have been carried out with sodium borohydride in dioxan: 3β-acetoxy-20-iso-5-cholenic acid chloride[50], *trans*-norpinic acid monomethyl ester chloride[51, 52], the acid chloride of ethyl *trans*-β(2-carboxycyclohexane) propionate[53], 3α-hydroxyolean-12-en-29-oic acid chloride[54], tetracosanedioic monomethyl ester acid chloride (11)[55], 5-ethylfluorenone-6-carboxylic acid chloride, 5-ethylfluorene-6-carboxylic acid chloride[56] and 2-phenyloxazole-4-carboxylic acid chloride[57]. Reduction of 3,4-dinitrobenzoyl chloride with sodium borohydride in dimethoxyethane gave a low yield of the carbinol[58]. 2,3,4-Trimethyl-6-nitrobenzoyl chloride[59] and 2-nitrohomoveratric acid chloride[60] were both reduced in tetrahydrofuran. Hemimelletic anhydride chloride (12) in diglyme gave 7-carboxyphthalide (13), the dihydroisobenzofuran (14) and hemimelletic acid. 4-Carboxyphthalide chloride (15) also gave 4-hydroxymethylphthalide, and the acid chloride of 7-carboxyphthalide formed the corresponding carbinol together with the dihydroisobenzofuran (16). Excess sodium borohydride afforded 1,2,3-trihydroxymethylbenzene in the case of 12[61]. A method for the degradation of peptides[62] involved preparing the hydantoin derivative (17), converting this into the acid chloride, which was then reduced by sodium borohydride and the product hydrolysed. Identification of the aminocarbinol (18) established the nature of the end group.

Substituted complex metal hydrides have also been used for the reduction of acid chlorides. Thus, 2-(2'-nitrophenyl)-phenylmethanol

$$CH_3O_2C(CH_2)_{22}COCl$$

(11)

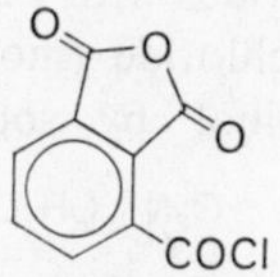

(12)

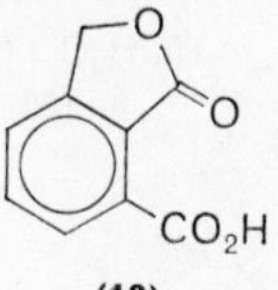

(13)

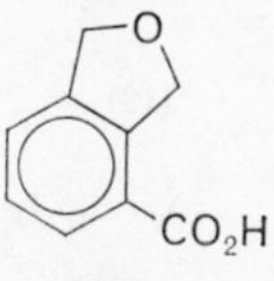

(14)

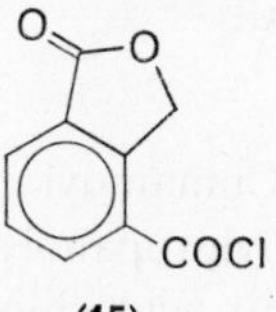

(15)

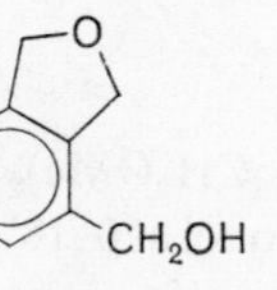

(16)

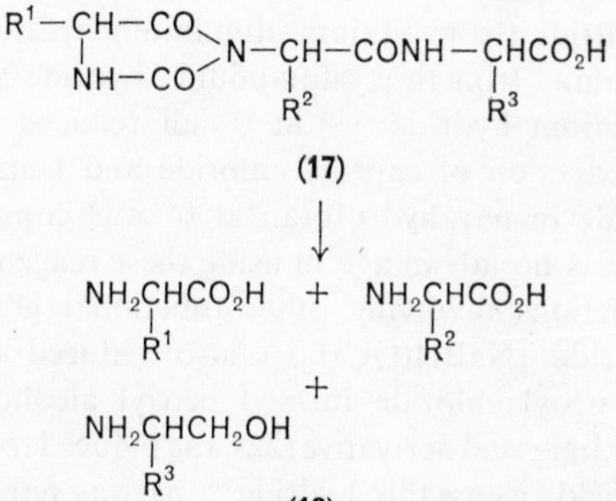

$$R^1-CH-CO \diagdown N-CH-CONH-CHCO_2H$$
$$NH-CO \diagup \underset{R^2}{|} \quad \underset{R^3}{|}$$

(17)

↓

$$NH_2CHCO_2H \quad + \quad NH_2CHCO_2H$$
$$\underset{R^1}{|} \qquad\qquad \underset{R^2}{|}$$

+

$$NH_2CHCH_2OH$$
$$\underset{R^3}{|}$$

(18)

(19)[63] and 2,2'-bis(hydroxymethyl)-6,6'-dinitrobiphenyl (20)[64] were prepared by reducing the corresponding acid chlorides with sodium trimethoxyborohydride in tetrahydrofuran at −80°. Tetraacetylquinic acid chloride (21) also produced the carbinol on reduction with the same hydride[65].

Reduction of α,α-diethylsuccinic acid monomethyl ester chloride with lithium tri-t-butoxyaluminium hydride gave the carbinol[66]. (Δ²-2,3-Diphenylcyclopropenyl) carbinol was also prepared by reducing the appropriate chloride[67]. The triterpene derivatives, dihydrocassaminic acid and erythrophlaminic acid were converted into carbinols by reducing their acid chlorides with LiAl(t-BuO)$_3$H[68].

Pelargonyl chloride (monanoyl chloride) was reduced to the corresponding carbinol by sodium dihydro-bis(2-methoxyethoxy)aluminate

O$_2$N CH$_2$OH

(19)

O$_2$N NO$_2$

HOH$_2$C CH$_2$OH

(20)

AcO OAc

OAc

ClOC OAc

(21)

(NaAlH$_2$(OCH$_2$CH$_2$OMe)$_2$) in benzene. Cinnamoyl chloride was reduced to cinnamyl alcohol at 0–18° and to 3-phenylpropanol at 70–100°[69]. A series of aliphatic (caproic, caprylic and lauric) acid chlorides gave yields of 88–99% of carbinols in benzene, toluene or xylene with this reagent[70].

Aluminium hydride (from lithium aluminium hydride and aluminium trichloride), lithium trimethoxyaluminium hydride[71, 72] and lithium tri-t-butoxyaluminium hydride[71, 73] at 0° all reduced acid chlorides to carbinols. The reduction of caproyl chloride and benzoyl chloride with aluminium hydride in tetrahydrofuran at 0° was complete in 15 min[71]. However, there was no advantage in using these reagents, except in cases of selective reductions involving other functional groups. Sodium trimethoxyborohydride (NaBH(OCH$_3$)$_3$) also reduced acid chlorides to alcohols, and benzoyl chloride formed benzyl alcohol in 66% yield[74]. However, the shikimic acid derivative (**22**) was reduced to the corresponding aldehyde (85% yield) using this hydride[75], as was tetraacetylglycoferulic acid chloride (**23**)[76]. Lithium borohydride reduction of α-tritio-(p-nitrophenyl)acetyl chloride afforded the tritiated carbinol[77]. Trifluoracetyl chloride was reduced to the carbinol with lithium aluminium hydride, but not with sodium borohydride or sodium hydride[23]. Lithium aluminium hydride also reduced 1,3,5-tricarboxymethylbenzene chloride to 1,3,5-tris-2'-(hydroxymethyl)benzene, and benzene-1,3,5-tricarbonyl chloride to the corresponding polycarbinol[78].

A great advance in the use of complex hydrides for selective reductions was the discovery that one equivalent of lithium tri-t-butoxyaluminium hydride at low temperature ($-78°$) reduced acid chlorides to aldehydes[79]. The reagent was simply prepared by adding three equivalents of t-butanol to one equivalent of lithium aluminium hydride in ether[80]. Benzoyl chloride was reduced to benzaldehyde in 78% yield, and p-nitrobenzoyl

(22)   (23)

chloride and terephthaloyl chloride to the corresponding aldehydes in 80% and 85% yield, respectively[79]. At 0° benzoyl chloride gave only 60% benzaldehyde, while lithium tri-t-amyloxyaluminium hydride afforded only 48% aldehyde at $-75°$[79]. *Meta-* and *para*-substituted aromatic aldehydes were formed in 60–90% yield and nitro, cyano and carbethoxy groups were not reduced by this hydride. *Ortho*-substituted benzoyl chlorides gave lower yields (o-nitro 77%, o-chloro 41% and o-methoxy, 27%), but α,β-naphthoyl, nicotinoyl (69%) and cinnamoyl chlorides (71%) were reduced in good yield. The yields of aldehydes from aliphatic and alicyclic acid chlorides were lower (isobutyryl 57%, adipyl 53%, cyclo-propane carboxylic 42%, fumaryl 59% and crotonyl 48%), and these yields are inferior to those given by Rosenmund reduction (see section II)[81]. However, α-fluoroaldehydes were obtained by the reduction of α-fluoro acyl chlorides with LiAl(t-BuO)$_3$H in tetrahydrofuran at $-70°$ (α-fluoro-acetaldehyde 52%, α-fluoropropionaldehyde 60% and α-fluorobutyr-aldehyde 65%[82]).

The reduction of diacetyltartaric acid methyl ester chloride[83] and of α,α-dimethylsuccinic acid monoethyl ester chloride[84] with LiAl(t-BuO)$_3$H at $-65°$ to $-75°$ afforded the corresponding ester aldehydes. However, pentyne-2-oylic and heptyne-2-oylic chlorides, in diethylcarbitol at $-60°$, gave a complex mixture of products in which the triple bond had been reduced[85]. The same hydride at low temperature reduced 1,3,5-tricarboxy-methyl benzene chloride and benzene-1,3,5-tricarbonyl chloride to their aldehydes[78]. 2-Nitro-3,4,5-trimethoxybenzoyl chloride gave the aldehyde on treatment with this hydride at $-70°$[86]. 4-Bromo-3-methylisothiazole-5-carboxyl chloride formed both 4-bromo-5-formyl-3-methylisothiazole (24%) and 4-bromo-5-hydroxymethyl-3-methylisothiazole (21%) at $-40°$, but gave only the aldehyde at $-70°$[87]. Pyrazinoyl chloride afforded a low yield of the aldehyde (20%) at $-70°$[88].

Sodium tri-t-butoxyaluminium hydride has been reported to give better yields in the reduction of acid chlorides to aldehydes than sodium aluminium hydride, sodium di-t-butoxyaluminium hydride. Reductions with sodium tri-t-butoxyaluminium hydride[89] were carried out in diglyme or tetrahydrofuran at $-60°$ to $-70°$. Benzaldehyde (86%) and o-chloro-(68%), o-bromo-(61%), m-nitro-(78%), p-chloro-(73%), p-bromo-(80%), p-nitro-(78%) and 3,5-dinitro-(87%) benzaldehydes were prepared. Cinnamaldehyde (69%) and phthalaldehyde (78%) were also prepared, as were furaldehyde (85%), thiophenecarboxaldehyde (63%), furylacrolein (72%), coumarincarboxaldehyde (68%), nicotinaldehyde (75%), n-butyr-aldehyde (48%), isobutyraldehyde (34%), n-hexaldehyde (45%) and pivalaldehyde (45%)[90].

In addition to directed reduction with metal hydrides, acid chlorides can also be converted to aldehydes by forming N-methylanilides and reducing them with lithium aluminium hydride[91, 92], by forming dimethyl-amides and reducing with lithium triethoxyaluminium hydride[93], and by reacting with ethylenimine in the presence of triethylamine to afford an acylaziridine, which is then reduced with lithium aluminium hydride[94]. The yields in the last case were higher with aliphatic acid chlorides (n-butyroyl 75% and pivaloyl 88%)[95]. Carboxylic acids can also be reduced directly to aldehydes (14–80% yield) with lithium in ethylamine[96], and diisobutylaluminium hydride reduces acids (40–70%)[97] and esters (85%)[98] to aldehydes[99].

Diborane does not reduce acid chlorides[100] although aldehydes and acids are reduced rapidly by this reagent[101]. However, t-butylamine-borane reduced benzoyl chloride to benzaldehyde in 85% yield, although esters, acids and amides were not affected[102]. On the other hand, dimethylamine-borane in ether at room temperature converted both benzoyl chloride and benzaldehyde to benzyl alcohol[103]. Ethane-1,2-diamine-borane—$(CH_2NH_2—BH_3)_2$—was also reported to reduce acetyl chloride to acetaldehyde in diglyme[104].

Triphenyltin hydride converted benzoyl chloride to benzaldehyde (equation 11)[105]. This reduction can also be carried out using tri-n-butyltin hydride[106]. However, this hydride at 40° without solvent reduced

$$PhCOCl + Ph_3SnH \longrightarrow PhCHO + Ph_3SnCl \qquad (11)$$

succinoyl chloride (24) to α-chloro-γ-butyrolactone (25), probably via the half aldehyde–half acid chloride (26). Phthaloyl chloride (27) also afforded phthalide (29), probably through 3-chlorophthalide (28). In ether solution benzoyl chloride formed 45% benzaldehyde, but in the absence of solvent the main product was benzyl benzoate[107, 108]. The relative yields of aldehyde and ester formed using tri-*n*-butyltin hydride depend on the acid chloride. Thus acetyl chloride gave 5% acetaldehyde and 95% ethyl

$$Bu_3SnH + RCOCl \longrightarrow RCHO + Bu_3SnCl$$

$$2\,Bu_3SnH + 2\,RCOCl \longrightarrow RCO_2CH_2R + 2\,Bu_3SnCl$$

acetate, while benzoyl chloride afforded 65% benzaldehyde and 35% benzyl acetate. Acid bromides also formed largely aldehydes. Thus,

CH$_2$COCl   CH$_2$CHO   CH$_2$CHCl
| $\longrightarrow$ | $\longrightarrow$ | $\diagdown$O
CH$_2$COCl   CH$_2$COCl   CH$_2$—CO$^{\diagup}$

(24)   (25)   (26)

[ortho-C$_6$H$_4$(COCl)$_2$] $\longrightarrow$ (28) $\longrightarrow$ (29)

(27)   (28)   (29)

propionyl bromide gave 79% propionaldehyde and 2% butyl propionate (propionyl chloride gave 87% ester, also in 2,3-dimethylbutane), and benzaldehyde was formed exclusively (99%) from benzoyl bromide[108]. The yield of aldehyde from acid chlorides increased with larger or branched-chain acyl chlorides, 2,2-dimethylpropionyl chloride giving 65% aldehyde and 33% ester. Whereas propionyl chloride formed only ester on treatment with tri-*n*-butyltin hydride without solvent, the use of either 2,3-dimethylbutane, toluene, *m*-xylene or methyl acetate as solvent reduced the ester yield to about 25%. If an aldehyde or ketone was added to the reaction mixture, a mixed ester resulted in which the aldehyde or ketone had formed the alkyl group. Thus the addition of benzaldehyde to a 1 : 1 mixture of propionyl chloride and tri-*n*-butyltin hydride produced 70% benzyl propionate, with propionaldehyde as a secondary product. In toluene or

$$PhCHO + C_2H_5COCl + Bu_3SnH \longrightarrow PhCH_2OCOC_2H_5 + Bu_3SnCl$$

2,3-dimethylbutane solution only 34% benzyl propionate was formed and cyclohexanone without solvent gave only 12% cyclohexylpropionate. The suggested mechanism involved the formation of alkyltin free radicals[109], initiated by free radicals from the acid chloride. These radicals remove chlorine from the acid chloride giving an acyl radical (equation 12), which

$$R^{\bullet} + Bu_3SnH \longrightarrow Bu_3Sn^{\bullet} + RH$$

$$Bu_3Sn^{\bullet} + RCOCl \longrightarrow Bu_3SnCl + R\overset{\bullet}{C}{=}O \qquad (12)$$

$$R\overset{\bullet}{C}{=}O + Bu_3SnH \longrightarrow RCHO + Bu_3Sn^{\bullet} \qquad (13)$$

then removes hydrogen from the hydride, producing the aldehyde and regenerating the alkyltin radical (equation 13). The addition of azobisisobutyronitrile as radical initiator leads to complete reduction of ethyl chloroformate (forming ethyl formate) in 3 h at 80°, whereas only 5% reduction occurred under the same conditions without the addition of initiator. Triphenylacetyl chloride afforded 90% triphenylacetaldehyde, but 5% triphenylmethane (and 10–12% carbon monoxide) was also formed through decomposition of the intermediate ($Ph_3C\dot{C}{=}O$ radical). Benzyl chloroformate also gave 39% toluene, together with 61% benzyl formate, and the acyloxy radical must be decarboxylated to form the benzyl radical (equation 14). The relative rates of reduction of acid chlorides

$$PhCH_2OCOCl \longrightarrow PhCH_2O\dot{C}{=}O \longrightarrow PhCH_2{}^{\bullet} + CO_2 \qquad (14)$$

have been measured by competition with benzyl bromide or 2-bromooctane for tri-n-butyltin hydride. Electron-withdrawing groups facilitate the reduction, the electropositive organotin radical acting as electron donor to the electronegative chlorocarbonyl group. The rates followed a Hammett correlation with ρ of 2·61 for a series of substituted benzoyl chlorides. Benzoyl bromide was 331 times as reactive as benzoyl chloride, whereas propionyl chloride was 5·63 times as reactive[111]. The mechanism of formation of the ester has been suggested to involve the reaction of the intermediate acyl radical with a second molecule of aldehyde forming an α-acyloxy radical (30) (equation 15). This radical then abstracts hydrogen

$$R\dot{C}{=}O + RCHO \longrightarrow \underset{\underset{O}{\|}}{RC}{-}O\dot{C}HR \qquad (15)$$
$$(30)$$

from the organic tin hydride giving the ester and an organic tin radical (equation 16)[110]. The rates of reaction of the acyl radical with substituted

$$\underset{\underset{O}{\|}}{RC}{-}O\dot{C}HR + Bu_3SnH \longrightarrow \underset{\underset{O}{\|}}{RC}{-}OCH_2R + Bu_3Sn{}^{\bullet} \qquad (16)$$

benzaldehydes follow the Hammett relation with ρ of 0·43. The acyl radical acts as a nucleophile. The relative rates of reaction of the propionyl radical in 2,3-dimethylbutane at 25° relative to benzaldehyde were cyclohexanone 0·093, n-heptaldehyde 1·62, p-chlorobenzaldehyde 1·26 and p-methoxybenzaldehyde 0·468[110].

Triphenyltin hydride, with or without a solvent, reduced benzoyl chloride to benzyl benzoate[111]. Propionyl chloride afforded 14% aldehyde and 50% ester, while phenacetyl chloride gave only 6·6% aldehyde with 90% ester. The bridged-ring acid chlorides, 1-apocamphane carbonyl chloride and 1-norbornane carbonyl chloride, gave both ester (76·6% and 80·1%, respectively) and lesser amounts of aldehyde (15·0% and 11·1%, respectively). However, triphenylacetyl chloride formed only aldehyde (61·9%), while ferrocenoyl chloride gave the corresponding carboxylic anhydride (70·4%). All these results refer to neat solution. 2-Thenoyl chloride in benzene formed 28·8% ester and 1·2% aldehyde[111]. Tribenzyl-silane, however, reduced benzoyl chloride in ether to benzaldehyde (equation 17). Triethylsilane, with a trace of aluminium chloride, also effected this reduction, but no reaction took place in the absence of aluminium chloride. Benzoyl bromide was reduced to benzaldehyde by tribenzylsilane, and also by triethylsilane without aluminium chloride. o- and p-Chlorobenzoyl chloride and p-ethoxybenzoyl chloride were

$$(PhCH_2)_3SiH + PhCOX \longrightarrow (PhCH_2)_3SiX + PhCHO \qquad (17)$$

reduced to the corresponding aldehydes with triethylsilane and a little aluminium trichloride. p-Ethoxybenzoyl bromide did not react with triethylsilane alone, although reduction occurred in ether solution[112].

Phenylacetyl chloride reacted with triphenylstannane to produce 2-phenylethyl phenylacetate[111]. Reaction of acid chlorides with tri-n-butylstannane deuteride (Bu_3SnD) in the presence of nickel and platinum produced deuterated aldehydes (RCDO) and esters (RCO_2CD_2R)[113].

Copper hydride, prepared by the reduction of copper sulphate with sodium hypophosphite, converts benzoyl chloride into benzoic acid (70%) and ethyl benzoate (10%)[114]. The benzoic acid arose from water of hydration in the hydride and the ester from the ethyl alcohol used to wash the hydride[115]. Ethyl benzoate is not formed if the hydride is dried with acetone. Acetyl chloride similarly formed acetic acid (80%)[114]. However, dry copper hydride, prepared by reducing copper iodide in pyridine with lithium aluminium hydride, did reduce benzoyl chloride to benzaldehyde[116]. Lithium hydride in petroleum ether reduced benzoyl chloride at 150–210° to benzyl benzoate (65%). Only a trace of benzaldehyde was formed and a residue consisted of lithium benzoate[117].

IV. REDUCTION BY DISSOLVING METALS

Reduction by dissolving metals in alcohols, in liquid ammonia, or in amines generally cannot be employed since the acid halides would be solvolysed. Although sodium amalgam in methanol converted palmitoyl chloride to hexadecanol, the reaction probably involved Bouveault–Blanc reduction of methyl palmitate[118]. Sodium amalgam also reduced succinyl chloride to butyrolactone[119]. However, sodium and moist ether reduced

9

acid chlorides to acyloins or diketones[120, 121]. Butyryl, isobutyryl and isovaleroyl chlorides were converted to the esters of enediols (33)[122, 123],

$$2\,RCOCl \;+\; 2\,Na \longrightarrow \begin{array}{c} R-C=O \\ | \\ R-C=O \end{array} \;+\; 2\,NaCl$$

(31)

$$\begin{array}{c} R-C=O \\ | \\ R-C=O \end{array} + 2\,Na \longrightarrow \begin{array}{c} R-C-ONa \\ \| \\ R-C-ONa \end{array} \xrightarrow{2\,ROCl} \begin{array}{c} R-C-OCOR \\ \| \\ R-C-OCOR \end{array}$$

(31)   (32)   (33)

and similar compounds were formed from lauryl, myristyl, palmitoyl and stearoyl chlorides in yields of 64–70%[124]. The reaction probably proceeds via the diketone (31), which is further reduced to the acyloin-enol (32)[121]. Trimethylacetyl chloride afforded hexamethylbiacetyl (34) (70%) and the trimethylacetyl ester of the acyloin (35)[125].

$$\begin{array}{c} (CH_3)_3C-C-C-C(CH_3)_3 \\ \| \; \| \\ O \; O \end{array} \qquad\qquad \begin{array}{c} (CH_3)C-C-C-C(CH_3)_3 \\ \| \; | \\ O \; OCOC(CH_3)_3 \end{array}$$

(34)   (35)

Univalent magnesium also causes bimolecular reduction of acid chlorides[126, 127]. The reagent is prepared from excess magnesium and iodine or bromine. Propionyl chloride and butyryl chloride in ether gave

$$Mg + MgX_2 \rightleftharpoons 2MgX$$

only low yields of hexan-3,4-dione and octan-4,5-dione[128]. However, hindered aromatic acyl chlorides were reduced in satisfactory yield. Thus, mesitoyl chloride afforded 2,2′,4,4′,6,6′-hexamethyl-α,α'-stilbenediol (36) (35%)[129] and mesitil (37) (34%)[130]. 2,4,6-Triethylbenzoyl chloride similarly gave the diol (30%)[131] and the diketone (18%)[94], as did 2,4,6-tri-isopropyl-benzoyl chloride[132].

4-Bromo-2,6-xyloyl chloride was reduced to 4,4′-dibromo-2,2′,6,6′-tetramethylbenzil (14%) and 4,4′-dibromo-2,2′,6,6′-tetramethylbenzoin (10%)[133]. 2,3,4,6-Tetramethyl-[134], and 2,3,5,6-tetramethyl-benzoyl chloride[135] and 2-methyl-1-naphthoyl chloride[136] were reduced to mixtures of the corresponding stilbenediols and benzoins.

Reaction with metals in the absence of hydrogen donors gave other products. Thus, treatment of acetyl chloride with zinc dust afforded 1,1,6,6-tetraacetyl-3,4-dimethyl-1,2,4,5-hexatetraene (38), probably formed from the enol of biacetyl[137, 138]. Sodium in anhydrous ether converted benzoyl chloride to ethyl benzoate, while sodium or potassium in xylene gave benzoic anhydride. Succinoyl chloride and phthaloyl chloride both formed the corresponding anhydrides (60% and 70% yields, respectively) with sodium or potassium in anhydrous ether[139]. Butyryl chloride and

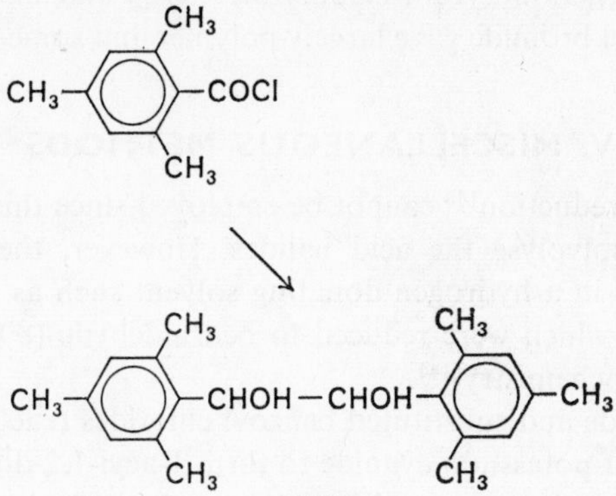

$$CH_3-\underset{\underset{CH_3}{|}}{\overset{\overset{CH_3}{|}}{C_6H_2}}-COCl$$

$$CH_3-\underset{\underset{CH_3}{|}}{\overset{\overset{CH_3}{|}}{C_6H_2}}-CHOH-CHOH-\underset{\underset{CH_3}{|}}{\overset{\overset{CH_3}{|}}{C_6H_2}}-CH_3$$

(36)

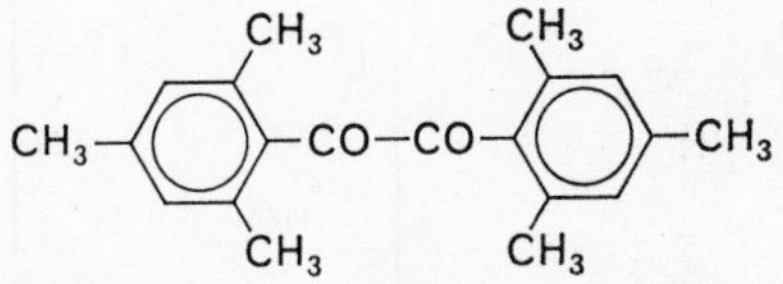

+

$$CH_3-\underset{\underset{CH_3}{|}}{\overset{\overset{CH_3}{|}}{C_6H_2}}-CO-CO-\underset{\underset{CH_3}{|}}{\overset{\overset{CH_3}{|}}{C_6H_2}}-CH_3$$

(37)

$$\underset{CH_3CO}{\overset{CH_3CO}{>}}C=C=\underset{CH_3}{\overset{CH_3}{C}}-\underset{CH_3}{\overset{CH_3}{C}}=C=C\underset{COCH_3}{\overset{COCH_3}{<}}$$

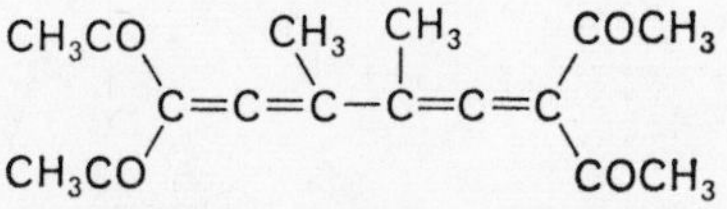

(38)

sodium in anhydrous ether formed the dibutyrate of 4-octene-4,5-diol[123]. However, benzoyl chloride vapour and sodium vapour reacted to produce benzil[140]. Zinc or iron with benzoyl chloride in ether afforded ethyl benzoate, but no benzil, although some benzyl benzoate was formed. In diisoamyl ether or dioxan, isoamyl benzoate or ethylene glycol dibenzoate were formed. No reaction took place in solutions of benzene, carbon disulphide or carbon tetrachloride[141].

Acetyl bromide with magnesium itself in ether formed ethyl acetate (35%) and biacetyl (30%). A violent reaction occurred in the absence of solvent. Isovaleroyl bromide similarly afforded after hydrolysis, isovaleraldehyde (27%), 2-isopropyl-6-methyl-hex-2-enal **(39)** (40%), isovaleroin

$$(CH_3)_2CHCH_2CH=C(CH(CH_3)_2)CHO$$

(39)

isovalerate (15%) and biisovaleryl (7%) formed via the magnesium (RCOMgBr) derivative. Isobutyryl bromide was similarly converted to isobutyroin (30%), isobutyroin isobutyrate (30%) and biisobutyryl (5%). However, benzoyl bromide gave largely polymer and some benzoic acid[142].

V. MISCELLANEOUS METHODS

Electrochemical reduction[143] cannot be employed since this requires polar solvents which solvolyse the acid halides. However, the irradiation of benzoyl bromide in a hydrogen-donating solvent such as ether produced benzoyl radicals which were reduced to benzaldehyde (80% in 30 h) (see chapter on Photochemistry)[144].

Benzoyl chloride and substituted benzoyl chlorides react with quinoline in the presence of potassium cyanide to form 1-acyl-1,2-dihydroquinaldo-nitriles (**40**, 'Reissert's compounds') which can be hydrolysed to benzalde-hyde and quinaldic acid (**42**). The reaction is known as Reissert reduction

(**42**)[92, 145]. *p*-Chlorobenzoyl chloride and cinnamoyl chloride were con-verted into the corresponding aldehydes in good yield[146], and *o*-nitro- and 2-nitro-3,4,5-trimethylbenzaldehyde were formed in 60–70% yield from the acid chlorides[147]. Aliphatic acid chlorides do not form Reissert's compounds under the normal conditions, but can be formed using hydro-gen cyanide in benzene[148]. The hydrolysis of the Reissert's compound probably proceeds through the quinoline derivative (**41**). The hydrolysis of 1-benzoyl-1,2-dihydroquinaldonitrile with hydrogen bromide in acetic acid afforded 87% benzaldehyde[149]. Benzaldehyde labelled with carbon-14 on the carbonyl group has been prepared via a Reissert compound from labelled benzoyl chloride[150].

A further method of converting an acid chloride into an aldehyde utilizes the Grundmann reaction[151], proceeding through a diazoketone (43), ketol acetate (44) and diol (45)[92]. Stearaldehyde, citronellal and benzalde-

$$RCOCl \xrightarrow{\;CH_2N_2\;} RCOCHN_2 \longrightarrow RCOCH_2OAc \longrightarrow$$
$$\quad\quad\quad\quad\quad (43) \quad\quad\quad\quad\quad\quad\quad (44)$$

$$RCHOHCH_2OH \xrightarrow{\;Pb(OAc)_4\;} RCHO$$
$$\quad\quad\quad\quad (45)$$

hyde were synthesized via this route in yields of 56%, 42% and 48%, respectively[145]. Oleic acid also afforded an aldehyde[152], and 3-benzyloxy-4-methoxyacetaldehyde was also prepared by this series of reactions[153].

Acid chlorides can also be converted to aldehydes by first forming the thiolesters, which are then desulphurized with Raney nickel[92,154,155]. The use of standard Raney nickel results in further reduction to the

$$RCOCl \longrightarrow RCOSR' \longrightarrow RCHO$$

carbinol[156,157], but desulphurization with deactivated Raney nickel produces the aldehyde[158].

Chromous chloride in aqueous potassium hydroxide reduced benzoyl chloride in a hydrogen atmosphere to benzyl alcohol (30%). However, 2,6-dichloroisonicotinyl chloride gave only a poor yield of 2,6-dichloro-4-pyridylcarbinol under these conditions[159].

n-Butylmagnesium chloride reduced pivaloyl chloride to neopentyl alcohol (27%) as well as forming n-butyl-t-butyl carbinol (69%). Acetyl chloride with n-butylmagnesium chloride gave ethanol (8%), n-hexanol (13%) and traces of butanol, dibutylmethyl carbinol and decenes[160].

VI. REFERENCES

1. E. Mosettig and R. Mozingo, *Org. Reactions*, **4**, 362 (1948).
2. N. O. V. Sonntag, *Chem. Rev.*, **52**, 237 (1953).
3. K. W. Rosenmund, *Ber.*, **51**, 585 (1918).
4. K. W. Rosenmund and F. Zetzsche, *Ber.*, **54**, 425 (1921).
5. E. B. Hershberg and J. Cason, *Org. Syn. Coll.* Vol. III, Wiley, New York, 1955, p. 627.
6. C. Weygand and W. Mensel, *Ber.*, **76**, 503 (1943).
7. F. Zetzsche, F. Enderlin, C. Flütsch and E. Menzi, *Helv. Chim. Acta*, **9**, 177 (1926).
8. F. Zetzsche and O. Arnd, *Helv. Chim. Acta*, **8**, 591 (1925).
9. F. Zetzsche and O. Arnd, *Helv. Chim. Acta*, **9**, 173 (1926).
10. K. W. Rosenmund, F. Zetzsche and F. Heise, *Ber.*, **54**, 638 (1921).
11. K. W. Rosenmund, F. Zetzsche and F. Heise, *Ber.*, **54**, 2038 (1921).
12. C. A. Rojahn and K. Fahr, *Ann. Chem.*, **434**, 252 (1923).
13. M. J. Egerton, G. I. Gregory and T. Malkin, *J. Chem. Soc.*, 2272 (1952).

14. Y. Sakurai and Y. Tanaka, *J. Pharm. Soc. Japan*, **64**, 25 (1944).
15. E. Spath, *Monatsh*, **40**, 129 (1919).
16. S. Daniloff and E. Venus-Danilova, *Ber.*, **59**, 377 (1926).
17. N. Fröschl, F. Maier and A. Heuberger, *Monatsh.*, **59**, 256 (1932).
18. F. Zetzsche, C. Flütsch, F. Enderlin and A. Loosli, *Helv. Chim. Acta*, **9**, 182 (1926).
19. S. A. Harris, D. E. Wolf, R. Mozingo, G. E. Arth, R. C. Anderson, N. R. Easton and K. Folkes, *J. Am. Chem. Soc.*, **67**, 2098 (1945).
20. H. Sobotka and F. E. Stynler, *J. Am. Chem. Soc.*, **72**, 5139 (1950).
21. Y. Egawa, M. Suzuki and T. Okuda, *Chem. Pharm. Bull. Japan*, **11**, 589 (1963).
22. F. Brown and W. K. R. Musgrave, *J. Chem. Soc.*, 5049 (1952).
23. A. L. Henne, R. L. Pelley and R. M. Alm, *J. Am. Chem. Soc.*, **72**, 3370 (1950).
24. J. W. Sellers and W. E. Bissinger, *J. Am. Chem. Soc.*, **76**, 4486 (1954).
25. R. H. Nesley and J. S. Driscall, *J. Heterocyclic Chem.*, **4**, 587 (1967).
26. T. Amakasu and K. Sato, *Bull. Soc. Chim. Japan*, **40**, 1428 (1967).
27. H. B. White, L. L. Sulya and C. E. Cain, *J. Lipid Res.*, **8**, 158 (1967).
28. A. F. Thomson and N. H. Cromwell, *J. Am. Chem. Soc.*, **61**, 1374 (1939).
29. C. B. Calleja and P. Rogers, *J. Labelled Compds.*, **6**, 135 (1970).
30. S. Affrossman and S. J. Thomson, *J. Chem. Soc.*, 2024 (1962).
31. R. Escourrou, *Bull. Soc. Chim. France*, **6**, 1173 (1939).
32. N. Fröschl and G. Danoff, *J. Prakt. Chem.*, **144**, 217 (1936).
33. K. V. Puzitskii, T. F. Bulanova, Y. T. Eidus, K. G. Ryabova and N. S. Sergeeva, *Zhur. Org. Khim.*, **3**, 785 (1967).
34. K. V. Puzitskii, Y. T. Eidus, T. F. Bulanova, K. G. Ryabova and N. S. Sergeeva, *Izv. Akad. Nauk SSSR, Ser. Khim.*, 117 (1968).
35. K. W. Rosenmund, *Angew. Chem.*, **38**, 145 (1925).
36. V. Grignard and G. Mingasson, *Compt. Rend.*, **185**, 1173 (1927).
37. A. Mailhe, *Compt. Rend.*, **180**, 111 (1925).
38. E. A. Braude, R. P. Linstead, P. W. D. Mitchell and K. R. H. Wooldrige, *J. Chem. Soc.*, **3**, 595 (1954).
39. W. G. Brown, *Org. React.*, **6**, 469 (1951).
40. N. G. Gaylord, *Reduction with Complex Metal Hydrides*, Interscience, New York, 1956.
41. R. F. Nystrom and W. G. Brown, *J. Am. Chem. Soc.*, **69**, 1197 (1947).
42. C. E. Sroog, C. M. Chih, F. A. Short and H. W. Woodburn, *J. Am. Chem. Soc.*, **71**, 1710 (1949).
43. A. L. Henne, R. M. Alm and M. Smook, *J. Am. Chem. Soc.*, **70**, 1968 (1948).
44. H. H. Zeiss, C. E. Slimowicz and V. Z. Pasternak, *J. Am. Chem. Soc.*, **70**, 1981 (1948).
45. H. C. Brown, P. M. Weissman and N. M. Yoon, *J. Am. Chem. Soc.*, **88**, 1458 (1966).
46. S. W. Chaikin and W. G. Brown, *J. Am. Chem. Soc.*, **71**, 122 (1949).
47. H. C. Brown, *U.S. Pat.* 3,277,178 October 4, 1966.
48. H. C. Brown and B. C. Subba Rao, *J. Am. Chem. Soc.*, **77**, 3164 (1955).
49. H. C. Brown and B. C. Subba Rao, *J. Am. Chem. Soc.*, **78**, 2582 (1956).
50. R. Hayazu, *Pharm. Bull. (Tokyo)*, **5**, 452 (1957).

51. R. Trave and L. Garanti, *Gazz. Chim. Ital.*, **90**, 597 (1960).
52. R. Trave and L. Garanti, *Gazz. Chim. Ital.*, **93**, 549 (1963).
53. L. A. Paquette and N. A. Nelson, *J. Org. Chem.*, **27**, 2272 (1962).
54. F. E. King and J. W. W. Morgan, *J. Chem. Soc.*, 4738 (1960).
55. L. Duhamel, *Ann. Chim. (Paris)*, **8**, 315 (1963).
56. S. Takahaski, *Agri. Biol. Chem. (Tokyo)*, **26**, 401 (1962).
57. A. B. A. Jansen and M. Szelke, *J. Chem. Soc.*, 405 (1961).
58. R. Fuchs and D. M. Carlton, *J. Org. Chem.*, **27**, 1520 (1962).
59. K. Hirai, *Yakugaku Zasshi*, **80**, 608 (1966).
60. P. K. Banerjee and D. N. Choudhury, *J. Org. Chem.*, **26**, 4344 (1961).
61. E. Wenkert, D. B. R. Johnston and K. G. Dave, *J. Org. Chem.*, **29**, 2534 (1964).
62. F. Wessely, K. Schlögl and G. Korger, *Nature*, **169**, 708 (1952).
63. D. C. Iffland and H. Siegel, *J. Am. Chem. Soc.*, **80**, 1947 (1958).
64. C. W. Muth, J. C. Ellers and O. F. Folmer, *J. Am. Chem. Soc.*, **79**, 6500 (1957).
65. R. Grewe, H. Büttner and G. Burmeister, *Angew. Chem.*, **69**, 61 (1957).
66. K. Nagarajan, C. Weismann, H. Schmid and P. Karrer, *Helv. Chim. Acta*, **46**, 1212 (1963).
67. R. Breslow, J. Lockhart and A. Small, *J. Am. Chem. Soc.*, **84**, 2793 (1962).
68. V. P. Arya and B. G. Engel, *Helv. Chim. Acta*, **44**, 1650 (1961).
69. V. Bazant, M. Capka, M. Cerny, V. Chvalovsky, K. Kochloefl, M. Kraus and J. Malek, *Tetrahedron Letters*, 3303 (1968).
70. M. Cerny, J. Malek, M. Capka and V. Chvalovsky, *Collection Czech. Chem. Commun.*, **34**, 1025 (1969).
71. H. C. Brown and N. M. Yoon, *J. Am. Chem. Soc.*, **88**, 1464 (1966).
72. H. C. Brown and P. M. Weissman, *J. Am. Chem. Soc.*, **87**, 5614 (1965).
73. H. C. Brown and P. M. Weissman, *Israel J. Chem.*, **1**, 430 (1963).
74. H. C. Brown and E. J. Mead, *J. Am. Chem. Soc.*, **75**, 6263 (1953).
75. R. Grewe and H. Büttner, *Chem. Ber.*, **91**, 2452 (1958).
76. K. Freudenberg and W. Fuchs, *Chem. Ber.*, **88**, 1824 (1955).
77. E. M. Hodnett and J. I. Sparapany, *Pure Appl. Chem.*, **8**, 385 (1964).
78. W. P. Cochrane, P. L. Pauson and T. S. Stevens, *J. Chem. Soc.*, C, 630 (1968).
79. H. C. Brown and R. F. McFarlin, *J. Am. Chem. Soc.*, **78**, 252 (1958).
80. H. C. Brown and R. F. McFarlin, *J. Am. Chem. Soc.*, **80**, 5372 (1958).
81. H. C. Brown and B. C. Subba Rao, *J. Am. Chem. Soc.*, **80**, 5377 (1958).
82. E. D. Bergmann and A. Cohen, *Tetrahedron Letters*, 1151 (1965).
83. H. J. Bestman and R. Schmiecken, *Chem. Ber.*, **94**, 751 (1961).
84. M. Julia, S. Julia and B. Cochet, *Bull. Soc. Chim. France*, 1487 (1964).
85. A. Vallet, J. Janin and R. Romanet, *J. Labelled Compds.*, **4**, 299 (1968).
86. E. Hardegger and H. Corradi, *Pharm. Acta Helv.*, **39**, 101 (1964).
87. D. Buttimore, D. H. Jones, R. Slack and K. R. H. Wooldridge, *J. Chem. Soc.*, 2032 (1963).
88. H. Rutner and P. E. Spoerri, *J. Org. Chem.*, **28**, 1898 (1963).
89. L. I. Zakharkin, D. N. Maslin and V. V. Gavrilenko, *Zh. Org. Khim.*, **2**, 2153 (1966).
90. L. I. Zakharkin, D. N. Maslin and V. V. Gavrilenko, *Zh. Org. Khim.*, **2**, 2197 (1966).

91. F. Weygand, G. Eberhardt, H. Linden, F. Schafer and I. Eigen, *Angew. Chem.*, **65**, 525 (1953).
92. E. Mosettig, *Org. React.*, **8**, 218 (1954).
93. H. C. Brown and A. Tsukamoto, *J. Am. Chem. Soc.*, **86**, 1089 (1964).
94. H. C. Brown and A. Tsukamoto, *J. Am. Chem. Soc.*, **83**, 2016 (1961).
95. H. C. Brown and A. Tsukamoto, *J. Am. Chem. Soc.*, **83**, 4549 (1961).
96. A. W. Burgstahler, L. H. Worden and T. B. Lewis, *J. Org. Chem.*, **28**, 2918 (1963).
97. L. I. Zakharkin and I. M. Khorlina, *Zh. Obshch. Khim.*, **34**, 1029 (1964).
98. L. I. Zakharkin and I. M. Khorlina, *Izvest. Akad. Nauk SSSR, Otdel, Khim. Nauk*, 316 (1961).
99. J. Carnduff, *Quart. Rev. Chem. Soc.*, **20**, 169 (1966).
100. H. C. Brown and B. C. Subba Rao, *J. Am. Chem. Soc.*, **82**, 681 (1960).
101. H. C. Brown and W. Korytnyk, *J. Am. Chem. Soc.*, **82**, 3866 (1960).
102. H. Noth and H. Beyer, *Chem. Ber.*, **93**, 1078 (1960).
103. D. L. Chamberlain and W. H. Schechter, *U.S. Pat.* 2, 898, 379, August 4, (1959).
104. H. C. Kelly and J. O. Edwards, *J. Am. Chem. Soc.*, **82**, 4842 (1960).
105. C. J. M. van der Kerk, J. G. Noltes and J. G. A. Luijten, *J. Appl. Chem.*, **7**, 356 (1957).
106. H. G. Kuivila, *Advances in Organometallic Chemistry* (Ed. F. G. A. Stone and R. West), Academic, New York 1964, p. 47.
107. H. G. Kuivila, *J. Org. Chem.*, **25**, 284 (1960).
108. H. G. Kuivila and E. J. Walsh, *J. Am. Chem. Soc.*, **88**, 571 (1966).
109. L. W. Menapace and H. G. Kuivila, *J. Am. Chem. Soc.*, **86**, 3047 (1964).
110. E. J. Walsh and H. G. Kuivila, *J. Am. Chem. Soc.*, **88**, 576 (1966).
111. E. J. Kupchik and R. J. Kiesel, *J. Org. Chem.*, **31**, 456 (1966).
112. J. W. Jenkins and H. W. Post, *J. Org. Chem.*, **15**, 552 (1950).
113. K. Kuehlein, W. P. Neumann and H. Mohring, *Angew. Chem. Intern. Ed. Engl.*, **7**, 455 (1968).
114. E. A. Braude, *J. Chem. Soc.*, 1940 (1949).
115. O. Neuhoeffer and F. Nerdel, *J. Prakt. Chem.*, **144**, 63 (1935).
116. E. Wiberg and W. Henle, *Z. Naturforsch.*, **7b**, 250 (1952).
117. A. Hodaghian and R. Levaillant, *Compt. Rend.*, **194**, 2059 (1932).
118. S. Sikata, N. Osaka and Y. Inoue, *U.S. Pat.* 2, 263, 195 (1941).
119. N. Fraschl, A. Maier and A. Neuberger, *Monatsh.*, **59**, 256 (1932).
120. W. H. Perkin and J. J. Sudborough, *Ber.*, **29**, R.662 (1896).
121. S. M. McElvain, *Org. React.*, **4**, 256 (1948).
122. H. Klinger and L. Schmitz, *Ber.*, **24**, 1271 (1891).
123. A. Basse and K. Klinger, *Ber.*, **31**, 1218 (1898).
124. A. W. Ralston and W. M. Selby, *J. Am. Chem. Soc.*, **61**, 1019 (1939).
125. V. I. Egorova, *J. Russ. Phys. Chem. Soc.*, **60**, 1199 (1928).
126. W. S. Ide and J. S. Buck, *Org. React.*, **4**, 269 (1948).
127. M. D. Rausch, W. E. McEwen and J. Kleinberg, *Chem. Rev.*, **57**, 417 (1957).
128. H. H. Inhoffen, H. Pommer and F. Bohlmann, *Chem. Ber.*, **81**, 507 (1948).
129. R. C. Fuson and J. Corse, *J. Am. Chem. Soc.*, **61**, 975 (1939).
130. R. C. Fuson, C. H. McKeever and J. Corse, *J. Am. Chem. Soc.*, **62**, 600 (1940).

131. R. C. Fuson, J. Corse and C. H. McKeever, *J. Am. Chem. Soc.*, **61**, 2010 (1939).
132. R. C. Fuson and E. C. Horning, *J. Am. Chem. Soc.*, **62**, 2962 (1940).
133. R. C. Fuson, S. L. Scott and R. V. Lindsey, *J. Am. Chem. Soc.*, **63**, 1679 (1941).
134. R. C. Fuson and S. C. Kelton, *J. Am. Chem. Soc.*, **63**, 1500 (1941).
135. R. C. Fuson, J. Corse and P. B. Welldon, *J. Am. Chem. Soc.*, **63**, 2645 (1941).
136. R. C. Fuson, C. H. McKeever and L. C. Behr, *J. Am. Chem. Soc.*, **63**, 2648 (1941).
137. D. A. Pospekhov, *Khim. Referat. Zhur.*, **4**, 30 (1941); *Chem. Abstr.*, **37**, 4359 (1943).
138. D. A. Pospekhov, *Sbornik Rabot Kiev. Tekh. Inst. Kozhevenno-Obuvnoi Prom.*, **3**, 268 (1940); *Chem. Abstr.*, **37**, 4359 (1943).
139. I. A. Pearl, T. W. Evans and W. M. Dehn, *J. Am. Chem. Soc.*, **60**, 2478 (1938).
140. H. V. Hartel, *Trans. Faraday Soc.*, **30**, 187 (1934).
141. G. A. Varvoglis, *Ber.*, **70**, 2391 (1937).
142. D. V. Tistchenko, *Bull. Soc. Chim. France*, **37**, 623 (1925).
143. F. D. Popp and H. P. Schultz, *Chem. Rev.*, **62**, 19 (1962).
144. U. Schmidt, *Angew. Chem.*, **77**, 169 (1965).
145. W. E. McEwen and R. L. Cobb, *Chem. Rev.*, **55**, 511 (1955).
146. W. E. McEwen, J. V. Kindall, R. N. Hazlett and R. H. Glazier, *J. Am. Chem. Soc.*, **73**, 4591 (1951).
147. G. L. Buchanan, J. W. Cook and J. D. Loudon, *J. Chem. Soc.*, 325 (1944).
148. J. M. Grosheintz and H. O. L. Fischer, *J. Am. Chem. Soc.*, **63**, 2021 (1941).
149. J. W. Davis, *J. Org. Chem.*, **24**, 1691 (1959).
150. D. G. Ott, *Organic Synthesis with Isotopes* (Ed. A. Murray and D. L. Williams), Interscience, New York, 1958, p. 626.
151. V. C. Grundmann, *Ann. Chem.*, **524**, 31 (1936).
152. H. K. Mangold, *J. Org. Chem.*, **24**, 405 (1959).
153. C. Schöpf, E. Brass, E. Jacobi, W. Jorde, W. Mocnik, L. Neuroth and W. Salzer, *Ann. Chem.*, **544**, 30 (1940).
154. M. L. Wolfrom and J. V. Karabinos, *J. Am. Chem. Soc.*, **68**, 724 (1946).
155. M. L. Wolfrom and J. V. Karabinos, *J. Am. Chem. Soc.*, **68**, 1455 (1946).
156. O. Jeger, J. Norymberski, S. Szpilfogel and V. Prelog, *Helv. Chim. Acta*, **29**, 684 (1946).
157. V. Prelog, J. Norymberski and O. Jeger, *Helv. Chim. Acta*, **29**, 360 (1949).
158. G. B. Spero, A. V. McIntosh and R. H. Levin, *J. Am. Chem. Soc.*, **70**, 1907 (1948).
159. R. Graf and M. Tatzel, *J. Prakt. Chem.*, 146, 198 (1936).
160. F. C. Whitmore, A. H. Popkin, J. S. Whitaker, K. F. Mattil and J. D. Zech., *J. Am. Chem. Soc.*, **60**, 2458 (1938).

CHAPTER 8

Rearrangements involving acyl halides

D. V. BANTHORPE
University College, London

and

B. V. SMITH
Chelsea College of Science and Technology, University of London

I. INTRODUCTION

Systematic investigations of the rearrangements of acyl halides are few in number and where such reactions are reported they are usually incidental to other studies. Consequently, references to this class of reactions are widely scattered throughout the literature and the present survey, which covers the period up to April 1971, cannot claim to be exhaustive or even to have unearthed every type.

Unlike the situation for alkyl halides, the generation of a cation from simple acyl or aroyl halides rarely leads to rearrangement of the ion itself. This contrast between the properties of a carbonium and an acylium ion is undoubtedly due to the smaller positive charge on the α-carbon atom

253

of the latter, and for the aryl analogues additionally to the delocalization of the charge into the aromatic ring. Unsolvated cations possessing enhanced tendencies for rearrangement such as are obtained on treatment of alkyl amines with nitrous acid are apparently not generated from carbonyl halides. Most of the reactions considered within the province of this discussion consequently involve isomerizations, or rearrangements coupled with extrusion of groups or atoms, that are not generally paralleled in the chemistry of alkyl and aryl halides, amines and such like.

In certain systems, where alternative formulations of the structure of a carbonyl halide are possible, it is not always clear whether isomerization precedes a particular sequence of reactions or occurs in its course, and often therefore the examples cited may be open to more than one mechanistic interpretation. It will also be apparent that the sub-divisions of the following sections are somewhat arbitrary and many examples could well be classified in a different, or in more than one, section. Conditions under which the cited reactions have been carried out vary widely and mention of these is only made when the point has some mechanistic significance.

Most of the experimental data cited is semi-quantitative in being based on isolation of products by classical techniques: few studies have been made using spectroscopic or chromatographic means of analysis. Likewise, although most of the structural assignments appear reasonable and are usually based on some chemical evidence, much of the work was carried out before the advent of spectroscopic methods of structure determination. These limitations may account for certain anomalies that have been observed.

II. REARRANGEMENTS OF HALF-ESTER CHLORIDES

A well-known synthetic route to ketones is the reaction of organometallic compounds with acid chlorides. One of its most important applications is the preparation of keto-esters by treatment of half-ester chlorides with organozinc or organocadmium compounds[1]; this enables elongation of a carbon chain whilst retaining the terminal functional group for further synthetic modification, reaction (1). A difficulty that has limited the use of

$$R_2Cd + ClCO(CH_2)_nCO_2Et \longrightarrow RCO(CH_2)_nCO_2Et + RCdCl \qquad (1)$$

unsymmetrical ester chlorides in this process is the frequent migration of the functional group in the products resulting in the formation of two keto-esters (reaction 2). In some examples the same product was obtained whichever of the isomeric half-ester chlorides was used and so rearrangement of one of the isomers must have been essentially complete. Cyclizations also occurred in certain suitably constructed compounds, whilst

$$
\begin{array}{ccc}
CO_2R' & CO_2R & COR'' \\
| & | & | \\
(CH_2)_m & (CH_2)_m & (CH_2)_m \\
| & | & | \\
CHR \xrightarrow[\text{or}]{R''_2Cd \atop R''Z_nI} & CHR & + \quad CHR \\
| & | & | \\
(CH_2)_n & (CH_2)_n & (CH_2)_n \\
| & | & | \\
COCl & COR'' & CO_2R
\end{array} \qquad (2)
$$

for others acylations that were catalysed by Lewis acids gave partly rearranged products. No rearrangement appears to have been reported on treatment with sodio-diethyl malonate.

The main examples of these types of reactions involve derivatives of succinic, glutaric and phthalic acids, and these together with some minor classes will be discussed in turn and finally the mechanism will be outlined.

(i) *Derivatives of succinic acid.* Although it has been shown[2] that the half-ester chloride (1) (R = alkyl) condensed with sodio-diethyl malonate

$$RO_2CCH_2CMe_2COCl \qquad PhCOCH_2CMe_2CO_2H \qquad EtO_2CCHMeCMe_2COCl$$
$$\text{(1)} \qquad\qquad\qquad \text{(2)} \qquad\qquad\qquad \text{(3)}$$

without rearrangement and also that the reaction of 1 (R = Et) with ethylzinc iodide also proceeded as expected[3], the acylation of benzene with the analogous methyl ester gave[4] only the rearranged keto-acid (2). Treatment of 3 with methylzinc chloride gave a mixture of keto-esters[5] in contrast with the result[3] for its lower homologue; whereas 4 gave only the rearranged acid 5 on reaction with benzene in the presence of aluminium trichloride[6, 7]. The isomeric ester chlorides of methylsuccinic acid were

$$\overset{+}{A}g\overset{\bar{}}{O}_2CCPh_2CH_2CO_2R$$

(4)   (5)   (6)

prepared using oxalyl chloride[8] and the derived anilides were shown to be well-characterized compounds that were each formed with no isomerization of the substrate: this result has a very significant bearing on the elucidation of the pathway to rearrangement in related compounds, for in most of the other examples that have been studied the acid chloride was prepared with the use of thionyl chloride.

Reactions of each isomeric methyl-ester chloride of phenylsuccinic acid with various organometallics were reported[9] to furnish separate ketones, but similar reactions of the α,α-diphenyl analogue gave extensive yields of rearranged product when the acid chloride was prepared using thionyl chloride[10]; although reaction of the dry silver salt (6) with the same chlorinating agent gave no rearranged chloride. A rearrangement that occurred during the actual formation of the acid chloride in the first example was later confirmed[11] and although the reaction of the α,α-diphenyl compound with diazomethane led to unrearranged products, reactions of the same substrate under Friedel–Crafts conditions gave rearranged products. Thus under these last conditions the ester chloride (7) cyclized and either 8 or 9 could be isolated depending on the

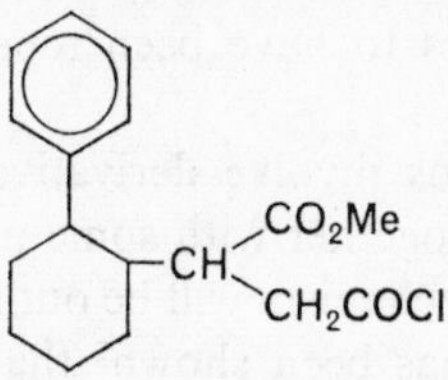

(7) 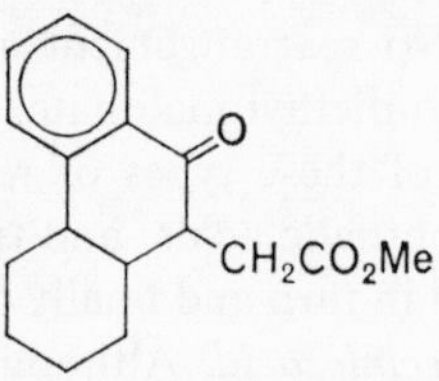   (8) 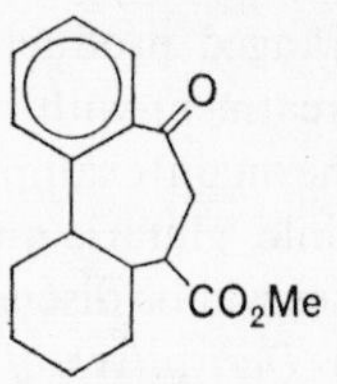   (9)

catalyst[12, 13]: aluminium trichloride and tribromide gave 8 or 9 respectively, but the action of either on 9 gave 7 in high (ca. 90%) yield.

A related cyclization[14] is the reaction of 10 with stannic chloride to give the keto-ester (11) accompanied by a small percentage of 12. The yield of the latter was increased either by warming the substrate before reaction or by preparing it with thionyl chloride rather than with other halogenating agents. It was concluded that the conversion of 10 into its isomeric ester-chloride and subsequent cyclization was responsible for the formation of 12.

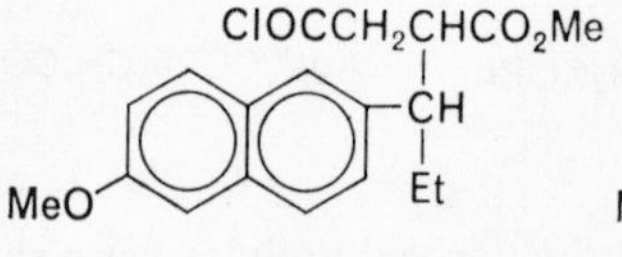

(10) 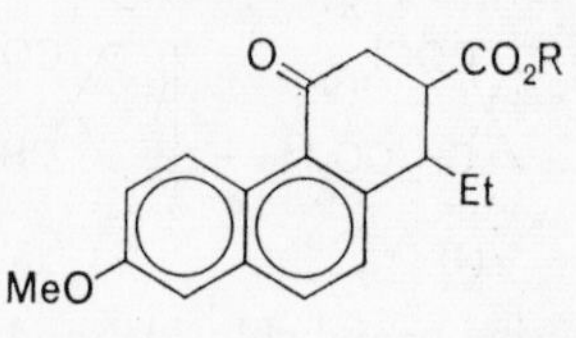   (11)

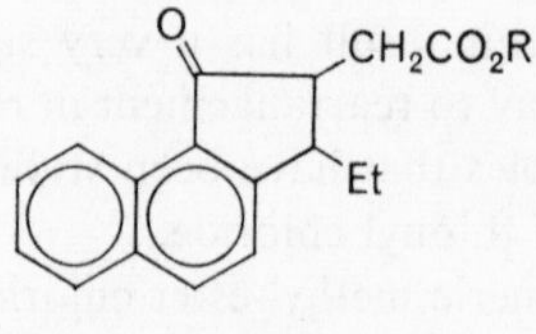

(12)

(ii) *Derivatives of glutaric acid and related compounds.* A well-authenticated example[15] of rearrangement in this series is the interconversion of the α- and β- half-esters of camphoric acid via the chlorides (reaction 3).

(3)

(α-half-ester)   (α)   (β)

Decomposition of the α-chloride with water gave mainly the β-acid together with some of the α-acid and the same mixture of products resulted from similar treatment of the β-chloride. Reaction of the α- or β-half-ester chlorides with ammonia also gave the same proportions of amides irrespective of which isomer was the starting material: this finding accounted for the unexpected difficulty[16] of preparing the amides as pure compounds. Treatment of either the α- or β-half-ester chlorides of camphoric acid with methylzinc chloride also gave[17] the same keto-ester and the substrates were considered to be tautomers. In these studies it was not clear whether the interchange of groups took place after the formation of the acid chlorides or during their preparation; but the acid chlorides of the half-esters were later shown to be separable and interconvertible, the view that tautomers were involved was shown to be erroneous, and what is now accepted as the mechanism was proposed[18] and later reiterated[15].

Another detailed study[19] of a substituted glutaric acid revealed that when **13** was converted into the acid chloride and reacted with *s*-tribromoaniline the same anilide was obtained as that prepared from the

$$ HO_2C-\underset{\underset{Et}{|}}{\overset{\overset{n\text{-}Bu}{|}}{C}}-(CH_2)_2CO_2Me $$

(13)

$$ MeO_2C-\underset{\underset{Et}{|}}{\overset{\overset{n\text{-}Bu}{|}}{C}}-(CH_2)_2CO_2H $$

(14)

$$ MeO_2C-\underset{\underset{Et}{|}}{\overset{\overset{n\text{-}Bu}{|}}{C}}-(CH_2)_3CO_2H $$

(15)

isomer **14**. The former acid chloride was stable to storage or heat, but underwent extensive (ca. 40%) rearrangement when treated with a trace of ferric chloride[20]. With zinc chloride in anhydrous ether a higher yield (86%) of **14** was isolated after work-up.

Homologues of **14** gave less rearrangement: e.g. the acid chloride of **15** gave approximately 32% of such products on treatment with ferric

$$\underset{\underset{\text{Et}}{|}}{\overset{\overset{\textit{n}\text{-Bu}}{|}}{MeO_2C-C}}-(CH_2)_4CO_2H$$

(16)

$$\underset{\underset{\text{Et}}{|}}{\overset{\overset{\textit{n}\text{-Bu}}{|}}{\textit{n}\text{-}C_5H_{11}-C}}-(CH_2)_2CO_2H$$

(17)

$$\underset{\underset{\text{Et}}{|}}{\overset{\overset{\textit{n}\text{-Bu}}{|}}{HO_2C-C}}-\textit{n}\text{-}C_7H_{11}$$

(18)

chloride whereas **16** was negligibly rearranged under the same conditions. The half-ester acid chlorides derived from **13** or **14** when treated with di-*n*-butylcadmium gave[20] a mixture of keto-esters which were reduced by the Wolff–Kishner method to form **17** (ca. 20%) and **18** (ca. 75%). As this system has one severely hindered carbonyl group, it was possible both to establish the homogeneity of the half-esters used as starting materials with some certainty and to assign the formation of isomeric products as a genuine rearrangement of the half-ester chlorides.

Reaction (4) led to a racemic acid when thionyl chloride was the chlorinating agent, although no racemization occurred when oxalyl

$$\underset{\underset{\text{CH}_2\text{CO}_2\text{H}}{|}}{\overset{\overset{\text{CH}_2\text{CO}_2\text{Me}}{|}}{^*\text{CHMe}}} \quad [\equiv RCO_2H] \xrightarrow{\text{SOCl}_2} RCOCl$$

$$\downarrow \text{H}_2\text{O}$$

$$(\pm)-RCO_2H \tag{4}$$

chloride was used[21]. However, racemization did not result when the thionyl chloride was purified[22] or even when up to 10% sulphuryl chloride was added to the pure reagent. The effective catalyst for the rearrangement that was manifested as racemization was proposed to be a metal salt present in commercial samples of thionyl chloride that was acting as a Lewis acid. In concord with this view, the addition of a few moles per cent of ferric chloride to the pure chlorinating agent and use of the mixture in the reaction led to very extensive racemization.

(iii) *Derivatives of phthalic acid.* The isomeric acid chlorides from the monomethyl esters of 3-nitrophthalic acid reacted[23] under Friedel–Crafts conditions to give rearranged products (reaction 5). In a related study[24], the chloride from **19** acylated chlorobenzene to give isomeric chloro-benzoylnitro–benzoic acids and a similar reaction with toluene was found[25]

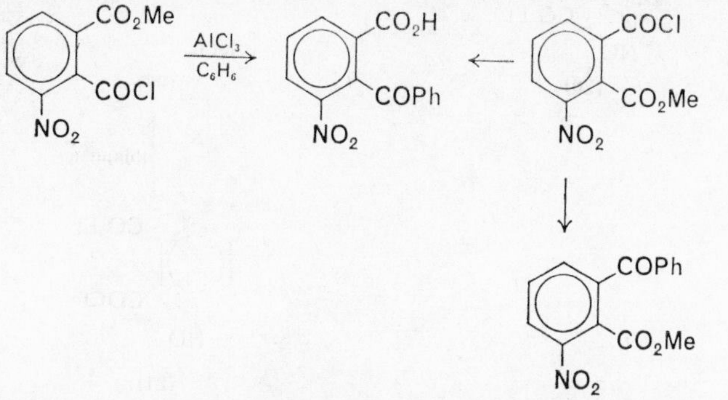

(5)

also to furnish two products despite an earlier report[26] to the contrary. The finding that **19** formed a peroxide that decomposed in benzene to yield two isomeric substituted biphenyls, reaction (6), also provided clear evidence that the acid chloride partly isomerized before reaction[27, 28]. A very detailed reinvestigation and reinterpretation[11] of the earlier results revealed that an equilibrium mixture was set up from either of the isomeric

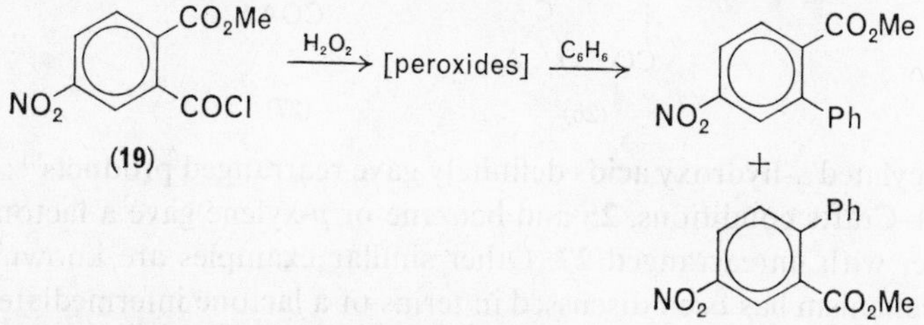

(6)

chlorides of 3-nitrophthalic acid in which the 1-methyl ester predominated, and the analogous 4-nitro-system gave a similar pattern.

Acylation of thiophene by **20** or its isomer **(21)** under catalysis by stannic chloride gave[29] only one keto-acid **(22)**, reaction (7). An independent experiment demonstrated that preliminary isomerization of **20** to the more stable isomer **(21)** had preceded reaction.

These and similar[11] rearrangements occurred with σ-phthalic acids and their derivatives: the absence of similar reactions for *iso* and *tere*-phthalic acids strongly suggests an intramolecular pathway for the first type of compounds.

(iv) *Half-ester chlorides of other dibasic acids and miscellaneous compounds.* Acylation of benzene using the chlorides of the acids **(23)** and **(24)** was reported[30] to give unrearranged products although these were not conclusively identified. However, the reactions of several acid chlorides derived

$$(7)$$

from acylated α-hydroxy acids definitely gave rearranged products[31]: under Friedel–Crafts conditions, 25 and benzene or p-xylene gave a lactone (26) together with unrearranged 27. Other similar examples are known[32] and the mechanism has been discussed in terms of a lactone intermediate[33]. An

attempt to convert 28 into an aryl ketone by a similar method was frustrated by fragmentation products[34] that were attributed to the decomposition of the lactone in feasible but unexpected ways.

The previously discussed rearrangement of the acid chloride of 15 is the only known example in which a derivative of adipic acid or an acid with a longer carbon chain undergoes rearrangement[35]: for example, the homocamphoric acid derivative (29) gave no rearranged products when treated with sodiodiethyl malonate[36].

(v) *Mechanism.* The studies outlined in the previous sections have been discussed, insofar as any mechanism has been assigned, in terms of three intermediates: an anhydride (30), an oxonium chloride (31) and an alkoxychlorolactone (32). The intermediacy of the first has been ruled out

(30)   (31)   (32)

for the particular example of the reaction of the chloride from 19 and by inference in the general case: the anhydride was prepared and shown to lead almost exclusively to the 1-methyl ester under the reaction conditions whereas the reaction of 19 led mainly to derivatives of the 2-methyl ester. Other evidence of a similar nature has also been presented against this mechanism[11].

The intervention of 31 is more reasonable but is less easy to define, although the indications are[11] that in circumstance where the species could be formed from a suitably constructed ester chloride little rearrangement does in fact occur. The consensus points to the involvement of 32, the formation of which is catalysed by Lewis acids. Evidence for the spontaneous conversion of the open-chain form into the lactone is sparse and inconclusive and the reaction requires the preliminary formation of an acylium ion[22]. No rearrangement occurs when half-ester chlorides react with alcohols or mercaptans[37] or with diazomethane in the first step of the Arndt–Eisert homologation[11], and in none of these processes are cationic intermediates believed to occur.

The accepted mechanism[11] is outlined in reaction (8); although a contrary view[8] envisages the intermediacy of an oxonium salt followed

(8)

by bimolecular reactions to give the diester (33) and the dichlorolactone (34), and similar subsequent processes (reaction 9). There appears to be neither direct evidence nor suitable analogues for these reactions.

III. REARRANGEMENTS OF UNSATURATED SYSTEMS

A. Positional Isomerization

In the course of investigations on isomeric α,β- and β,γ-unsaturated acids it was found[38] that some acid chlorides rearranged by migration of the double bond. 35 (R = Me) showed no such tendency to isomerize as judged by the recovery of unrearranged acid on hydrolysis but homologues

(R = Me, Et, n-Pr, i-Pr) and β,γ-unsaturated halides (36) (R', R'' = H or alkyl) gave up to 80% rearranged products, e.g. 37, from the α,β-compound in similar reactions. Such isomerizations occurred either in the syntheses of the acid chlorides using thionyl chloride or by the use of other chlorinating agents or when the chlorides thus formed were heated.

No rearrangement was detected when the isomeric chlorides (38) and (39) (R = Me, Et) were treated under conditions similar to those used for

$$\text{MeCH=CRCH}_2\text{COCl} \qquad \text{MeCH}_2\text{CR=CHCOCl}$$
$$\textbf{(38)} \qquad\qquad\qquad \textbf{(39)}$$

the phenyl-substituted compounds. In aliphatic compounds, the α,β- and β,γ-unsaturated enes are probably stabilized by hyperconjugation to a similar extent and there is presumably little driving force for rearrangement whereas, in the aromatic series, the β,γ-unsaturated isomer has a greater stabilization than its α,β-isomer. Reaction (10), which occurs during the formation of the acid chloride by the action of thionyl chloride[39], has some analogy with the foregoing examples. All presumably proceed by the general scheme of reaction (11).

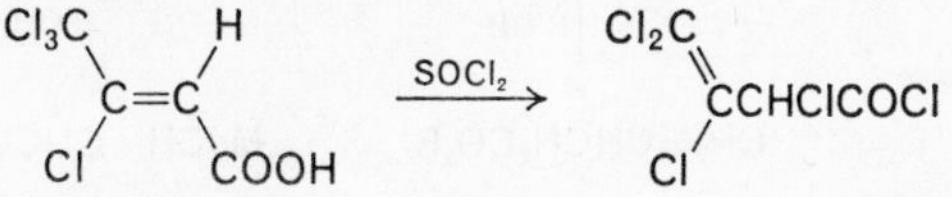

(10)

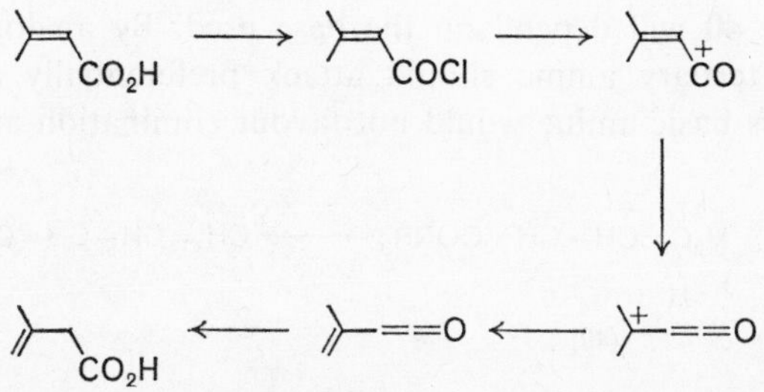

(11)

Treatment of an acid chloride with an alcohol in the presence of a strongly basic amine (the Einhorn Reaction) leads, in some cases, to isomerizations similar to those just described[40], e.g. reaction (12), although

$$\text{MeCH=CHCOCl} \xrightarrow[\text{R}_3'\text{N}]{\text{ROH}} \text{CH}_2\text{=CHCH}_2\text{CO}_2\text{R}$$
(major)
$$+$$
$$\text{MeCH=CHCO}_2\text{R}$$
(minor)

(12)

now these occur in purely aliphatic systems. Use of tertiary amines (Et$_3$N, n-Bu$_3$N, etc.) led to extensive double-bond migration (70–96%)

with a variety of alcohols but the unrearranged product predominated when pyridine or other relatively weak amines were used, or when sodium hydroxide was employed as catalyst. In the absence of amine no rearrangement was observed. A more detailed study[41] revealed that the basicity of the amine was indeed a crucial factor in controlling products: with catalysts of pK_a ca. 10 rearrangement predominated (ca. 90%), whereas for amines with pK_a less than 7·2 the unrearranged product was uniquely formed. Amines of intermediate basicity led to the formation of mixtures of comparable quantities of rearranged and unrearranged products.

The mechanism suggested is outlined in reaction (13). It is known[42] that α,β-unsaturated acid halides form ketenes with strong bases such as tertiary alkylamines and the relative proportions of the ketenes generated

$$\text{MeCH=CHCOCl} \xrightarrow{\text{NR}_3} \text{MeCH=CHCON}\overset{+}{R}_3$$

$$\downarrow \text{base}$$

$$\text{CH}_2\text{=CH–CH=C=O} \qquad \text{MeCH=C=C=O} \tag{13}$$

$$\downarrow \text{R'OH} \qquad\qquad\qquad \downarrow \text{R'OH}$$

$$\text{CH}_2\text{=CHCH}_2\text{CO}_2\text{R}' \qquad \text{MeCH=CHCO}_2\text{R}'$$

by attack on **40** will depend on the base used. By analogy with other reactions, a tertiary amine should attack preferentially at position γ whereas a less basic amine would not favour elimination and the species

$$\overset{\gamma\quad\beta\quad\alpha\qquad+}{\text{H}_2\text{C–CH=CH–CONR}_3} \longrightarrow \text{CH}_2\text{=CH–CH=C=O}$$

$$\text{B:}\overset{}{\curvearrowright}\overset{|}{\underset{\text{H}}{}}$$

$$\textbf{(40)}$$

40 would decompose to form unrearranged products. The kinetic forms of these different modes of reactions have not been checked, however.

Under certain conditions a related process (reaction 14), can be identified, which leads to unrearranged products[43] via the preliminary addition of the catalyst to the double bond.

$$\text{RCH=CHCOCl} + \text{R'OH} + \text{NR}_3'' \longrightarrow \text{R}_3''\text{NCHRCH}_2\text{CO}_2\text{R}'$$

$$\tag{14}$$

$$\downarrow$$

$$\text{RCH=CHCO}_2\text{R}'$$

B. Geometrical Isomerization

The interconversion of maleic and fumaric acids by means of formation of their respective chlorides has been long known. The cis-dichloride (41) can also be isomerized into an unsymmetrical form (42) and although this

$$
\begin{array}{cc}
\begin{array}{c}
\text{CHCOCl} \\
\parallel \\
\text{CHCOCl}
\end{array}
&
\begin{array}{c}
\text{CH—CO} \\
\parallel \qquad \diagdown \text{O} \\
\text{CH—C—(Cl)}_2
\end{array}
\\
(41) & (42)
\end{array}
$$

type of reaction is not strictly geometrical isomerization, it is conveniently discussed in this section.

The principal transformations[44-47] of the maleic–fumaric acid system are shown in reaction (15). The cyclic compound (42) was considered to

$$
\begin{array}{c}
\begin{array}{c}
\text{HO}_2\text{CCH} \\
\parallel \\
\text{HCCO}_2\text{H}
\end{array}
\xrightarrow{(a)}
\begin{array}{c}
\text{ClOCCH} \\
\parallel \\
\text{HCCOCl}
\end{array}
\underset{(c)}{\overset{(b)}{\rightleftharpoons}}
\begin{array}{c}
\text{HCCOCl} \\
\parallel \\
\text{HCCOCl}
\end{array}
\underset{}{\overset{(g)}{\rightleftharpoons}}
\begin{array}{c}
\text{HC—CO} \\
\parallel \qquad \diagdown \text{O} \\
\text{HC—C—(Cl)}_2
\end{array} \\
(42)
\end{array}
$$

$$
\begin{array}{c}
\begin{array}{c}
\text{CH—CO} \\
\parallel \qquad \diagdown \text{O} \\
\text{CH—CO}
\end{array}
\xleftarrow{(a)}
\begin{array}{c}
\text{HCCO}_2\text{H} \\
\parallel \\
\text{HCCO}_2\text{H}
\end{array}
\end{array} \tag{15}
$$

Key   (a) Phthaloyl chloride   (e) Thionyl chloride/zinc chloride
      (b) Aluminium trichloride   (f) Phosphorous pentachloride
      (c) Storage or distillation   (g) Phosphorous trichloride
      (d) Phthaloyl chloride/zinc chloride

be the best representation of the structure of maleyl chloride for, whereas fumaroyl chloride (which is open-chain of necessity) reacted rapidly with aniline or methanol, the reaction of the maleyl compound was very slow.

The isomerizations of substituted maleic and fumaric acid chlorides have also been well studied, albeit a considerable time ago[48]. One set of reactions is in reaction (16): the uncharacterized complex (43) had not undergone rearrangement since on treatment with ice water it reverted to the fumaroyl derivative, whereas 44 under similar conditions gave rearranged products that were considered to have cyclic structures as on pyrolysis they decomposed to give phosgene, among other products. In analogous experiments, dibromofumaroyl dibromide was similarly isomerized by either aluminium trichloride or tribromide and cyclic structures analogous to 42 were proposed and supported by the finding that dehalogenation of the maleyl products gave lactones[48, 49].

$$
\begin{array}{c}
\text{ClCCOCl} \\
\parallel \\
\text{ClOCCH}
\end{array}
\xrightarrow{\text{AlCl}_3}
[\text{complex}]_1
\underset{100^\circ/\text{AlCl}_3}{\overset{100^\circ}{\rightleftharpoons}}
[\text{complex}]_2
$$

$$(43) \qquad\qquad\qquad\qquad (44)$$

$$\searrow \; \text{H}_2\text{O}, 0^\circ \qquad\qquad\qquad (16)$$

$$
\begin{array}{c}
\text{ClCCOCl} \\
\parallel \\
\text{HCCOCl}
\end{array}
\;\text{or}\;
\begin{array}{c}
\text{ClC--CO} \\
\parallel \quad\;\; \diagdown \\
\quad\quad\quad \text{O} \\
\parallel \quad\;\; \diagup \\
\text{HC--C--(Cl)}_2
\end{array}
\;\text{or}\;
\begin{array}{c}
\text{ClC--C--(Cl)}_2 \\
\parallel \quad\quad\;\; \diagdown \\
\quad\quad\quad\quad \text{O} \\
\parallel \quad\quad\;\; \diagup \\
\text{HC--CO}
\end{array}
$$

Several examples of geometrical isomerization of alkylfumaroyl acid chlorides that are catalysed by heat, aluminium trichloride or phosphorous pentachloride have been reported[50–53]. An interesting point is that although the ester-chloride (45) is readily isomerized (reaction 17), the dibromo-analogue was not, even when heated with aluminium trichloride[54]. An

$$
\begin{array}{c}
\text{ClOCCMe} \\
\parallel \\
\text{MeCCO}_2\text{Me}
\end{array}
\xrightarrow{\text{AlCl}_3}
\begin{array}{c}
\text{MeC--CO} \\
\parallel \quad\;\; \diagdown \\
\quad\quad\quad \text{O} \\
\parallel \quad\;\; \diagup \\
\text{MeC--C--(Cl)}_2
\end{array}
\xrightarrow{[\text{H}]}
\begin{array}{c}
\text{MeC--CO} \\
\parallel \quad\;\; \diagdown \\
\quad\quad\quad \text{O} \\
\parallel \quad\;\; \diagup \\
\text{MeC--CO}
\end{array}
$$

$$(45) \qquad\qquad\qquad\qquad\qquad\qquad\qquad\qquad (17)$$

earlier report[55] had mentioned that the isomeric ester chlorides (46) and (47) formed different anilides and no evidence of rearrangement was found.

$$
\begin{array}{cc}
\begin{array}{c}
\text{Me} \quad\quad \text{CO}_2\text{R} \\
\diagdown \quad\;\; \diagup \\
\text{C}{=}\text{C} \\
\diagup \quad\;\; \diagdown \\
\text{ClCO} \quad\quad \text{H}
\end{array}
&
\begin{array}{c}
\text{Me} \quad\quad \text{COCl} \\
\diagdown \quad\;\; \diagup \\
\text{C}{=}\text{C} \\
\diagup \quad\;\; \diagdown \\
\text{RO}_2\text{C} \quad\quad \text{H}
\end{array}
\\
(46) & (47)
\end{array}
$$

There is also no evidence, as far as we are aware, for geometrical isomerization of the acid chlorides of unsaturated monocarboxylic acids of the oleic, linoleic and similar types[56].

IV. REACTIONS OF DIBASIC ACID CHLORIDES

In this section acid chlorides derived from dibasic acids other than maleic and fumaric will be discussed. Often it is not clear whether the isomerization of such compounds is spontaneous or occurs under conditions of acylation in the presence of Lewis acids. Accordingly some of the material under this general heading will be discussed in section V.

(i) *Reduction.* As previously mentioned, the formation of a lactone on reduction of the diacid chloride of a dibasic acid has been held to indicate the predominant extreme of the latter in a ring form such as 42, although it is quite feasible that cyclization occurs after reduction of one acid chloride group (reaction 18), perhaps on the catalytic surface. The various

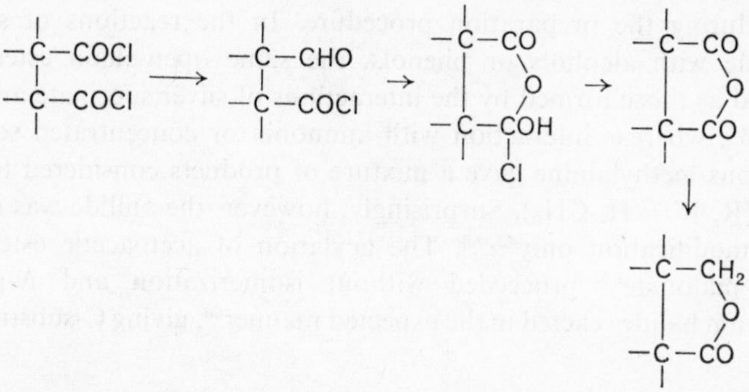

$$
\begin{array}{c}
-\overset{|}{\underset{|}{C}}-COCl \\
-\overset{|}{\underset{|}{C}}-COCl
\end{array}
\longrightarrow
\begin{array}{c}
-\overset{|}{\underset{|}{C}}-CHO \\
-\overset{|}{\underset{|}{C}}-COCl
\end{array}
\rightarrow
\begin{array}{c}
-\overset{|}{C}-CO \\
\quad\;\; O \\
-\overset{|}{C}-\underset{\underset{Cl}{|}}{C}OH
\end{array}
\rightarrow
\begin{array}{c}
-\overset{|}{C}-CO \\
\quad\;\; O \\
-\overset{|}{C}-CO
\end{array}
$$

(18)

$$
\begin{array}{c}
-\overset{|}{C}-CH_2 \\
\quad\;\; O \\
-\overset{|}{C}-CO
\end{array}
$$

and often rather ill-defined conditions of different sets of experiments make it difficult to assess the likelihood of such a pathway or to compare the behaviour of different chlorides.

Reduction of the diacid chloride of malonic acid gave[57] an open-chain acid whereas the corresponding derivative of succinic acid gave the lactone (48) and occasionally the acid (49). It was inferred that the latter dichloride

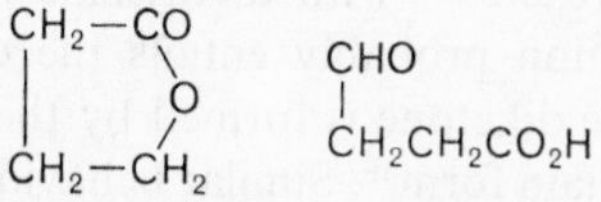

$$
\begin{array}{c}
CH_2-CO \\
\quad\;\;\; O \\
CH_2-CH_2
\end{array}
\qquad
\begin{array}{c}
CHO \\
CH_2CH_2CO_2H
\end{array}
$$

(48)   (49)

existed in both open-chain and cyclic structures. In a related experiment dibromosuccinyl dichloride gave a lactone together with succinic anhydride, the latter presumably arising from the partially reduced substrate, cf. reaction (18).

Phthaloyl dichloride was reduced to phthalide (50) and this was presumed[58] to have arisen by prior isomerization of the symmetrical form (51): under more forcing conditions some ring-opened products

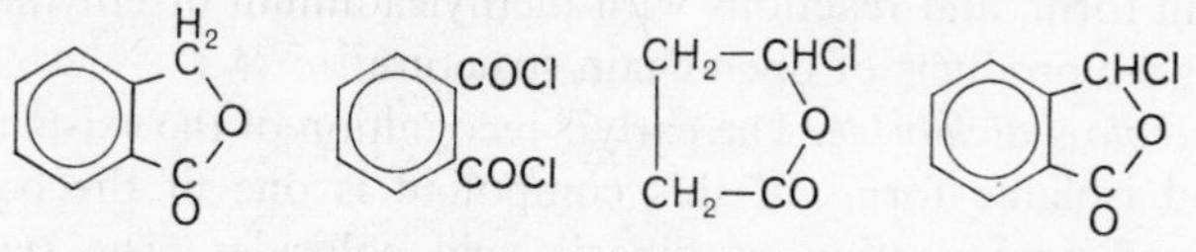

(50)   (51)   (52)   (53)

were also isolated[59]. The reduction of succinyl and phthaloyl dichlorides with tributyltin hydride gave[60] chlorolactones (52) and (53) respectively.

(ii) *Succinyl dichloride*. This substrate apparently did not isomerize in the presence of aluminium trichloride[48] and it was concluded that the equilibrium between the open-chain and cyclic forms had been rapidly set up during the preparation procedure. In the reactions of succinyl dichloride with alcohols or phenols, the same open-chain esters were produced as those formed by the interactions of silver succinate and these reagents[61], whereas interaction with ammonia or concentrated solutions of aqueous methylamine gave a mixture of products considered to be **54** and **55** (R, R' = H, CH$_3$). Surprisingly, however, the anilide was isolated in one modification only[62, 63]. The acylation of acetoacetic ester[64] and diethyl malonate[65] proceeded without isomerization and N-pyrrolyl magnesium halide reacted in the expected manner[66], giving C-substitutions.

CH$_2$CONHR

CH$_2$CONHR'

(54)

CH$_2$—CO

| \\ O

CH$_2$—C—NHR'

|

NHR

(55)

CH$_2$—C—(Et)$_2$

| \\ O

CH$_2$—CO

(56)

CH$_2$—C—(OEt)$_2$

| \\ O

CH$_2$—CO

(57)

Succinyl dichloride reacted[67] with diethylzinc at low temperatures to form **56** and this reaction probably entails the cyclic form whereas at higher temperatures the diketone is formed by the prior isomerization of the cyclic to the open-chain form[68]. Similar behaviour has been suggested[69] in the reaction between the dichloride and anisole.

The i.r. spectrum of the dichloride is consistent with an open-chain form[70] and the observation[71] that reaction of the half-ester chloride with ethanol containing suspended sodium carbonate led to **57** lends some support to the suggestion that the chloride cyclized after one acid chloride group had reacted or had been modified by interaction with a Lewis acid. The reactions of certain α- and β-substituted succinyl dichlorides are also best accommodated on the basis of an open-chain structure[70].

(iii) *Glutaryl dichloride*. The i.r. spectrum suggests that this exists in the open-chain form, and reactions with diethylcadmium or ethylmagnesium bromide gave products of open-chain structure[71, 72].

(iv) *Phthaloyl dichloride*. The early[73] recognition of the existence of two stable and isolable forms of this compound is one of the best-known examples of isomerization of dibasic acid chlorides. The symmetrical form (**51**) can be converted into the unsymmetrical form by mineral or Lewis acids[73-76], the isolation and interconversions have been discussed[48, 77] and the proposed mechanism[78] has been opposed[79].

Heating phthalic anhydride with phosphorus pentachloride leads to the symmetrical form of the dichloride[80] and whereas this reacted in general faster than the cyclic form[81] (cf. the situation for maleyl chloride[48]) the two forms were not distinguishable by their reactions with ammonia or with compounds possessing active methylene groups. It is not possible to deduce from these observations whether any equilibrium between the different forms is set up under these reaction conditions, but with N-pyrrolyl magnesium halide both open-chain (58) and cyclic (59) products were isolated[82], although rapid interconversions between these were not ruled out.

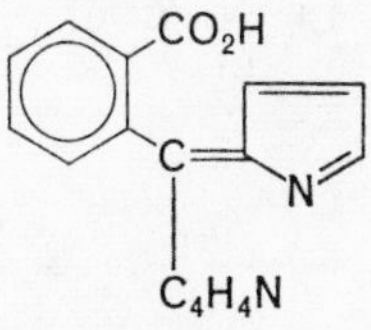

(58) 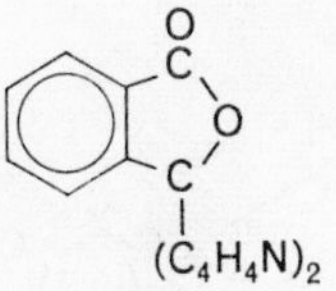   (59)

Halogen-exchange in the dichloride was achieved by heating with sodium fluoride at 250° and the product was considered to be a mixture of open-chain and cyclic fluorides[83] but no clear evidence for the latter was put forward. Treatment of the mixture with aluminium trichloride formed the cyclic dichloride. One of the few other well-documented examples of acid halide exchange with rearrangement is reaction (19): the

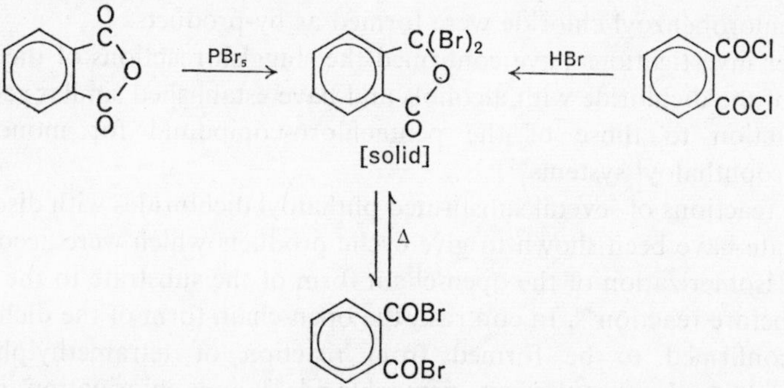

(19)

cyclic structure was assigned[84] to the solid product by reason of the value of its reaction with isopropanol and this compound was claimed to be converted into the open-chain form on melting or on heating to 90°.

Chloro-substituted phthaloyl dichlorides have been extensively studied[85] and the earlier work[86] has been extended and revised. Clear evidence was obtained for the existence and interconversion of open-chain and cyclic forms and reaction (20) shows the rearrangements that have been characterized in this system. Particular points of interest are: (i) the interconversions show the greater stability of the cyclic compound; (ii)

(20)

Key   (a) PCl$_5$ (d) OH$^-$ (g) heat
          (b) EtOH/CaCO$_3$ (e) EtOH   (h) silver oxide
          (c) EtI   (f) spontaneous

the rates of interconversions were greatly enhanced in the presence of a suspension of charcoal which points to the heterogeneous nature of some of these steps; and (iii) disruptive rearrangement must have occurred when the cyclic form was further heated since hexachlorobenzene and pentachlorobenzoyl chloride were formed as by-products.

Later investigations have confirmed the sluggish reactions of the cyclic form of the dichloride with alcohols and have established similar schemes of reaction to those of the pentachloro-compound for mono and dichlorophthaloyl systems[87].

The reactions of several substituted phthaloyl dichlorides with disodium salicylate have been shown to give cyclic products which were accounted for by isomerization of the open-chain form of the substrate to the cyclic form before reaction[88]. In contrast, the open-chain form of the dichloride was confirmed to be formed from reaction of tetramethylphthalic anhydride with phosphorous pentachloride[89] and chlorination of the thio-analogue of phthalic anhydride or its tetramethyl derivative gave the symmetrical form of the dichloride: the latter was characterized in the usual manner as well as by the finding that the velocity of its reaction with aniline was the same regardless of the mode of preparation.

(v) *Miscellaneous.* Treatment of quinolinic acid (60) with phosphorus pentachloride gave a dichloride that showed no tendency to isomerize to a cyclic form[90], and formed the same dimethyl ester as that obtained from reaction of the silver salt and methyl iodide. On treatment of the dichloride with aqueous ammonia **61** was found. The adduct (**62**) was formed with

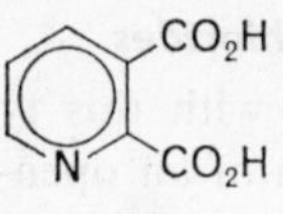

(**60**) 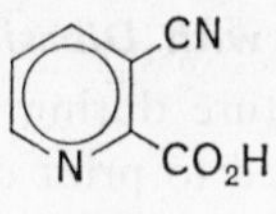　　　　　　　(**61**)

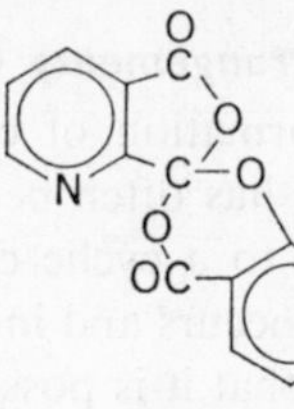

(**62**)

disodium salicylate, but this was rightly considered no proof of a cyclic structure for the chloride.

Although not a dibasic carboxylic acid, the compound (**63**) has been shown to give a chloride that exists in open-chain (**64**) and cyclic (**65**) forms, both of which are appreciably stable[90-92].

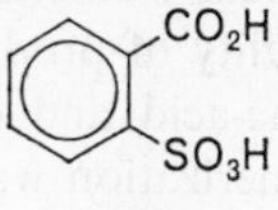

(**63**) 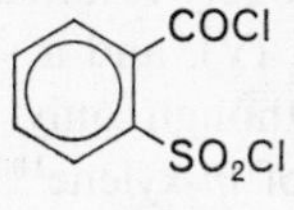　　　　　　　(**64**) 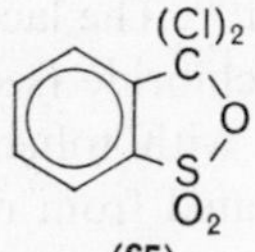　　　　　　　(**65**)

V. REARRANGEMENTS ASSOCIATED WITH FRIEDEL–CRAFTS REACTIONS

The rearrangement of an alkyl group (R) occasioned by the interaction of RX and a Lewis acid is a familiar aspect of Friedel–Crafts processes of alkylation. The corresponding rearrangement of an acyl group (RCO), when an acyl halide is used, is rare[93] even when branched or cyclic substrates are involved. Thus tertiary structures of the type R_3CCOCl react normally in almost all cases and the acid chloride derived from cyclopropane carboxylic acid acylates benzene in good yield to give the expected ketone[94].

This difference in behaviour between alkyl and acyl halides can probably be attributed to the nature of the intermediates that are formed[95]. Spectral evidence indicates that the complexes between acyl halides and aluminium trihalides are either tightly bonded ion pairs or oxonium complexes, e.g. **66** or **67**, and the free acylium ion is present in trace amounts only[96-100]: the alkyl cations are probably more loosely bonded in an ion pair and, in any event, have a greater density of positive charge on the α-carbon that evokes rearrangement very readily.

Despite this situation, several types of rearrangement and accompanying cyclizations have been observed for the reactions of acyl halides in the course of typical Friedel–Crafts reactions.

A.　Rearrangements Occurring with Dibasic Acid Chlorides

The formation of cyclic structure during acylation with this type of substrate has often been attributed to prior cyclization of an open-chain structure to a cyclic chloride. In some examples there is good evidence that this occurs and in others analogy with other reactions (cf. section IV) suggests that it is possible.

The 'normal' acylation reaction has been noted in many reactions of malonyl and substituted malonyl-chlorides with aromatics[101, 102] but dimethylmalonyl dichloride gave some lactone (68) on reaction with benzene, especially when the latter was in excess[102].

Succinyl chloride gave a similar lactone[103] which became the main product of the acylation reaction if reverse addition of the reactants was employed[104]. The lactone probably resulted from a facile isomerization of the acid chloride (see section IV), and a similar duality of products was obtained with toluene[105], although only the ketone-acid and diketone were isolated from reaction of m-xylene[106]. No isomerization was found when meso- or racemic-dihalogenosuccinyl chlorides were used to

$$\overset{+}{R C O} {-} \overset{-}{A l C l_4} \qquad \overset{+}{R C O A l C l_3} {-} \overset{-}{C l} \qquad Me_2C \overset{\overset{Ph_2}{C}}{\underset{\underset{O}{C}}{\big\langle}} O$$

$$\text{(66)} \qquad\qquad\qquad \text{(67)} \qquad\qquad\qquad \text{(68)}$$

acylate benzene; a lactone analogous to 68 was formed, especially if the conventional order of mixing the reactants was reversed[104].

These studies suggest that either the open-chain form is cyclized prior to the acylation reaction or the cyclization occurs after reaction of one of the functional groups has taken place. Repetition of earlier work[107] led to the recognition of two products[108]—the expected diketone and the lactone—from acylation of anisole by glutaryl dichloride. With adipyl dichloride as reactant the diketone was the only product[109].

Friedel–Crafts reactions of substituted benzenes with the acid chlorides derived from fumaric acid and its methylated derivatives gave the expected trans-diketones[110]. On the other hand, those chlorides known to be either cyclic or readily to assume this structure during reaction, e.g. 69, 70 and 71, gave other products[110-116]. Acylation of benzene with 69 was anomalous in yielding 72 in over 70% yield rather than the expected lactone (73). The product of reaction of 71 and bromobenzene in the presence of aluminium tribromide was an open-chain keto-acid[110] that differed from that obtained from the methyl fumaroyl analogue and presumably was a geometrical isomer.

$$
\begin{array}{ccc}
\underset{\|}{\overset{\overset{\displaystyle Cl}{|}}{BrC-C-Ph}} & \underset{\|}{\overset{\overset{\displaystyle Cl}{|}}{MeC-C-C_{12}H_9}} & \underset{\|}{MeC-C(Cl)_2} \\
BrC-CO & MeC-CO & HC-CO \\
(69) & (70) & (71)
\end{array}
$$

(72)   (73)

Similar rearrangements of the acid chlorides of phthalic acid and its derivatives have been carefully studied. In view of the ready interconversion (cf. section IV) of the two forms of these dichlorides it is not surprising that the pattern of products is quite complex, and reaction of phthaloyl dichloride with benzene in the presence of aluminium trichloride gave anthraquinone, σ-benzoylbenzoic acid, σ-(triphenylmethyl)benzoic acid and 9,9-diphenylanthrone depending on the conditions and on the order of addition of the reactants[117–120]. Low temperatures and a short reaction time favoured the keto acid, and the addition of the acid chloride to the other premixed reactants also minimized rearrangement[121, 122], whereas at higher temperatures and at high substrate concentrations the proportion of rearrangement products was greater[123]. Analogous reactions of toluene[124], pyrogallol trimethylether[125] and 1-methyoxynaphthalene[126] gave substantial yields of rearranged products whereas m-xylene and mesitylene furnished diketones[127].

In general the pattern of products is as in reaction (21). The lack of such products when the corresponding derivatives of iso- and tere-phthalic acids are used[128] confirms the intramolecularity of the changes, but it is less certain to what extent all or some of the suggested steps are necessary for reaction to occur. The reported[129] rearrangement of isophthaloyl dichloride to the para-isomer was traced[128] to the presence of the latter as an impurity in the reactant.

The dichloride of naphthalic acid has been assigned[130] the symmetrical structure 74 and an attempt[131] to obtain isomeric forms was unsuccessful. Nevertheless, Friedel–Crafts reactions led to 75 in addition to the major diketone product[132, 133].

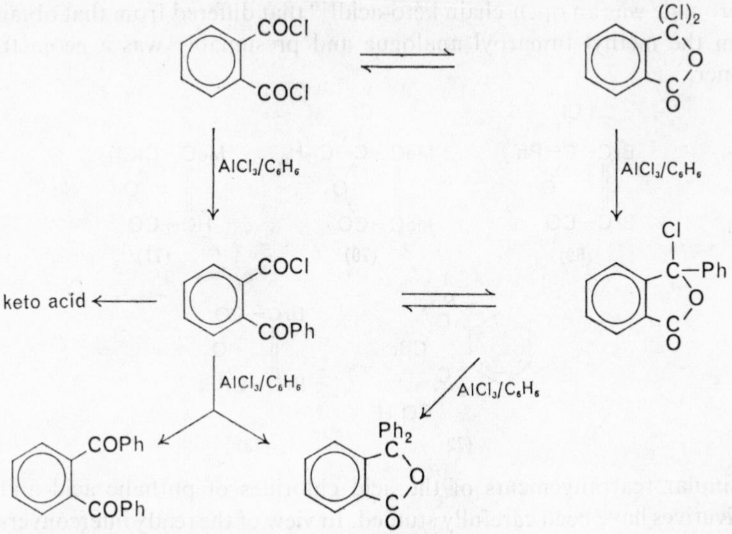

COCl

COCl

(Cl)₂
C
O
C
O

⇌

AlCl₃/C₆H₆

AlCl₃/C₆H₆

(21)

keto acid ⟵

COCl

COPh

⇌

Cl
C—Ph
O
C
O

AlCl₃/C₆H₆

AlCl₃/C₆H₆

COPh

COPh

Ph₂
C
O
C
O

Several of the keto-acid chlorides which are possible intermediates in the reactions of dibasic acid chlorides have been shown[134, 135] to yield cyclic products under Friedel–Crafts conditions and so to react in the

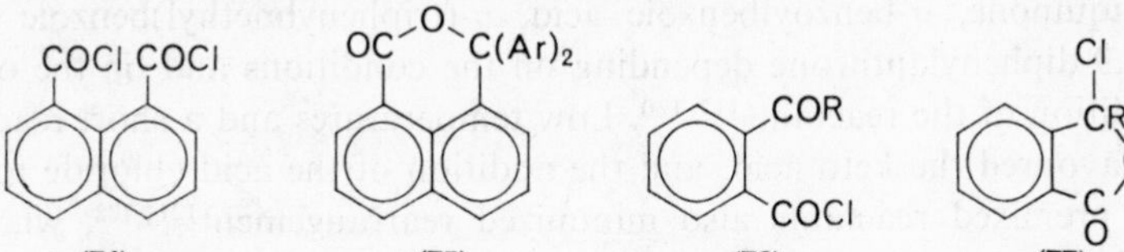

COCl COCl

(74)

OC—O—C(Ar)₂

(75)

COR

COCl

(76)

Cl
CR
O
C
O

(77)

cyclic (77) rather than the symmetrical form (76). There is independent evidence[136] that such compounds exist preferentially in the cyclic form but the variation of proportions of products derived from 76 and 77 caused by difference such as the use of thionyl chloride or phosphorus pentachloride as chlorinating agents suggests that an equilibrium between the two forms of the reactant may be rapidly set up under the conditions for acylation.

Detailed studies[137-139] of products of Friedel–Crafts reactions using the chlorides of the anthraquinone dicarboxylic acids (78) and (79) and

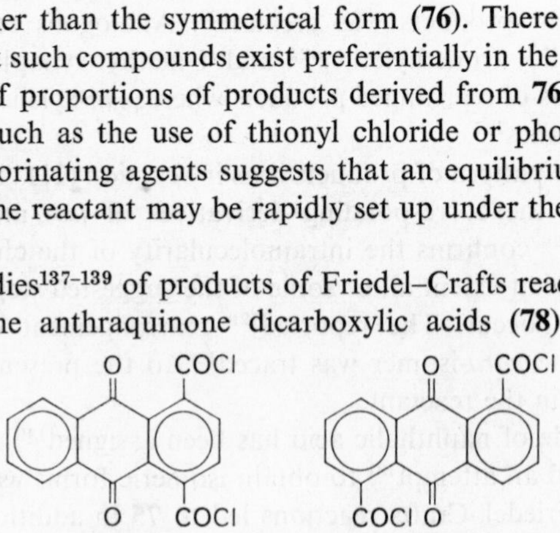

O   COCl

O   COCl

(78)

O   COCl

COCl O

(79)

substituted derivatives showed that the keto group interacted with the chlorocarbonyl group: this influence, which was found to be modified by ring-substitution or variation in catalysts and solvents, altered the ratio of dilactone to diketone in products in a predictable way.

Other miscellaneous rearrangements in similar reactions have been recorded[140, 141].

B. Rearrangements Occurring with Hydride or Alkyl Shifts

Under certain conditions the well-established[142] tendency of acid chlorides of the pivaloyl type to lose carbon monoxide in the presence of a Lewis acid can be suppressed. Pivaloyl chloride itself thus can react as in reaction (22) in the presence of a hydride donor such as isopentane[143, 144].

$$\text{Me}_3\text{COCl} \xrightarrow{\text{AlCl}_3} \text{Me}_3\overset{+}{\text{C}}\text{CO} \longrightarrow \text{Me}_2\overset{+}{\text{C}}\text{COMe} \longrightarrow \text{Me}_2\text{CHCOMe} \qquad (22)$$

In the absence of such a donor only products derived from t-butyl carbonium ion could be detected[142, 145]. A similar reaction may account for the formation[146] of ethyl isopropyl ketone as well as the expected acid (80) in the treatment of n-pentane with carbon monoxide and aluminium trichloride, reaction (23).

$$\text{CH}_3(\text{CH}_2)_3\text{CH}_3 \xrightarrow[\text{AlCl}_3]{\text{CO}} \text{CH}_3\text{CH}_2\text{CHCH}_2\text{CH}_3 \underset{}{\overset{}{\rightleftharpoons}} \text{CH}_3\text{CH}_2\overset{+}{\text{CO}}\text{CHCH}_2\text{CH}_3$$

$$\text{CH}_3\text{CH}_2\overset{+}{\text{CO}}\text{C}(\text{CH}_3)_2 \qquad (23)$$

$$\underset{\substack{|\\ \text{CO}_2\text{H}\\ (80)}}{\text{CH}_3\text{CH}(\text{CH}_2)_2\text{CH}_3} \qquad\qquad \text{derivatives}$$

The addition of acyl halides to olefins in the presence of Lewis acids has also been found to lead to rearranged products. Electrophilic attack of an acylium ion is believed[147] to be followed by stepwise hydride shifts which transfer the positive centre to different sites where it is neutralized by chloride ion, reaction (24). By these means, the product of 1,2-addition of cyclohexene is often accompanied by up to 10% of products of abnormal addition. When solvents capable of interacting with the intermediate cationic species are used, additional products can be obtained. Thus addition of acetyl chloride to cyclohexene in cyclohexane and benzene respectively gave good yields of acetylcyclohexane and 1-acetyl-4-phenyl-cyclohexane[148, 149]. Cyclopentane reacts similarly in inert solvents to give

10

$$(24)$$

81 and **82** in 86 and 4% yields[150], but cycloheptene undergoes extensive ring contraction to form methylcyclohexyl ketones[151].

The occurrence of rearrangement during acylation of both straight-forward and alicyclic saturated hydrocarbons is well known: *n*-alkanes and cyclohexanes (but not cyclopentanes) react with acetyl chloride in the presence of a Lewis acid as in reactions (25) and (26) and variable amounts

$$CH_3(CH_2)_3CH_3 \xrightarrow{MeCOCl} (CH_3)_2CCHCH_3 \atop COMe \qquad (25)$$

$$(26)$$

of unsaturated ketones are also formed[152–160]. The interaction of benzoyl chloride and cyclohexane in the presence of aluminium trichloride led to reduction of the former to give benzaldehyde and 1-methyl-2-benzoyl cyclopentane as an unexpected side-product. These reactions undoubtedly involve the preliminary isomerization of the hydrocarbon under the influence of the Lewis acid: a typical proposed scheme is reaction (27) and the chloroketone has been detected as an intermediate in certain cases. In similar conditions cyclopropanes undergo ring-opening (reaction 28), presumably by a similar mechanism[161, 162].

$$(27)$$

$$\triangle \xrightarrow{\text{RCOCl}} RCO(CH_2)_2Cl + RCOCHMeCH_2Cl$$

$$\left.\begin{array}{l}\\ \end{array}\right\} \quad (28)$$

$$\bowtie \xrightarrow{\text{RCOCl}} RCOCHMeCMe_2Cl$$

C. Miscellaneous Reactions

It is difficult to systematize the other examples of rearrangements observed under Friedel–Crafts conditions, and some reactions have been mentioned in earlier sections: e.g. the isomerizations of certain α-acyl acid halides.

(i) *Reactions of oxalyl halides.* Reaction of oxalyl chloride with aromatic substrates proceeded[163] according to reaction (29) although the bromide

$$ArH + ClCOCOCl \longrightarrow [ArCOCOCl] \longrightarrow ArCOCl \qquad (29)$$
$$(83)$$

yielded some benzil and benzophenone[164]. The formation of **83** and thence of the derived acid has also been recorded for aromatic substrates[165–169]: with naphthalene both α- and β-naphthoic acids were isolated[169]. The mechanism in reaction (30) has been proposed[167], and by analogy the

$$ClCOCOCl + 2 AlCl_3 \rightleftharpoons \left[\overset{\delta-}{AlCl_4}\cdots\overset{\delta+}{CO}-\overset{\delta+}{CO}\cdots\overset{\delta-}{AlCl_4}\right]$$

$$\downarrow ArH$$

$$Ar\overset{+}{CO}CO\overline{A}lCl_4 \qquad (30)$$

$$Ar\underset{O}{\overset{+}{C}}-C=O \longrightarrow \underset{O}{\overset{||}{C}} + Ar-\overset{+}{C}=O \xrightarrow{Ar'H} Ar'COAr'$$
$$\searrow{AlCl_4}$$
$$ArCOCl$$

rearrangement of oxalyl chloride that was catalysed by aluminium trichloride to form phosgene is written as in reaction (31).

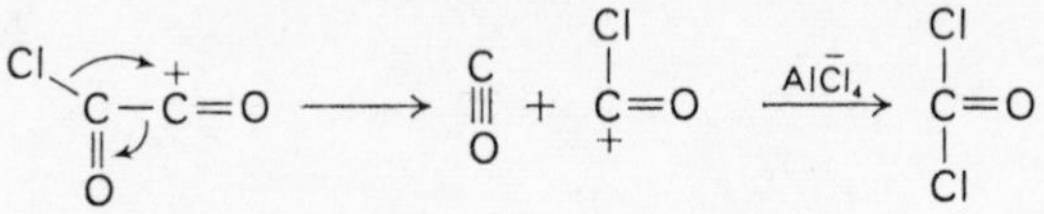

$$\underset{\underset{O}{\parallel}}{Cl}\overset{+}{\underset{}{C}}-C=O \longrightarrow \underset{O}{\overset{\parallel}{C}} + \overset{Cl}{\underset{+}{C}}=O \xrightarrow{A l\bar{C}l_4} \overset{Cl}{\underset{Cl}{C}}=O \tag{31}$$

(ii) *Alkylchloroformates*. Under Friedel–Crafts conditions, alkylchloroformates furnish alkylated products[170] although the aryl analogues react differently[171]. More detailed study showed[172] the decomposition generally to be as in reaction (32) and the mechanism is doubtless similar to that proposed for oxalyl chloride. A similar process[173-175] is reaction (33).

$$ROCOCl \longrightarrow RCl + CO_2 \tag{32}$$

$$ClCOOCCl_3 \xrightarrow{AlCl_3} CO_2 + CCl_4 \tag{33}$$

The decompositions of certain aryl chlorocarbonates in the presence of aluminium trichloride (reaction 34) are also similar to these processes[176, 177]:

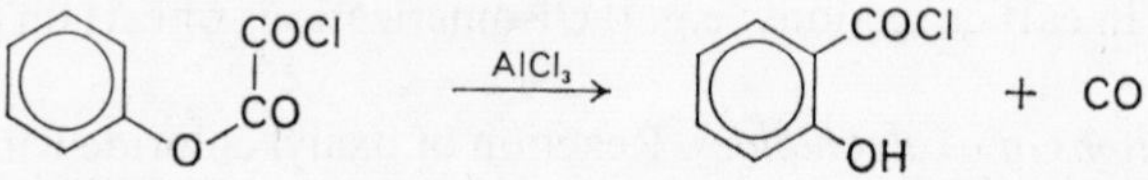

$$\tag{34}$$

when an activating substituent was present in the *para*-position, ring closure occurred to give **84**.

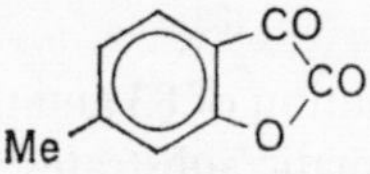

(84)

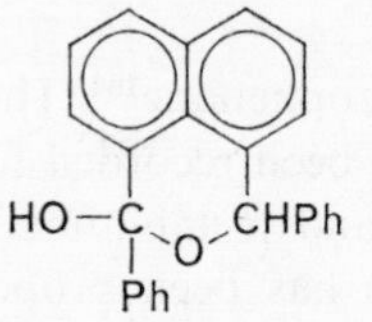

(85)

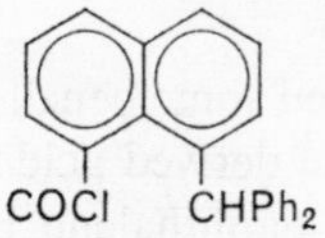

(86)

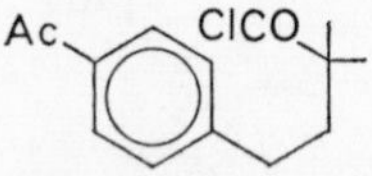

(87)

VI. REACTIONS LEADING TO CYCLIZATIONS

Again, some examples of this type overlap with those cited in other sections, notably V.

(i) *Friedel–Crafts and related reactions*. The cyclization of aryl-substituted aliphatic acid halides (reaction 35) is a well-documented reaction

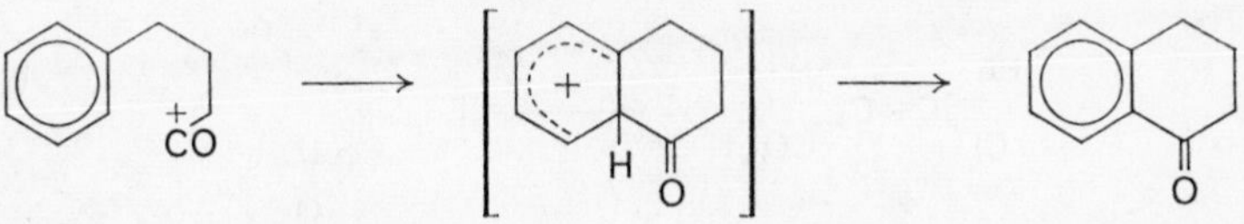

(35)

with regard to the influence of ring-size, steric factors, substituent effects and catalysts[178, 179]. Formally, the reaction may be classified with those processes in which a new bond is formed by a migration to a cationic centre, although since the changes occur in an aryl system in which a proton may be expelled the product is not rearranged in the strict sense. Cyclization of **86** led[180] to an accompanying reaction to form **85**, and deacylation of **87** occurred during attempted ring-closure[181]. Similar treatment of **88** with stannic chloride gave **89** whereas **90** formed **91**[182, 183].

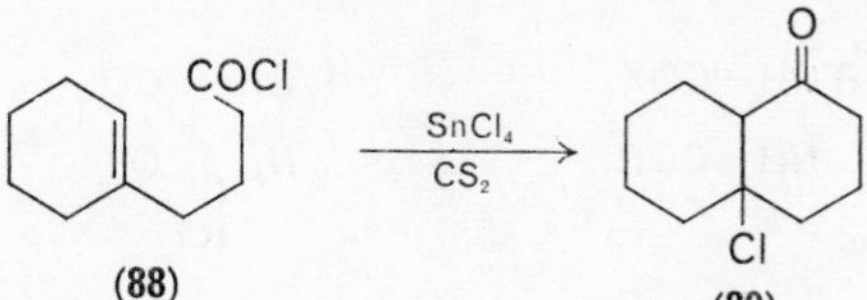

(88) $\xrightarrow[CS_2]{SnCl_4}$ (89)

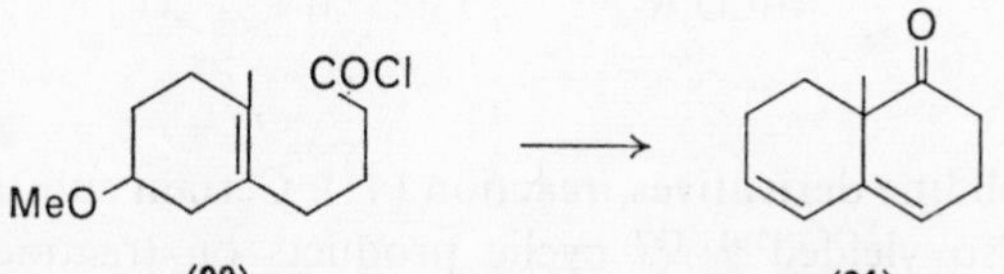

(90)   →   (91)

Certain reactions of cinnamoyl chloride and some of its derivatives have been shown[184] to cause cyclization, although the phenyl and chloro-carbonyl group were *trans*-related[185, 186], cf. reaction (36). The α- and β-chloro-analogues also gave[187] products of ring-closure, reactions (37) and (38), and reaction (39) is similar[184]. Surprisingly, use of the dimethyl-ated compound in the last reaction gave only open-chain products[188].

$\xrightarrow[AlCl_3]{C_6H_6}$ PhCOCH₂CHPh₂ +   (36)

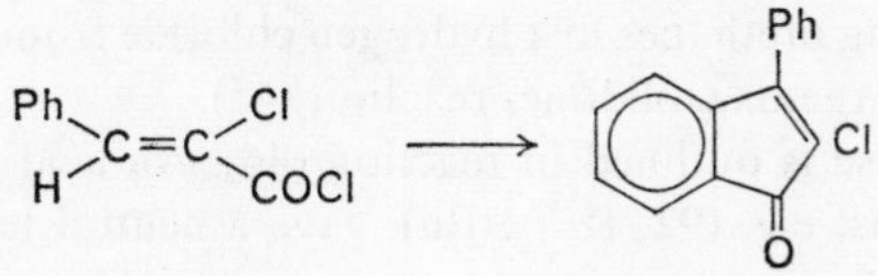

(37)

$$\underset{Cl}{\overset{Ph}{>}}C=C\underset{COCl}{\overset{H}{<}} \longrightarrow \text{[indanone structure with Ph, Ph and O]} \tag{38}$$

$$\text{[benzene]} + CH_2{=}CHCOCl \longrightarrow \text{[indanone structure with O]} \tag{39}$$

(ii) *Other types*. Several examples in which a cyclic structure may result from reaction of a carbonyl halide are known and the subject has been discussed[189].

Acid chlorides derived from α-acylamino acids rearrange rapidly into oxazolones (reaction 40) and urethane-type structures similarly cyclize to

$$\underset{NH-COR}{\overset{R'CH-COX}{|}} \longrightarrow \underset{\underset{R}{N{\underset{C}{\diagdown}}O}}{\overset{R'CH-CO}{|}} + HX \tag{40}$$

$$\underset{NHCO_2R}{\overset{R'CH-COX}{|}} \longrightarrow \underset{HN{\underset{\underset{O}{C}}{\diagdown}}O}{\overset{R'CH-CO}{|}} \tag{41}$$

form[190] oxazolidine derivatives, reaction (41). Certain substituted hydroxamic acids also yielded[191, 192] cyclic products on treatment with ethyl

$$\text{[phenol with OH and CONHOH, R]} \xrightarrow{ClCO_2C_2H_5} \text{[cyclic product with O, CO, N—OH]} \tag{42}$$

$$\text{[phenol with OH and CONHPh]} \xrightarrow{ClCO_2C_2H_5} \text{[cyclic product with O, CO, NPh]} \tag{43}$$

chloroformate, reactions (42) and (43). The interaction of phosgene and β-hydroxyamides resulted in a cyclic product formed by a neighbouring group interaction and this product rearranged on heating[193], reaction (44). The corresponding urethanes lost hydrogen chloride from the intermediate and thus formed the oxazolidone, reaction (45).

An unusual case is outlined in reaction (46). An acid chloride without α-hydrogen atoms, e.g. (92, R = *t*-Bu), gave a neutral heterocycle via an

$$HO(CH_2)_2NHCOR \xrightarrow{COCl_2} ClCOO(CH_2)_2NHCOR$$

$$\downarrow \Delta \qquad (44)$$

$$Cl(CH_2)_2NHCOR \xleftarrow{\Delta} \left[\begin{array}{c} CH_2-CH_2 \\ HN \overset{+}{=} \underset{R}{\overset{|}{C}} O \end{array} \right] Cl^- + CO_2$$

$$HO(CH_2)_2NHCO_2R \xrightarrow{COCl_2} ClCO_2(CH_2)_2NHCO_2R$$

$$\downarrow \Delta \qquad (45)$$

$$RCOO-N \begin{array}{c} CH_2-CH_2 \\ \underset{C}{\overset{|}{O}} \\ O \end{array} + HCl$$

$$\overset{+}{K}Et_3\bar{B}CN + Cl-\underset{O}{\overset{|}{C}}-R \xrightarrow{ether} Et_3\bar{B}-C\equiv\overset{+}{N}-\underset{\underset{O^-}{\overset{|}{C}}}{\overset{Cl}{\overset{|}{C}}}-R$$

(92)

$$\left. \begin{array}{c} \overset{Et_2}{\underset{O-CR}{EtB}} \overset{C}{\underset{N}{\diagdown}} \xleftarrow{} \overset{Et}{\underset{O=C}{\overset{B-C}{\diagdown}}} \overset{Et}{\underset{R}{\diagup}} N \xleftarrow{} \overset{Et}{\underset{O}{Et_2\bar{B}-C\equiv\overset{+}{N}-\underset{O}{\overset{|}{C}}-R}} \end{array} \right\} \quad (46)$$

N-acyl nitrilium intermediate, migration of an alkyl group and ring closure. If α-hydrogen was present, a betaine resulted from formal addition of hydrogen cyanide[194], reaction (47).

$$\overset{Et_2}{\underset{O-CR}{EtB}} \overset{C}{\underset{N}{\diagdown}} + HCN \longrightarrow \overset{Et}{\underset{NC}{\overset{Et_2}{\overset{C}{\diagup}}}} \overset{+}{\underset{O-CR}{\overset{NH}{\diagdown}}} \quad (47)$$

VII. MISCELLANEOUS REACTIONS

(i) *The bromination of carboxylic acids.* The suggestion[195] that α-bromination of carboxylic acids under the conditions of the Hell–Volhard–Zelinsky reaction involves a rate-determining enolization of the acid, or the acid bromide, has been rendered improbable by the demonstration[196-199] that bromination of acetyl bromide definitely involved the ketonic form of the substrate. The evidence for prototropic change preceding reaction is thus inconclusive and the course of the reaction probably is dependent on the formation of the acid bromide.

(ii) *The reaction of carbonyl halides with Grignard reagents.* Certain benzylic and heterocyclic Grignard reagents react with acyl halides or alkyl chloroformates to give products in which rearrangement to an *ortho*- and exceptionally to *para*-position has taken place, reaction (48) (R = Me,

(48)

(93)

(94)

Et, Ph, OMe, OEt)[200-203]. The amount of these products varied from substrate to substrate but was usually about 20% of the reaction: although the thiophene derivative **93** gave 72% reaction by this route[204]; and the product of acylation of 2,6-dichlorobenzyl magnesium chloride was apparently exclusively derived[200] from reaction at position 4. Several similar rearrangements have been reported[205-208] and it is interesting that whereas 1-naphthylmethyl magnesium chloride reacted with ethyl chloroformate to give 41% of the 2-product, its naphthyl parent gave only 1-benzoylnaphthalene on treatment with benzoyl chloride[209]. These rearrangements have been discussed[208] in terms of a benzylic anion or its analogue in which a contributing structure is **94**. The anion presumably exists in an aggregate in the solvents of low polarity that are typically used to effect the reaction, but the equilibria associated with Grignard reagents in solution made it difficult to identify the actual species involved. There is no reference to any rearrangement in a study of the reactions of a series of substituted aliphatic acyl halides with Grignard reagents, even when branched, tertiary structures were present in the acyl radical[210]. Presumably the charge on the α-carbon in the latter species is now small and the normal process of reaction is followed.

(iii) *Pyrolysis of acyl halides.* Decarbonylation as in reaction (49) is often observed[211] and sometimes more complex disruption occurs[212] but the mechanism is obscure and few kinetic studies are available. The rearranged olefins that are also formed may result from direct decomposition of the substrate or may be derived from the alkyl halide.

$$RCOCl \longrightarrow RCl + CO \qquad (49)$$

$$ROCOCl \longrightarrow RCl + CO_2 \qquad (50)$$

(iv) *Reactions of alkyl and aralkyl chloroformates*. Kinetic studies of the decomposition of the former class in the vapour phase (reaction 50) led to the suggestion [213, 214] that loss of carbon dioxide occurred in the rate-determining step, reaction (51).

$$
\begin{array}{c}
Cl-C=O \\
\quad | \\
\quad OR
\end{array}
\rightleftharpoons
\left[
\begin{array}{c}
\overset{+}{C}=O \\
\quad | \\
\overset{-}{C}lOR
\end{array}
\right]
\longrightarrow CO_2 + RCl \qquad (51)
$$

The postulated migration to oxygen did not apparently occur for aryl compounds[215] and phenyl chloroformate was shown to undergo bimolecular decomposition (reaction 52).

$$2(PhOCOCl) \longrightarrow (PhO)_2CO + COCl_2 \qquad (52)$$

The decomposition on an optically active aralkyl substrate **95** (reaction 53) gave[216] a higher degree of retention of configuration than when the

$$
\begin{array}{c}
(-)\ PhCHMe \\
\quad | \\
\quad O-COCl \\
\quad \textbf{(95)}
\end{array}
\overset{\Delta}{\longrightarrow}
\begin{array}{c}
(-)\ PhCHMe + CO_2 \\
\quad | \\
\quad Cl
\end{array}
\qquad (53)
$$

chlorosulphite (**96**) was so reacted[217] (reaction 54). It was concluded[216] that the mechanism of the former process resembled that in reaction (51),

$$
\begin{array}{c}
(-)\ PhCHMe \\
\quad | \\
\quad OSOCl \\
\quad \textbf{(96)}
\end{array}
\overset{\Delta}{\longrightarrow}
\begin{array}{c}
(-)\ PhCHMe + SO_2 \\
\quad | \\
\quad Cl
\end{array}
\qquad (54)
$$

but that a 1,3-positional shift of the aralkyl group took place within an ion pair (reaction 55), without leading to racemization. The effect of

$$
\begin{array}{c}
R-O-C-Cl \\
\quad\quad \| \\
\quad\quad O
\end{array}
\longrightarrow
\left[
\begin{array}{c}
\overset{+}{R}O-\overset{-}{C}-Cl \\
\quad\quad \| \\
\quad\quad O
\end{array}
\right]
\rightleftharpoons
\left[
\begin{array}{c}
\overset{-}{O}-C-Cl\overset{+}{R} \\
\quad\quad \| \\
\quad\quad O
\end{array}
\right]
$$

$$\qquad (55)$$

$$\downarrow$$

$$CO_2 + RCl$$

substituents in the aryl ring on enthalpies and entropies of activation were also discussed in terms of the details of this route.

An attempt to test the 'degree of freedom' of the proposed carbonium ion in this scheme was made using neopentyl chloroformate. A relatively slow decomposition took place to yield olefins with skeletal rearrangement and it was concluded that the initial product was neopentyl chloride. However, a more recent study[218] of the thermal decomposition of a series of alkyl chloroformates revealed extensive rearrangement in the formation of the chlorides comparable in extent to that encountered in the nitrous acid deamination of the corresponding alkyl amines. Thus *n*-butyl chloroformate decomposed to form *sec*-butyl chloride (78%); neopentyl chloroformate gave *t*-pentyl chloride (100%); and *iso*-butyl chloroformate yielded almost entirely *sec*- and *tert*-butyl chlorides. The mechanism in reaction (55) may accommodate these observations (which were obtained under slightly different conditions to those of the earlier experiments) if the cationic species has a greater degree of freedom during migration. Alternatively, a cyclic transition state, cf. reaction (56), may be invoked[218] to account for some of the rearranged products.

$$
(H)R \overset{\displaystyle -\overset{|}{C}-O}{\underset{\displaystyle -\overset{|}{C}\diagdown_{Cl}}{\diagup}} Cl{=}O \longrightarrow \overset{\displaystyle R{-}\overset{\diagup}{\underset{|}{C}}}{\underset{\displaystyle \overset{|}{C}{-}Cl}{}} + CO_2 \tag{56}
$$

A detailed kinetic study of the decomposition of various chloroformates in the gas phase elucidated[219] the pattern of reactions (57). Heterogeneous components were present but the rates were unaffected by the addition of nitric oxide and so radical chains could probably be ruled out. A somewhat

$$
\begin{array}{ccc}
MeCH{=}CHCH_2OCOCl & \rightleftharpoons & MeCHCH{=}CH_2 \\
 & & \overset{|}{O}COCl \\
\downarrow & \times & \downarrow \\
MeCH{=}CHCH_2Cl & \rightleftharpoons & MeCHCH{=}CH_2 \\
 & & \overset{|}{C}l \\
 & CH_2{=}CHCH{=}CH_2 &
\end{array} \tag{57}
$$

similar reaction is that of oxalyl chloride and the diazoketone (97). Reaction (58) has been proposed[220] but no intermediates have been isolated.

The ionization of haloformates in suitable media (e.g. SbF_5-SO_2) has been observed[221] and forms a convenient route for the generation of the

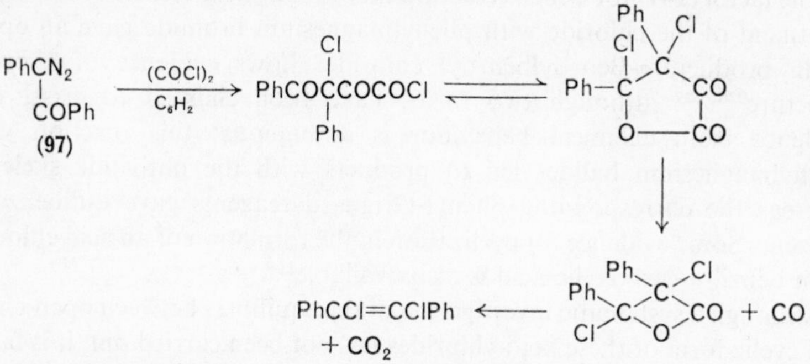

(58)

appropriate carbonium ion (reaction 59). It was not stated, however, whether isomerization of the cation was observed.

$$ROH \xrightarrow{COX_2} ROCOX \xrightarrow{SbF_5} \overset{+}{R}SbF_5\bar{X} + CO_2 \quad \{X = Cl, F\} \tag{59}$$

(v) *Acid halides derived from* ortho-*carbonyl-substituted benzoic acids.* 2-Formylbenzoic acid (98) and its derivatives form acid chlorides that have been shown to exist in the cyclic form (99) by chemical and physical evidence[222, 223]. The isomerization may occur during the formation of the compound or by spontaneous interconversion of first-formed 98. An attempt[224] to prepare the acid bromide by reaction of o-phthaldehyde with bromine gave only the bromolactone which could be reduced to phthalide. Reaction of 99 with an alcohol in the presence of potassium carbonate gave a ψ-ester, isomerized by dilute acid to the normal form. The reactions of opianic acid (100) were similar and the formation of a ψ-chloride and ψ-esters were well established[225].

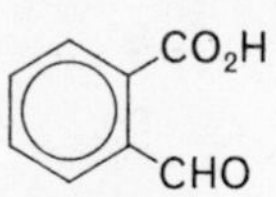

(98) 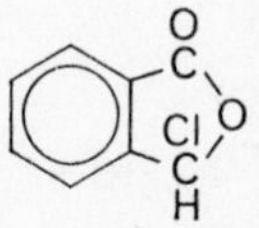 (99) 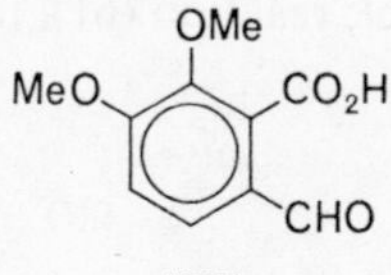 (100)

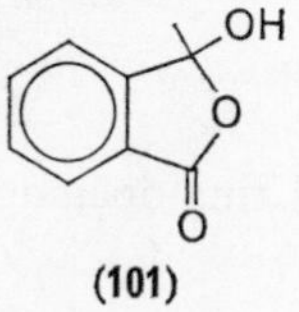

(101)

σ-Acetylbenzoic acid also forms a chloride which is cyclic: this may be formed either by cyclization of the open-chain chloride or by replacement on the lactol (101) for which structure there is physical evidence. However, treatment of the chloride with phenylmagnesium bromide gave an open-chain product. σ-Benzoylbenzoyl chloride shows evidence of a cyclic structure[226, 227] although two forms have been claimed to exist. The evidence from chemical behaviour is ambiguous: this reaction with methylmagnesium halides led to products with the phthalide skeleton whereas the corresponding phenyl-Grignard reagents gave σ-dibenzoyl-benzene. Some evidence for cyclization in the formation of an acid chloride from benzil-σ-carboxylic acid is also available[228].

Although a systematic investigation of the equilibria between open-chain and cyclic form of these keto-chlorides has not been carried out, it is fairly clear that the relationship between reactant and product is complicated by solvent and steric effects.

(vi) *Other reactions.* A variety of other rearrangements that do not fit easily into the above classes fall within our terms of reference and can be conveniently grouped together. Thermal isomerization of 4-ethoxy-butyryl and 5-ethoxyvaleryl chlorides led to the respective ω-chloroesters. The reaction showed first-order kinetic behaviour at 100–150°, proceeded in the absence of catalyst, and was presumably an intramolecular process (reaction 60), similar to that observed with half-ester chlorides[229].

$$\qquad\qquad (60)$$

Ethyl chloroformate and ethyleneimine reacted[230] to form the expected product, reaction (61), in the presence of an alkylamine. In the absence of

$$\qquad\qquad (61)$$

the latter, the hydrogen chloride produced catalysed ring-opening in the reaction.

Other processes studied[231-234] with varying degrees of thoroughness are in reactions (62–68).

$$\text{CNCH}_2\text{COCl} \longrightarrow$$

(62)

$$\text{PhCH(OH)COMe} + \text{PhCOCl} \longrightarrow \underset{\underset{\text{OCOPh}}{|}}{\text{PhCHCOMe}} + \underset{\underset{\text{OCOPh}}{|}}{\text{PhCOCHMe}}$$
(63)

$$\xrightarrow[150-175°]{\text{AlCl}_3}$$
(64)

$$\xrightarrow[]{\text{Fe, 200°}}$$
(65)

$$\xrightarrow[\text{or Fe III}]{\text{Fe}}$$
(66)

$$\text{Cl}_2\text{C=CClCCl=CClOEt} \longrightarrow$$
(67)

$$\xrightarrow[\text{Fe III}]{\text{Fe}}$$
(68)

VIII. REFERENCES

1. D. A. Shirley, *Org. Reactions*, **8**, 23 (1954).
2. W. N. Haworth and A. T. King, *J. Chem. Soc.*, 1348 (1914).
3. C. Rosanbo, *Ann. Chim.*, **19**, 335 (1923).
4. E. Rothstein and M. A. Saboor, *J. Chem. Soc.*, 427 (1943).
5. J. Bardhan, *J. Chem. Soc.*, 2604 (1928).
6. J. Bardhan, *J. Chem. Soc.*, 2593 (1928).
7. S. C. Sengupta, *J. Ind. Chem. Soc.*, **11**, 392 (1934).
8. J. E. Hancock and R. P. Linstead, *J. Chem. Soc.*, 3490 (1953).
9. R. Anshutz, *Ann. Chem.*, **343**, 139 (1907).
10. F. Salmon-Legagneur and F. Soudan, *Compt. Rend.*, **218**, 681 (1944).

11. B. H. Chase and D. H. Hey, *J. Chem. Soc.*, 553 (1952).
12. C. D. Gutsche and K. C. Seligman, *J. Am. Chem. Soc.*, **75**, 2579 (1953).
13. C. D. Gutsche and A. J. Lanck, *Chem. Ind. (London)*, 116 (1955).
14. B. L. Turner, B. I. T. Bhattacharyya, R. P. Graber and W. S. Johnson, *J. Am. Chem. Soc.*, **72**, 5654 (1950).
15. F. Salmon-Legagneur, *Bull. Soc. Sci. Bretagne*, **15**, 189 (1938); *Chem. Abstr.*, **34**, 5833 (1940).
16. A. Haller, *Compt. Rend.*, **141**, 697 (1905).
17. M. Qudrat-I-Khuda, *J. Chem. Soc.*, 206 (1930).
18. J. Bredt, *J. Prakt. Chem.*, **137**, 89 (1932).
19. J. Cason, *J. Org. Chem.*, **13**, 227 (1948).
20. J. Cason, *J. Am. Chem. Soc.*, **69**, 1548 (1947).
21. S. Stallberg-Stenhagen, *J. Am. Chem. Soc.*, **69**, 2568 (1947).
22. J. Cason and R. D. Smith, *J. Org. Chem.*, **18**, 1201 (1953).
23. M. Hayashi, S. Tsuruoka and A. Nakayama, *J. Chem. Soc. Japan*, **56**, 1086 (1935).
24. M. Hayashi and A. Nakayama, *J. Soc. Chem. Ind. Japan*, **36**, 203 (1933).
25. M. Hayashi and A. Nakayama, *J. Soc. Chem. Ind. Japan*, **37**, 238 (1934).
26. R. Lawrance, *J. Am. Chem. Soc.*, **43**, 2579 (1921).
27. F. F. Blicke and P. Castro, *J. Am. Chem. Soc.*, **63**, 2437 (1941).
28. D. H. Hey and J. A. Walker, *J. Chem. Soc.*, 2214 (1948).
29. R. Goncalves and E. V. Brown, *J. Org. Chem.*, **19**, 4 (1954).
30. A. Kirpal, *Monatsh.*, **31**, 295 (1910).
31. E. E. Blaize and R. Herzog, *Compt. Rend.*, **184**, 1332 (1927).
32. A. V. de Maatschappy, *Fr. Pat.*, 790,002; *Chem. Abstr.*, **30**, 2987 (1936).
33. M. Hayashi, S. Tsuruoka and A. Nakayama, *J. Chem. Soc. Japan*, **56**, 1096 (1935).
34. A. McKenzie and M. S. Lesslie, *Ber.*, **61**, 153 (1928).
35. S. Grateau, *Compt. Rend.*, **191**, 947 (1930).
36. F. Litvan and R. Robinson, *J. Chem. Soc.*, 1998 (1938).
37. A. von Leeman and G. K. Grant, *J. Am. Chem. Soc.*, **72**, 4073 (1950).
38. J. D. A. Johnson and G. A. R. Kon, *J. Chem. Soc.*, 2748 (1927).
39. G. Maahs, *Ann. Chem.*, **686**, 55 (1965).
40. T. Ozeki and M. Kusaka, *Bull. Chem. Soc. Japan*, **39**, 1995 (1966).
41. T. Ozeki and M. Kusaka, *Bull. Chem. Soc. Japan*, **40**, 1232 (1967).
42. G. B. Payne, *J. Org. Chem.*, **31**, 718 (1966).
43. P. W. Hickmott, *J. Chem. Soc.*, 883 (1964).
44. R. Anschutz and Q. Wurtz, *Ann. Chem.*, **239**, 137 (1887).
45. W. H. Perkin, *Ber.*, **14**, 2540 (1881).
46. L. Vanino and E. Thiele, *Ber.*, **29**, 1724 (1896).
47. L. P. Kyrides, *J. Am. Chem. Soc.*, **59**, 206 (1937).
48. E. Ott, *Ann. Chem.*, **392**, 245 (1912).
49. H. B. Hill and R. W. Cornelison, *Am. J. Chem.*, **16**, 278 (1894).
50. R. E. Lutz and R. J. Taylor, *J. Am. Chem. Soc.*, **55**, 1585 (1933).
51. R. E. Lutz and R. J. Taylor, *J. Am. Chem. Soc.*, **55**, 1593 (1933).
52. D. T. Mowry, *J. Am. Chem. Soc.*, **66**, 372 (1944).
53. R. E. Lutz, *J. Am. Chem. Soc.*, **52**, 3409 (1930).
54. R. E. Lutz and W. M. Eisner, *J. Am. Chem. Soc.*, **56**, 2700 (1934).
55. R. Anschutz and J. Drugman, *Ber.*, **30**, 2651 (1897).

56. T. R. Wood, F. L. Jackson, A. R. Baldwin and H. E. Longenecker, *J. Am. Chem. Soc.*, **66**, 287 (1944).
57. N. Fröschl, A. Maier and A. Henberger, *Monatsch.*, **59**, 256 (1932).
58. K. W. Rosenmund and F. Zetzsche, *Ber.*, **54**, 2888 (1921).
59. F. Zetzsche, C. Flütsch, F. Enderlin and A. Loosli, *Helv. Chim. Acta*, **9**, 9182 (1926).
60. H. G. Kuivila, *J. Org. Chem.*, **25**, 284 (1960).
61. R. Meyer and K. Marx, *Ber.*, **41**, 2459 (1908).
62. V. Auger, *Ann. Chim. Phys.*, **22**, [6], 326 (1891).
63. G. F. Morell, *J. Chem. Soc.*, **105**, 1733 (1914).
64. J. Scheiber, *Ber.*, **44**, 2422 (1911).
65. J. Scheiber, *Ber.*, **42**, 1318 (1909).
66. W. Tschelinzeff and A. Skwozoff, *J. Russ. Phys. Chem. Soc.*, **47**, 170 (1915).
67. J. Cason and E. Reist, *J. Org. Chem.*, **23**, 1492 (1958).
68. E. Buchta and G. Schaeffer, *Ann. Chem.*, **579**, 129 (1955).
69. W. Borsche, *Ann. Chem.*, **426**, 1 (1936).
70. J. A. McRae, R. A. Barnard and R. B. Ross, *Can. J. Res.*, **28B**, 73 (1950).
71. J. Cason and E. Reist, *J. Org. Chem.*, **23**, 1675 (1958).
72. J. Cason and E. Reist, *J. Org. Chem.*, **23**, 1668 (1958).
73. A. von Baeyer, *Ber.*, **4**, 658 (1871).
74. G. Stadnikoff and I. Goldfarb, *Ber.*, **61**, 2341 (1928).
75. L. P. Kyrides, *Org. Syn.*, **11**, 88 (1931).
76. E. Clar and D. G. Stewart, *J. Am. Chem. Soc.*, **76**, 3504 (1954).
77. W. Csányi, *Monatsh.*, **40**, 81 (1919).
78. G. Baddeley, *Quart. Revs. (London)*, **8**, 355 (1954).
79. Y. K. Syrkin, *Izvest. Akad. Nauk S.S.S.R.*, 389–400 (1959); *Chem. Abstr.*, **53**, 21603 (1959).
80. L. P. Kyrides, *J. Am. Chem. Soc.*, **59**, 209 (1937).
81. J. Scheiber, *Ber.*, **46**, 2366 (1913).
82. B. Oddo and L. Tognacchini, *Gazz. Chim. Ital.*, **53**, 265 (1929).
83. A. T. Down, A. N. Hambly, R. E. Paul and G. S. C. Semmens, *J. Chem. Soc.*, 15 (1933).
84. W. Davies, A. N. Hambly and G. S. C. Semmens, *J. Chem. Soc.*, 1309 (1933).
85. A. Kirpal and H. Kunze, *Ber.*, **62**, 2102 (1929).
86. C. Graebe, *Ann. Chem.*, **238**, 318 (1887).
87. A. Kirpal, A. Gabinschka and E. Lassak, *Ber.*, **68**, 1330 (1935).
88. H. P. Kaufmann and H. Voss, *Ber.*, **56**, 2513 (1923).
89. E. Ott, A. Langenohl and W. Zerweck, *Ber.*, **70**, 2360 (1937).
90. J. Scheiber and M. Knothe, *Ber.*, **45**, 2252 (1912).
91. R. List and M. Stein, *Ber.*, **31**, 1648 (1898).
92. I. Remsen, *Am. J. Chem.*, **18**, 806 (1896); **30**, 247 (1903).
93. P. H. Gore, *Chem. Rev.*, **55**, 229 (1955).
94. N. O. Calloway, *Chem. Rev.*, **17**, 327 (1935).
95. G. A. Olah (Ed.) *Friedel–Crafts and Related Reactions*, Ch. XI, Interscience, New York, 1963.
96. G. Baddeley and D. Voss, *J. Chem. Soc.*, 418 (1954).
97. V. Gold and D. Bethell, *Carbonium Ions, An Introduction*, Academic, New York, 1967, p. 283.

98. B. P. Susz and I. Cooke, *Helv. Chim. Acta*, **37**, 1273 (1954).
99. B. P. Susz, I. Cooke and C. Herschmann, *Helv. Chim. Acta*, **37**, 1280 (1954).
100. D. Cook, *Can. J. Chem.*, **37**, 48 (1959).
101. A. Behal and V. Auger, *Bull. Soc. Chim. France*, **9** [3], 696 (1889).
102. M. Freund and K. Fleischer, *Ann. Chem.*, **399**, 182 (1913).
103. V. Auger, *Ann. Chim. Phys.*, **22** [6], 310 (1891).
104. R. E. Lutz, *J. Am. Chem. Soc.*, **49**, 1106 (1927).
105. H. Limpricht, *Ann. Chem.*, **312**, 110 (1900).
106. A. Claus, *Ber.*, **20**, 1374 (1887).
107. S. Skraun and S. Guggenheimer, *Ber.*, **58**, 2488 (1925).
108. S. G. Plant and M. Tomlinson, *J. Chem. Soc.*, 856 (1935).
109. R. C. Fuson and J. T. Walker, *Org. Syn.*, **13**, 32 (1934).
110. J. B. Conant and R. E. Lutz, *J. Am. Chem. Soc.*, **45**, 1303 (1923).
111. J. B. Conant and R. E. Lutz, *J. Am. Chem. Soc.*, **47**, 881 (1925).
112. R. E. Lutz, *J. Am. Chem. Soc.*, **52**, 3405 (1930).
113. R. E. Lutz, *J. Am. Chem. Soc.*, **55**, 1178 (1933).
114. R. E. Lutz, *J. Am. Chem. Soc.*, **56**, 445 (1934).
115. R. E. Lutz, *J. Am. Chem. Soc.*, **56**, 1378 (1934).
116. R. E. Lutz, *J. Am. Chem. Soc.*, **56**, 2698 (1934).
117. C. Friedel and J. M. Crafts, *Ann. Chim. Phys.*, **6** [1], 449 (1884).
118. C. Deichler and C. Weizmann, *Ber.*, **36**, 547 (1903).
119. M. Copisarow, *J. Chem. Soc.*, 10 (1917).
120. A. Heitler, *Bull. Soc. Chem. France*, **17** [3], 873 (1897).
121. M. Copisarow and C. Weizmann, *J. Chem. Soc.*, 878 (1915).
122. J. Scheiber, *Ann. Chem.*, **389**, 121 (1912).
123. M. Copisarow, *J. Chem. Soc.*, 209 (1920).
124. H. Limpricht, *Ann. Chem.*, **299**, 286 (1898).
125. W. H. Perkin and C. Weizmann, *J. Chem. Soc.*, 1657 (1906).
126. W. Schulenberg, *Ber.*, **53**, 1445 (1920).
127. R. C. Fuson, *J. Org. Chem.*, **10**, 55 (1945).
128. F. F. Blicke and R. A. Patelski, *J. Am. Chem. Soc.*, **60**, 2283 (1938).
129. R. Weiss and L. Chedlowski, *Monatsh.*, **65**, 357 (1935).
130. F. A. Mason, *J. Chem. Soc.*, 2119 (1924).
131. W. Davies and G. W. Leeper, *J. Chem. Soc.*, 1124 (1927).
132. F. A. Mason and W. R. Brown, *J. Chem. Soc.*, 651 (1934).
133. H. E. French and J. E. Kircher, *J. Am. Chem. Soc.*, **63**, 3270 (1941).
134. H. Meyer, *Monatsh.*, **25**, 1171 (1904).
135. E. Clar, *Ber.*, **63**, 112 (1930).
136. M. Renson, *Bull. Soc. Chim. Belges*, **70**, 77 (1961).
137. R. Scholl, J. Donat and O. Böttger, *Ann. Chem.*, **512**, 112 (1934).
138. R. Scholl, J. Donat and O. Böttger, *Ann. Chem.*, **512**, 124 (1934).
139. R. Scholl, K. Meyer and A. Keller, *Ann. Chem.*, **513**, 295 (1934).
140. R. E. Lutz and F. S. Palmer, *J. Am. Chem. Soc.*, **57**, 1947 (1935).
141. W. Borsche and H. Schmidt, *Ber.*, **72**, 1827 (1939).
142. E. Rothstein and R. W. Saville, *J. Chem. Soc.*, 1961 (1949).
143. A. T. Balaban and C. D. Nenitzescu, *Ann. Chem.*, **625**, 66 (1959).
144. H. Hopff, C. D. Nenitzescu, D. A. Isasescu and I. P. Cantuniari, *Ber.*, **69**, 2244 (1936).
145. V. Boeseken, *Rec. Trav. Chim.*, **29**, 85 (1910).

146. A. T. Balaban and C. D. Nenitzescu, *Tetrahedron*, **10**, 55 (1960).
147. C. L. Stevens and A. Farkas, *J. Am. Chem. Soc.*, **85**, 3306 (1953).
148. C. D. Nenitzescu and E. Cioranescu, *Ber.*, **69**, 1920 (1936).
149. C. D. Nenitzescu and I. Gavat, *Ann. Chem.*, **519**, 260 (1935).
150. C. D. Nenitzescu, *Acad. Rep. Populaire Romaine, Studii Cercetari Chim.*, **6**, 375 (1958).
151. S. L. Friess and R. Pinson, *J. Am. Chem. Soc.*, **73**, 3512 (1951).
152. H. Hopff, *Ber.*, **64**, 2739 (1931).
153. C. D. Nenitzescu and C. N. Ionescu, *Ann. Chem.*, **491**, 189 (1931).
154. C. D. Nenitzescu and C. N. Ionescu, *Ann. Chem.*, **491**, 1451 (1931).
155. F. Unger, *Ber.*, **65**, 467 (1932).
156. C. D. Nenitzescu and I. P. Cantuniari, *Ber.*, **65**, 807 (1932).
157. C. D. Nenitzescu and I. P. Cantuniari, *Ber.*, **65**, 1449 (1932).
158. C. D. Nenitzescu and I. Chicos, *Ber.*, **66**, 969 (1933).
159. C. D. Nenitzescu and I. Chicos, *Ber.*, **68**, 1584 (1935).
160. C. D. Nenitzescu and I. P. Cantuniari, *Ann. Chem.*, **510**, 269 (1934).
161. H. Hart and D. E. Curtis, *J. Am. Chem. Soc.*, **79**, 931 (1957).
162. H. Hart and G. Levitt, *J. Org. Chem.*, **24**, 1261 (1959).
163. H. Staudinger, *Ber.*, **41**, 3558 (1908).
164. H. Staudinger, *Ber.*, **45**, 1594 (1912).
165. A. Schonberg and O. Kraemer, *Ber.*, **55**, 1174 (1922).
166. C. Liebermann, *Ber.*, **45**, 1186 (1912).
167. G. Varvoglis and N. Alexandrou, *Chim. Chronika Athens*, **26**, 137 (1961).
168. S. L. Silver and A. Lowy, *J. Am. Chem. Soc.*, **56**, 2429 (1934).
169. C. Liebermann and M. Zsuffa, *Ber.*, **44**, 204 (1911).
170. C. Friedel and J. M. Crafts, *Compt. Rend.*, **84**, 1456 (1877).
171. W. H. Coppok, *J. Org. Chem.*, **22**, 325 (1957).
172. N. P. Bui-Hoi and J. Jamicand, *Bull. Chem. Soc. France*, **12**, 640 (1945).
173. J. Hentschel, *J. Prakt. Chem.*, **36** [2], 308 (1887).
174. V. Grignard, *Ann. Chem.*, **12** [9], 229 (1920).
175. H. C. Ramsperger and G. Waddington, *J. Am. Chem. Soc.*, **55**, 214 (1933).
176. R. Stolle, *Ber.*, **47**, 1130 (1914).
177. R. Stolle and E. Knebel, *Ber.*, **54**, 1213 (1921).
178. W. S. Johnson, *Org. Reactions*, **2**, 114 (1944).
179. S. Sethna in G. A. Olah (Ed.) *Friedel–Crafts and Related Reactions*, Ch. 35, Interscience, New York, 1966.
180. P. T. Lansbury and R. L. Letsinger, *J. Am. Chem. Soc.*, **81**, 941 (1959).
181. G. Baddeley and R. Williamson, *J. Chem. Soc.*, 4647 (1956).
182. J. W. Cook and C. A. Lawrence, *J. Chem. Soc.*, 1637 (1935).
183. J. W. Cook and C. A. Lawrence, *J. Chem. Soc.*, **58** (1938).
184. E. P. Kohler, *Am. Chem. J.*, **42**, 375 (1909).
185. E. P. Kohler, G. L. Heritage and M. C. Burnley, *Am. Chem. J.*, **44**, 60 (1910).
186. K. von Auwers and E. Risse, *Ann. Chem.*, **502**, 282 (1933).
187. K. von Auwers and R. Hugel, *J. Prakt. Chem.*, **143**, 157 (1935).
188. G. Darzens, *Compt. Rend.*, **189**, 766 (1929).
189. H. E. Carter and J. W. Hinsman, *J. Biol. Chem.*, **178**, 303 (1949).
190. J. S. Fruton, *Adv. Protein Chem.*, **5**, 21 (1949).
191. W. B. Wright, *U.S. Pat.*, 2,714,105; *Chem. Abstr.* **50**, 12120 (1956).

192. B. S. Joshi, R. Srinivasan, R. Talaydekar and K. Venkataraman, *Tetrahedron*, **11**, 133 (1960).
193. D. Ben Ishai, *J. Am. Chem. Soc.*, **78**, 4962 (1956).
194. E. Brehm, A. Haag, G. Hesse and H. Wilte, *Ann. Chem.*, **737**, 70 (1970).
195. A. Ward, *J. Chem. Soc.*, **123**, 2207 (1923).
196. H. B. Watson, *J. Chem. Soc.*, 2067 (1925).
197. H. B. Watson, *J. Chem. Soc.*, 2103 (1925).
198. H. B. Watson, *J. Chem. Soc.*, 1137 (1928).
199. H. B. Watson, *J. Chem. Soc.*, 2779 (1928).
200. P. R. Austin and J. R. Johnson, *J. Am. Chem. Soc.*, **54**, 647 (1935).
201. H. Gilman, *J. Am. Chem. Soc.*, **51**, 3475 (1929).
202. H. Gilman, *J. Am. Chem. Soc.*, **54**, 345 (1932).
203. F. C. Whitmore, *J. Am. Chem. Soc.*, **64**, 2968 (1942).
204. R. Gaertner, *J. Am. Chem. Soc.*, **72**, 4326 (1950).
205. V. Chelintzev and S. Karmanov, *J. Russ. Phys. Chem. Soc.*, **47**, 161 (1915).
206. R. Gaertner, *J. Am. Chem. Soc.*, **73**, 3934 (1951).
207. R. Gaertner, *J. Am. Chem. Soc.*, **74**, 760 (1952).
208. R. Gaertner, *J. Am. Chem. Soc.*, **74**, 2185 (1952).
209. F. Acree, *Ber.*, **37**, 2573 (1904).
210. F. C. Whitmore, *J. Am. Chem. Soc.*, **64**, 1618 (1942).
211. A. Bistrzycki and A. Landtwing, *Ber.*, **41**, 686 (1908).
212. A. Maihle, *Compt. Rend.*, **180**, 111 (1925).
213. A. R. Choppin and G. F. Kirby, *J. Am. Chem. Soc.*, **62**, 1591 (1940).
214. A. R. Choppin and E. L. Compere, *J. Am. Chem. Soc.*, **70**, 3797 (1948).
215. S. T. Bowden, *J. Chem. Soc.*, 310 (1939).
216. K. B. Wiberg and T. M. Shryne, *J. Am. Chem. Soc.*, **77**, 2774 (1955).
217. A. McKenzie and G. W. Clough, *J. Chem. Soc.*, 687 (1913).
218. P. W. Clinch and H. R. Hudson, *Chem. Comm.*, 925 (1968).
219. E. S. Lewis and K. Witte, *J. Chem. Soc.* (*B*), 1198 (1968).
220. H. Staudinger, *Ber.*, **49**, 1969 (1916).
221. G. A. Olah and J. A. Olah in *Carbonium Ions* (Eds. G. A. Olah and P. von R. Schleyer), Vol. 2, Wiley–Interscience, New York, 1970, p. 715.
222. S. Gabriel, *Ber.*, **49**, 1612 (1916).
223. A. Kirpal and K. Zeigler, *Ber.*, **62**, 2106 (1929).
224. H. Simonis, *Ber.*, **45**, 1584 (1912).
225. A. Kirpal, *Ber.*, **60**, 382 (1927).
226. M. S. Newman, *J. Am. Chem. Soc.*, **67**, 254 (1945).
227. H. Meyer, *Monatsh.*, **28**, 1235 (1907).
228. A. Hantzsch, *Ber.*, **49**, 213 (1916).
229. V. Prelog and S. Heimbach-Jushasz, *Ber.*, **74**, 1702 (1941).
230. H. Bestian, *Ann. Chem.*, **566**, 210 (1950).
231. G. Schroeter, C. Serdler, M. Sulzbacher and R. Kanetz, *Ber.*, **65**, 432 (1932).
232. T. I. Temnikova, *Vestnik Leningrad Univ.*, 138 (1947); *Chem. Abstr.*, **42**, 4156 (1948).
233. G. Maahs, *Ann. Chem.*, **688**, 53 (1965).
234. G. Maahs, *Ann. Chem.*, **676**, 303 (1964).

CHAPTER **9**

Photochemistry and radiation chemistry of carbonyl halides

U. SCHMIDT

Organisch-chemisches Institut, University of Vienna, Austria

and

H. EGGER

Sandoz Forschungsinstitut, Vienna, Austria

I. INTRODUCTION

In striking contrast to the numerous studies on the photochemistry and radiation chemistry of many classes of compounds containing a carbonyl group, our knowledge of the behaviour of acyl halides in photochemical and radiation-induced processes is very limited. Apart from a few reactions of simple compounds such as phosgene and oxalyl chloride, which have also been the subject of kinetic studies, mechanistic concepts relating to more complex processes are largely speculative, based on product distribution and bond energies. For example, the gas-phase formation and photolysis of phosgene was the subject of a classical series of studies by Bodenstein and co-workers. Later on, reactions were increasingly investigated in the liquid phase. Since very complex reaction mixtures are the rule rather than the exception, elucidation of such reaction pathways had to await the development of modern methods of separation and identification, in which, at present, gas chromatography in combination with mass spectrometry and n.m.r. plays the crucial role.

The primary photochemical step on irradiation of an acyl halide is probably almost invariably the formation of an acyl–halogen radical pair. Complex reaction mixtures arise through secondary radical reactions, either of the radical pair itself or through reactions of the separated radicals following dissociation. Which of the secondary reactions predominates depends on temperature, concentration and solvent (separation of the radical pair; possible diffusion of the radicals out of a solvent cage). In addition, non-photochemical reactions of the reaction products and substrates with acyl halide, hydrogen halide and halogen occur.

The multiplicity of possible pathways is largely responsible for the fact that only a very few of the known photochemical or radiation-induced reactions of acyl halides have attained limited preparative significance. Examples will be dealt with explicitly in subsequent sections.

II. ELECTRON ABSORPTION SPECTRA OF ACYL HALIDES

As this topic is treated elsewhere in this volume, only the essential features of the u.v. absorption spectra of acyl halides will be dealt with in this chapter; a knowledge of the band positions and intensities forms the starting point for any photochemical experiment.

Corresponding to the relatively small number of photochemical studies of acyl halides, only a limited number of absorption measurements in this class are available. The conflicting data to be found in the literature concerning such simple compounds as acetyl chloride[1,2,3,4] are a further limitation, and may in some cases be due to considerable difficulties encountered in purification (e.g. of the extremely corrosive oxalyl chloride) and to photochemical decomposition observed[1] in the course of measurement.

Comparison of the u.v. spectrum of aliphatic acyl chlorides with those of related classes of compounds of the type $R-CO-X$ (where $X = OH$, OR, NHR, R)[3, 5] shows that all possess a band in the region between 200 and 300 mμ with ε from approximately 10 to 100, which is assigned to the $n \rightarrow \pi^*$ transition of a lone electron in the carbonyl group. As would be expected, the exact position of the maximum depends greatly on the nature of the group X, whereby the transition to heavier halogens (Cl$\rightarrow$I) in the acyl halides is accompanied by a marked bathochromic shift with concurrent increase in extinction. Nagakura[3] gives the values $\lambda_{max} = 220$ mμ with $\varepsilon_{max} \sim 100$ for acetyl chloride in heptane.

This $n \rightarrow \pi^*$ band shows a fairly gradual decline in intensity towards the long wavelength region of the spectrum, so that in the range of the Hg emission line at 254 mμ, which is important as a photochemical radiation source, an extinction sufficient for excitation is still available. The energy of this radiation corresponds to 111 kcal, which suffices to rupture the weakest bond in the acyl halide molecule, normally the C$-$X bond; its dissociation energy is of the order of 70–80 kcal for $X = Cl$, and about 60 kcal for $X = Br$, to which is added a variable amount expended on photochemical excitation of the fragments.

The u.v. spectra of perfluoroacyl halides were recorded in the context of photochemical studies[6]. Here, too, a marked bathochromic shift of the maximum of the $n \rightarrow \pi^*$ transition was observed on passing from the fluorides through the chlorides to the bromides. The measurements for the perfluorobutyryl halides are recorded in Table 1.

TABLE 1. Absorption bands of perfluorobutyryl halides in the u.v. region[6]

n-C$_3$F$_7$COX X	Gaseous phase λ_{max} mμ (ε)	Cyclohexane λ_{max} mμ (ε)
F	215 (66)	—
Cl	258 (38)	259 (46·5)
	266 (37)	266 (46)
Br	—	217 (692)
		273 (56)

Aromatic acyl chlorides also show great similarity in position, intensity and substituent shift of corresponding bands to related compounds ArCOX, especially to the corresponding aldehydes ($X = H$) and acetophenones ($X = CH_3$)[7]. Benzoyl chloride in cyclohexane, with the longest wavelength maximum at 242 mμ ($\varepsilon = 14,500$), serves as an example[7].

In the case of phosgene[8], oxalyl chloride[9] and oxalyl bromide[10], compounds frequently the subject of photochemical studies, an absorption which continuously increased towards the vacuum u.v. without pronounced maxima was observed (in part with fine structure, as the measurements were made in the gaseous phase).

III. PHOTOCHEMICAL AND RADIATION-CHEMICAL FORMATION OF ACID HALIDES

A. Photochemical Formation of Phosgene

The formation of phosgene from carbon monoxide and chlorine under the influence of light belongs to the topics of classical photochemistry. Extensive fundamental knowledge was gained by study of the kinetics of this reaction, which was investigated principally by Bodenstein and co-workers[11, 12] and by Rollefson[13]. The reaction is reversible under irradiation, with attainment of a photostationary state.

Phosgene formation at pressures greater than 200 mm at room temperature obeys the experimental rate law:

$$\frac{d[COCl_2]}{dt} = \kappa_1 I_{abs}{}^{\frac{1}{2}}[Cl_2][CO]^{\frac{1}{2}} \tag{1}$$

According to Bodenstein[11] this kinetic behaviour is consistent with and can be derived quantitatively from the following kinetic scheme:

$$Cl_2 + h\nu \longrightarrow 2\,Cl^{\cdot} \tag{2}$$

$$^{\cdot}Cl + CO + M \longrightarrow {}^{\cdot}COCl + M \tag{3}$$

$$^{\cdot}COCl + M \longrightarrow CO + {}^{\cdot}Cl + M \tag{4}$$

$$^{\cdot}COCl + Cl_2 \longrightarrow COCl_2 + {}^{\cdot}Cl \tag{5}$$

$$^{\cdot}COCl + {}^{\cdot}Cl \longrightarrow CO + Cl_2 \tag{6}$$

Equation (2): initiation
Equations (3) and (4): equilibrium state (M: three-body collision partner; e.g. collision with wall)
Equation (6): termination

It follows that

$$\kappa_1 = \frac{k_4}{k_5^{\frac{1}{2}} K_{\cdot COCl}{}^{\frac{1}{2}}} \quad \text{where} \quad K_{\cdot COCl} = \frac{[^{\cdot}Cl][CO]}{[^{\cdot}COCl]} \tag{7}$$

For the dissociation of $^{\cdot}COCl$ into $CO + {}^{\cdot}Cl$ an activation energy of only 6 kcal was calculated.

B. Radiation-chemical Formation of Acyl Chlorides

The system phosgene/cyclohexane was subjected to γ-radiolysis with a ^{60}Co radiation source[14]. For the formation of cyclohexane carboxylic acid chloride a G value of about 70 was obtained, which supports a chain reaction.

The radiochemical chlorocarbonylation of cyclohexane in the system CCl_4/CO was investigated in greater detail[15]. Reaction (9) competes with the principal reaction (8):

$$RH + CCl_4 + CO \xrightarrow{\gamma(^{60}Co)} RCOCl + CHCl_3 \tag{8}$$

$$RH + CCl_4 \xrightarrow{\gamma(^{60}Co)} RCl + CHCl_3 \tag{9}$$

A quantitative relationship between the rates of these two processes and the concentration variables can be formulated from the following reaction scheme:

$$RH + {}^{\bullet}CCl_3 \xrightarrow{k_1} R^{\bullet} + CHCl_3 \tag{10}$$

$$R^{\bullet} + CCl_4 \xrightarrow{k_2} RCl + {}^{\bullet}CCl_3 \tag{11}$$

$$R^{\bullet} + CO \underset{k_{-3}}{\overset{k_3}{\rightleftharpoons}} RCO^{\bullet} \tag{12}$$

$$RCO^{\bullet} + CCl_4 \xrightarrow{k_4} RCOCl + {}^{\bullet}CCl_3 \tag{13}$$

Steps (10) and (11) are familiar from the free-radical chlorination of hydrocarbons with CCl_4.

The formation of dicarboxylic acid chlorides cannot be achieved, since the acid chloride acts as an auto-inhibitor of the reaction. Thaler[15] assumes a competing attack of the chain-propagating trichloromethyl radical on the acid chloride with formation of an (unknown) radical species which is too stable to continue the chain. By suitable choice of the reaction parameters (as derived from the results of the kinetic study) conversions approaching 100% with RCOCl as the greatly predominant product can be achieved.

C. Acyl Chlorides by Photooxidation

Haszeldine and Nyman[16] found that individual fluorinated acyl halides are formed in good yields by photooxidation of suitable halogenated hydrocarbons. Trifluoroacetyl chloride is accessible by the following path:

$$Cl_2C{=}CHCl \xrightarrow{+HF} F_3C{-}CH_2Cl \xrightarrow{+Cl_2} F_3C{-}CHCl_2 \xrightarrow[Cl_2/O_2]{h\nu} F_3CCOCl \tag{14}$$

The photooxidation proceeds with Cl_2 as sensitizer in 90% yield. A chain mechanism with primary attack by chlorine radicals was advanced[16].

Chlorodifluoroacetyl fluoride has been prepared similarly[17]:

$$F_2C{=}CFCl \xrightarrow[Cl_2/O_2]{h\nu} ClF_2CCOF \tag{15}$$

IV. PHOTOLYSIS, PHOTOCHEMICAL OXIDATION AND REDUCTION OF ACYL HALIDES

A. Phosgene

The photochemical decomposition of phosgene is caused by light of wavelengths less than 270 mμ (predissociation boundary[18]) with formation of the ·COCl radical as intermediate[11, 19]. As mentioned in section III, A, a photostationary state can be reached from both sides:

$$COCl_2 \underset{}{\overset{h\nu}{\rightleftharpoons}} CO + Cl_2 \tag{16}$$

The oxidation of phosgene under the action of light follows the equation[20]

$$2\,COCl_2 + O_2 \xrightarrow{h\nu} 2\,CO_2 + 2\,Cl_2 \tag{17}$$

The course of the reaction can readily be followed manometrically in the gaseous phase owing to the change in molar number (3 moles → 4 moles). Accepting the familiar photochemical primary reaction of phosgene, the reaction scheme can be formulated as follows[20]:

Initiation: $\quad COCl_2 \xrightarrow{h\nu} ·COCl + ·Cl \tag{18}$

$·COCl + O_2 \longrightarrow CO_2 + ·ClO \tag{19}$

$COCl_2 + ·ClO \longrightarrow CO_2 + Cl_2 + ·Cl \tag{20}$

$·COCl + Cl_2 \longrightarrow COCl_2 + ·Cl \tag{21}$

Termination: $2·Cl + M \longrightarrow Cl_2 + M \tag{22}$

The rate law derived therefrom,

$$\frac{d[CO_2]}{dt} = \frac{kI_0[COCl_2]}{1 + k'[Cl_2]/[O_2]} \tag{23}$$

is closely followed experimentally. I_0 means intensity of light. The quantum yield of 2 demanded by the mechanism was also confirmed by the experimental value of 1·8[8].

The photochemical reduction of phosgene in the gaseous phase also proceeds smoothly and has been studied in detail[8]:

$$COCl_2 + H_2 \xrightarrow{h\nu} CO + 2\,HCl \tag{24}$$

As the reaction proceeds, the experimentally determined quantum yield decreases from 3·8 to 2·4. Montgomery and Rollefson[8] have proposed a plausible kinetic scheme.

The photolysis of phosgene in the presence of ethylene, which was studied by Wijnen[21], is considerably more complex than the reactions discussed so far. It was found that all ·COCl radicals formed decay into

$CO + \cdot Cl$ (activation energy only 6 kcal). The mechanism suggested by Wijnen interprets the formation of the major products 1-chlorobutane and 1,4-dichlorobutane as due to a complex reaction sequence, details of which are given in the original publication[21]. The chloroethyl radical $\cdot CH_2CH_2Cl$ plays the essential part in chain propagation.

B. Oxalic Acid Halides

The gas-phase photolysis of oxalyl chloride[9] and oxalyl bromide[10] has been studied by Rollefson and co-workers. As the two acid halides behave quite differently they will be discussed separately.

$$(COCl)_2 \xrightarrow[\lambda < 380\ m\mu]{h\nu} [CO + Cl_2] \longrightarrow \underset{\text{(major products)}}{COCl_2 + CO} \qquad (25)$$

Determination of the quantum yield at the wavelengths 365 $m\mu$ and 254 $m\mu$ yielded the admittedly uncertain values

$$\phi \sim 2\ (\lambda = 365\ m\mu)$$
$$\phi \sim 1\ (\lambda = 254\ m\mu)$$

This result, together with the kinetic treatment, led to the assumption of two photolytic primary steps, namely rupture of the C—Cl bond (which predominates at long λ) and rupture of the C—C bond (predominant at short λ):

$$Cl \overset{O}{\underset{A}{\overset{\|}{C}}} \overset{O}{\underset{B}{\overset{\|}{C}}} Cl \qquad \begin{array}{l} A\ (COCl)_2 \xrightarrow{h\nu} \cdot COCOCl + \cdot Cl \qquad (26) \\[6pt] B\ (COCl)_2 \xrightarrow{h\nu} 2\ \cdot COCl \qquad (27) \end{array}$$

The primary radicals thus formed decay in succeeding steps in the familiar manner with separation of CO.

Although the kinetic behaviour and quantum yield are compatible with these assumptions, from estimates of the various dissociation energies a preferred rupture of the C—Cl bond when using low-energy, long-wavelength light is by no means self-evident. The two dissociation energies in question should not differ greatly from each other. For the roughly comparable diacetyl, values of 72[22] and 60[23] kcal for the dissociation energy of the central C—C bond have been given, which lie beneath the energy expected for the C—Cl bond (approx. 75 kcal). However, there are indications that the primary step in the photolysis of oxalyl chloride differs from that in the photolysis of acetyl chloride, which certainly involves C—Cl rupture. Namely, photoreactions with hydrogen donors proceed much faster and to a greater extent with oxalyl chloride and oxalic acid ethyl ester chloride than with acetyl chloride (cf. section V, B and V, D).

In addition, the reaction can be carried out with both compounds as a non-photochemical process with a peroxide initiator. All these findings are indications that the photoreactions of oxalyl halides proceed principally—in the sense of Kharasch's interpretation[38] (cf. section V, D)—via rupture of the C—C bond.

In contrast, the photochemical decomposition of oxalyl bromide[10] takes place in a uniform manner throughout the entire range investigated (436–265 mμ) with a virtually constant quantum yield $\phi \leqslant 1$, which is evidence against a chain reaction. Furthermore, addition of oxygen has almost no influence on the reaction rate.

$$(COBr)_2 \xrightarrow{h\nu} 2\,CO + Br_2 \qquad (28)$$

The relationship between initial and final pressure required by this stoicheiometry, $p_\infty = 3p_0$, is closely followed. At longer wavelengths the bromine formed functions by absorption as a filter and slows down the reaction. The presence of free bromine atoms was demonstrated by addition of hexane to the reactant with formation of alkyl bromides.

The thermal decomposition of oxalyl bromide ($> 200°C$) was also investigated by Rollefson[10], and differs interestingly from the photochemical reaction in that bromophosgene $COBr_2$ is formed as an intermediate ($E_A = 32$ kcal), which decays further in a heterogeneous reaction to carbon monoxide and bromine.

C. Acyl Halides

The decomposition of illuminated acetyl chloride vapour[4] affords methyl chloride and carbon monoxide together with methane, vinyl chloride and much polymeric material.

$$CH_3COCl \xrightarrow[\substack{40-200\ mm \\ vapour \\ (100-200°)}]{h\nu} CH_3Cl + CO + CH_4 + CH_2{=}CHCl + polymers \qquad (29)$$

The partial pressure of methyl chloride is always proportional to the acetyl chloride pressure. In no experiment could ethane be identified. The authors conclude from these findings that the photochemical primary process in methyl chloride formation is *molecular*, since all radicals are trapped by acetyl chloride with formation of polymers. For the two molecular processes (30) and (31) activation energies of 15·5 and 20·4 kcal respectively were obtained.

$$CH_3COCl \xrightarrow{h\nu} CH_3Cl + CO \qquad (30)$$

$$CH_3COCl \xrightarrow{h\nu} CH_2{=}CHCl + O \qquad (31)$$

A rather different situation obtains in the gaseous phase photolysis of acetyl bromide. Etzler and Rollefson[24] found that the decomposition takes place with about 90% formation of methyl bromide and CO. In addition, the reaction mixture was shown to contain about 10% methane and more than 0·5% Br_2; H_2, C_2H_6, C_2H_4 and diacetyl are not present. The product ratio remains constant throughout the reaction, and the pressure increase $p_\infty = 2p_0$ is observed, as anticipated.

$$CH_3COBr \xrightarrow[\substack{\text{vapour} \\ \text{(15–55°C)}}]{h\nu} CH_3Br + CO (+ \text{ by-products: } CH_4, Br_2, C_2H_4Br_2) \quad (32)$$

The reaction rate is directly proportional to the quantity of light absorbed; a value of about 0·5 was determined for the quantum yield.

As methane formation can be totally suppressed by a sufficient amount of NO as a radical trap, the quantity of methane can be regarded as a measure of the proportion of CH_3COBr which yields methyl radicals.

As a result of their experiments Etzler and Rollefson[24] postulate three different primary steps:

$$\text{Direct dissociation:} \quad CH_3COBr \xrightarrow{h\nu} CH_3Br + CO \quad (33)$$

Dissociation via radicals:
$$\begin{cases} CH_3COBr \xrightarrow{h\nu} {}^\bullet CH_3 + {}^\bullet COBr & (34) \\ CH_3COBr \xrightarrow{h\nu} CH_3CO^\bullet + Br^\bullet & (35) \end{cases}$$

The authors have described the photolysis of acetyl bromide with a set of secondary processes undergone by the radicals thus formed.

Irradiation of perfluoroacyl chlorides and bromides in the liquid phase affords perfluoroalkyl radicals, which predominantly lose carbon monoxide. The perfluoroalkyl radicals subsequently dimerize or yield perfluoroalkyl halides.[6]

$$n\text{-}C_3F_7\text{-}COCl \xrightarrow{h\nu} \underset{(81\%)}{n\text{-}C_3F_7Cl} + \underset{(4\%)}{n\text{-}C_6F_{14}} + CO \quad (36)$$

$$ClCO\text{-}(CF_2)_3\text{-}COCl \xrightarrow{h\nu} \underset{(71\%)}{Cl(CF_2)_3Cl} + \underset{(11\%)}{Cl(CF_2)_6Cl} + CO \quad (37)$$

$$n\text{-}C_3F_7\text{-}COBr \xrightarrow{h\nu} \underset{(94\%)}{n\text{-}C_3F_7Br} + CO \quad (38)$$

In the photolysis of carboxylic acid fluorides, rupture of the α-C—C bond is favoured owing to the high bond energy of the C—F bond. Harris[6] has observed decarbonylation and dimerization of the perfluoropropyl radical on irradiation of perfluorobutyryl fluoride:

$$n\text{-}C_3F_7COF \xrightarrow[\substack{\text{(low-pressure} \\ \text{lamp)}}]{h\nu} \underset{(58\%)}{n\text{-}C_6F_{14}} + \underset{n\text{-}C_3F_7}{\overset{n\text{-}C_3F_7}{>}}CF\text{-}OC_3F_7$$

$$+ \quad CO + CO_2 + COF_2 \quad (39)$$

In a special photoreactor it was also possible to photodimerize per-fluorodiacyl fluorides[25] in good yields. In accordance with the primary cleavage discussed above, the reaction conforms to the scheme:

$$2 \; \overset{O}{\underset{\|}{F C}}-(CF_2)_n-\overset{O}{\underset{\|}{C F}} \xrightleftharpoons[\substack{\text{(medium-}\\\text{pressure}\\\text{lamp)}}]{h\nu} 2 \; \overset{O}{\underset{\|}{F C}}-(CF_2)_n^{\cdot} + 2 \; {}^{\cdot}\overset{O}{\underset{\|}{C F}} \xrightleftharpoons{h\nu} \overset{O \quad O}{\underset{\| \quad \|}{F-C-C-F}}$$

$$\downarrow$$

$$\overset{O}{\underset{\|}{F C}}-(CF_2)_{2n}-\overset{O}{\underset{\|}{C F}}$$

(40)

Ethers of the type $FCO(CF_2CF_2-O-CF_2CF_2)_nCOF$ react similarly.

D. Aroyl Bromides

The benzoyl radical $C_6H_5CO^{\cdot}$ was first isolated in substance as an orange-red condensate by freezing out the vapour pumped off after subjecting benzoyl bromide to gas-phase photolysis[26]. The radical was characterized by its e.s.r. spectrum which has a signal width of 15 Gauss (peak–peak) and a g-factor of 2·007. On warming from the temperature of liquid nitrogen to 125°K, colour and e.s.r. signal disappear simultaneously. A radical with identical properties can be obtained by photolysis of benzaldehyde in the same experimental arrangement.

$$\left.\begin{array}{l} C_6H_5COBr \xrightarrow{h\nu} \\[2mm] C_6H_5CHO \xrightarrow{h\nu} \end{array}\right\} \longrightarrow \underset{\text{orange-red}}{C_6H_5CO^{\cdot}}$$

(41)

$$\underset{\text{green}}{CH_3O\langle\bigcirc\rangle CO^{\cdot}} \qquad \underset{\text{red-brown}}{Cl\langle\bigcirc\rangle CO^{\cdot}}$$

The green p-methoxybenzoyl radical and the red-brown p-chlorobenzoyl radical were obtained similarly.

In the warmed-up condensate of the benzoyl radical, benzil and benzophenone were identified as recombination products.

V. PHOTOLYSIS OF ACYL HALIDES IN THE PRESENCE OF HYDROGEN DONORS, ALKANES AND ALKENES

In the case of aliphatic and aromatic carboxylic acid halides there is no doubt that the photochemical primary step is excitation, involving

$n \rightarrow \pi^*$ transition of an electron of the carbonyl group to the singlet state with subsequent transformation (intersystem crossing) into a triplet. As the energy absorbed is large compared with the dissociation energy of the acyl–chlorine and acyl–bromine bonds (60–80 kcal)[27, 28], dissociation of carboxylic acid chlorides and bromides in the singlet or triplet state to acyl and halogen radicals occurs. In the case of aroyl bromides this could be proved by direct methods (cf. section IV, D).

A. Acyl Cyanides and Fluorides

The acyl–cyanide and acyl–fluorine bonds in acyl cyanides and fluorides respectively are so strong that homolysis of these bonds in the excited state does not occur, and consequently the familiar reactions of the carbonyl triplet state are observed.

Irradiation of acetyl cyanide in olefins affords oxetanes in good yields[29] and thus behaves quite analogously to the carbonyl compounds ('Paterno–Büchi reaction'[30]).

The direction of addition conforms to the principle of addition via the most stable diradical.

$$CH_3COCN \; + \; (C_6H_5)_2C{=}CH_2 \; \xrightarrow{h\nu} \; \overset{Ph_2C-CH_2}{\underset{CH_3 \; CN}{|\;\;\;\;\;\;|}} \; \longrightarrow \; \underset{CH_3\,CN}{\overset{Ph \quad H}{Ph-\!\!\!-\!\!\!-H}} \tag{42}$$

52% yield

As a mixture of *cis–trans* oxetane isomers is obtained from sterically pure olefins, it must be assumed that the addition proceeds via the triplet of the acyl cyanide.

Smooth addition of perfluoroacyl fluorides to olefins was also observed[6, 31]:

$$n\text{-}C_3F_7COF \; \xrightarrow{h\nu,\; CF_2{=}CF\text{-}CF_3} \; \underset{F}{\overset{O\;\;\;\;\;C_3F_7}{|\;\;\;\;\;\;\;|}} \tag{43}$$

Sensitized irradiation of aroyl cyanides in hydrogen donors results in dehydrogenation of the solvent and formation of cyanohydrin radicals which dimerize. The resulting benzil dicyanohydrin loses hydrogen cyanide to form benzil[32]:

$$ArCOCN + (C_2H_5)_2O \xrightarrow{h\nu,\ sens.} Ar-\overset{\overset{\displaystyle OH}{|}}{\underset{\displaystyle \cdot}{C}}-CN + C_2H_5-O\overset{\cdot}{C}H-CH_3 \qquad (44)$$

$$2\ Ar-\overset{\overset{\displaystyle OH}{|}}{\underset{\displaystyle \cdot}{C}}-CN \longrightarrow Ar-\overset{\overset{\displaystyle CN}{|}}{\underset{\displaystyle |}{\underset{\displaystyle OH}{C}}}-\overset{\overset{\displaystyle CN}{|}}{\underset{\displaystyle |}{\underset{\displaystyle OH}{C}}}-Ar \longrightarrow ArCOCOAr + 2\ HCN \qquad (45)$$

B. Acyl Chlorides and Bromides

The photoreactions of acyl chlorides in hydrogen donor solvents result in the formation of an acyl–halogen radical pair in the immediate neighbourhood of the C—H bond of a solvent molecule. The radical pair either reacts at once or becomes separated, followed by reaction of the individual separated radicals. In the first case the course of the subsequent reaction can be deduced from the energies of the bonds subjected to opening and closure (46) or (47):

$$\qquad\qquad \longrightarrow R-\overset{\overset{\displaystyle O}{\|}}{C}-C\!\!\diagup \!\!\!\!\diagdown + \ H\!-\!Hal \qquad (46)$$

$$\qquad\qquad \longrightarrow R-\overset{\overset{\displaystyle O}{\|}}{C}-H + \ Hal\!-\!C\!\!\diagup \!\!\!\!\diagdown \qquad (47)$$

Path (46) is always observed when the H—Hal bond is the strongest of the bonds to be formed, for example on irradiating acyl chlorides in hydrogen donors such as cyclohexane or ether. Path (47) represents the course of reactions in which the dissociation energy of the H—halogen bond is comparable with that of RCO—H, as in the irradiation of aroyl bromides in diethyl ether.

Irradiation of an acyl halide–olefin mixture leads to the addition of one of the radicals formed. The product composition can be interpreted as resulting from addition of the primary radical formed without assuming significant allylic substitution.

The irradiation of acetyl chloride and cyclohexane[33] for 24 h yielded 10% acetylcyclohexane and about 3% chlorocyclohexane. The formation of the identified products **1–3** is compatible with the following mechanism:

$$CH_3COCl + \bigcirc \xrightarrow[254m\mu]{h\nu} \overset{COCH_3}{\bigcirc} + \overset{Cl}{\bigcirc} + \bigcirc\!\!-\!\!\bigcirc \tag{48}$$

$$(\sim 10\%) \qquad (\sim 3\%)$$

$$\textbf{(1)} \qquad\quad \textbf{(2)} \qquad\qquad \textbf{(3)}$$

$$R-\overset{O}{\overset{\|}{C}}-Cl \xrightarrow{h\nu} R-\overset{O}{\overset{\|}{C}}{}^\bullet + {}^\bullet Cl \tag{49}$$

$$Cl^\bullet + \bigcirc \longrightarrow \bigcirc^\bullet + HCl \tag{50}$$

$$\bigcirc^\bullet + R-\overset{O}{\overset{\|}{C}}{}^\bullet \longrightarrow \bigcirc-CO-R \tag{51}$$

$$\textbf{(1)}$$

$$2 \bigcirc^\bullet \longrightarrow \bigcirc\!\!-\!\!\bigcirc \tag{52}$$

$$\textbf{(3)}$$

$$\bigcirc^\bullet + R-\overset{\|}{\underset{O}{C}}-Cl \longrightarrow \bigcirc-CO-R + Cl^\bullet \tag{53}$$

$$\bigcirc^\bullet + R-\overset{\|}{\underset{O}{C}}-Cl \longrightarrow \bigcirc-Cl + R-\overset{O}{\overset{\|}{\underset{\bullet}{C}}} \tag{54}$$

$$\textbf{(2)}$$

The reaction proceeds slowly. Short chains represented by the transfer steps (53) and (54) seem to play no important part. Biacetyl is not formed in identifiable amounts; chlorine is present in a minute quantity during the irradiation. Chlorocyclohexane may be formed either from the cyclohexyl radical and chlorine or according to (54).

Considerably more products result, and the extent of conversion is greater in the irradiation of a mixture of acetyl chloride and cyclohexene[33] than with cyclohexane. Six products (**1**), (**2**), (**4**)–(**7**) of the reaction have been identified, and point to the mechanism overleaf described by (56)–(63); **4** and **5** may be formed by the transfer steps (62) and (63).

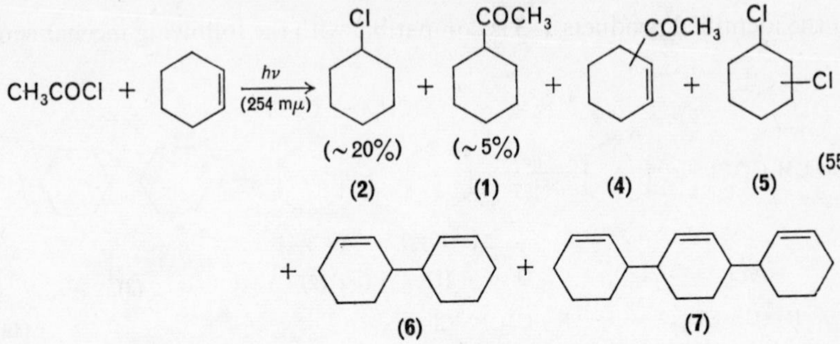

$$CH_3COCl + \text{(cyclohexene)} \xrightarrow[\text{(254 m}\mu\text{)}]{h\nu} \text{(2)} + \text{(1)} + \text{(4)} + \text{(5)} \quad (55)$$

(~20%)   (~5%)

(2)   (1)   (4)   (5)

$$+ \text{(6)} + \text{(7)} \quad$$

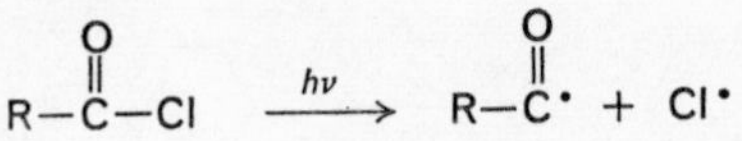

(6)   (7)

$$R-\overset{O}{\underset{\|}{C}}-Cl \xrightarrow{h\nu} R-\overset{O}{\underset{\|}{C}}{}^{\bullet} + Cl^{\bullet} \quad (56)$$

$$R-\overset{O}{\underset{\|}{C}}{}^{\bullet} + \text{(cyclohexene)} \longrightarrow \text{(R-C=O cyclohexyl radical)} \quad (57)$$

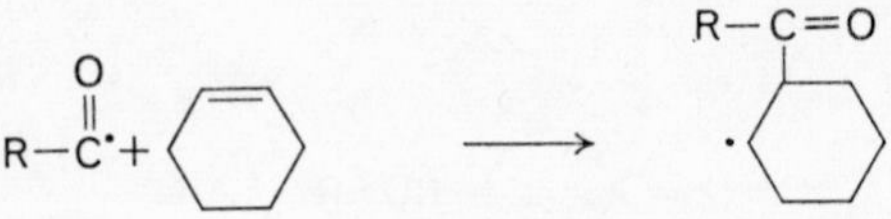

$$\text{(R-C=O cyclohexyl radical)} + \text{(cyclohexene)} \longrightarrow \text{(1)} + \text{(cyclohexenyl radical)} \quad (58)$$

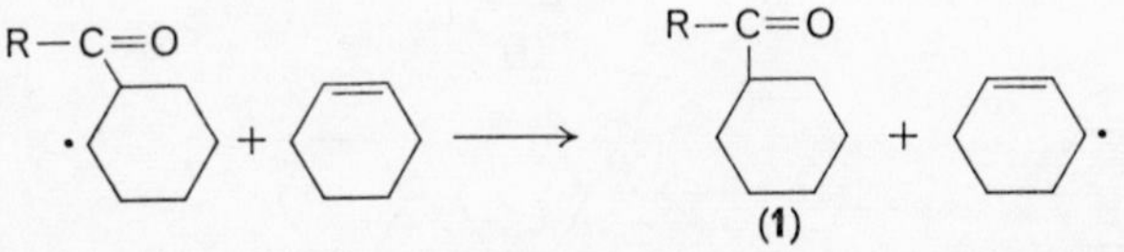

(1)

$$2 \text{(cyclohexenyl radical)} \longrightarrow \text{(6)} \left[\text{(7)} \right] \quad (59)$$

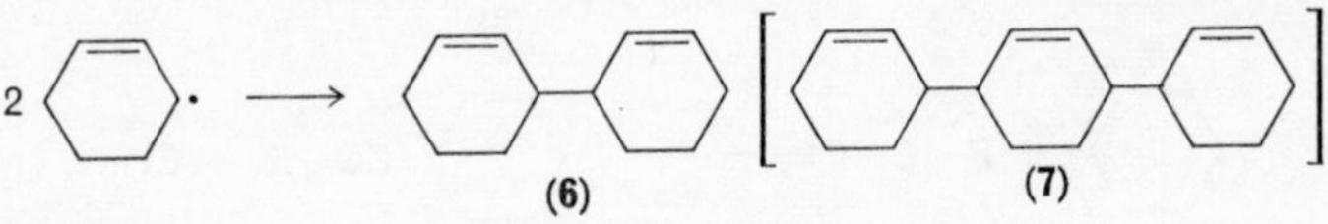

(6)   (7)

$$Cl^{\bullet} + \text{(cyclohexene)} \longrightarrow \text{(Cl cyclohexyl radical)} \quad (60)$$

$$\text{(Cl cyclohexyl radical)} + \text{(cyclohexene)} \longrightarrow \text{(2)} + \text{(cyclohexenyl radical)} \quad (61)$$

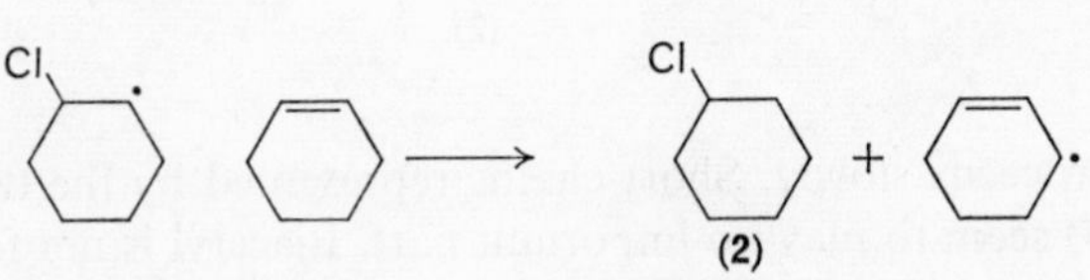

(2)

$$\text{(Cl cyclohexyl radical)} + RCOCl \longrightarrow \text{(5)} + R-\overset{O}{\underset{\|}{C}}{}^{\bullet} \quad (62)$$

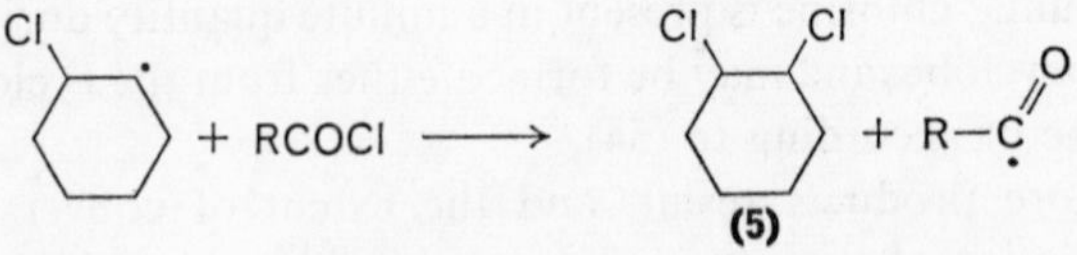

(5)

$$\text{(COCH}_3 \text{ cyclohexyl radical)} + RCOCl \longrightarrow R-\overset{O}{\underset{\|}{C}}{}^{\bullet} + \text{(CH}_3\text{CO, Cl)} \xrightarrow{-HCl} \text{(4)} \quad (63)$$

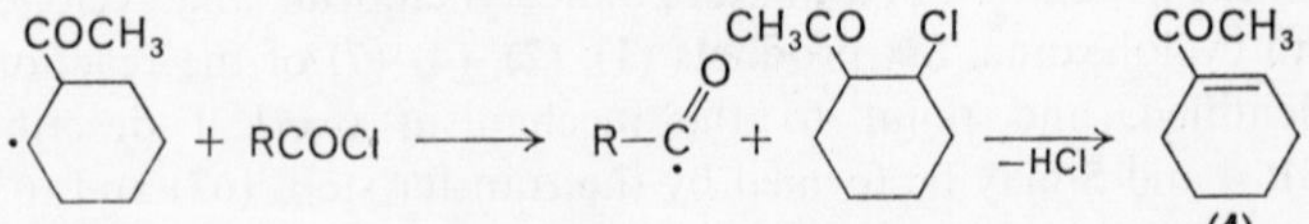

(4)

Irradiation of acyl halides in ether[33, 34] results in considerably improved yields and conversions. Large amounts of hydrogen halide are formed and remain dissolved in the ether, which undergoes non-photolytic fission by the acid halide to the ethyl ester of the corresponding acid (8). For this reason the light source used should be as strong as possible in order to accelerate the photoreaction and limit the extent of the competing non-photochemical reaction. Up to 65% 3-ethoxy-2-butanone (9) is formed from diethyl ether and acetyl chloride[34]:

$$CH_3COCl + (C_2H_5)_2O \xrightarrow{h\nu\ (254\ m\mu)} C_2H_5OCH(COCH_3)CH_3 + HCl \qquad (64)$$
$$(65\%)$$
$$(9)$$

With increasing temperature and increasing acetyl chloride concentration a much greater proportion of by-products is formed. These consist largely of 3-acetoxy-2-butanone (10), which is probably formed non-photochemically from 3-ethoxy-2-butanone by ether fission. The photochemical reaction between acetyl chloride and diethyl ether is considerably faster than that between cyclohexane and acetyl chloride, which very likely reflects the lower dissociation energy of the C—H bond in the α-position of the ether. It is also possible that, owing to loose acetyl chloride–ether oxonium salt formation, the radical pair after photolysis is very close to the C—H bond to be broken. The following mechanism is compatible with products (9), (11), (12), found in addition to 3-acetoxy-2-butanone and ethyl acetate (8):

$$CH_3COOC_2H_5$$
$$(8)$$

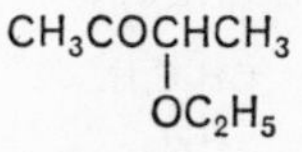

$$CH_3COCHCH_3$$
$$|$$
$$OC_2H_5$$
$$(9)$$

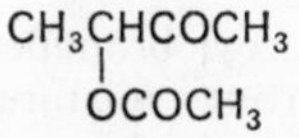

$$CH_3CHCOCH_3$$
$$|$$
$$OCOCH_3$$
$$(10)$$

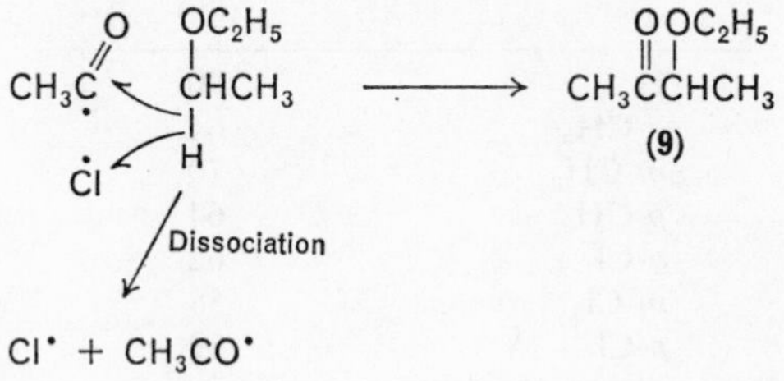

$$(65)$$

$$Cl^{\cdot} + CH_3CO^{\cdot}$$

11

$$Cl^{\bullet} + (C_2H_5)_2O \longrightarrow HCl + CH_3\overset{\bullet}{C}HOC_2H_5 \qquad (66)$$

$$2\ CH_3\overset{\bullet}{C}HOC_2H_5 \longrightarrow (C_2H_5-O-\underset{|}{C}H-CH_3)_2 \qquad (67)$$

(11)

$$2\ CH_3\overset{\bullet}{C}O \longrightarrow CH_3COCOCH_3 \qquad (68)$$

(12)

On photolysis of phenylacetyl halides[35], decarbonylation of the phenyl-acetyl radical (equation 70) or direct scission of the benzyl–chlorocarbonyl bond occurs, forming benzyl radicals. The generation of small amounts of 1,4-diphenylbutane-2,3-dione (16), proves primary formation of the phenylacetyl–radical (equation 69) at least to some extent. Therefore, on reaction in cyclohexane and cyclohexene, the major products found are the combination products of the benzyl radical (1,2-diphenylethane (13), and 1,2,3-triphenylpropane (14)) and the combination products of the 'solvent radicals', which are formed by abstraction of hydrogen by the chlorine radicals (dicyclohexenyl, dicyclohexyl). On irradiation of phenyl-acetyl chloride in diethyl ether, α-benzyldiethyl ether (15), formed by reaction of benzyl radicals with diethyl ether, was identified in addition to large amounts of 13 and traces of the α-diketone (16).

$$C_6H_5CH_2COCl \xrightarrow{h\nu(254m\mu)} C_6H_5CH_2CO^{\bullet} + Cl^{\bullet} \qquad (69)$$

$$C_6H_5CH_2CO^{\bullet} \longrightarrow C_6H_5CH_2^{\bullet} + CO \qquad (70)$$

$$C_6H_5CH_2CH_2C_6H_5 \qquad\qquad C_6H_5CH_2\underset{|}{C}HCH_2C_6H_5$$
(13) C_6H_5

(14)

$$CH_3\underset{|}{C}HOC_2H_5 \qquad\qquad C_6H_5CH_2\overset{\|}{C}-\overset{\|}{C}CH_2C_6H_5$$
$$CH_2C_6H_5 \qquad\qquad\qquad O\quad\ O$$
(15)   (16)

TABLE 2. Photoreduction of aroyl bromides RC$_6$H$_4$COCl; 0·1 mole starting mixture, molar ratio ArCOBr/ether 1 : 3

R	Yield of ArCH=O (%)
H	70
o-CH$_3$	64
m-CH$_3$	70
p-CH$_3$	64
o-Cl	62
m-Cl	58
p-Cl	69

Irradiation of aroyl bromides in diethyl ether[36] leads to smooth photo-reduction of the bromides to aldehydes. About 20% of the ethyl ester of the aromatic acid is formed by a concurrent non-photochemical reaction.

$$ArCOBr + (C_2H_5)_2O \longrightarrow ArCHO + C_2H_5OCHBrCH_3 \quad (71)$$
$$\textbf{(17)}$$

The α-bromoether (**17**) very probably formed in accordance with the above equation could not be isolated. Its formation was deduced from the identification of large quantities of acetaldehyde. 2,3-Diethoxybutane was not identified.

C. Ethyl Chloroformate

The photolysis of ethyl chloroformate was studied in cyclohexane, cyclohexene, 1-octene and without solvent[37]. A mercury high-pressure lamp was used as radiation source. The course of the reaction is very complex and the product distribution depends to a large degree on the reaction conditions. A chain mechanism is not involved. The ethoxy-carbonyl radicals formed in the primary step can substitute hydrocarbons and add to olefins, the yields being very variable.

The authors assume the primary process to be the selective fission of the C—Cl bond (equation 72).

$$Cl-COOC_2H_5 \xrightarrow{h\nu} {}^{\bullet}COOC_2H_5 + {}^{\bullet}Cl \quad (72)$$
$${}^{\bullet}COOC_2H_5 \longrightarrow {}^{\bullet}C_2H_5 + CO_2 \quad (73)$$

The reactions resemble those of acyl halides with alkanes and alkenes discussed in greater detail above, but the yields are poorer owing to the lability of the ethoxycarbonyl radical, which dissociates into the ethyl radical and CO_2 (equation 73) with the extremely low activation energy of about 1–2 kcal/mole.

$$Cl-COOC_2H_5 + \bigcirc$$

$$\downarrow h\nu \qquad (74)$$

$$C_2H_6 + CO_2 + \bigcirc{-}COOC_2H_5 + \bigcirc{-}\bigcirc$$
(main products)

$$Cl-COOC_2H_5 + \bigcirc \xrightarrow{h\nu} \bigcirc{-}COOC_2H_5 + \bigcirc{-}\bigcirc$$
(main products)

$$+ C_2H_6 + CO_2 + C_2H_4 \quad (75)$$

$$Cl-COOC_2H_5 + CH_2{=}CHC_6H_{13} \xrightarrow{h\nu} CH_3(CH_2)_7COOC_2H_5 + C_2H_6$$
$$+ CO_2 + C_2H_4 \qquad (76)$$

The olefins thus yield the corresponding saturated ethyl ester in a radical addition reaction.

Conversions of only a few per cent are obtained in these reactions, with irradiation times of up to 14 h. They therefore have no preparative significance.

D. Oxalic Acid Halides

Aliphatic and alicyclic acyl halides are formed on light-induced or peroxide-initiated reaction of oxalyl chloride with aliphatic and alicyclic hydrocarbons. The reaction was discovered by Kharasch and Brown[38] and later re-examined by Runge[39]:

$$\text{(cyclohexane)} + (COCl)_2 \xrightarrow{h\nu} \text{(cyclohexyl-COCl)} + CO + HCl \qquad (77)$$

The irradiation was conducted in the liquid phase at 30–60° with various radiation sources. Conversions of up to 60% were achieved. From hydrocarbons with non-equivalent C—H bonds mixtures of products were formed as expected, but were not analysed. If an excess of oxalyl chloride is used, it reacts further non-photochemically at the boiling point with the monocarboxylic acid chloride formed to a substituted malonyl chloride:

$$\qquad (78)$$

Mixed aliphatic–aromatic hydrocarbons such as toluene do not react; benzene acts as an inhibitor. This may be due to absorption of the photolytically active radiation by the aromatic system.

The smooth course of photochemical 'chlorocarbonylation' makes the reaction preparatively useful in principle, although deposition of strongly light-absorbing decomposition products on the wall of the reaction vessel frequently causes premature cessation of the reaction.

Compared with the photochemical processes discussed in the preceding sections, this reaction proceeds much more quickly, with significantly higher conversions. In addition, the reaction can also be conducted with peroxide catalysis instead of light. These are indications that the reaction proceeds by way of short chains and differs fundamentally from irradiation of acetyl chloride or chloroformic ester in hydrocarbons. Kharasch and Brown[38] advanced the most plausible interpretation, assuming a transfer step (83) to follow the two primary processes (79), (80) by cleaving the CO—CO bond:

$$(COCl)_2 \xrightarrow{h\nu} 2\ ^{\bullet}COCl \tag{79}$$

$$(COCl)_2 \xrightarrow{h\nu} {}^{\bullet}Cl + {}^{\bullet}COCOCl \tag{80}$$

$$^{\bullet}COCl \longrightarrow CO + {}^{\bullet}Cl \tag{81}$$

$$^{\bullet}Cl + RH \longrightarrow {}^{\bullet}R + HCl \tag{82}$$

$$^{\bullet}R + (COCl)_2 \longrightarrow RCOCl + {}^{\bullet}COCl \tag{83}$$

The photochemical reaction of ethyl chloroglyoxylate with hydrocarbons[40] also proceeds as a chain reaction. The yields relate to ethyl chloroglyoxylate converted.

$$\tag{84}$$

$$\tag{85}$$

Thus, in the case of cyclohexene, addition and allylic substitution proceed concurrently.

VI. ACKNOWLEDGMENT

The authors are much indebted to Dr. A. Stephen of Sandoz Research Institute, Vienna, for translating the German manuscript and Dr. G. Wagner-Sääf for preparing the manuscript.

VII. REFERENCES

1. H. Ley and B. Arends, *Z. Physik. Chem.*, **17B**, 177 (1932).
2. B. D. Saksena and R. E. Kagarise, *J. Chem. Phys.*, **19**, 994 (1951).
3. S. Nagakura, *Bull. Chem. Soc. Japan.*, **25**, 164 (1952).
4. W. D. Capey, J. R. Majer and J. C. Robb, *J. Chem. Soc.* (B), 447 (1968).

5. K. Bowden, E. A. Braude and E. R. H. Jones, *J. Chem. Soc.*, 948 (1946).
6. J. F. Harris, Jr., *J. Org. Chem.*, **30**, 2182 (1965).
7. W. F. Forbes and J. J. J. Myron, *Can. J. Chem.*, **39**, 2452 (1961).
8. C. W. Montgomery and G. K. Rollefson, *J. Am. Chem. Soc.*, **55**, 4025 (1933).
9. K. B. Krauskopf and G. K. Rollefson, *J. Am. Chem. Soc.*, **58**, 443 (1936).
10. J. E. Tuttle and G. K. Rollefson, *J. Am. Chem. Soc.*, **63**, 1525 (1941).
11. M. Bodenstein, W. Brenschede and H.-J. Schumacher, *Z. Physik. Chem.*, **40B**, 121 (1938).
12. M. Bodenstein, S. Lenher and C. Wagner, *Z. Physik. Chem.*, **3B**, 459 (1929).
13. S. Lenher and G. K. Rollefson, *J. Am. Chem. Soc.*, **52**, 500 (1930).
14. M. J. Grosmangin, *World Petrol. Congr., Proc. 5th New York 1959*, No. 10, 85; *Chem. Abstr.*, **56**, 9606 (1962).
15. W. A. Thaler, *J. Am. Chem. Soc.*, **90**, 4370 (1968).
16. R. N. Haszeldine and F. Nyman, *Proc. Chem. Soc. (London)*, 146 (1957); R. N. Haszeldine and F. Nyman, *J. Chem. Soc.*, 387 (1959).
17. R. N. Haszeldine and F. Nyman, *J. Chem. Soc.*, 1084 (1959).
18. V. Henri, *Proc. Roy. Soc. (London)*, **A128**, 178 (1930).
19. C. W. Montgomery and G. K. Rollefson, *J. Am. Chem. Soc.*, **56**, 1089 (1934).
20. G. K. Rollefson and C. W. Montgomery, *J. Am. Chem. Soc.*, **55**, 142 (1933).
21. M. H. J. Wijnen, *J. Am. Chem. Soc.*, **83**, 3014 (1961).
22. J. G. Calvert and J. T. Gruver, *J. Am. Chem. Soc.*, **80**, 1313 (1958).
23. M. Szwarc, *Chem. Rev.*, **47**, 114 (1950).
24. D. H. Etzler and G. K. Rollefson, *J. Am. Chem. Soc.*, **61**, 800 (1939).
25. R. A. Mitsch, P. H. Ogden and A. H. Stoskopf, *J. Org. Chem.*, **35**, 2817 (1970).
26. U. Schmidt, K. H. Kabitzke and K. Markau, *Angew. Chem.*, **77**, 378 (1965); U. Schmidt, K. H. Kabitzke and K. Markau, *Monatsh.*, **97**, 1000 (1966).
27. A. S. Carson and H. A. Skinner, *J. Chem. Soc.*, 936 (1949).
28. J. R. Majer, C. R. Patrick and J. C. Robb, *Trans. Faraday Soc.*, **57**, 14 (1961).
29. Y. Shigemitsu, Y. Odaira and S. Tsutsumi, *Tetrahedron Letters*, 55 (1967).
30. A. Schönberg, *Preparative Organic Photochemistry*, Springer Verlag, Berlin, Heidelberg, New York, 1968, p. 414.
31. J. F. Harris, Jr. and D. D. Coffman, *J. Am. Chem. Soc.*, **84**, 1553 (1962).
32. V. F. Raaen, *J. Org. Chem.*, **31**, 3310 (1966).
33. U. Schmidt, *Angew. Chem.*, **77**, 216 (1965); U. Schmidt, H. Egger and A. Nikiforov, unpublished results.
34. U. Schmidt and A. Nikiforov, unpublished results.
35. K. Schnass, Dissertation, Vienna (1971).
36. U. Schmidt, *Angew. Chem.*, **77**, 169 (1965); U. Schmidt and W. Silhan, *Monatsh.*, **102**, in press (1971).
37. C. Pac and S. Tsutsumi, *Bull. Chem. Soc. Japan*, **36**, 234 (1963); C. Pac and S. Tsutsumi, *Bull. Chem. Soc. Japan*, **38**, 1916 (1965).
38. M. S. Kharasch and H. C. Brown, *J. Am. Chem. Soc.*, **64**, 329 (1942).
39. F. Runge, *Z. Elektrochemie*, **56**, 779 (1952); F. Runge, *Z. Elektrochemie*, **60**, 956 (1956).
40. C. Pac and S. Tsutsumi, *Tetrahedron Letters*, 2341 (1965); C. Pac and S. Tsutsumi, *Bull. Chem. Soc. Japan*, **39**, 1926 (1966).

CHAPTER 10

Biological reactions of carbonyl halides

SASSON COHEN

Israel Institute for Biological Research, Ness-Ziona, Israel

I. INTRODUCTION

Chemists are familiar with the primary irritation of eyes, skin or respiratory tract occasionally incurred in handling acyl halides. These unpleasant effects are commonly attributed to the high reactivity of the compounds. There has, however, been little investigation of the specific action of acyl halides on tissues or biochemical systems. Most texts[1] refer to this action as strong irritation or severe tissue damage. Also, the relatively high vapour pressure of most members of this class entails a serious inhalation hazard.

The feeling of irritation or damage and destruction of tissues are only the ultimate physiological expression of chemical changes which occur in the tissue affected. The changes may be widespread and indiscriminate with the more reactive agents, confined and specific with others. Hence carbonyl halides do not necessarily form a homogeneous toxicological or pharmacological group.

313

The purpose of this chapter is a balanced presentation of the effects of the carbonyl halides at the molecular and biochemical levels and the final physiological or pathological implications of these reactions. The term 'carbonyl halide' will be henceforth used to denote all members of the general formula $R-CO-X$ where X is always a halogen and R may be a halogen, an alkyl, aryl, alkoxy, aryloxy or amino group.

II. ACTUAL AND POTENTIAL CARBONYL HALIDES

Any presentation of the biological effects of carbonyl halides would be lacking if it were limited to a discussion of compounds of RCOX structure only. Other types of compounds that do not formally classify as carbonyl halides are powerful, though latent, acylating agents. These potential carbonyl halides are capable of participating in chemical reactions which would eventually lead to products similar to or identical with those expected from the analogous reaction with the corresponding carbonyl halide.

To illustrate this, consider for example the case of a compound RCX_3, where the carbon–halogen bond in $-CX_3$ is responsive to heterolysis in protic media under relatively mild conditions[2]; then,

$$RCX_3 \xrightarrow{\text{H}_2\text{O}} (RCOX + 2\,HX) \xrightarrow{\text{HB}} RCOB + 3\,HX$$

where B represents base, e.g. OH^-, NH_2^-, SH^- or their alkyl congeners. It must be understood, however, that the carbonyl halide RCOX in the above reaction is implied only and may not necessarily arise stepwise in the course of the reaction. Yet the end products RCOB and HX are those expected from its reaction with the substrate HB.

The highly-electrophilic olefins $XYC=CXY$ where X and Y may be the same or different halogens are another example of potential carbonyl halides. These compounds, especially when fluorine is one of the halogens, readily undergo nucleophilic addition reactions, yielding ultimately an acylated product and a halogen acid. Thus, the reaction of perfluoro-propene and perfluoroisobutene with hydroxylamine, reported by Knunyants and co-workers[3], led eventually to hydrogen fluoride and a fluorinated carboxylic acid:

$$CF_3CR=CF_2 \xrightarrow{\text{NH}_2\text{OH/MeOH}} CF_3CHRCF=NOH + HF$$

$$\xrightarrow{\text{H}_2\text{O}} CF_3CHRCOOH + HF$$

$$(R = F;\ CF_3)$$

A related case is the air oxidation of compounds such as $CFBr=CFBr$[4]

and $CHF{=}CBr_2{}^5$ which are then converted into acid halides by rearrangement:

$$CFBr{=}CFBr \xrightarrow{O_2} (CFBr-CFBr) \longrightarrow CFBr_2COF$$
$$\underset{O_2}{\diagdown\diagup}$$

$$CHF{=}CBr_2 \xrightarrow{O_2} (CHF-CBr_2) \longrightarrow CHBrFCOBr$$
$$\underset{O_2}{\diagdown\diagup}$$

Since the oxidation of these olefins occurs spontaneously at ambient temperature, they should also be of concern as potential carbonyl halides.

Presentation in this review will be limited to acylating agents not involved in normal biochemical processes which produce unusual toxicological or biological effects. Acyl transfer reactions of general biochemical interest have been reviewed by Bruice and Benkovic[6].

III. POTENTIAL REACTION SITES IN BIOLOGICAL MEDIA

The end products of the reaction of carbonyl halides with a wide range of representative substrates have been given in a review article by Sonntag[7]. This may be taken as a general guideline to anticipate the nature of products that would result from the reaction of carbonyl halides with the constituents of body fluids and tissues. Some potential reaction sites are given in Table 1.

The list of these is by no means comprehensive, but should convey a fair impression of the possible reaction pathways open to carbonyl halides. Also, the figures for molarity of the various functional groups should be taken with reserve, because most of these are part of highly-organized structures rather than uniformly distributed as in a perfect solution.

The products of the reaction of carbonyl halides with such substrates are a halogen acid HX, on the one hand, and a carboxylic acid RCOOH (or CO_2 if R is a halogen), its amide $RCONHR^1$ or imide $RCON{<}$, an oxygen ester $RCOOR^1$ or a sulphur ester $RCOSR^1$, on the other, depending on the conditions and site of the reaction and on the nature and reactivity of the reactants. In this context, it is useful to remember that the reaction of acyl halides and anhydrides with nucleophiles in protic media possesses both an S_N1 and S_N2 character[8]. The relative contribution of each of these two mechanisms should be of consequence in determining the course of the reaction in the presence of a wide range of competing substrates. With the exception of water, which constitutes by far the most abundant

TABLE 1. Potential reaction sites for carbonyl halides in biological media

Reactive group	Carrier molecule	pK_a range	Relative nucleophilicity[a]	Molar concentration[b] $\times 10^3$	Reference
—OH	H_2O (free and combined)	−1·7	−6·20 (H_2O)	40,000 (blood)	9
	Free glucose	14–15	4·10 (pentaerythritol)	20 (blood)	9
	Glycogen			150 (muscle)	9
	RNA-pentose	15–16	4·18 (ethanol)	17 (liver)	10
	Free cholesterol	13–14	3·95 (N-acetylserinamide)	0·7 (blood)	9
	Serine, threonine residues, in				
	Serum albumin			20 (blood)	11
	Haemoglobin			140 (blood)	12
	Tyrosine residues, in	10–11	1·76 (phenolate)		
	Serum albumin			7 (blood)	11
	Haemoglobin			25 (blood)	12
	Brain proteins			18 (brain)	9
—NH₂	Free amino acids	9–10	2·19 (glycine)	0·4 (blood)	9
	Lysine residues, in	9–10	2·62 (ethylenediamine)		
	Serum albumin			22 (blood)	11
	Haemoglobin			100 (blood)	12
	Brain proteins			30 (brain)	9
	Spermine			15 (semen)	13
	Cytosine, guanine bases, in	4–5	−1·82 (aniline)		
	RNA			13 (liver)	14
	DNA			13 (liver)	14

TABLE 1. (continued)

Nucleophile	pK_a	Relative nucleophilicity[a]	Approximate molar concentration[b]	Reference
>NH	6–7			
Histidine residues, in		1·54 (imidazole)		
Serum albumin			6 (blood)	11
Haemoglobin			86 (blood)	12
Brain proteins			22 (brain)	9
≫N	7–8			
Purine bases, in		1·30 (N-methylimidazole)		
RNA, DNA			13 (liver)	14, 15
—SH	8–9			
Glutathione		2·58 (glutathione)	1·6 (muscle)	9
			1 (blood)	9
			0·7 (blood)	9
Ergothioneine	8–9	2·63 (cysteine)		
Cysteine residues, in				
Serum albumin			1·5 (blood)	11
Fibrinogen			0·05 (blood)	16
P—OH	6–7	−2·13 (HPO_3^{2-})	3 (blood)	9
Inorganic phosphate			2 (blood)	9
Phospholipids			3 (muscle)	9
ATP				
—COOH	4–5	−3·29 (acetate)	10 (blood)	9
Fatty acids				
Glutamic acid residue, in				
Acid glycoprotein			0·4 (blood)	17

[a] Relative nucleophilicity is defined as $\log k_n$ (M^{-1} min^{-1}) in the reaction of p-nitrophenyl acetate in water with nucleophiles (shown in parentheses) that are closely related in structure and basicity to the biological system under consideration. Data are from a compilation by Johnson[8].

[b] Approximate molar concentrations refer to human tissue indicated and have been calculated from data published in references.

molecule, the concentration of most other possible reactants lies within the rather narrow limits 10^{-2} to 10^{-4} M and, therefore, is not expected to have far-reaching effects on the course of the reaction. Much more pronounced are the differences in nucleophilicities of the various substrates, the range being -6 to 4 log units. Finally, as most reaction sites are parts of organized structures, steric factors assume great importance. For example, the $-NH_2$ group which is highly reactive towards carbonyl halides, might not be accessible to the reagent if 'submerged' within the tertiary structure of a native protein. This situation will be considered in greater detail in the following sections.

A common feature to all possible reaction pathways, however, is formation of one mole of acid HX for each mole of reacting RCOX. In view of this, it is convenient to consider the effects of carbonyl halides under two categories: (1) toxicological effects due to acid formation, mostly HX, in tissues and (2) fine biological effects due to the coupling of the residue RCO— to a molecule of biological importance.

IV. TOXICOLOGICAL EFFECTS DUE TO GENERATION OF ACID

Few organisms are able to withstand a high concentration of hydrogen ions*. This is because the conformation of most biologically important macromolecules becomes altered, resulting in the reduction or complete cessation of specialized biological functions. According to Steinhardt and Beychok[19], acid- or base-denaturation of proteins may be ascribed to loss of solubility at the isoelectric point, loss of specific properties such as enzymatic activity, loss of compact structure with accompanying changes in the electrostatic term w, and uncovering of all or most of the previously inaccessible prototropic groups. The limits of reversibility given by these authors for various proteins, i.e. the range of pH within which conformational changes are reversible but beyond which irreversible denaturation occurs, are far too wide to have a practical meaning *in vivo*. Control of the pH of tissue fluids and of the blood within narrow limits is necessary for normal physiological function. For human blood, the normal range is 7·35–7·45; values below 7·0 are extremely rare and usually imply a critical condition such as advanced diabetes.

Peters and Van Slyke[20] gave an estimate of the total buffer capacity of the blood. For instance, when the pH of the blood falls from 7·4 to 7·0

* Notable exceptions are the cells of the alga *Cyanidium caldarium* which withstand 1N sulphuric acid; the pH inside the cells may be much higher than that of the external medium[18].

as may happen in severe diabetic acidosis, the contribution of the various buffer systems in neutralizing excess acid is as follows:

Buffer system	*mmoles acid neutralized*
Bicarbonate/carbonate	18
Haemoglobinate/haemoglobin	8
Proteinate/protein	1·7
Others	2

Since one mole of carbonyl halide generates at least one mole of acid, irrespective of the site or nature of its reaction, then exposure to RCOX is equivalent to exposure to HX, the severity of action depending on the systemic or local concentrations of HX generated. Low systemic concentrations of HX lead to a condition clinically described as primary alkali deficit. A rise in the hydrogen ion concentration of the blood moves the equilibrium in the bicarbonate/carbonate buffer system to the right, resulting in an increase in the amount of exhaled carbon dioxide:

$$HCO_3^- + H^+ \rightleftharpoons H_2CO_3 \longrightarrow H_2O + CO_2 \text{ (exhaled)}$$

The alkali reserve of the blood, i.e. HCO_3^-, would eventually be exhausted if the increment in halide ion, X^-, were not removed from the body and replaced by HCO_3^-. This is accomplished by the kidneys which also eliminate cations of sodium and potassium to compensate for halide ion withdrawn from circulation[21].

The clinical symptoms associated with systemic acidosis are thirst, rapid weak pulse, slow shallow breathing, twitching of muscles, convulsions and collapse[22]. To a large measure, however, the gross toxicological effects following exposure to carbonyl halide depend on the site or organ that has been first exposed to the direct action of the chemical. For example, ingestion produces burns on lips and mouth, severe pain in throat and stomach, nausea and vomiting[22]. Of all sites, the lungs are the most vulnerable because, in addition to the special properties of the

TABLE 2. Estimated total surface area of various organs in human adults

Organ	Surface area (m²)
Eyes, lips	0·001
Skin	1·8
Small intestine	10
Lungs	80

alveolar membrane which permits rapid exchange of material, they also offer the largest area of the vertebrate body to external conditions (Table 2).

In view of this, the more volatile carbonyl halides became the focus of interest as chemical warfare agents. A few of them, like phosgene, were used in actual combat.

V. PHOSGENE AND RELATED COMPOUNDS

The clinical symptoms of phosgene poisoning in man, following use of this compound by the Germans as chemical warfare agent in 1915, were first given by Gilchrist, quoted by Prentiss[23]. In moderate cases of exposure, slight symptoms such as coughing and a sensation of irritation in the trachea and bronchi may appear immediately after inhalation of phosgene, though not to the extent of interfering with a man's normal activities. However, the condition of the victim may become suddenly worse, with signs of extreme cyanosis, followed by collapse and death. Pulmonary oedema appears very early and is the outstanding condition at the height of the illness. After the second or third day, if death does not occur, the oedema fluid is reabsorbed and recovery follows unless complications intervene.

A number of theories have been advanced to explain the action of phosgene. The LCT_{50}* of phosgene is 5000, which means that exposure for 10 minutes to an atmosphere containing phosgene at a concentration of 500 mg per cubic metre would be fatal to 50% of the individuals exposed. Under these conditions, the total amount of inhaled phosgene by a person breathing air at the rate of 20 litres per minute would be about 100 mg which, in turn, is equivalent to 70 mg of hydrogen chloride, assuming 100% retention in the lungs. This amount of acid would have insignificant effects were it administered parenterally, or by inhalation. Doses of phosgene and diphosgene far above those that are lethal on inhalation (see Table 3) have been administered intravenously or intra-peritoneally to experimental animals with, at most, only some local damage[24]. It is now generally accepted that inhaled phosgene acts immediately and strictly locally on the lung tissues. The agent as such does not appear past the lungs and has no effect on the rest of the body. Thus, when a plug was placed in a main bronchus of a dog and the animal exposed to a dose far exceeding the normal lethal dose, then the plug removed from the protected lung, the animal could live indefinitely,

* Concentration in air (mg/m³) × Time of exposure (min) that would be Lethal to 50% of the individuals exposed.

despite extreme pathology in the exposed lung and retention in the body of many times the lethal amount of phosgene[24, 24a]. Another line of evidence corroborating the above findings was obtained by crossing the circulation between two dogs of which only one was exposed to phosgene. After some hours the dogs were separated and the exposed animal died but the other survived undamaged[24, 24a].

Lungs removed from experimental animals immediately after exposure fail to collapse as do normal ones. Emphysema is the first pathological condition. There is a steady accumulation of water and chloride in the lungs, till these become waterlogged and may have four or five times their normal weight. A frothy oedema fluid finally obstructs the larger air channels and death follows severe interference with blood oxygenation[24].

The far-reaching systemic consequences of phosgene poisoning are most probably triggered by a local reaction that destroys the semipermeability of alveolar capillary membrane. The normal lung is lined with a thin film of lipoprotein, rich in lecithin, which fulfils the function of lowering the tension of the sharply curved alveolar surface. Otherwise, the latter would set up a negative pressure resulting in transudation from the capillaries, collapse and compensatory emphysema[25]. There is no evidence, however, that the 'lining film', which is freely permeable to gas, is damaged by phosgene.

Phosgene is not very reactive in the gas phase, even in the presence of a strong nucleophile such as NH_3[26]. This casts serious doubts on the assumption that the agent undergoes rapid hydrolysis to hydrogen chloride in the humid space of the air channels. To achieve lung damage comparable to that caused by phosgene in experimental animals by use of hydrogen chloride from an extraneous source, about 800 equivalents of HCl were required[26].

A more likely course is reaction of phosgene in the condensed phase with a nucleophilic group (see Table 1) in the alveolar capillary membrane leading to protein or other macromolecule analogue of carbamoyl chloride or alkyl chloroformate:

$$\text{Protein}-\text{NH}-\text{COCl} \qquad \text{Protein}-\text{O}-\text{COCl}$$

accompanied by primary release of HCl at the site of reaction. The consecutive reaction leading to the release of a second mole of HCl is most probably of a different nature and is expected to proceed at a slower rate. An example is N,N-dimethylcarbamoyl chloride, $(CH_3)_2NCOCl$, which is known to hydrolyse by an ionization mechanism, as indicated by the insensitivity of its rate of disappearance to the concentration of strong nucleophiles like hydroxide. The S_N1 mechanism here, as in the case of alkyl chloroformate, could be due to the stabilized acylium ions[27, 28].

$$R_2\overset{+}{N}-C=O \longleftrightarrow R_2\overset{+}{N}=C=O \qquad R\overset{+}{O}-C=O \longleftrightarrow R\overset{+}{O}=C=O$$

Whether the secondary release of acid, *in situ*, is more damaging than the primary one cannot be said at the present state of knowledge. There is, however, adequate support of the view that the reaction proceeds in two steps, along the lines given above.

The first supporting evidence comes from preparative organic chemistry where examples of preparation of alkyl chloroformate[29] and *N*-substituted carbamoyl chloride[30] are common procedures. It is perhaps significant that most of the so-called 'asphyxiating agents'[26, 31] are able to generate two or more moles of acid in reaction with water, in contrast to the less toxic but lachrymatory halides which generate only one mole of halogen acid:

$$COCl_2 \xrightarrow{H_2O} CO_2 + 2\,HCl$$
Phosgene

$$ClCOOCCl_3 \xrightarrow{2\,H_2O} 2\,CO_2 + 4\,HCl$$
Diphosgene

$$Cl_3COCOOCCl_3 \xrightarrow{3\,H_2O} 3\,CO_2 + 6\,HCl$$
Triphosgene

$$ClCOOCH_2Cl \xrightarrow{H_2O} CO_2 + CH_2O + 2\,HCl$$

$$ClCH_2OCH_2Cl \xrightarrow{H_2O} 2\,CH_2O + 2\,HCl$$

⎫ asphyxiating agents

$$ClCOOCH_3 \xrightarrow{H_2O} CO_2 + CH_3OH + HCl$$

$$CH_3COCl \xrightarrow{H_2O} CH_3COOH + HCl$$

⎫ irritating or lachrymatory agents

Recently, however, a lethal case of chloroformate intoxication has been reported[31a]. Some toxicological data on phosgene and related compounds are given in Table 3.

There are no publications on the biological reactions of $COBr_2$[32], COF_2[33], $COClF$[34], $COBrF$[35] and $COIF$[36]. The last four compounds should be of special interest because their reaction with amines stops at the carbamoyl fluoride stage, $RNHCOF$ or R_2NCOF, these being relatively resistant to further displacement of fluoride by nucleophiles[34, 37].

Finally, inhalation of sub-micron particles of zinc chloride, $ZnCl_2$, which is a major component of the smoke generated by smoke-grenades, is known to lead to a pathological condition similar to that caused by phosgene[38], implicating a slow and localized release of HCl at the alveolar level. An unusual but perhaps related case of intoxication resulted from the inhalation of masonry dust containing adsorbed phosgene[38a].

TABLE 3. Toxicological data of some carbonyl halides[a]

Compound	LCT_{50}	MAC	LIR	IC
$COCl_2$	5000 (man)	0·5, 4	5, 12	20
	500 (cat)			
	2000 (rat)			
	3000 (mouse)			
	10,000 (dog)			
$ClCOOCCl_3$	5000		5	40
$Cl_3COCOOCCl_3$	5000			
$ClCOOCH_2Cl$	10,000		50	2
$ClCOOCH_3$				45
$ClCH_2OCH_2Cl$	4700		14	40
$BrCH_2OCH_2Br$	4000			
$HCOF$	10,000[b]			
CH_3COF	20,000[b]			
$ZnCl_2$	2000			

[a] Compilation of data from references 22, 23, 24, 26, 31 and 38.

LCT_{50}: concentration $(mg/m^3) \times$ time (min) that would cause death to 50% of the individuals exposed. The values given are for different animals under different experimental conditions and may not be directly comparable.

MAC: maximum allowable concentration in mg/m^3.

LIR: minimum irritating concentration in mg/m^3.

IC: intolerable concentration in mg/m^3.

[b] Estimated from data in reference 31.

The premise that phosgene or diphosgene reacts with a nucleophilic group in lung tissue suggested the use of an effective competing nucleophile as a potential prophylactic agent against phosgene poisoning. In a preliminary screening test using baker's yeast as a trial organism, it was found that a number of amines afforded complete protection against the action of diphosgene in aqueous medium[24, 24b]. Concurrently, the effectiveness of the nucleophile in competing with water for the reaction with phosgene was determined by measuring the volume of CO_2 evolved from the solution of the nucleophile. The less CO_2 is evolved, the more $-CO-$ is trapped by the nucleophile:

$$COCl_2 \quad \xrightarrow{H_2O} \quad CO_2 + 2\,HCl$$
$$COCl_2 \quad \xrightarrow{2\,NH_2R} \quad CO(NHR)_2 + 2\,HCl$$

No quantitative correlation could be found between the ability to protect yeast on the one hand and that to bind the $-CO-$ group on the other (Table 4); nor could a simple structure–activity relationship be deduced from the results.

TABLE 4. Effectiveness of some nucleophiles in protecting yeast against the action of diphosgene in aqueous medium[a]

Compound	Protection[b]	CO binding capacity[c]
Hexamethylene tetramine	complete	80 (1); 88 (10)
p-Aminobenzoic acid	complete	65 (1); 80 (10)
Ethyl p-aminobenzoate	none	20
2-Naphthylamino-6-sulphonic acid	complete	50 (2)
Aniline-3,4-disulphonic acid	complete	20
Coagulin	complete	20
Glycine ethyl ester HCl	considerable	60 (1); 80 (10)
Glycine, pH 10·5	fair	30 (10)
Glycine, pH 7·0	none	0 (10)
1,2-diaminobenzene-4-sulphonic acid	considerable	80 (2)
Taurine	fair	30 (10)
Sulphanilic acid	fair	45 (1); 70 (10)
Plasma	fair	10
Diethylamine	none	0 (2)
Aniline	none	70 (1); 85 (10)
Cysteine	none	80 (2)
Glutathione	none	60 (2)
Serine	none	40 (2)
Cholesterol	none	0 (1)

[a] Data from references 24 and 24b.

[b] Protection is expressed as a qualitative measure; 10 moles of protective agent were used for 1 equivalent of diphosgene.

[c] Percentage of CO_2 that failed to evolve on addition of diphosgene to water containing nucleophile; equivalents of nucleophile used for each equivalent of diphosgene are shown in parentheses.

A possible reaction that could account for some of the ambiguous results consists in the formation of a carbamic acid derivative rather than an N,N′-substituted urea which would eventually decompose into CO_2 and the original amine:

$$RNH_2 \xrightarrow{\ COCl_2\ } RNHCOCl \xrightarrow{\ H_2O\ } RNHCOOH \xrightarrow{\ H^+\ } RNH_2 + CO_2$$

The more active compounds conferred a fair degree of protection on experimental animals when they had been injected intraperitoneally prior to exposure to a lethal concentration of diphosgene, but none had any effect when administered after exposure (Table 5)[24].

TABLE 5. Effectiveness of some prophylactic agents in animals against diphosgene poisoning[a]

Agent	Dose[b] (mg/kg)	Time[c] (h)	Survivors[d] (%)
Hexamethylene tetramine	2 or less	0–6	42
None			3
Hexamethylene tetramine	2	0·25	100
None			0
Taurine	2–6	0·5–1·0	31
None			5
1,2-Diaminobenzene-4-sulphonic acid	2–4	0·25–6	29
None			13
2-Naphthylamine-6-sulphonic acid	0·8	0·5	12
None			0
Coagulin	atomized suspension	3–6	20
None			0

[a] Data from reference 24.
[b] Intraperitoneally to rats and mice.
[c] Time interval between administration of prophylactic agent and exposure to diphosgene.
[d] Indefinitely, usually after 24 h.

Outstanding among the prophylactic agents is hexamethylene tetramine, the activity of which was also confirmed by Porembskii[38b]. Some electrolytes also proved beneficial, e.g. the subcutaneous injection of a soluble calcium salt proved effective in preventing reactions induced by sub-lethal concentrations of phosgene for as long as 24 h after administration[39]. Also, death in frogs, the skin respiration of which had been impaired after exposure to phosgene, could be prevented by parenteral administration of sodium chloride or by maintaining the animals in a NaCl bath[40]. The molecular processes underlying these effects have so far remained unexplored.

Another line of approach for the study of the reaction of phosgene with lung tissue is based on an immunochemical investigation of proteins derived from these organs. It has been found that lung-protein extracts derived from various animals which had been exposed to phosgene contain an immunochemically related antigen and elicit in rabbits antibodies directed against a common molecular grouping[41]. This occurs to such an extent that differentiation between proteins from normal lungs and proteins from lungs exposed to phosgene is possible with the aid of the anaphylactic reaction[42]. That is, animals sensitized to phosgene-treated proteins respond violently to an inoculum of phosgene-treated proteins but not to normal proteins. Most probably, any modification of protein due to cross-linking of adjoining chains through —NH—CO—NH—, —NH—CO—O —or —O—CO—O— groups is of a temporary nature, since no sensitization to phosgene occurred in guinea-pigs that had been subjected to repeated exposure to low concentrations of the agent[43]. It is also remarkable that in repeated exposure of cats to sub-lethal concentrations of phosgene, those cats that had been exposed more frequently than others showed no greater respiratory impairment than those exposed only a few times[44].

From a physiological standpoint, Enikeeva[45, 46] showed that irritation of tracheal or bronchial mucosa with 20–30% sulphuric acid causes emphysema in dogs and cats but not if preceded by vagotomy. However, such treatment as vagotomy, atropine-block of the vagus nerve or respiratory narcosis with procaine failed to prevent the occurrence of emphysema in cats that had been exposed for 15 min to diphosgene at a concentration of 0.2 mg per litre air. The author concluded that emphysema due to mere acid irritation of these organs is not of the same origin as in phosgene poisoning.

Artificial oxygenation of the blood of victims suffering from anoxia following phosgene poisoning is not recommended. Motylev[47] reported that prolonged inhalation of oxygen-enriched air (30–45%) lessened the severity of cyanosis and dyspnoea in cats and mice that had been exposed for 15–30 min to diphosgene at a concentration of 0.26–1.44 mg per litre air. However, neither the course of poisoning nor the mortality rate was significantly affected. This is not surprising in view of the observation by Ohlssen[48] that the lesion produced by phosgene is localized in the alveolar wall, while oxygen poisoning may affect the capillary wall. Thus, the combined action of the two agents could be more damaging than the separate effect of phosgene alone.

VI. HALIDES OF TOXIC CARBOXYLIC ACIDS

Certain acid halides owe their action not to the carbonyl halide function as such, but to toxic carboxylic acids which they eventually generate *in vivo*, either through direct hydrolysis or by the metabolic breakdown of amides and esters derived from their interaction with various substrates. In this case, the toxicity of the halide is qualitatively and quantitatively equivalent to that of the parent acid itself and is of systemic nature rather than confined locally to the organ or tissue directly exposed to the action of the agent.

Classical examples are the halides of fluoroacetic acid, a powerful systemic poison which acts by blocking the conversion of citrate to *cis*-aconitate in the tricarboxylic acid cycle[49, 50, 51]. That the action of these halides resides in the FCH_2CO- group rather than in the carbonyl halide function has been proved unequivocally by Saunders and Stacey[52] after submitting test animals to the action of compounds embodying the FCH_2CO- group on the one hand and $R-COX$ (X = F, Cl, Br) on the other. The results confirmed that acyl halides derived from fluoroacetic acid were far more toxic than corresponding compounds derived from acetic or chloroacetic acid, the magnitude of the effect being almost independent of the method of administration. Some results are shown in Table 6. This subject has been adequately covered in a comprehensive monograph by Pattison[53].

TABLE 6. Toxicities of some derivatives of fluoroacetic and formic acids

Compound	Toxicity (reference)	
	LD_{50} (mice) (mg/Kg)	$LC_{50}{}^a$ (mg/l/10 min)
FCH_2COCl	6–10 [52]	0·1 [52]
FCH_2COBr	3–25 [54, 55]	—
FCH_2COF	6–10 [52]	0·1 [52]
FCH_2COOCH_3	6–10 [52, 56, 57]	0·1 [52]
$FCOOC_2H_5$	insignificant [52]	
$FCOCH_3$	insignificant [52]	
$FCOCH_2Cl$	insignificant [52]	

^{a} LC_{50} is the lethal concentration for a 10 min exposure.

VII. FLUORO-ALKENES AS LATENT CARBONYL HALIDES

Certain fluoro-alkenes which are extremely toxic when inhaled appear to exert their action *in vivo* by a mechanism analogous to the one attributed to carbonyl halides. Indeed, a common structural feature to both classes of compounds is a highly electrophilic sp^2 or trigonal carbon atom which is highly susceptible to attack by nucleophiles:

$$X-\overset{\overset{\displaystyle Y}{\|}}{C}-X$$

X = F, Cl, Br

Y = O in carbonyl halides

Y = $CR^1 R^2$, where R^1 and R^2 are electron-withdrawing groups in fluoro-alkenes

In both cases, the end products of the reactions with nucleophiles are an acyl derivative of the latter on the one hand and one or, more often, two moles of halogen acid on the other. Presumably therefore, the local toxic action of fluoro-alkenes may be due to the *in situ* generation of acid.

There has been no systematic investigation of the action of the fluoro-alkenes on biological systems at the molecular level. However, an indication of probable reaction pathways may be obtained from a large volume of data on the reactivity of these compounds *in vitro* with selected substrates.

For example, in the reaction sequence of a fluoroalkene, $CF_2{=}CFX$ (X = F or Cl) with a primary amine[58] two moles of hydrogen fluoride are formed in the course of the reaction, first following addition of the

$$CF_2{=}CFX \xrightarrow{+RNH_2} (RNHCF_2-CFXH) \xrightarrow{-HF} RN{=}CF-CFXH$$

$$\xrightarrow{+RNH_2, -HF} \underset{\underset{\displaystyle NHR}{|}}{RN{=}C}-CFXH$$

primary amine to $CF_2{=}CFX$ to yield an imidyl fluoride, then following addition of a second mole of primary amine to the latter to yield an amidine. Thus, both fluorine atoms in the terminal CF_2 group have been substituted in reasonable analogy with the reaction of a carbonyl halide, COX_2, with amines.

In the presence of water, an amide may be obtained as in the following cases[59, 60]:

$$C_4H_9NH_2 + CF_2{=}CF_2 \xrightarrow{H_2O} CHF_2CONHC_4H_9 + 2\ HF$$

$$(CF_3)_2C{=}CF_2 \xrightarrow{NH_3\ (Et_2O,H_2O)} (CF_3)_2CHCN + (CF_3)_2CHCONH_2 + 2\ HF$$

In general, the reaction with amines proceeds in the cold and without the aid of a basic catalyst[58]. It is reasonable to assume, therefore, that similar reactions occur at nucleophilic sites *in vivo*.

The reactions of fluoro-alkenes with nucleophiles were reviewed by Chamber and Mobbs[61], according to whom the order of chemical reactivity increases along the series $CF_2{=}CFX < CF_2{=}CF_2 < CF_3CF{=}CF_2 < CF_2$ $={C(CF_3)_2}$.

The paucity of data on the biological reactions of fluoro-alkenes makes any correlation between toxicity and chemical reactivity necessarily speculative. Even so, it may be stated that the more reactive compounds are also the more toxic ones. This may be inferred from a comparison between toxic fluoro-alkenes and closely related non-toxic analogues, with the understanding that the demarcation line between toxic and non-toxic compounds is a diffuse one, depending on the relative concentration of the agent in air and on the length of exposure. Some fluoro-alkenes reported to be toxic are shown in Table 7.

TABLE 7. Some toxic and non-toxic fluoro-alkenes

Toxic	Reference	Non-toxic	Reference
$CF_2{=}CF_2$	53, 62	$CF_2{=}CH_2$	53, 62
$CF_2{=}CFCl$	53, 63, 64	$CF_2{=}CCl_2$	53
$CF_3{-}CF{=}CF_2$	53	$CF_3{-}CF{=}CF{-}CF_3$	65
$(CF_3)_2C{=}CF_2$	53, 65	$CF_3{-}CF_2{-}CF{=}CF_2$	65
$CF_3{-}\underset{\diagdown\diagup}{\underset{CF_2}{CF}}{-}CF_2$	66	$\underset{\diagdown\diagup}{\underset{CF_2}{CF_2}}{-}CF_2$	66
$CHF_2{-}\underset{\diagdown\diagup}{\underset{CF_2}{CF}}{-}CF_2$	66	$\begin{array}{c}CF_2{-}CF_2\\ \mid\qquad\mid\\ CF{=}CF\end{array}$	66
$CClF_2{-}\underset{\diagdown\diagup}{\underset{CF_2}{CF}}{-}CF_2$	66	$\begin{array}{c}CF{=}CF\\ \mid\qquad\mid\\ CF_2\quad CF_2\\ \diagdown\diagup\\ CF_2\end{array}$	66

Perfluoroisobutene, $(CF_3)_2C{=}CF_2$, the most reactive compound in the series[60, 65], has been reported to be ten times more toxic than phosgene[53]. It appears to be a constituent of the dangerous gases arising from the accidental or intentional pyrolysis of perfluoro-ethylene polymer[67-70]. The symptoms following exposure include respiratory distress, dyspnoea, gasping and prostration. Acute pulmonary irritation, characterized by haemorrhage and oedema, was revealed by post-mortem examination of test animals. Perfluorisobutene itself is exceptional among the perfluorobutenes in that it will react with alcohols even in neutral or slightly acidic media[65, 65a].

A 4 h exposure of dogs, rabbits or rats to air containing 0·5% by volume of $CF_2{=}CFCl$ resulted in the death of all animals from pulmonary hypostatic congestion and oedema[53]. As in the case of phosgene poisoning, no immediate signs of intoxication could be observed, death occurring within a delay of a few hours. Ambros[63] showed that $CF_2\text{=-}CFCl$ undergoes oxidation with molecular oxygen at $30°$, yielding a polymeric peroxide $(C_2F_3ClOO)_n$ and CF_2ClCOF as main products, with COF_2 and $COFCl$ in lesser amounts. A 3 h exposure of rats to $CF_2\text{=}CFCl$ led to the death of all animals within $3\cdot5$ h when the concentration of the agent was 250 mg per litre. At a concentration of 30 mg per litre death occurred within one week. At concentrations less than 15 mg per litre no death occurred[64].

In an investigation of respiratory irritants, Gerard and co-workers[24, 24b] found that ketene acts much like phosgene, being even more toxic. In the absence of halogens in the molecule, one has to assume that it has properties not shared by other acylating agents. For example, it could acylate free —COOH residues in proteins to yield anhydrides which may then react with neighbouring nucleophiles to yield cross-linked proteins:

$$\text{Protein—COOH} \xrightarrow{\ CH_2=C=O\ } \text{Protein—} \overset{O}{\overset{\|}{C}}\text{—O—}\overset{O}{\overset{\|}{C}}\text{—CH}_3$$

$$\xrightarrow{\ Protein-NH_2\ } \text{Protein—CO—NH—protein} + CH_3COOH$$

These views, however, have not yet been substantiated by adequate experimentation.

The LCT of ketene to mice is about 750 which means that its toxicity by inhalation is eight times greater than that of phosgene. After a $1\cdot5$ min exposure to air containing $0\cdot5$ mg ketene per litre, seven out of eight mice died[24b]. Like phosgene, ketene considerably inhibits enzymatic activity in lung homogenates and completely inhibits that of baker's yeast (Table 8).

TABLE 8. Effect of equivalent amounts of ketene, hydrochloric acid and diphosgene on enzymatic activity of lung homogenates and yeast [a]

Material	Measurement	Substrate	Inhibition (%)		
			Ketene	Diphosgene	HCl
Lung homogenate	methylene blue reduction	none	80	70	—
		succinate	70	65	35
		glycero-phosphate	70	40	15
Yeast	CO₂ production	glucose	100	100	0

[a] Data from reference 24b.

VIII. FINE BIOLOGICAL EFFECTS OF CARBONYL HALIDES

A protein molecule may be modified by the incorporation of an acyl residue in its structure. The properties of the modified protein may differ from those of the native one, as surface charge, hydration, topology or tertiary structure depart from the original. If modification has not been so extensive as to result in the complete and irreversible loss of original character, then some qualitative and quantitative changes in the function of the molecule will occur.

In general, reaction of carbonyl halides with proteins consists of a fast and non-selective acylation of free amino groups and thiols, which are followed in reactivity by aromatic and aliphatic hydroxyl groups of tyrosine, serine and threonine[71, 72]. Usually however, use of the less-reactive acid anhydrides has been preferred by protein chemists. A review on this subject has been given recently by Stark[73].

Limited success has been achieved by the use of carbonyl halides. An example is the controlled treatment of insulin with phosgene. The derivative produced possessed a short or moderate insulin effect when administered subcutaneously, but larger amounts of phosgene led to inactivation[74]. Sluyterman[75] showed that the acetylation of insulin with acetic anhydride, depending on the relative amounts, could give partly acetylated insulins, containing 33–93% of the theoretical amount of possible N-acetyl residues.

Bier, Ram, Nord and Terminiello found that whilst the incorporation of acyl residues in the trypsin molecule had little effect on the proteolytic activity of the enzyme with respect to its substrate, acylated trypsin was not subject to self-digestion as in the case of native trypsin[76, 77]. In particular, acylated trypsin showed a diminished tendency to bind to inhibitory factors such as ovomucoid or inhibitors normally present in human serum[78]. Some quantitative results are shown in Table 9.

Provided the integrity of the tertiary structure is preserved and the reactive centre in enzymes not blocked, the net effect of carbonyl halides should be expressed in a reduction of positive charge on the protein surface:

$$\text{Protein}-\overset{+}{\text{NH}_3} \xrightarrow{\text{RCOX}} \text{Protein}-\text{NHCOR} + \text{HX} + \text{H}^+$$

In view of this, the capacity of acylated proteins to participate in ionic interactions should be impaired; otherwise, acylation will have little effect on interaction requiring hydrophobic bonding. Thus, the binding of testosterone by bovine serum albumin (BSA) is unaltered, even after extensive acetylation of the protein[79].

TABLE 9. Effect of the incorporation of acyl residues on some properties of trypsin[a]

Acyl residue	Free—NH_2/ 10^4 g protein	Substitution (%)	Relative activity[b]	K^c $\times 10^7$ M
None	8·8	0	100	$2·2 \times 10^{-2}$
Acetyl	1·3	85	75	$5·2 \times 10^{-2}$
n-Propionyl	2·9	67	92	$1·5 \times 10^{-1}$
n-Butyryl	1·3	85	70	1·1
Citraconyl	4·5	49	71	1·4
Itaconyl	2·2	75	68	1·5

[a] Data from references 76–78.
[b] With respect to casein as substrate.
[c] K is defined as the dissociation constant of enzyme–human serum complex.

This contrasts with the finding of Theorell and Nygaard[80, 81] that acetylation of flavoprotein (old yellow enzyme) prevents the interaction of the apoenzyme with flavin mononucleotide which requires the presence of free amino and hydroxyl groups for attachment:

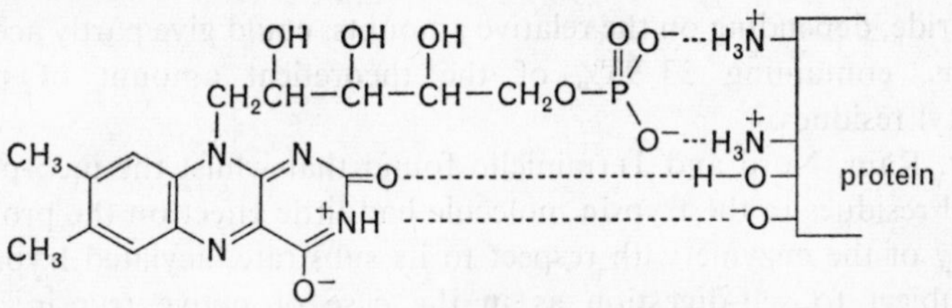

On the other hand, the interaction between antigens and their corresponding antibodies seems to be of the non-ionic type, since the presence of free-amino groups is unessential. This conclusion has been borne out by the results of Marrack and Orlans[81a] and Nisinoff and Pressman[81b] who subjected either antigen or antibody to acetylation, thereby blocking the ε-amino groups of lysine. Acetylation did not result in any appreciable loss of hapten-binding capacity. Loss of charge, however, led to a loss in the precipitating capacity of the antigen–antibody complex.

Numerous examples exist of the modification of proteins with acylating agents[71-73], but it is not clear to what extent the analogous reactions would occur *in vivo*. One difficulty lies in the very low probability of achieving selective acylation with the extremely reactive carbonyl halides or anhydrides when these are applied to the whole organism. Also, the fate of ε-N-acyl-lysyl residues which would arise after such an application is unknown. Of relevance to this problem is the discovery by Paik and co-workers of an acyl–lysine deacylase that was isolated from rat kidney. This enzyme hydrolyses ε-N-acetyl-L-lysine and ε-N-chloroacetyl-L-lysine but is ineffective towards ε-N-carbobenzoxy-L-lysine, ε-N-acetyl-D-lysine and δ-N-chloroacetylornithine[82].

The effect of acid chlorides on the rate of tumour induction in mice by 3,4-benzpyrene was investigated by Crabtree[82a]. All proved inhibitory when applied to the skin, but the degree of inhibition could not be correlated with any physical or chemical parameter of the compounds tested (Table 10). Remarkably, the most effective agent was benzenesulphonyl chloride, the hydrolysis rate of which is considerably lower than that of the aliphatic carbonyl chlorides used (Table 15). The author concluded that the action of acid chlorides dwells on a non-specific impairment of cell-functions. For example, the rate of glycolysis in a slice of tissue is reduced equally whether treated with an acid chloride or the equivalent amount of HCl.

TABLE 10. Average induction time for papillomas in mice treated with benzpyrene and acid chlorides[a]

Acid chloride	Average induction time (days)	
	0·3% benzpyrene	0·1% benzpyrene
None	79	99
Stearoyl chloride	103	121
Palmitoyl chloride	103	115
Valeroyl chloride	112	160
Acetyl chloride	122	156
Benzenesulphonyl chloride	177	186

[a] Data from reference 82a.

Encouraging results were obtained with latent carbonyl halides which would generate the halide or an intermediate of equivalent reactivity under normal physiological conditions. An example is 6-trichloromethylpurine which acts as though it were the chloride of purinoic acid[83]:

(R = alkyl, aryl or protein residue)

The acylation reaction may proceed by two different pathways, depending on the pH of the medium[84]. Since the pK_a of 6-trichloromethylpurine is 7·70, it will exist as an anion to the extent of 24% at pH 7·2. The latter species will expel chloride in a rate-determining step to yield a highly reactive intermediate which would add H_2O, ROH or RNH_2 to yield in one or more steps the corresponding derivative of purinoic acid. At low pH where the concentration of the anion would be negligible, or in a case where the proton in position nine has been replaced with an alkyl group, acylation may still proceed but at a considerably slower rate:

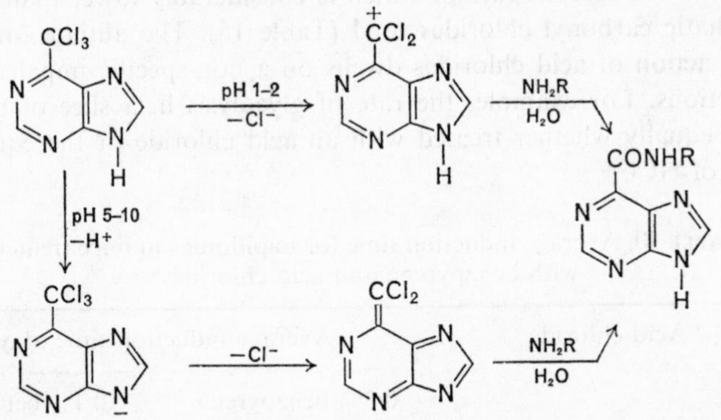

Judged by its rate of hydrolysis, 6-trichloromethylpurine anion ranks amongst the least reactive carbonyl halides (Table 11). This relative stability may account, at least partly, for the selective acylation of amino groups even in very dilute aqueous solution.

TABLE 11. Rates of hydrolysis of some acid chlorides

Acid chloride	Conditions	$10^4 k_1$ (s^{-1})	Ref
p-Methoxybenzoyl chloride	30% H_2O/acetone, 0°	7·03	85
p-Methylbenzoyl chloride	33% H_2O/dioxan, 25°	14·83	86
Benzoyl chloride	50% H_2O/dioxan, 25°	26	85
p-Nitrobenzoyl chloride	40% H_2O/acetone, 0°	50·16	85
Acetyl chloride	25% H_2O/acetone, 25°	8600	87
6-Trichloromethylpurine	H_2O, pH 1–2, 25°	8×10^{-3}	84
6-Trichloromethylpurine anion	H_2O, pH 5–10, 25°	10·9	84

6-Trichloromethylpurine displays an interesting range of biological properties which could not be accounted for exclusively by the purine structure were it not able to 'couple' to some biologically important molecule via the carbonyl group. For example, it has a marked anti-neoplastic activity against a number of experimental tumours[88].

The reaction of 6-trichloromethylpurine with various proteins has been studied *in vitro*. In the pH range 7–9 it acylates the ε-amino groups of

TABLE 12. Activity of 6-trichloromethylpurine against mouse tumours[a]

Tumour	Result
Sarcoma 180 (solid)	no effect
Sarcoma 180 (ascitic)	complete inhibition
Ehrlich carcinoma (ascitic)	complete inhibition
Adenocarcinoma EO771	slight inhibition
Carcinoma 1025	slight inhibition
Lewis lung carcinoma	no effect
Wagner osteogenic sarcoma	no effect
Ridgway osteogenic sarcoma	marked inhibition
Mecca lymphosarcoma	complete inhibition
Harding–Passey melanoma	slight inhibition
Glioma 26	no effect
Friend virus leukaemia	slight inhibition

[a] Data from reference 88. Dose of 6-trichloromethylpurine was 60 mg/kg/day.

lysine residues without much interference with the tertiary structure of the recipient protein or its activity if it were an enzyme[89, 90]. Thus, bovine pancreatic ribonuclease containing three purinoyl residues per mole protein was still enzymatically active, proving that ε-amino group of lysine in position 41, known to be essential for enzymatic activity, has remained unaffected[90]. In general, purinoyl proteins prepared under mild conditions are homogeneous, contain a constant number of purinoyl residues per mole protein and possess higher T_m values than the corresponding native proteins (Table 13).

An important property of purinoyl proteins is their ability to elicit purine-specific antibodies in rabbits[91], mice[89, 89a] and hamsters[92]. These antibodies cross-react with single-stranded deoxyribonucleic acid. Some of the purinoyl proteins are antineoplastic, inducing regression of Ehrlich ascites tumour in mice[89, 89a] and SV-40 induced tumour in hamsters[92]. The antigenicity of the purinoyl residue and the antineoplastic activity of the protein seem to be interdependent properties with a common origin. The properties of some purinoyl-protein conjugates are given in Table 13.

TABLE 13. Some N-acylated proteins and their biological activity[a]

Acylating agent	Protein[b]	Acyl residues/ mole protein	T_m[c]	Relative anti- neoplastic activity[d]
None	HSA	0	73°	0
	RNase	0	62·5°	0
6-trichloromethylpurine (CCl₃)	HSA	14	84°	100
	RSA	14		100
	B-γ-G	12		75
	RNase	3	65°	
	RNase	8		20
9-methyl-6-trichloromethylpurine (CCl₃, CH₃)	HSA	14	79°	100
	B-γ-G	12		30
	RNase	8		30
naphthalene COCl	HSA	12	77·5°	10
2-amino-6-trichloromethylpurine (CCl₃, H₂N)	HSA	14	71°	0
	RNase	8		0

[a] Data from references 89 and 90.

[b] HSA: human serum albumin; RSA: rabbit serum albumin; B-γ-G: bovine γ-globulin; RNase: bovine pancreatic ribonuclease.

[c] T_m is the temperature at which 50% transition occurs from the native helical structure to the random coil structure, measured in 0·01M phosphate buffer, pH 7·0.

[d] Against Ehrlich ascites tumour, by intraperitoneal injection of 100 μg protein per mouse.

N-Acylation of proteins with 6-trichloromethylpurine derivatives bearing various substituents in position two, eight or nine of the purine ring[93, 94] proceeds under similar conditions. Among the other halogen analogues, 6-tribromomethylpurine[2] is more reactive and, therefore, less suitable for acylation reactions in aqueous media, whilst 6-trifluoromethyl-purine[95] has proved singularly unreactive. In contradistinction, 5-tri-fluoromethyluracil (pK_a, 7·35) undergoes rapid hydrolysis in water to yield iso-orotic acid and will acylate glycine or glycylglycine in protic media and under extremely mild conditions to yield the N-iso-orotyl derivatives[96]:

(R = CH₂COOH; CH₂CONHCH₂COOH)

The facile heterolysis of the C—F bond in —CF$_3$ groups has been reported by several groups[97-99], but little effort seems to have been made so far to apply the acylating properties of such compounds with molecules of biological interest.

IX. SELECTIVITY VERSUS REACTIVITY

With few exceptions, little success has been achieved in using the carbonyl halide group, —COX, for the design of active site-directed inhibitors of enzymes. The mechanism of action of such inhibitors is based on the following concept, formulated by Baker[100]: 'The macromolecular enzyme has functional groups on its surface which logically could be attacked selectively in the tremendously accelerated neighbouring group reactions capable of taking place within the reversible complex formed between the enzyme and an inhibitor substituted with a properly placed neighbouring group.'

The apparent lack of selectivity that characterizes the carbonyl halides stems to a large extent from their intrinsic reactivity which is the highest in a given series of halides of carbon, phosphorus and sulphur acids. The less reactive a halide, the more it will discriminate among the various possible substrates, structures or functions. Added selectivity could be conferred on the halide by designing it to conform to the molecular structure and shape which makes the closest and best-oriented fit with the biological function to be modified.

Thus chymotrypsin could be selectively acylated or alkylated at each of three different sites, depending on the type of active site-directed inhibitor used[101]:

Inhibitor	Formula	Reaction (site of specific attack)
N, N-Diphenylcarb-amoyl chloride		Acylates —OH in serine (195)
3-Phenoxy-1, 2-epoxy-propane		Alkylates methionine (192)
N-Tosyl-L-phenyl-alanyl-chloromethane		Alkylates histidine (57)

The relationship between *in vitro* reactivity on the one hand and selectivity on the other is best exemplified in a series of sulphonyl, phosphoryl and carbonyl halides, certain members of which react preferentially with the —OH group of serine in the active centre of hydrolytic enzymes[102]. The reactivity of these halides in non-catalysed hydrolysis in protic media follows the sequence:

$$RCOX > (R)_2NCOX > ROCOX > (RO)_2P(O)X > RSO_2X$$

where R is alkyl or aryl and X is chlorine or fluorine, correspondingly[103, 104].

Maximum biological activity is reached with the acid fluorides which are also the most stable under 'neutral' conditions (Table 14).

The remarkable selective activity of the sulphonyl and phosphoryl fluorides is also shared by N,N-dimethylcarbamoyl fluoride which owes its high toxicity to the inhibition of the enzyme acetylcholinesterase, with consequent accumulation of acetylcholine. The symptoms of acute intoxication with such agents are probably due chiefly to acetylcholine itself, the normal removal of which has been blocked by inhibition of the enzyme. All anti-cholinesterase agents which are active *in vivo* produce the so-called 'muscarinic effects' which include bronchoconstriction, sweating, salivation, anorexia, nausea, abdominal cramps, vomiting, diarrhoea, involuntary defaecation and increased urination. The 'nicotinic effects' of these agents include muscular twitching and fasciculation, increased fatigability and weakness of skeletal muscles. Death occurs

TABLE 14. Reactivity of halides of sulphur, phosphorus and carbon acids in non-catalysed hydrolysis at 25° [a]

Acid halide	Solvent	$10^4 k$ (s^{-1})	Relative rate	Reference
$C_6H_5SO_2F$	50% H_2O/acetone	0·0005	1	87
(iso-PrO)$_2$POF	H_2O	0·017	$3·4 \times 10^1$	103
C_6H_5COF	25% H_2O/acetone	0·082	$1·6 \times 10^2$	87
CH_3COF	25% H_2O/acetone	1·1	$2·2 \times 10^3$	87
C_2H_5OCOCl	H_2O	(20°) 2·15	$4·3 \times 10^3$	28
$C_6H_5SO_2Cl$	50% H_2O/acetone	2·4	$4·8 \times 10^3$	87
$(CH_3O)_2POF$	H_2O	4·0	$8·0 \times 10^3$	104
C_6H_5COCl	25% H_2O/acetone	7·2	$1·4 \times 10^4$	87
$(CH_3)_2NCOCl$	H_2O	(8·5°) 56	$1·1 \times 10^5$	28
$(CH_3O)_2POCl$	ethanol	($-8·5°$) 60	$1·2 \times 10^5$	104
(iso-PrO)$_2$POCl	H_2O	81	$1·6 \times 10^5$	103
CH_3COCl	25% H_2O/acetone	8600	$1·7 \times 10^7$	87

[a] Most data from compilation by Halman[103].

TABLE 15. Biological activity of some halides of sulphur, phosphorus and carbon acids

Acid halide	$-\log k_1$[a]	pI$_{50}$[b]				LD$_{50}$[c] (mg/kg)	Reference
		A	B	C	D		
$(CH_3)_2NCOF$	7–8*	6·4	5·2	5·5	5·7	7	105
CH_3SO_2F	6–7*	5·0	3·9	4·2	3·3	3	105
$C_6H_5SO_2F$	7·3	3·2	3·1	3·6	4·9	100	105
(iso-PrO)$_2$POF	5·8	6·7	5·0	7·1	7·2	4	105
$(CH_3O)_2POF$	3·4	—	—	7·0	—	30	106 107
$(Pr^nO)_2POCl$	2·0	—	—	—	—	>20	108

[a] k_1 = first-order rate constant for hydrolysis under 'neutral' conditions[103]; figures marked * have been derived by estimation.

[b] pI$_{50}$ = $-\log$ molar concentration required to produce 50% inhibition of A, rat brain cholinesterase; B, rat brain tributyrin esterase; C, serum pseudocholinesterase; D, serum tributyrin esterase.

[c] For rats, by injection.

following paralysis of the respiratory muscles. A detailed account of these effects has been given in numerous texts[109].

In principle, all carbamates where X in $R^1R^2N—CO—X$ is a potential leaving group are capable of acylating the hydroxyl group of activated

12

serine in the esteratic site of the enzyme[110]. Like acetylated cholinesterase, the carbamoylated enzyme may recover its original activity by losing the carbamoyl residue, but at a much slower rate which is also independent of the nature of the leaving group X[111, 111a, 112].

The inhibition of acetylcholinesterase with carbamates has been rationalized[111a, 113] in terms of the structural similarities between the natural substrate, acetylcholine, on the one hand, and the carbamates on the other. A presentation of the structure–activity relationship in carbamates has been given by Long[114] and by Casida and co-workers[115].

In the carbamoyl halides proper, Metzger and Wilson[116] found that in all cases the fluorides were more powerful inhibitors towards a number of hydrolytic enzymes than the corresponding chlorides (Table 16). In view

TABLE 16. Specific rate constants for the reaction of hydrolytic enzymes with carbamoyl halides[a]

Inhibitor	k (l/mole s)			
	AcChE	Human serum ChE	Chymo-trypsin	Trypsin
$(C_6H_5)_2NOCF$	347	12,900	3790	58
$(C_6H_5)_2NCOCl$	2·2	145	480	7
$\dfrac{k_{RCOF}}{k_{RCOCl}}$	160	89	8	8·3
$C_6H_5\underset{\underset{CH_3}{\mid}}{N}COF$	12,300	2070	18	0·14
$C_6H_5\underset{\underset{CH_3}{\mid}}{N}COCl$	2300	163	3·5	0
$\dfrac{k_{RCOF}}{k_{RCOCl}}$	5·4	13	5	—

[a] Data from Metzger and Wilson[116].

of these results, the authors concluded that an electrophilic mechanism is involved in the process of acylation. A change from Cl to F in —COX entails a promotion in the electrophilicity of the carbonyl carbon which becomes more susceptible to attack by nucleophiles. The change is accompanied by a transition from an S_N1 to an S_N2 mechanism, in conformity with the findings of Swain and Scott[87] that the basic hydrolysis of C_6H_5COF proceeds 40% faster than that of C_6H_5COCl while the rate of hydrolysis of C_6H_5COCl under neutral conditions is 100 times faster than that of C_6H_5COF.

Another line of approach dwells on a rational choice of the groups R^1 and R^2 so that they are able to conform to specific structural requirements of the enzyme. An example is the discovery by Erlanger and co-workers[117–119] that diphenylcarbamoyl chloride is a specific inhibitor of chymotrypsin.

TABLE 17. Specific rate constants of the inactivation of chymotrypsin and trypsin with specific inhibitors[a]

Inhibitor	k (l/mole s)	
	Chymotrypsin	Trypsin
$(C_6H_5)_2NCOCl$	610	8·2
$(iso-PrO)_2POF$	317	
$C_6H_5—CH_2CHCOCH_2Br$ | NH—tosyl	2·37	

[a] Data from Erlanger, Cooper and Cohen[119].

The reaction of diphenylcarbamoyl chloride with chymotrypsin proceeds to completion in a mole/mole ratio and may be competitively inhibited with indole. These results strongly suggest the involvement of the active site of chymotrypsin on the one hand and at least one of the two aromatic rings of the inhibitor molecule on the other. A similar inhibitor of chymotrypsin is N-chloroformylphenothiazine[120], in which the free movement of the —COCl group relative to the aromatic structure is subject to considerable constraint. In these carbamoyl chlorides the aromatic ring is believed to fulfil an active role in positioning the inhibitor on the enzyme surface as may be inferred by comparing the structure of the inactivator molecules with that of specific substrates such as 1-keto-3-carbomethoxy-1,2,3,4-tetrahydroisoquinoline and benzoyl-L-phenylalanine methyl ester. In three-dimensional models, one could see that the orientation and distance of the —COOMe group with respect to the phenyl ring is similar to that of the —COCl group in the inhibitor molecules[120].

In vitro, diphenyl carbamoyl chloride N-acylates a wide range of amino acids and peptides under relatively mild conditions[121].

Recently, however, Makinen[122] reported that the hydrolysis of L-arginyl-2-naphthylamide by human foetal liver homogenate is subject to a specific

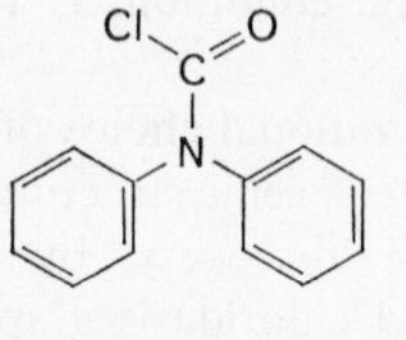

Diphenylcarbamoyl chloride

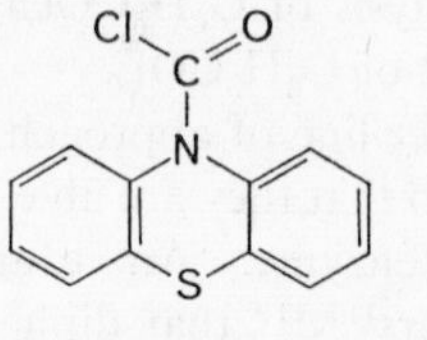

N-chloroformylphenothiazine

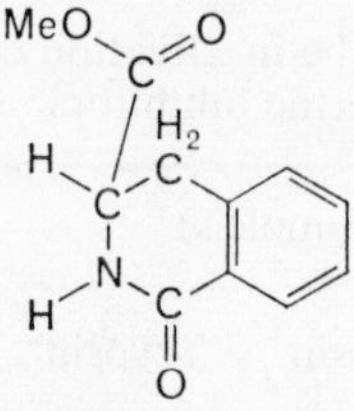

1-Keto-3-carbomethoxy-1,2,3,4-
tetrahydroisoquinoline

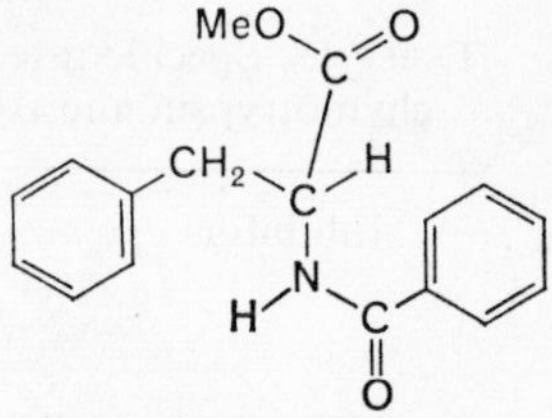

Benzoyl-L-phenylalanine
methyl ester

and non-competitive inhibition by diphenylcarbamoyl chloride. The following $K_i(M)$ values were reported (where $K_i(M)$ is the molar constant of inhibition):

Inhibitor	$K_i(M)$
$C_6H_5CH_2SO_2F$	$2 \cdot 5 \times 10^{-4}$
$(C_6H_5)_2NCOCl$	3×10^{-5}
$C_6H_5CH_2CHCOCH_2Cl$ $\quad\quad\mid$ $\quad$ NH—tosyl (L-isomer)	1×10^{-6}

The enzyme inactivated by these agents, tentatively identified as amino-peptidase-B, is believed to react with the inhibitors at a nucleophilic group at the catalytic site rather than at the substrate-binding site proper.

X. CONCLUDING NOTES

In retrospect, the evolution of thought on the possible biological effects of the acid halides has undergone a profound change with the advent of the more specific agents, first made public in 1946[123, 124]. The early pre-occupation with highly reactive agents such as phosgene gave way to a more rational concept of biological activity. Specificity is also a function of chemical reactivity, but while, for example, ethyl chloroformate appears less reactive than dimethyl phosphorofluoridate or dimethylcarbamoyl chloride, both of which represent groups of highly specific agents, the chloroformates are just lachrymatory. Neither has there been mention of any specific activity on behalf of the far more stable fluoroformates[125]. Formyl fluoride, the only stable halide of formic acid, is not endowed with any particular property in biological media that could be an extension of the range of its reactions with simpler substrates[126].

Finally, and because the biological activity of the carbonyl halides depends in the first place on the creation of a covalent bond between the carbonyl group and any of the functions exemplified in Table 1, their relative value as potential chemotherapeutic or investigational agents should be weighed in the light of knowledge of compounds known to exert their specific effect by a similar process. Surprisingly enough, the alkylating agents[127] lack specificity in their therapeutic activity or toxicity[128] and, like the carbonyl halides, they react rather indiscriminately with most of the nucleophilic sites in the macromolecular constituents of the cell. However, one particular reaction among a multitude of possibilities seems to confer on the alkylating agents their antineoplastic property—namely the alkylation of N-7 in guanine or N-3 in adenine bases of deoxyribonucleic acid[129]. No such common property has been found yet for the carbonyl halides.

XI. REFERENCES

1. D. W. Fassett, 'Organic acids, anhydrides, lactones, acid halides and amides, thioacids' in *Industrial Hygiene and Toxicology*, Vol. II, 2nd rev. ed. (Ed. F. A. Patty), Interscience, New York, 1963, p. 1826.
2. S. Cohen, E. Thom and A. Bendich, *J. Org. Chem.*, **27**, 3545 (1962).
3. I. L. Knunyants, E. G. Bykhovskaya and V. N. Frosin, *Dokl. Akad. Nauk S.S.S.R.*, **127**, 337 (1959); *Chem. Abstr.*, **54**, 259 (1960).
4. F. Swarts, *Bull. Acad. Roy. Belg.*, **35**(3), 849 (1898).
5. F. Swarts, *Mem. Couronnés Acad. Roy. Belg.*, **61**, 94 (1901).
6. Th. C. Bruice and S. Benkovic, *Bioorganic Mechanisms*, Vol. I, W. A. Benjamin, New York, 1966, p. 2.
7. N. O. V. Sonntag, *Chem. Rev.*, **52**, 237 (1953).
8. S. L. Johnson, 'General base and nucleophilic catalysis of ester hydrolysis and related reactions' in *Adv. Phys. Org. Chem.*, Vol. 5 (Ed. V. Gold), Academic, London, 1967, p. 321.
9. E. S. West and W. R. Todd, *Textbook of Biochemistry*, 4th ed., Macmillan, New York, 1966.
10. J. N. Davidson, I. Leslie and J. C. White, *Lancet*, **260**, 1287 (1951).
11. R. A. Phelps and F. W. Putnam, *The Plasma Proteins* (Ed. F. W. Putnam), Academic, New York, 1960, p. 143.
12. R. J. Hill, R. W. Konigsberg, G. Guidotti and L. C. Craig, *J. Biol. Chem.*, **237**, 1549 (1962).
13. C. Huggins, *Physiol. Rev.*, **25**, 281 (1945).
14. E. Chargaff and R. Lipshitz, *J. Am. Chem. Soc.*, **75**, 3658 (1953).

15. E. Chargaff, B. Magasanik, E. Vischen, C. Green, R. Doningen and D. Elson, *J. Biol. Chem.*, **186**, 51 (1950).
16. F. W. Putnam, 'Structure and function of the plasma proteins' in *The Proteins*, Vol. III, 2nd ed. (Ed. H. Neurath), Academic, New York, 1965, p. 153.
17. A. Gottschalk and E. R. Bruce Graham, 'The basic structure of glyco-proteins', in *The Proteins*, Vol. IV, 2nd ed. (Ed. H. Neurath), Academic, New York, 1966, p. 95.
18. M. B. Allen, *Arch. Mikrobiol.*, **32**, 270 (1959).
19. J. Steinhardt and S. Beychok, 'Interaction of proteins with hydrogen ions and other small ions and molecules', in *The Proteins*, Vol. II, 2nd ed. (Ed. H. Neurath), Academic, New York, 1964, p. 139.
20. J. P. Peters and D. D. Van Slyke, *Quantitative Clinical Chemistry*, Vol. I, William and Wilkins, Baltimore, 1931, p. 894.
21. A. Cantarow and M. Trumper, *Clinical Biochemistry*, 6th ed., W. B. Saunders, Philadelphia, 1962.
22. V. J. Jacobs and M. B. Jacobs, *Poisons*, 2nd ed., D. Van Nostrand, London, 1958
23. A. M. Prentiss, *Chemicals in War*, McGraw–Hill, New York, 1937, p. 154.
24. R. W. Gerard, 'Recent research on respiratory irritants' in *Advances in Military Medicine*, Vol. II, Little, Brown, Boston, 1948, p. 565.
24a. J. M. Tobias, S. Postel, H. M. Patt, C. C. Lushbaugh, M. N. Swift and R. W. Gerard, *Am. J. Physiol.*, **158**, 173 (1949).
24b. A. M. Potts, F. P. Simon and R. W. Gerard, *Arch. Biochem.*, **24**, 329 (1949).
25. R. E. Pattle and F. Burgess, *J. Pathol. Bacteriol.*, **82**, 315 (1961).
26. S. Franke, *Lehrbuch der Militarchemie*, Vol. I, Deutscher Militarverlag, Berlin, 1967, pp. 132, 140.
27. H. K. Hall, *J. Am. Chem. Soc.*, **77**, 5993 (1955).
28. H. K. Hall and C. H. Lueck, *J. Org. Chem.*, **28**, 2818 (1963).
29. J. B. A. Dumas, *Ann.*, **15**, 39 (1835).
30. J. N. Tilley and A. A. R. Sayigh, *J. Org. Chem.*, **28**, 2076 (1963).
31. M. Sartori, *The War Gases*, J. and A. Churchill, London, 1939.
31a. A. M. Thiess and W. Hey, *Zentralbl. Arbeits-med. Arbeitsschutz*, **18**, 141 (1968); *Chem. Abstr.*, **69**, 79970 (1968).
32. H. Schumacher and S. Lenher, *Ber.*, **61**, 1671 (1928).
33. O. Ruff and G. Miltschitzky, *Z. Anorg. Chem.*, **221**, 154 (1934).
34. H. J. Emeléus and J. F. Wood, *J. Chem. Soc.*, 2183 (1948).
35. G. A. Olah and S. J. Kuhn, *J. Org. Chem.*, **21**, 1319 (1956).
36. G. Brauer, *Handbuch der präparativen organischen Chemie*, Vol. I, Enke Verlag, Stuttgart, 1960, p. 201.
37. G. D. Buckley, H. A. Piggott and A. J. E. Welch, *J. Chem. Soc.*, 864 (1945).
38. H. Cullumbine, *J. Roy. Army Med. Corps*, **103**, 118 (1957).
38a. A. M. Thiess and P. J. Goldman, *Zentralbl. Arbeits-med. Arbeitsschutz*, **18**, 132 (1968); *Chem. Abstr.*, **69**, 79969 (1968).
38b. B. M. Porembskii, *Farmakol. i Toksikol.*, **3**, 73 (1940); *Chem. Abstr.*, **36**, 2931 (1940).
39. E. Rothlin, *Schweiz med. Wochschr.*, **70**, 641 (1940); *Chem. Abstr.*, **35**, 5192 (1941).

40. E. M. Boyd and W. C. Stewart, *J. Pharm. Pharmacol.*, **5**, 39 (1953).
41. S. G. Ong, *Verslag Gewone Vergad. Afdeel Natuurkunde*, **52**, 270 (1943); *Chem. Abstr.*, **38**, 5929 (1944).
42. S. G. Ong, *Proc. Nederland Akad. Wetensch.*, **45**, 774 (1942); *Chem. Abstr.*, **38**, 5571 (1944).
43. D. Cordier and G. Cordier, *Compt. Rend. Soc. Biol.*, **147**, 327 (1953).
44. D. Cordier and G. Cordier, *J. Physiol (Paris)*, **45**, 421 (1953).
45. S. I. Enikeeva, *Farmakol. i Toksikol.*, **6**, 54 (1943); *Chem. Abstr.*, **39**, 1470 (1945).
46. S. I. Enikeeva, *Farmakol. i Toksikol.*, **7**, 51 (1944); *Chem. Abstr.*, **40**, 5153 (1946).
47. V. G. Motylev, *Farmakol. i Toksikol.*, **8**, 61 (1945); *Chem. Abstr.*, **40**, 7410 (1946).
48. W. T. L. Ohlssen, *Acta Med. Scand.*, **128**, suppl. 190 (1947); *Chem. Abstr.*, **41**, 7427 (1947).
49. C. Liebecq and R. A. Peters, *Biochim. Biophys. Acta*, **3**, 215 (1949).
50. C. Martius, *Ann.*, **561**, 227 (1949).
51. W. B. Elliott and G. Kalnitsky, *J. Biol. Chem.*, **186**, 487 (1950).
52. B. C. Saunders and G. J. Stacey, *J. Chem. Soc.*, 1773 (1948).
53. F. L. M. Pattison, *Toxic Aliphatic Fluorine Compounds*, Elsevier, Amsterdam, 1959.
54. E. Gryszkiewicz-Trochimowski, A. Sporzynski and J. Wnuk, *Rec. Trav. Chim.*, **66**, 419 (1947).
55. F. L. M. Pattison, R. R. Fraser, G. J. O'Neill and J. F. K. Wilshire, *J. Org. Chem.*, **21**, 887 (1956).
56. F. L. M. Pattison, S. B. D. Hunt and J. B. Stothers, *J. Org. Chem.*, **21**, 883 (1956).
57. M. B. Chenoweth, *J. Pharmacol. and Exptl. Therapy*, **97**, 383 (1949).
58. D. C. England, L. R. Melby, M. A. Dietrich and R. V. Lindsey, *J. Am. Chem. Soc.*, **82**, 5116 (1960).
59. D. D. Coffman, M. S. Raasch, G. W. Rigby, P. L. Barrick and W. E. Hanford, *J. Org. Chem.*, **14**, 747 (1949).
60. I. L. Knunyants, L. S. German and B. L. Dyatkin, *Izv. Akad. Nauk S.S.S.R., otdel Khim. Nauk*, 1353 (1956); *Chem. Abstr.*, **51**, 8037 (1957).
61. R. D. Chambers and R. H. Mobbs, 'Ionic interactions of fluoro-olefins' in *Advances in Fluorine Chemistry*, Vol. IV (Ed. M. Stacey, J. C. Tatlow and A. G. Sharpe), Butterworths, London, 1965, p. 50.
62. D. Lester and L. A. Greenberg, *Arch. Indust. Hyg. Occupational Med.*, **2**, 335 (1950).
63. D. Ambros, *Chem. Průmysl*, **11**, 60 (1960); *Chem. Abstr.*, **55**, 18556 (1961).
64. J. Kopečný and D. Ambros, *Chem. Průmysl*, **11**, 366 (1961); *Chem. Abstr.*, **55**, 27641 (1961).
65. T. J. Brice, J. D. LaZerte, L. J. Hals and W. H. Pearlson, *J. Am. Chem. Soc.*, **75**, 2698 (1953).
65a. R. J. Koshar, Th. C. Simmons and F. W. Hoffmann, *J. Am. Chem. Soc.*, **79**, 1741 (1957).
66. J. D. Park, A. F. Benning, F. B. Downing, J. F. Laucius and R. C. Mettarness, *Ind. Eng. Chem.*, **39**, 354 (1947).
67. D. R. Harris, *Lancet*, **261**, 1008 (1951).

68. D. R. Hagenmeyer and W. Stubbeline, *Mod. Plastics*, **31**, 136, 219 (1954).
69. J. F. Treon, J. W. Cappel, F. P. Cleveland, E. E. Larson, R. W. Atchley and R. T. Denham, *Am. Ind. Hyg. Assoc. Quart.*, **16**, 187 (1955).
70. P. J. R. Challen, R. J. Sherwood and J. Bedford, *Brit. J. Ind. Med.*, **12**, 177 (1955).
71. L. A. Cohen, *Ann. Revs. Biochem.*, **37**, 695 (1968).
72. B. L. Vallee and J. F. Riordan, *Ann. Revs. Biochem.*, **38**, 733 (1969).
73. G. R. Stark, 'Recent developments in chemical modification and sequential degradation of proteins' in *Advances in Protein Chemistry*, Vol. 24, Academic, New York, 1970, p. 261.
74. G. Mohnike, G. Schnuchel, L. Kuppfer and W. Langenbeck, *Hoppe-Seylers Z. Physiol. Chem.*, **294**, 12 (1953).
75. L. A. AE. Sluyterman, *Biochim. Biophys. Acta*, **17**, 169 (1955).
76. J. S. Ram, L. Terminiello, M. Bier and F. F. Nord, *Arch. Biochem. Biophys.*, **52**, 451, 464 (1954).
77. L. Terminiello, J. S. Ram, M. Bier and F. F. Nord, *Arch. Biochem. Biophys.*, **57**, 252 (1955).
78. M. Bier, J. S. Ram and F. F. Nord, *Nature*, **176**, 789 (1955).
79. B. H. Levedahl and H. Bernstein, *Arch. Biochem. Biophys.*, **52**, 353 (1954).
80. A. P. Nygaard and H. Theorell, *Acta Chem. Scand.*, **8**, 1489 (1954).
81. H. Theorell and A. P. Nygaard, *Acta Chem. Scand.*, **8**, 1649 (1954).
81a. J. R. Marrack and E. S. Orlans, *Brit. J. Exptl. Pathol.*, **35**, 389 (1954).
81b. H. Nisinoff and D. Pressman, *J. Immunol.*, **83**, 138 (1959).
82. W. K. Paik, L. Bloch-Frankenthal, S. M. Birnbaum, M. Winitz and J. P. Greenstein, *Arch. Biochem. Biophys.*, **69**, 56 (1957).
82a. H. G. Crabtree, *Cancer Res.*, **1**, 40 (1941).
83. S. Cohen, E. Thom and A. Bendich, *Biochem.*, **2**, 176 (1963).
84. S. Cohen and N. Dinar, *J. Am. Chem. Soc.*, **87**, 3195 (1965).
85. D. A. Brown and R. F. Hudson, *J. Chem. Soc.*, 3352 (1953).
86. C. A. Bunton and T. A. Lewis, *Chem. Ind.*, 180 (1956).
87. C. G. Swain and C. B. Scott, *J. Am. Chem. Soc.*, **75**, 246 (1953).
88. K. Sugiura, private communication. Work done at Sloan–Kettering Institute for Cancer Research, New York (1963).
89. Ch. Lachman and S. Cohen, *Cancer Res.*, **30**, 439 (1970).
89a. Y. Sinai, Ch. Lachman and S. Cohen, *Nature*, **205**, 192 (1965).
90. Ch. Lachman, 'Physico-chemical and biological properties of purine and pyrimidine conjugates of proteins', Ph.D. Thesis, Hebrew University, Jerusalem 1969.
91. V. Butler, Jr., S. M. Beiser, B. F. Erlanger, S. W. Tanenbaum, S. Cohen and A. Bendich, *Proc. Nat. Acad. Sci.*, **48**, 1597 (1962).
92. I. Hellering, Ch. Lachman, S. Ben-Ephraim and W. Klingberg, unpublished results (1970).
93. S. Cohen and A. Vincze, *Israel J. Chem.*, **2**, 1 (1964).
94. A. Vincze, 'Adducts of purines and pyrimidines with amino acids, peptides polypeptides and proteins', Ph.D. Thesis, Hebrew University, Jerusalem 1968.
95. A. Giner-Sorolla and A. Bendich, *J. Am. Chem. Soc.*, **80**, 5744 (1958).
96. C. Heidelberger, D. Parsons and D. C. Remy, *J. Med. Chem.*, **7**, 1 (1964).

97. J. D. Roberts, R. L. Webb and E. A. McElhill, *J. Am. Chem. Soc.*, **72**, 408 (1950).

98. J. Bornstein, S. A. Leone, W. F. Sullivan and O. F. Bennett, *J. Am. Chem. Soc.*, **79**, 1745 (1957).

99. Y. Ashani and S. Cohen, *Israel J. Chem.*, **3**, 101 (1965).

100. B. R. Baker, *J. Pharm. Sci.*, **53**, 347 (1964).

101. J. R. Brown and B. S. Hartley, *Abstr. 1st Meeting Federation European Biochem. Soc.*, **A29**, 25 (1964).

102. R. A. Oosterban, 'Constitution of DFP sensitive enzymes' in *Proceedings of the Conference on Structure and Reactions of DFP Sensitive Enzymes* (Ed. E. Heilbronn), Research Institute of National Defence, Stockholm, 1967, p. 25.

103. M. Halman, *J. Chem. Soc.*, **305**, (1959).

104. I. Dostrovsky and M. Halman, *J. Chem. Soc.*, 502, 516 (1953).

105. D. K. Myers and A. Kemp, Jr., *Nature*, **173**, 33 (1954).

106. B. A. Kilby and M. Kilby, *Brit. J. Pharmacol.*, **2**, 234 (1947).

107. J. F. Mackworth and E. C. Webb, *Biochem. J.*, **42**, 91 (1948).

108. D. R. Davies, P. Holland and M. J. Rumens, *Brit. J. Pharmacol.*, **15**, 271 (1960).

109. K. P. Dubois, 'Toxicological evaluation of the anticholinesterase agents' in *Cholinesterases and Anticholinesterase Agents* (Ed. G. B. Koelle), Springer Verlag, Berlin, 1963, p. 833.

110. R. D. O'Brien, *Ann. N.Y. Acad. Sci.*, **160**, 204 (1969).

111. I. B. Wilson, M. Hatch and S. Ginsburg, *J. Biol. Chem.*, **235**, 2312 (1960).

111a.I. B. Wilson, M. A. Harrison and S. Ginsburg, *J. Biol. Chem.*, **236**, 1498 (1961).

112. I. B. Wilson and M. A. Harrison, *J. Biol. Chem.*, **236**, 2292 (1961).

113. E. Reiner and V. Simeon-Rudolf, *Biochem. J.*, **98**, 501 (1966).

114. J. P. Long, 'Structure activity relationship of the reversible anticholinesterase agents' in *Cholinesterase and Anticholinesterase Agents* (Ed. G. B. Koelle), Springer Verlag, Berlin, 1963, p. 374.

115. J. E. Casida, K. B. Augustinson and G. Jonsson, *J. Econ. Entomol.*, **53**, 205 (1960).

116. H. P. Metzger and I. B. Wilson, *Biochem.*, **3**, 926 (1964).

117. B. F. Erlanger and W. Cohen, *J. Am. Chem. Soc.*, **85**, 348 (1963).

118. B. F. Erlanger, H. Castleman and A. G. Cooper, *J. Am. Chem. Soc.*, **85**, 1872 (1963).

119. B. F. Erlanger, A. G. Cooper and W. Cohen, *Biochem.*, **5**, 190 (1966).

120. B. F. Erlanger, 'Inactivation–reactivation studies on the topography of the active site of chymotrypsin', in *Proceedings of the Conference on Structure and Reactions of DFP Sensitive Enzymes* (Ed. E. Heilbronn), Research Institute of National Defence, Stockholm, 1967, p. 143.

121. D. E. Rivett and J. F. K. Wilshire, *Australian J. Chem.*, **18**, 1667 (1965).

122. K. K. Makinen, *Arch. Biochem. Biophys.*, **126**, 803 (1968).

123. H. McCombie and B. C. Saunders, *Nature*, **157**, 287 (1946).

124. G. Schrader, *Brit. Intell. Obj. Sub-com. Report No.* 714, item 8 (1947). See also G. Schrader, *Die Entwicklung neuer Insektizide auf Grundlage Organischen Fluor- und Phosphorverbindungen*, Verlag-Chem., Weinheim, 1951.

125. S. Nakanishi, T. C. Myers and E. V. Jensen, *J. Am. Chem. Soc.*, **77**, 3099 (1955).
126. G. A. Olah and S. J. Kuhn, *J. Am. Chem. Soc.*, **82**, 2380 (1960).
127. M. Ochoa, Jr. and E. Hirschberg, 'Alkylating agents' in *Experimental Chemotherapy* (Ed. R. J. Schnitzer and F. Hawking), Vol. V, Academic, New York, 1967, p. 1.
128. L. H. Schmidt, R. Fradkin, R. Sullivan and A. Flowers, 'Comparative pharmacology of alkylating agents', *Cancer Chemotherapy Rept.*, Suppl. **2**, 1 (1965).
129. P. D. Lawley and P. Brookes, *Nature*, **206**, 480 (1965).

CHAPTER 11

Thiocarbonyl halides

K. T. Potts *and* C. Sapino*

Rensselaer Polytechnic Institute, Troy, New York

*Present address: *Bristol Laboratories, Syracuse, New York.*

I. INTRODUCTION

The thiocarbonyl halides comprise a class of extremely reactive organic molecules which, in contrast to their oxygen analogues, have received relatively little attention largely on account of their inaccessibility. The counterparts of the simple acyl halides are still unknown except for poly-fluorinated derivatives and most attention has been focused on the thioaroyl halides. Because of this unavailability, many alternative ways of introducing the thioacyl or thioaroyl group have been devised. Attention is drawn to these alternatives where appropriate in this chapter.

A large body of literature deals with thiocarbonyl chloride (thiophosgene) and the ready availability of this product has resulted in its use in numerous synthetic sequences for the introduction of the thiocarbonyl group. Derivatives of thiocarbonyl chloride where one of the chlorine atoms has been replaced by amino, alkoxy or thioalkoxy groups comprise an important group of compounds containing the thiocarbonyl chloride group. Their properties are modified by the presence of these groups but they behave as thiocarbonyl chlorides and are discussed in this chapter.

II. PREPARATION OF THIOCARBONYL HALIDES

A. General Considerations

The reactivity of the thiocarbonyl group has made the most direct route to the various thiocarbonyl halides inapplicable. Thus, reaction of the corresponding acyl halides with various sulphur-containing reagents has not met with success. Similarly, attempts to utilize thiocarbonyl chloride as a source of the chlorothiocarbonyl function have been thwarted when it was hoped to form a carbon bond with the thiocarbonyl group. Dicon-densation products were formed under these Friedel–Craft type reaction conditions. However, when nucleophilic displacements were utilized, control of the reaction conditions readily resulted in formation of the desired thiocarbonyl halides.

B. Thioaroyl Halides

Thiobenzoyl chloride is the best known of the aryl-thioacid chlorides and may be prepared[1] from sulphur-containing derivatives of benzoic acid[2] and halogenating agents such as chlorine/thionyl chloride. Under these conditions various sulphur-containing by-products were obtained[1]. Refinement of the dithiobenzoic acid (1)–thionyl chloride method[3a, b] has resulted in reproducible yields (80%) of thiobenzoyl chloride (2), a violet oil, b.p. 60–65°/0·1 mm. This procedure has been extended satis-factorily[4] to the preparation of various substituted thiobenzoyl chlorides

(2; R = H; o-, m-, p-CH$_3$; m-, p-Cl; p-OCH$_3$) as well as thiophen-2-thiocarbonyl chloride* (3). Thiobenzoyl chloride is not especially stable

(1)   (2)   (3)

thermally, polymerizing to a black product above 130° and, in presence of excess copper powder, it forms cupric sulphide, cupric chloride, and a small amount of tetraphenylthiophene[3a]. The action of Raney nickel in boiling toluene quickly brings about complete decomposition, leading to trace amount of stilbene[3a].

C. Thioacyl Halides

Thioacetyl chloride, the simplest thioacyl halide corresponding to its oxygen analogue, is as yet unknown†. However, various fluorinated and chlorinated derivatives are available and they make up an extremely interesting and reactive class of compounds.

A general route to these products involves reaction of various chloro- and bromofluoroethylenes with boiling sulphur[6]. Thus, passage of chloro-trifluoroethylene (4) through boiling sulphur yields chlorodifluorothio-acetyl fluoride (5). Bromotrifluoroethylene also reacts with boiling sulphur,

$$CFCl=CF_2 + S \xrightarrow{445°} CClF_2CF\overset{\overset{\displaystyle S}{\|}}{}$$

(4)   (5)

in this case bromodifluorothioacetyl fluoride being formed. When 1,1-dichloro-2,2-difluoroethylene (6) is reacted under these conditions the product obtained is chlorodifluorothioacetyl chloride (8). A possible intermediate[6] in this reaction is the episulphide (7) which could undergo rearrangement with migration of a chlorine atom to give the product obtained.

(6)   (7)   (8)

* Alternative IUPAC name (Rule C-543.5) is 2-thiophenecarbothioyl chloride.

† An unsuccessful attempt[5] involved reaction of acetyl chloride with P$_4$S$_{10}$.

Alternative syntheses of thioacid fluorides have been described. A convenient procedure involved reaction of perfluoroalkyl mercurials with sulphur[6]. Thus, perfluorothioacetyl fluoride (11) was obtained from both bis(perfluoroethyl)mercury (9) and bis(1-chloroperfluoroethyl)mercury (10). Similarly, reaction of bis(1,1-dichloroperfluoroethyl)mercury with sulphur under these conditions gave trifluorothioacetyl chloride.

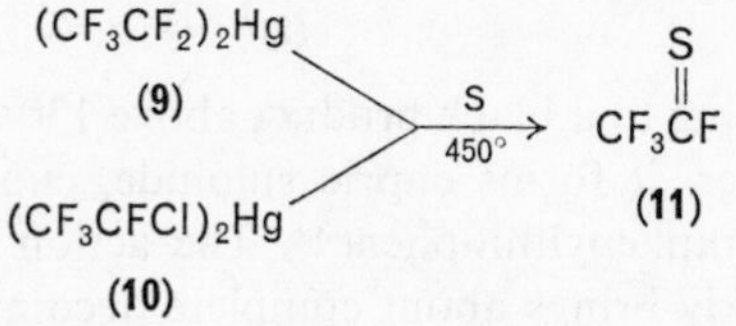

$$(CF_3CF_2)_2Hg \atop (9) \qquad\qquad \xrightarrow[450°]{S} \qquad CF_3CF \atop (11) \atop (CF_3CFCl)_2Hg \atop (10)$$

The yields obtained in the above reactions varied from 37% to 67%, and the hazards associated with the toxicity of these compounds require the utmost care in their manipulations.

The perfluoroalkylmercurials were readily prepared from reaction of mercuric fluoride with fluoroalkenes at 50–100° under autogenous pressure. Thus, from tetrafluoroethylene, trifluoroethylene and 1,1-difluoroethylene, yields of the order of 56–66% of bis(pentafluoroethyl)mercury, bis(1,2,2,2-tetraethyl)mercury and bis(2,2,2-trifluoroethyl)mercury, respectively, were obtained[7].

Other methods available[6] for the synthesis of perfluorothioacid fluorides involve reaction of phosphorus pentasulphide at 550° with perfluoroethyl iodide. In this case, perfluorothioacetyl fluoride is obtained and, under these conditions, 1-iodoperfluoropropane (12) is converted into perfluoro-thiopropionyl fluoride (13).

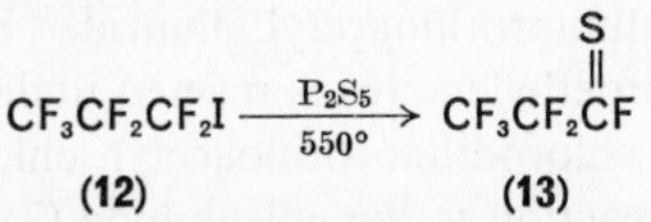

$$CF_3CF_2CF_2I \xrightarrow[550°]{P_2S_5} CF_3CF_2CF \atop \overset{\displaystyle S}{\|} \atop (12) \qquad\qquad (13)$$

An interesting variation[6] of the above types of procedure is that involving reaction of 4,4-diiodoperfluoro-1-butene[8] (14) with sulphur at 450°. In this case sulphur reacted principally at the position of the iodine atoms with the formation of perfluorothio-3-butenoyl fluoride (15) in 70% yield.

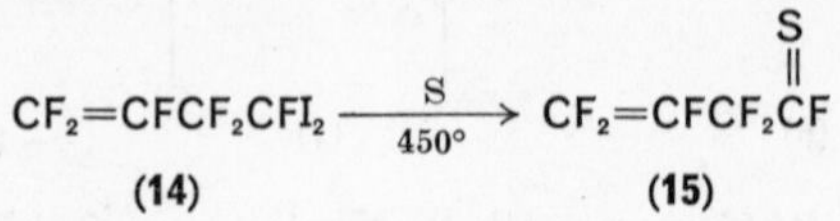

$$CF_2{=}CFCF_2CFI_2 \xrightarrow[450°]{S} CF_2{=}CFCF_2CF \atop (14) \qquad\qquad (15)$$

Other methods which yield thioacyl fluorides involve the elimination of hydrogen fluoride from fluorinated ethanethiols with fluorine atoms on the carbon containing the —SH group[9]. Thus, 1,1,2-trifluoro-2-chloro-ethanethiol (16), prepared by X-ray-initiated addition of hydrogen sulphide to 1,1,2-trifluoro-2-chloroethene, when passed in the vapour phase diluted with nitrogen through a tube packed with sodium fluoride pellets gave in good yield (62%) chlorofluorothioacetyl fluoride (17).

$$\text{HClCFCF}_2\text{SH} \xrightarrow{\text{NaF}} \overset{\displaystyle S}{\overset{\displaystyle \|}{\text{HClCFCF}}} + \text{NaF.HF}$$
$$\quad\quad (16) \quad\quad\quad\quad\quad\quad (17)$$

Under these conditions 2-*H*-tetrafluoroethanethiol (18) and 2-*H*-2-methoxytrifluoroethanethiol (19) gave difluorothioacetyl fluoride (20) and methoxyfluorothioacetyl fluoride (21), the latter being extremely unstable and polymerizing at −80°.

$$\text{HCF}_2\text{CF}_2\text{SH} \longrightarrow \overset{\displaystyle S}{\overset{\displaystyle \|}{\text{HCF}_2\text{CF}}}$$
$$\quad (18) \quad\quad\quad\quad\quad (20)$$

$$\text{CH}_3\text{OCHFCF}_2\text{SH} \longrightarrow \overset{\displaystyle S}{\overset{\displaystyle \|}{\text{CH}_3\text{OCHFCF}}}$$
$$\quad\quad (19) \quad\quad\quad\quad\quad\quad (21)$$

Reduction of 2-chlorotetrafluoroethanesulphenyl chloride (22), prepared by chlorination of dichlorooctafluorodiethyl sulphide[10] with tin and 28% hydrochloric acid, has been reported[11] to give chlorodifluorothioacetyl fluoride (23). A by-product in this reaction was 2,2-dichlorooctafluoro-diethyl disulphide.

$$\text{CF}_2\text{ClCF}_2\text{SCl} \xrightarrow[\text{HCl}]{\text{Sn}} \overset{\displaystyle S}{\overset{\displaystyle \|}{\text{CF}_2\text{ClCF}}}$$
$$\quad\quad (22) \quad\quad\quad\quad\quad (23)$$

Though of no practical significance, an example of a stable thioacyl bromide has been reported[12]. Reaction of ethylene trithiocarbonate (24) with bromocyanoacetylene yielded a product to which the structure α-cyano-1,3-dithiolane-Δ[2, α]-thioacetyl bromide (25) was assigned. The

$$(24) \quad\quad\quad\quad\quad\quad (25)$$

brilliant-violet, crystalline product fumed in moist air and the n.m.r. spectrum (CH$_2$Cl$_2$) showed a symmetrical pentuplet centred at δ 3·68.

Reaction of this thioacetyl bromide with potassium fluoride in refluxing acetonitrile gave the corresponding thioacetyl fluoride, similar in colour to the bromide above. The ^{19}F n.m.r. spectrum had a sharp singlet at -110.2 p.p.m. from $Cl_2FCCFCl_2$, consistent with the reported resonances of thioacyl fluorides between -107 and -162 p.p.m. from $Cl_2FCCFCl_2$[6]. The unusual possibilities for electron delocalization in **25** probably account for its stability.

D. Thiocarbonyl Chloride

The simplest chlorothioacid chloride, thiocarbonyl chloride (thio-phosgene)* (**26**), has been prepared in a large variety of ways, most of which are of no interest from a preparative point of view. In addition to poor yields, the thiophosgene is usually admixed with impurities[13]. The reduction of trichloromethanesulphenyl chloride (perchloromethyl mercaptan) (**27**) with a wide variety of reducing agents provides the most convenient route to thiophosgene. A procedure utilizing an organic reducing agent such as tetralin[14] is the most promising, with yields of the

$$2\ Cl_3CSCl\ +\ \underset{(27)}{\qquad} \xrightarrow{200°}\ 2\ CSCl_2\ +\ \underset{(26)}{\qquad}$$
$$+$$
$$4\ HCl$$

order of 80% being obtained. Other organic substrates such as tetrahydro-acenaphthene, cyclohexylxylene, naphthalene, anthracene and dodeca-hydroacenaphthene were found to be effective. The reduction also occurred readily (65–82%) when refluxing xylene containing ferric chloride as catalyst was used[15] and variations have included the use of chlorobenzene, toluene, benzene, naphthalene and biphenyl and catalysts such as aluminium chloride, zinc chloride and stannic chloride. A wide variety of reducing agents[16, 17, 18, 19] has been used to convert trichloromethanesulphenyl chloride into thiophosgene and none of these procedures offers any advantage over those described above.

E. Thiocarbonyl Fluoride

A recent synthesis[6] has now made the simplest fluoroacid fluoride, thiocarbonyl fluoride (**30**), conveniently available. The three-step process involved conversion of thiophosgene (**26**) into its dimer[20] (**28**), fluorination of the dimer with antimony trifluoride to give 2,2,4,4-tetrafluoro-1,3-dithietane (**29**), and pyrolysis of **29** at 475–500°. As would be expected in

* IUPAC name (Rule C-545.1) is thiocarbonyl dichloride.

the fluorination process, 2,2,4,4-tetrafluoro-1,3-dithietane (29) was contaminated with small amounts of the mixed chlorofluorodithietanes, the
2-chloro-2,4,4-trifluoro- and the 2,2-dichloro-4,4-difluoro products. The

$$CSCl_2 \longrightarrow \quad (28) \longrightarrow \quad (29)$$

(26)

$$\underset{(31)}{\overset{S}{\underset{\|}{FCCl}}} \qquad \underset{(30)}{\overset{S}{\underset{\|}{FCF}}}$$

former, on pyrolysis at 450°, gave a mixture of thiocarbonyl chlorofluoride
(31) and thiocarbonyl fluoride (30), whereas the latter formed thiocarbonyl
fluoride and thiophosgene.

Reaction of tetrafluoroethylene with sulphur at 500–600° may also be
used as a route to thiocarbonyl fluoride[6]. Appreciable amounts of trifluorothioacetyl fluoride (11) and bis(trifluoromethyl)disulphide (32) were
formed in the reaction but could be separated by distillation. It has been
reported that the composition of the product mixture is affected noticeably
by the reaction conditions. Thus, when tetrafluoroethylene undergoes

$$CF_2{=}CF_2 \xrightarrow[\;500\text{--}600°\;]{S} \overset{S}{\underset{\|}{FCF}} + CF_3\overset{S}{\underset{\|}{CF}} + CF_3SSCF_3$$

$$\quad\quad\quad\quad\quad\quad (30)\quad\quad (11)\quad\quad\quad (32)$$

reaction with sulphur vapour over a bed of activated charcoal, the major
product formed is trifluorothioacetyl fluoride[21].

Another procedure for obtaining thiocarbonyl fluoride involves reaction
of thiocarbonyl chloride and molten alkali metal fluorides[22]. Other
procedures described in the literature are of little consequence in light of
the above syntheses and a compilation of references to this topic is
available[22].

F. Thiocarbonyl Bromide

Thiocarbonyl bromide (33) was unknown until the work of the DuPont
group in 1965. It was prepared[6] by the addition of hydrogen bromide to
thiocarbonyl fluoride (30), and mercaptan intermediates that eliminate
hydrogen fluoride were thought to be involved in this reaction. Thiocarbonyl bromide, though it distils as a heavy orange-red liquid, is quite
unstable, readily decomposing in sunlight or on standing, largely to

bromine. These addition characteristics of the thiocarbonyl group are of considerable interest and are discussed in greater detail in section IV, F.

$$\underset{(30)}{\overset{\overset{\displaystyle S}{\parallel}}{FCF}} \xrightarrow{HBr} [BrCF_2SH] \xrightarrow{-HF} \left[\overset{\overset{\displaystyle S}{\parallel}}{BrCF} \right]$$

$$\underset{(33)}{\overset{\overset{\displaystyle S}{\parallel}}{BrCBr}} \xleftarrow{-HF} [Br_2CFSH] \overset{HBr}{\swarrow}$$

G. Mono- and Disubstituted Aminothiocarbonyl Halides

Representative examples of this type of substituted thiocarbonyl halide have been known for many years. In the case of the N,N-disubstituted thiocarbamoyl bromides*, a recent procedure for their synthesis involving bromination of the corresponding thioformanilides with bromine in refluxing carbon tetrachloride is now the method of choice[23]. A wide variety of substituted thioformanilides (34), available from the amide and phosphorus pentasulphide in boiling benzene[24], have been converted into the corresponding substituted thiocarbamoyl bromides (35).

$$\underset{(34)}{Aryl-N\overset{\displaystyle R}{\underset{\displaystyle}{C}}\overset{\displaystyle H}{\underset{\displaystyle S}{}}} \longrightarrow \underset{(35)}{Aryl-N\overset{\displaystyle R}{\underset{\displaystyle}{C}}\overset{\displaystyle Br}{\underset{\displaystyle S}{}}}$$

The N,N-dialkylthiocarbamoyl bromides are apparently relatively unstable. They have been prepared[25] by reaction of dialkylthioformamides with disulphur dibromide and the unstable bromide was then allowed to react with ethanol to form the corresponding thiourethane.

In contrast, the N,N-disubstituted thiocarbamoyl chlorides are stable, useful entities and are readily prepared from thiocarbonyl chloride and a variety of amines. The great reactivity of thiocarbonyl chloride makes the choice of experimental conditions critical in reactions of this type. With secondary amines such as dimethylamine, the reaction is carried out in anhydrous ether at ca. 5° using 2 moles of the amine and 1 mole of thiocarbonyl chloride. The resulting N,N-dimethylthiocarbamoyl chloride (37) is formed in good yield[26]. An alternative procedure, resulting in yields of product varying from 60% to 85%, is to treat a disubstituted thioamide with sulphur chloride in pyridine. Thus, when N,N-dimethylthioformamide (36) was treated with sulphur chloride in dichloromethane in the presence

* IUPAC name (Rule C-431.2).

of an equivalent amount of pyridine, N,N-dimethylthiocarbamoyl chloride (37) was formed in 73% yield[25]. Sulphur dichloride may also be used in

$$(CH_3)_2NC\begin{matrix}H\\ \diagdown\\ S\end{matrix} \quad + \quad S_2Cl_2 \quad \xrightarrow{\text{pyridine}} \quad (CH_3)_2NC\begin{matrix}Cl\\ \diagdown\\ S\end{matrix}$$

(36) (SCl₂) (37)

this reaction sequence which is also applicable to N-alkyl-N-arylthioamides.

Another procedure[27] which should be capable of wide application involves reaction of the trimethylsilyl derivative (38) of the secondary amines with carbon disulphide. The resulting trimethylsilyldithiocarbamate (39) is then treated with phosphorus pentachloride or thionyl chloride to give 37. Secondary amines such as dimethylamine, diethylamine, pyrrolidine, piperidine and morpholine have been used in this sequence, the yields of products being over 80%.

Another procedure involves treating the sodium salt of dimethyldithiocarbamic acid with perchloromethyl mercaptan, in which case a 33% yield of 37 was obtained along with other products[28].

$$(CH_3)_2NSi(CH_3)_3 \quad \xrightarrow{CS_2} \quad (CH_3)_2N\overset{\displaystyle S}{\overset{\displaystyle \|}{C}}SSi(CH_3)_3 \quad \xrightarrow[\substack{\text{or}\\ \text{SOCl}_2}]{PCl_5} \quad \text{(37)}$$

(38) (39)

Primary aromatic amines react readily in ether solution with thiocarbonyl chloride with formation of a monosubstituted thiocarbamoyl chloride, an isothiocyanate or a disubstituted thiourea. Heating, or addition of water, favoured formation of the isothiocyanate and the symmetrical diarylthiourea was invariably formed when 2 moles of the amine were used[29]. The effect of substituents in the aromatic ring on the course of the reaction is discussed in these publications.

However, the thiocarbamoyl chlorides derived from secondary aromatic amines such as N-methyl-, N-ethyl- and N-n-propylaniline are much more stable and may be prepared in the presence of protic solvents. Reaction of N-methylaniline in chloroform in the presence of an equimolar amount of the corresponding amine hydrochloride followed by 2 moles of aqueous sodium hydroxide gave the corresponding N-methyl-N-phenylthiocarbamoyl chloride in good yield[30]. An alternative synthesis of equal merit involves reaction of the secondary amine in chloroform in the presence of a tertiary amine to act as the hydrogen chloride acceptor[31].

Thiocarbamoyl chlorides derived from primary amines are too unstable to be isolated, usually undergoing elimination of hydrogen chloride to form the isothiocyanate or reacting with a second mole of amine to give the thiourea.

H. Alkoxy- and (Alkylthio)thiocarbonyl Chlorides

This variety of thiocarbonyl halide may be regarded as a derivative of thiocarbonic acid and may be described by the trivial names alkoxy or aryloxythiocarbonyl chlorides, (alkylthio) or (arylthio)thiocarbonyl chlorides. Alternatively, *Chemical Abstracts* treats compounds of this type as derivatives of formic acid, e.g. chlorothioformic acid, *O*-ethyl ester (**40**). The most direct route for their synthesis involves reaction of an equivalent amount of the sodium salt of the alcohol, phenol or thiol with thiocarbonyl chloride. Thus, reaction[32] of alcoholic sodium ethoxide with thiocarbonyl chloride in chloroform gave ethoxythiocarbonyl chloride (**40**). With phenol, addition of a chloroform solution of thiocarbonyl chloride to a 5% sodium hydroxide solution of the phenol gave[33] phenoxythiocarbonyl chloride (**41**) in 70% yield together with *O,O*-diphenylthiocarbonate (**42**).

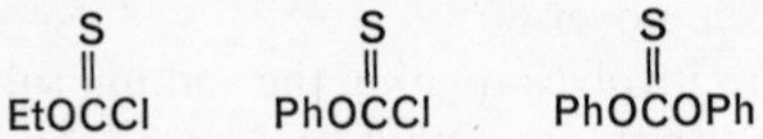

$$
\begin{array}{ccc}
\text{S} & \text{S} & \text{S} \\
\parallel & \parallel & \parallel \\
\text{EtOCCl} & \text{PhOCCl} & \text{PhOCOPh} \\
(\textbf{40}) & (\textbf{41}) & (\textbf{42})
\end{array}
$$

Variation of the above reaction conditions has been reported[34] to yield a variety of products and several organic solvents have been utilized in the reaction[35]. These procedures have been reported to be satisfactory for higher and more complex alcohols[36].

The corresponding *S*-alkyl compounds are also readily available from the appropriate thiol and thiocarbonyl chloride. Thus, methanethiol and thiocarbonyl chloride at $-18°$ over several days gave[37] (methylthio)thiocarbonyl chloride (**43**); the corresponding ethylthio compound (**44**) was formed slowly when a mixture of ethanethiol and thiocarbonyl chloride in carbon disulphide was allowed to stand[34]. (Phenylthio)thiocarbonyl chloride (**45**) was obtained[38] in 80% yield when thiocarbonyl chloride and thiophenol in chloroform were mixed with aqueous sodium hydroxide solution. In addition, the dicondensation product, diphenyltrithiocarbonate (**46**), was also obtained.

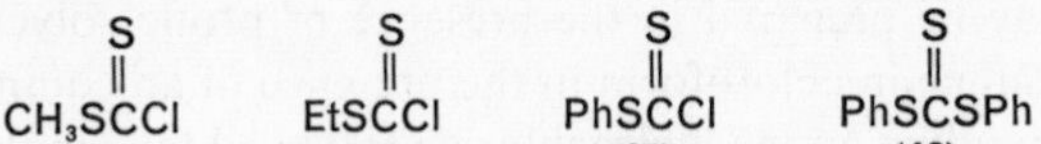

$$
\begin{array}{cccc}
\text{S} & \text{S} & \text{S} & \text{S} \\
\parallel & \parallel & \parallel & \parallel \\
\text{CH}_3\text{SCCl} & \text{EtSCCl} & \text{PhSCCl} & \text{PhSCSPh} \\
(\textbf{43}) & (\textbf{44}) & (\textbf{45}) & (\textbf{46})
\end{array}
$$

III. PHYSICAL PROPERTIES OF THIOCARBONYL HALIDES

A. General Characteristics

The majority of the arylthiocarbonyl halides described in the literature are highly coloured, labile oils with pronounced pungent odours. Purification is usually effected by repeated distillation in high vacuum. The various halogen-substituted alkylthiocarbonyl halides are low-boiling, toxic liquids varying in colour from yellow to red. Extreme care is advised in handling these products.

Incorporation of an oxygen, sulphur or nitrogen atom into the system increases the stability of the halides and has a marked effect on their properties. The oxygen compounds are yellow oils which, with increasing molecular weight, are obtained as yellow, crystalline products. The intensity of the colour is increased in the analogous sulphur products. On the other hand, the nitrogen-containing products are either colourless or, with those containing aromatic groups attached to the nitrogen atom, pale-yellow crystalline materials. These characteristics are summarized in Table 1 for representative thiocarbonyl halides.

TABLE 1. Physical properties of some thiocarbonyl halides

Structure	B.p. (m.p.) °C	Form	d_{20}^4	n_D^{25}	Reference
C_6H_5CSCl	60–65°/0·1 mm	violet oil	1·338		4
$o\text{-}CH_3C_6H_4CSCl$	76°/0·1 mm	violet oil	1·212		4
$m\text{-}CH_3C_6H_4CSCl$	85°/0·1 mm	violet oil	1·194		4
$p\text{-}CH_3C_6H_4CSCl$	75–76°/0·4 mm (12–13°)	violet oil	1·209		4
$m\text{-}ClC_6H_4CSCl$	99–100°/1·4 mm	violet oil	1·362		4
$p\text{-}ClC_6H_4CSCl$	91°/1·2 mm	violet oil	1·385		4
	82°/1·5 mm	blue-red oil			4
CF_3CSF	−22°	yellow oil			6
CF_3CSCl	28–29°	bright-red oil			6
CF_3CF_2CSF	9°				6
$CF_2{=}CFCF_2CSF$	45–46°	bright-yellow oil			6
$BrCF_2CSF$	41–42°	bright-yellow oil			6
$ClCF_2CSF$	23°, 36°	bright-yellow oil	1·5183ᵃ		6, 11
$ClCF_2CSCl$	−10°	deep-red oil		1·4465	6
F_2CS	−54°	colourless liquid			6
Cl_2CS	73°	red oil	1·5085ᵇ		13
$ClFCS$	6–7°	bright-yellow oil			6
Br_2CS	142–144°	orange-red oil		1·6015	6

TABLE 1. (*cont.*)

Structure	B.p. (m.p.) °C	Form	d_{20}^4	n_D^{25}	Reference
HClCFCSF	56–57°	bright-yellow oil		1·4182	9
HCF$_2$CSF	14–16°	bright-yellow oil			9
CH$_3$OCHFCSF	− 13° to − 11°/ 0·9 mm	bright-yellow oil			9
(CH$_3$)$_2$NCSCl	88–89°/ 0·1 mm (42°)	colourless needles			25, 26, 27 30
(Et)$_2$NCSCl	63–65°/ 0·1 mm (49°)				27
C$_4$H$_8$NCSCl	(96°) (97·5– 98·5°)				25, 27, 39
C$_5$H$_{10}$NCSCl	82–83°/0·1 mm 92–94°/0·4 mm			1·5960 1·5937^c	25, 27
C$_4$H$_8$ONCSCl	(66·5°) 100– 101°/10·7 mm (59–60°)				25, 27
N-CH$_3$, N-2,6- (CH$_3$)$_2$C$_6$H$_4$CSBr	(68–72°)	colourless needles			23
N-CH$_2$C$_6$H$_4$, N- 2,6-Cl$_2$C$_6$H$_4$CSBr	(121–121·5°)	colourless needles			23
C$_2$H$_5$OCSCl	(126–127°)	yellow oil			32
C$_6$H$_5$OCSCl	91°/10 mm	yellow oil			33
C$_6$H$_5$SCSCl	135°/15 mm	orange-red oil	1·331^b		38

^{a} d^0. ^{b} d_4^{15}. ^{c} 20°.

B. U.v. Spectra

Table 2 lists the main u.v. absorption bands for a series of substituted thiobenzoyl chlorides[4]. All show a characteristic long-wavelength absorption between 532 and 520 nm in cyclohexane which is shifted about 11 nm to shorter wavelength in acetonitrile. The low intensity (log ε 1·78–2·11) and the negative solvent effect indicate that this band is due to a symmetry forbidden $n \rightarrow \pi^*$ transition of the thiocarbonyl group. An intense absorption between 350 and 304 nm, which shows a positive solvent effect (ca. 3 nm) is most likely a $\pi \rightarrow \pi^*$ absorption. Using the various substituted thiobenzoyl chlorides described in Table 2, good agreement has been found for the energy involved in this $\pi \rightarrow \pi^*$ transition.

TABLE 2. U.v. absorption spectra of some substituted thiobenzoyl chlorides[4]

Substituent	Solvent	Absorption[a] λ_{max} nm (log ε)
H	cyclohexane	224[a] (3·90), 228[a] (3·90), 243[a] (3·70), 260[a] (3·6), 272[a] (3·5), 313 (4·14), 530 (1·82)
	acetonitrile	225 (3·74), 248[a] (3·5), 317 (4·08), 518 (1·78)
o-CH$_3$	cyclohexane	230 (3·97), 310 (3·89), 520 (2·19)
	acetonitrile	228 (3·90), 312 (3·79), 509 (2·11)
m-CH$_3$	cyclohexane	232 (3·90), 243[a] (3·74), 251[a] (3·69), 263[a] (3·55), 315 (4·09), 525 (1·88)
	acetonitrile	230 (3·75), 251[a] (3·43), 319 (4·08), 515 (1·93)
p-CH$_3$	cyclohexane	231 (3·85), 236 (3·86), 250 (3·69), 328 (4·17), 527 (1·85)
	acetonitrile	231[a] (3·78), 236 (3·79), 259 (3·66), 332 (4·17), 576 (1·92)
m-Cl	acetonitrile	233, 245, 307, 522
p-Cl	cyclohexane	233 (4·02), 239 (4·01), 254[a] (3·83), 324 (4·15), 532 (1·83)
	acetonitrile	232 (3·89), 255 (3·71), 327 (4·21), 522 (1·85)
p-OCH$_3$	cyclohexane	223, 248, 273, 282, 350, 522

[a] Shoulder.

From a consideration of the absorption spectra of a large number of various thiocarbonyl compounds, it has been shown[40] that the values of absorption bands calculated by means of absorption increments are in good agreement with the experimental results. This led to the prediction that in the unknown thioacetyl chloride the colour-imparting first absorption band would occur at 480 ± 9 nm (empirical method) or 491 ± 15 nm (quantum chemical method) and that the first intense band would be observed at 241 ± 8 nm (quantum chemical method). It is interesting to note that perfluorothioacetyl fluoride and difluorothioacetyl fluoride are yellow in colour and show absorption maxima at 410 nm and 220 nm[9].

C. I.r. Spectra

Thiocarbonyl chloride, dissolved in carbon disulphide, has been reported[41] to show a strong C=S stretching absorption at 1120 cm^{-1} in its i.r. spectrum. Strong absorptions at 810 cm^{-1} and 780 cm^{-1} were assigned to C—Cl stretching vibrations. In thiocarbonyl fluoride an observed i.r. frequency at 1375 cm^{-1} was assigned[42] to the C=S stretching absorption.

In more complex structural systems the C=S i.r. absorption is of limited use in structural diagnosis because of the large number of absorptions which occur in the above areas. This is clearly shown in the spectrum[6] of perfluorothioacetyl fluoride which contained absorption bands at 1351, 1242, 1183, 1055 and 1049 cm^{-1}. However, on the basis of values reported in the literature for ν_{C-Cl} in thiophosgene[41] (812 cm^{-1}), benzoyl chloride[43] (880 cm^{-1}) and phosgene[44] (849 cm^{-1}) it is possible to assign a strong and wide absorption at 850 cm^{-1} in thiobenzoyl chloride to the C—Cl vibration. Similarly, on the basis of a $\nu_{\gamma-CH}$ vibration at 779 cm^{-1} in benzoyl chloride, an analogous vibration at 770 cm^{-1} in thiobenzoyl chloride is most likely due to the C—H vibration. In thiophosgene and thiobenzophenone a band at 1121 cm^{-1} and 1207 cm^{-1}, respectively, has been assigned to the C=S group and a strong and wide absorption at 1253 cm^{-1} in thiobenzoyl chloride has accordingly been assigned to this group[3a]. Similar absorption in ethyl thiobenzoate and phenyl thiobenzoate has been noted in the region 1200–1300 cm^{-1}.

D. N.m.r. Spectra

The thiocarbonyl group exerts a marked de-shielding influence in ^{1}H n.m.r. spectra, much more so than the corresponding carbonyl group[45]. However, the major use of the n.m.r. technique in this area involves fluorine n.m.r. spectra. The appearance of peaks at low field for fluorine attached to C=S was observed[6, 9] in a series of thioacid fluorides. Thus, in perfluorothioacetyl fluoride the ^{19}F n.m.r. spectrum showed a quartet ($J = 10$ cps) centred at -120.9 p.p.m.* of relative area 1 and a doublet ($J = 10$ cps) centred at $+7.66$ p.p.m. of relative area 3. Similarly, in thiocarbonyl chlorofluoride, the ^{19}F n.m.r. spectrum consisted of a single, unsplit line at -16.8 p.p.m. which in thiocarbonyl difluoride occurred at -108 p.p.m.[6, 42].

IV. CHEMICAL PROPERTIES OF THIOCARBONYL HALIDES

A. Reaction with Nucleophiles

I. Amines

The thiocarbonyl halides all react readily with a variety of nucleophiles. Reaction with amines such as piperidine provides a convenient way of characterizing the labile thioaroyl halides. The thiobenzpiperidides are

---

* Spectra measured at 56·4 Mc/sec were calibrated in terms of displacement in p.p.m. from the ^{19}F resonance of 1,2-difluoro-1,1,2,2-tetrachloroethane used as an external standard.

usually pale-yellow, crystalline products easily purified by recrystallization from alcohol[4]. Aromatic amines such as aniline are also useful for characterization purposes and the majority of aromatic and heteroaromatic amines could be utilized for this purpose. Phenylhydrazine likewise forms a thiohydrazide[3a].

N-Alkylthiobenzamides, prepared from thiobenzoyl chloride and the aliphatic amine in ether, are usually obtained as low-melting solids or yellow oils in good yields[3a]. Thus, N,N-dimethylthiobenzamide, obtained in 69% yield, melts at 67° and distils at 180–186°/18 mm, whereas the N-n-butyl derivative (85%) distils at 155–164°/0·6–0·7 mm.

An acceptable alternative to the use of the thioacid chloride as an acylation agent is to utilize, for example, thiobenzoylthioglycollic acid[46]. Thioacylation of amines occurs very readily with this reagent[47] and various other alternatives have also been described[48]. In this context the use of ethyl dithioacetate, S-carboxymethyldithioacetate and S-carbomethoxymethyldithioacetate provide the alternative for the unknown thioacetyl chloride[49].

2. Alcohols, phenols, thiols and thiophenols

Alcohols also react readily with aromatic thioacid chlorides forming the corresponding thioesters[3a]. A comparison of the rate constants for the methanolysis of a series of substituted thiobenzoyl chlorides (o-CH$_3$, m-CH$_3$, o-Cl, m-Cl, p-Cl) with the corresponding benzoyl chlorides indicate that the thio compounds react more slowly with methanol[50] except for o-chlorothiobenzoyl chloride. Activation energies and entropies indicate that different reaction mechanisms are involved with the two classes of acid chlorides (Table 3).

With sodium alcoholates, formation of the corresponding thiocarboxylate is very rapid. Thus, ethyl thiobenzoate was obtained as a yellow

TABLE 3. Kinetic data for the methanolysis of substituted thiobenzoyl chlorides

Thiobenzoyl chlorides	k (min^{-1})				E_a (kcal/mol)	$\Delta S_{20}^{\ddagger}$ (e.u.)
	0°	10°	15°	20°		
Unsubstituted	0·0022	0·0064	0·0110	0·0179	16·9	−17·5
o-CH$_3$	>1					
m-CH$_3$	0·0033	0·0099	0·0171	0·0281	17·4	−14·9
p-CH$_3$	0·0040	0·0165	0·0322	0·0619	21·9	+2·1
p-Cl	0·00135	0·0034	0·0056	0·0086	14·8	−26·1

oil, b.p. 56–65°/0·8 mm, from the reaction of thiobenzoyl chloride and sodium ethoxide in ethanol[3a], the same product also being obtained from absolute ethanol and the thiobenzoyl chloride by refluxing under nitrogen.

Similar reactions have been reported with phenols. A pyridine solution of phenol was found[3a] to give phenylthiobenzoate, a yellow oil distilling at 92°/0·1 mm, whereas the corresponding thiophenol reacted with thiobenzoyl chloride in the absence of a hydrogen chloride scavenger. Phenyldithiobenzoate was obtained as a red oil which distilled at 145–150°/1 mm with some decomposition.

The presence of an electron-releasing group attached to the chlorothiocarbonyl function imparts greater stability to the thiocarbonyl chloride. However, alkoxy-, thioalkoxy- and aminothiocarbonyl chlorides react with alcohols, thiols and amines. Thus, in the reaction of phenol with thiocarbonyl chloride a small amount of the dicondensation product, O,O-diphenylthiocarbonate (42) was always obtained[33]. This arose by reaction of phenol with the phenoxythiocarbonyl chloride (41), the major product in the reaction. Variation of the reaction conditions in terms of temperature and molar proportion of reactants enables either the mono- or dicondensation product to be formed in greater amount[51]. This method leads to a simple O,O-ester but alternative routes are available for preparation of the mixed O,S-ester.

Reaction of the potassium salt of an alkoxy dithiocarboxylic acid (47) (a potassium xanthogenate) with alkyl halides[52], alkyl sulphate[53] or diazonium salts[54] provides a ready route to 48.

$$\underset{\textbf{(47)}}{\overset{\displaystyle S}{\overset{\displaystyle \|}{ROCSK}}} + RX \longrightarrow \underset{\textbf{(48)}}{\overset{\displaystyle S}{\overset{\displaystyle \|}{ROCSR}}}$$

The reaction of N,N-dimethylthiocarbonyl chloride with phenol is the basis of a very convenient synthesis of thiophenols. Thermolysis of the initially formed O-aryldimethylthiocarbamate (49) results in high yields of the S-aryldimethylthiocarbamate (50) which is easily hydrolysed to the

$$ArOH + \underset{}{\overset{\displaystyle S}{\overset{\displaystyle \|}{(CH_3)_2NCCl}}} \rightarrow \underset{\textbf{(49)}}{\overset{\displaystyle S}{\overset{\displaystyle \|}{(CH_3)_2NCOAr}}} \rightarrow \underset{\textbf{(50)}}{\overset{\displaystyle O}{\overset{\displaystyle \|}{(CH_3)_2NCSAr}}}$$

ArSH

thiophenol. This procedure may be extended to suitable hydroxyheterocycles and, as the thiols are readily desulphurized with Raney nickel, it provides an attractive alternative for replacing aromatic hydroxyl groups with hydrogen[55].

Aminothiocarbonyl chlorides (51) react readily with amines to give substituted thioureas which are also available from an amine and thiocarbonyl chloride. Utilization of the aminothiocarbonyl chlorides enables an unsymmetrical thiourea (52) to be prepared by condensation with another amine.[56]

With sodium alcoholates, thiocarbamic acid-*O*-esters (53) are formed. This method is not of great importance[57] as this class of compounds is very conveniently prepared in good yield from alkoxythiocarbonyl chlorides (54) and amines[58].

$$\underset{(51)}{\overset{S}{\underset{\|}{R_2NCCl}}} \xrightarrow{NaOR} \underset{(53)}{\overset{S}{\underset{\|}{R_2NCOR}}} \xleftarrow{R_2NH} \underset{(54)}{\overset{S}{\underset{\|}{ROCCl}}}$$

$$\Bigg\downarrow {\scriptstyle R'NH_2} \qquad \searrow {\scriptstyle RSH}$$

$$\underset{(52)}{\overset{S}{\underset{\|}{R_2NCNHR^1}}} \qquad \underset{(55)}{\overset{S}{\underset{\|}{R_2NCSR}}} \xleftarrow{R_2NH} \underset{(56)}{\overset{S}{\underset{\|}{ClCSR}}}$$

Though reaction of the aminothiocarbonyl chloride with thiols or their sodium salts leads[59] to the dithiocarbamic ester (55), numerous other advantageous methods of synthesis of 55 are available. Thioalkoxythiocarbonyl chlorides (56) and secondary amines yield the desired product which, when primary amines are used, is always a non-uniform product[60]. Other routes involve the reaction of isothiocyanates with thiols[61] and amines with trithiocarbonic acid diester[62].

3. Cyanides

Thiobenzoyl chloride reacts[3a] with cuprous cyanide, or potassium cyanide in aqueous acetone, to give thiobenzoyl cyanide (57) together with its dimer (58), represented as 1,4-dicyano-1,4-diphenyl-1,3-dithia-cyclobutane in analogy with the dimer of thiocarbonyl chloride[20], and some dicyanostilbene (59). Thiobenzoyl cyanide cannot be distilled and,

$$\underset{(57)}{\overset{S}{\underset{\|}{PhCCl}}} \rightarrow \underset{(57)}{\overset{S}{\underset{\|}{PhCCN}}} + \underset{(58)}{\underset{NC}{\overset{Ph}{\diagdown}}\underset{S}{\underset{}{\diagup}}\underset{CN}{\overset{Ph}{\diagdown}}} \qquad \underset{(59)}{\underset{NC}{\overset{Ph}{\diagdown}}C=C\underset{CN}{\overset{Ph}{\diagup}}}$$

on heating at 150° in an oil-bath, breaks down to sulphur and dicyano-stilbene, especially in the presence of copper salts.

4. Sodium azide

Thiocarbonyl chloride, in water at −5°, and sodium azide, gave rise to 5-chloro-1,2,3,4-thiatriazole (60). Caution must be maintained in reactions of this type and occasional violent detonations of an unpredictable nature were observed[63, 64]. N,N-Disubstituted thiocarbonyl chlorides also react readily with sodium azide in aqueous medium, giving 5-(di-substituted-amino)-1,2,3,4-thiatriazoles (61). The nature of the groups

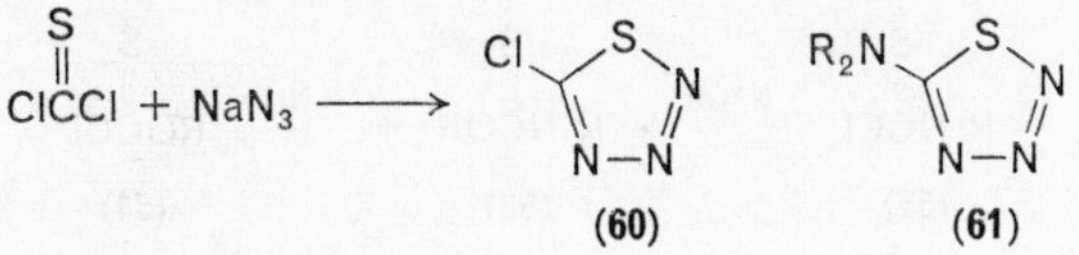

$$\underset{\underset{\text{ClCCl}}{\|}}{\overset{\text{S}}{}} + \text{NaN}_3 \longrightarrow$$

(60)   (61)

attached to the nitrogen had considerable influence on the reaction conditions with a reaction temperature of 100° and a reaction time of 2 hours being required for N,N-dimethylthiocarbonyl chloride[64].

Thiobenzoyl chloride and sodium azide also undergo ready reaction to form 5-phenyl-1,2,3,4-thiatriazole[65].

5. Heterocyclic tertiary amines

Ethoxythiocarbonyl chloride and phenoxythiocarbonyl chloride have been found[45a] to react with quinoline or isoquinoline and potassium cyanide to give Reissert compounds of type 62 and 63. It was not possible to utilize thiocarbonyl chloride in this reaction; instead o-isothiocyanato-trans-cinnamaldehyde (64) and 3-oxoimidazo[1,5-a]quinoline (65) were obtained in very small amounts when quinoline and potassium cyanide

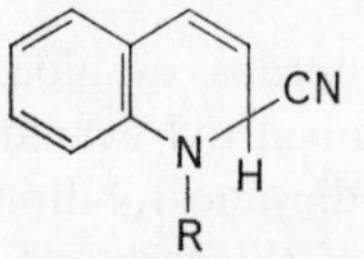

(62) 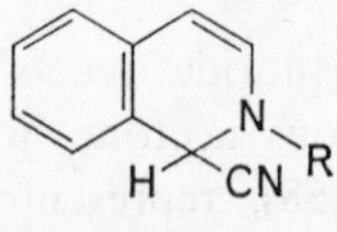   (63)

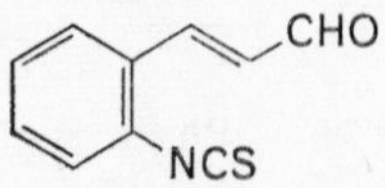

(64)

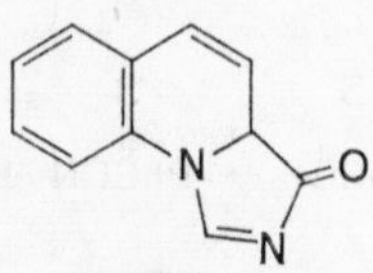

(65)

were used. With isoquinoline and thiocarbonyl chloride two products were also obtained. These were identified as **66** and **67**. *p*-Chlorothiophenoxythiocarbonyl chloride has also been reacted with isoquinoline[66] and in this case the product obtained was identified as **68**.

(66)   **(67)**

(68)   **(69)**

It is interesting to note[45a] that thiocarbonyl chloride and alkali (either 2*N*-sodium hydroxide or better, barium carbonate) and pyridine resulted in the formation of *trans,trans*-1-formyl-4-isothiocyanatobuta-1,3-diene **(69)**.

B. Reaction with Diazoalkanes

Diazomethane has been shown to react with thiobenzoyl chloride to give[65] 2-phenyl-1,3,4-thiadiazole **(70)**. Ring-closure in the manner envisaged was established by synthesis of **(70)** from *N*-benzoyl-*N'*-formylhydrazine and phosphorous pentasulphide[67]. Use of diazoacetic ester in the reaction leads[68] to 2-carbethoxy-5-phenyl-1,3,4-thiadiazole **(71)**.

(70)   **(71)**

It has also been established[67] that thiocarbonyl chloride and diazo ketones result in formation of the 1,3,4-thiadiazole nucleus rather than the 1,2,3-thiadiazole system[69]. Thus, from *p*-nitrobenzoyl diazomethane (72) and thiocarbonyl chloride, the major product obtained was 2-chloro-5-*p*-nitrobenzoyl-1,3,4-thiadiazole (73) and some ω-chloro-*p*-nitroaceto-phenone. Similarly, *p*-chlorobenzoyl diazomethane and thiocarbonyl chloride gave rise to 2-chloro-5-*p*-chlorobenzoyl-1,3,4-thiadiazole,

$$p\text{-NO}_2\text{C}_6\text{H}_4\text{COCHN}_2 + \overset{\overset{\text{S}}{\|}}{\text{ClCCl}} \rightarrow p\text{-NO}_2\text{C}_6\text{H}_4\text{CO}$$

(72)   (73)

(74)

ω-chloro-*p*-chloroacetophenone and the dithiadiene (74). This last product undergoes an interesting extrusion reaction when heated at 190° for 90 min. Sulphur is lost and 2,4-di-(*p*-chlorophenyl)thiophene is formed.

C. Reaction with 5-Phenyltetrazole and other Heterocycles

Thiobenzoyl chloride and 5-phenyltetrazole (75) react in the presence of pyridine to give 5-phenyl-2-thiobenzoyltetrazole (76). On warming, or when its preparation is carried out above 70°, nitrogen is lost[70]. This may be interpreted in terms of a *cis*-elimination, the thermal instability being a consequence of the formation of stable, molecular nitrogen on decomposition of 76 and the formation of a resonance-stabilized intermediate (78) via a species such as 77. The final product is 2,5-diphenyl-1,3,4-thiadiazole (79) which is formed in 50% yield. It has been suggested that

(75)   (76)   (77)

(79)   (78)   (77)

this rearrangement be regarded as a 1,5-dipolar cyclization involving a conjugated nitrilimine[71].

Under these conditions 2-benzoyl-5-phenyltetrazole results in 2,5-diphenyl-1,3,4-oxadiazole. It has been shown that the decomposition is a first-order reaction, independent of excess of acid chloride, indicating that the ring-cleavage must be the rate-determining step[72].

1-Thioacylazetidines show a ready tendency to isomerize when treated with concentrated hydrochloric acid[47]. Thus, 1-thiobenzoylazetidine (80) gave 2-phenyl-4,5-dihydro-6H-1,3-thiazine (81) on standing with acid at room temperature for several days.

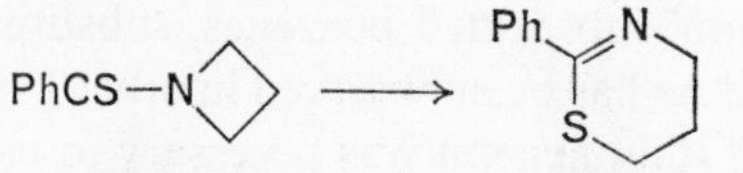

(80)   (81)

The thiobenzoylazetidines show a greater tendency to isomerize than do the corresponding 1-acylazetidines. 1-(Aryloxythiocarbonyl)- and 1-(dithio-p-chlorophenyloxycarbonyl)azetidine as well as 1-(N-phenyl-thiocarbamyl)-2-methylazetidine all undergo isomerization.

D. Friedel–Crafts Reactions

A wide variety of thiocarbonyl chlorides undergo reaction with aromatic substrates in the presence of Friedel–Crafts catalysts[73]. Thiobenzoyl, o-thiotoluoyl and p-thiotoluoyl chlorides (82) react under mild conditions (40° in 1,2-dichloroethane) with benzene, toluene, anisole or biphenyl in the presence of aluminium chloride to give the corresponding thiobenzophenone (83). This reaction is analogous to the earlier condensation of

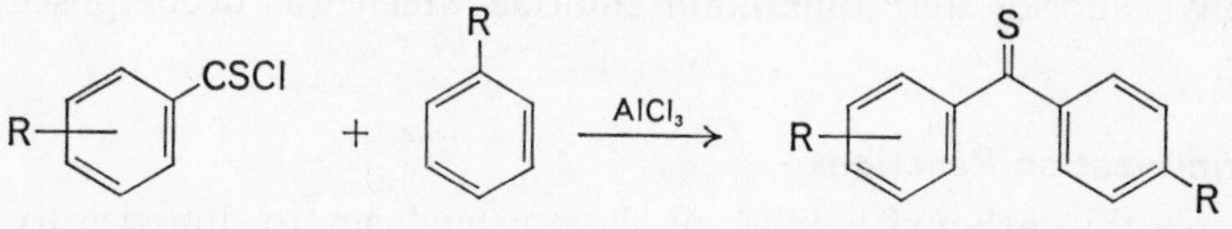

(82)   (83)

thiocarbonyl chloride with aromatic hydrocarbons to form thioketones[74]. Phenoxythiocarbonyl chloride also reacts[73, 75] in the presence of aluminium chloride with a variety of substituted benzenes to form thiobenzoic acid-O-aryl esters (84). The corresponding thiobenzoic acid-O-alkyl esters are not so readily accessible by this procedure due to elimination of carbonyl sulphide from the intermediate complex with aluminium chloride.

(Alkylthio)- or (arylthio)thiocarbonyl chlorides, in the presence of aluminium chloride or stannic chloride, react with a wide variety of aromatic hydrocarbons (or heterocycles such as furan and thiophene) to give in good-to-excellent yield the corresponding dithiobenzoic acid

$$
\underset{\text{S}}{\overset{\text{S}}{\parallel}}\text{PhOCCl} + \text{ArH} \xrightarrow[\text{or} \atop \text{SnCl}_4]{\text{AlCl}_3} \underset{(84)}{\overset{\text{S}}{\overset{\parallel}{\text{PhOCAr}}}}
$$

$$
\underset{\text{S}}{\overset{\text{S}}{\parallel}}\text{RSCCl} + \text{ArH} \xrightarrow[\text{or} \atop \text{SnCl}_4]{\text{AlCl}_3} \underset{(85)}{\overset{\text{S}}{\overset{\parallel}{\text{RSCAr}}}}
$$

esters (85). With monosubstituted benzenes, substitution always occurred in the *para*-position. As has been observed in other Friedel–Crafts reactions involving furan and thiophene, it was necessary to use the more moderate catalyst stannic chloride.

Under analogous conditions, *N,N*-disubstituted thiocarbonyl chlorides also react with similar aromatic hydrocarbons forming the corresponding *N,N*-disubstituted thioamides (86) in good yield[73]. However, benzene

$$
\overset{\text{S}}{\overset{\parallel}{(\text{CH}_3)_2\text{NCCl}}} + \text{ArH} \longrightarrow \overset{\text{S}}{\overset{\parallel}{(\text{CH}_3)_2\text{NCAr}}}
$$
$$
(86)
$$

$$
\overset{\text{S O}}{\overset{\parallel\ \ \parallel}{(\text{CH}_3)_2\text{NCSCN(CH}_3)_2}}
$$
$$
(87)
$$

itself does not give the expected thioamide but rather a product identified as 87. This arose from the intermediate complex formed from the thiocarbonyl chloride and aluminium chloride which was decomposed with water.

E. Dimerization Reactions

Simple thiocarbonyl halides all show a tendency to dimerize to some degree, especially fluorothioacyl fluorides. Thiocarbonyl chloride dimer is well known[20] and an analogous structure has been proposed for the dimer of thiocarbonyl fluoride. I.r. and Raman spectral data are consistent with structure (88) and it was concluded from the lack of coincident frequencies that the molecule has a centre of symmetry and belongs to the point group D_{2h}[76]. The various other mixed thiocarbonyl halides all form dimers of the same type[6].

Trifluorothioacetyl fluoride (11) does not dimerize spontaneously[6] but when irradiated with u.v. light at temperatures below 0°, it was transformed into approximately equal amounts of *cis*- and *trans*-2,4-difluoro-2,4-bis(trifluoromethyl)-1,3-dithietane (**89**). U.v. light irradiation also caused

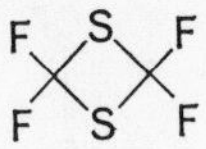

(88)

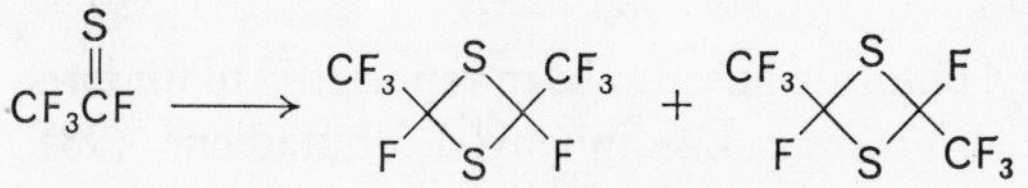

(11) **(89)**

trifluorothioacetyl chloride and chloro- and bromodifluorothioacetyl fluorides to dimerize to the corresponding dithietanes, whereas chlorodifluorothioacetyl chloride dimerized spontaneously at room temperature.

F. Cycloaddition Reactions

The simplest cycloaddition of a thiocarbonyl halide is of the [2+1] type in which thiocarbonyl fluoride adds carbenes[6]. Diazomethane (and diphenyldiazomethane) reacted to form the unstable thiirane (**90**) which immediately eliminated sulphur to give the olefin (**93**). In a [4+2] cyclo-

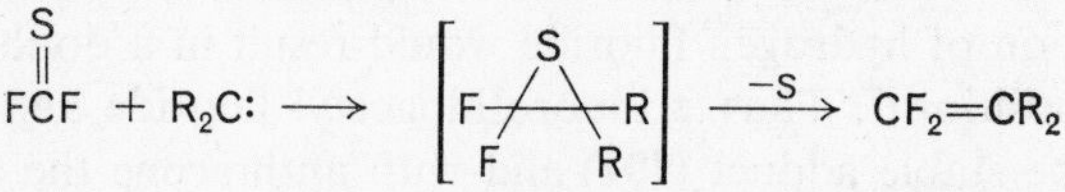

$$FCF + R_2C: \longrightarrow \left[F \overset{S}{\diagup} R \right] \overset{-S}{\longrightarrow} CF_2=CR_2$$

(90) **(93)**

addition, thiocarbonyl fluoride reacted as a dienophile with cyclopentadiene to give a relatively unstable adduct (**94**). Removal of the double bond in the cycloadduct by addition of bromine imparts stability to the product which otherwise must be kept in the cold. The corresponding adduct (**95**) from hexachlorocyclopentadiene had good stability[77].

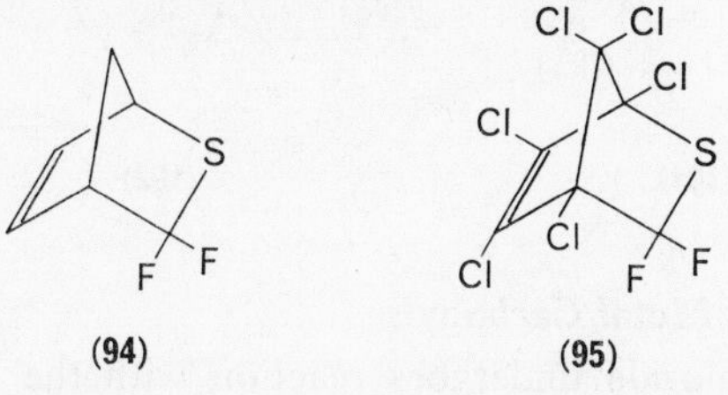

(94) **(95)**

Thioacyl fluorides are also exceedingly active dienophiles[77], though they are less reactive than hexafluorothioacetone. Trifluorothioacetyl fluoride, at −78°, reacted with butadiene (and dimethylbutadiene)

13

yielding the corresponding dihydrothiopyrans (96). These cycloadducts readily lost hydrogen fluoride to form the thiopyran (97).

(96)   (97)

Thiocarbonyl chloride has also been reported[78] to undergo an analogous reaction with *trans,trans*-1,4-diphenyl-1,3-butadiene (98). In this case 2-chloro-3,6-diphenylthiopyrylium chloride (100) was formed, most likely from the initial cycloadduct (99).

(98)   (99)   (100)

In the reaction of trifluorothioacyl fluorides with dienes, stability is imparted to the initial cycloadduct if the diene component is chosen so that elimination of hydrogen fluoride would result in a double bond at a bridgehead position[77]. Thus, trifluorothioacetyl fluoride and cyclopentadiene gave the stable adduct (101) and with anthracene the adduct (102) was obtained. An analogous stable adduct also resulted from the reaction of anthracene with pentafluorothiopropionyl fluoride.

(101)   (102)

G. Reaction with Metal Carbonyls

Thiobenzoyl chloride undergoes reaction with the strong Lewis-base sodium manganese pentacarbonyl at −78° in tetrahydrofuran[79]. The product initially formed, represented by 103, lost carbon monoxide (at ca. −60°) on warming the reaction to room temperature, giving brown trimeric, tetrameric and polymeric tetracarbonyl(thiobenzoyl)manganese complexes represented by the general expression **104**. These complexes

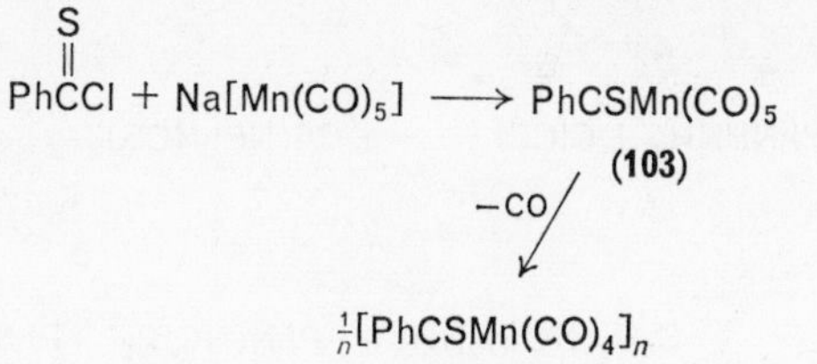

$$
\overset{\overset{\displaystyle S}{\|}}{PhCCl} + Na[Mn(CO)_5] \longrightarrow PhCSMn(CO)_5
$$

(103)

$$
\frac{1}{n}[PhCSMn(CO)_4]_n
$$

(104)

are thermally stable and diamagnetic. I.r. data, especially a ν_{C-S} absorption at 630 cm^{-1} and a series of bands attributable to the CO group in the regions 2100, 2020, 2011 and 1955 cm^{-1}, indicate that these products may be represented as either a thioketone (**105**) or a thiocarbene complex (**106**).

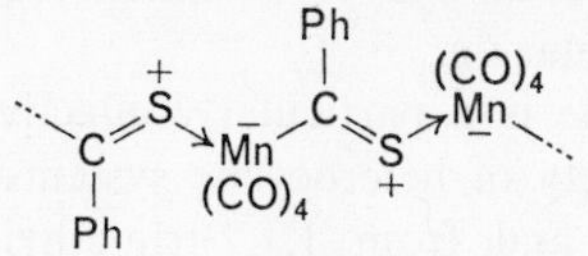

(105)

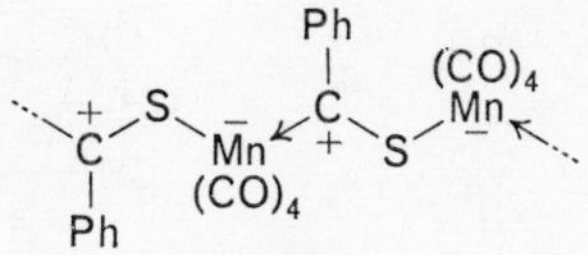

(106)

H. Miscellaneous Reactions

Esters of hydroxymethylphosphonic acid (**107**) react readily with N,N-dimethylthiocarbonyl chloride to give[80] esters of the type (**108**) in yields of 68–79%.

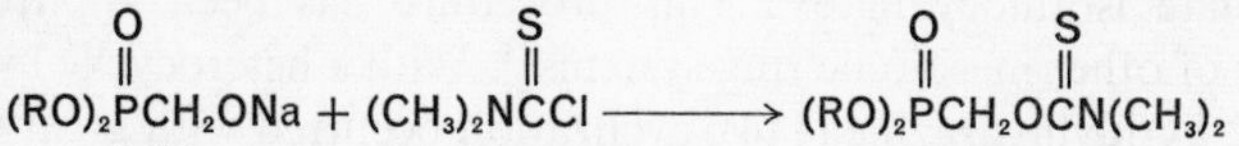

$$
(RO)_2PCH_2ONa + (CH_3)_2NCCl \longrightarrow (RO)_2PCH_2OCN(CH_3)_2
$$

(107)   **(108)**

The group R has been varied between ethyl, propyl, isopropyl, butyl and isobutyl, all of which yielded **108** as oils.

In addition to the normal reaction of thiocarbonyl chloride with hydrazines in which the chlorines are displaced by the amine function, a new type of reaction occurs when the reaction medium is dilute hydrochloric acid[81]. Thus, from phenylhydrazine and thiocarbonyl chloride, phenyl isothiocyanate and 1,2,4-triisothiocyanatobenzene (109) were obtained. These products have been rationalized in terms of the following reactions:

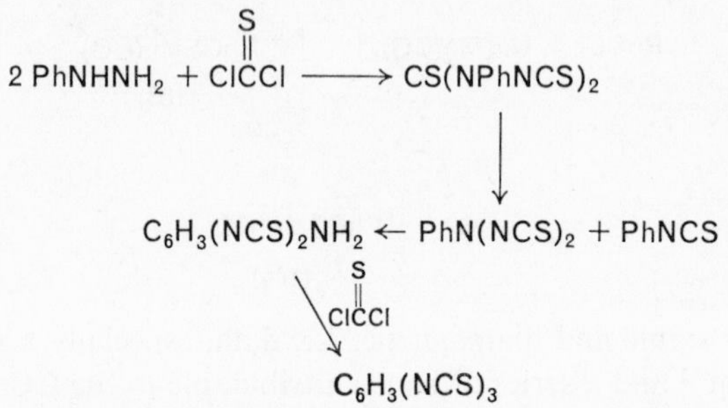

$$2\ PhNHNH_2 + Cl\overset{\overset{\displaystyle S}{\|}}{C}Cl \longrightarrow CS(NPhNCS)_2$$

$$C_6H_3(NCS)_2NH_2 \leftarrow PhN(NCS)_2 + PhNCS$$

$$\overset{\overset{\displaystyle S}{\big\backslash Cl\overset{\|}{C}Cl}}{\downarrow}$$

$$C_6H_3(NCS)_3$$

$$\text{(109)}$$

Related reactions were also observed with semicarbazide, thiosemicarbazide and semioxamazide.

Thiocarbonyl chloride is a particularly effective cyclization agent for the synthesis of a variety of heterocyclic systems[13]. Substituted 2-amino alcohols react readily and from 1,1,2-trimethyl-2-aminoethanol (110), 2-mercapto-4,5,5-trimethyl-2-oxazoline (111) is obtained[82].

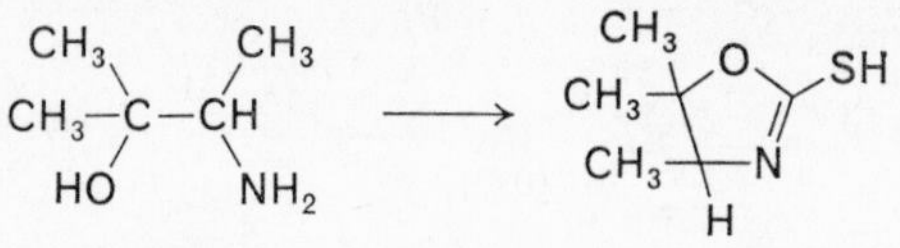

$$\text{(110)} \qquad\qquad \text{(111)}$$

As would be anticipated, thiocarbonyl chloride has found considerable use in cyclizations of this type. Reaction of 2-thiobenzoylmethylhydrazine (112) with thiocarbonyl chloride yielded the mesoionic 1,3,4-thiadiazole derivative (113), a ring-closure which probably involves formation of the intermediate isothiocyanate[83]. This procedure has been applied to the synthesis of other mesoionic ring systems[84]. With a heterocyclic hydrazine, such as 2-pyridylhydrazine (114), cyclization occurred with great ease[85] to the s-triazolo[4,3-a]pyridine-3-thiol (115). Numerous other examples of this type of utilization of thiocarbonyl chloride have been described in an earlier review[13].

A particularly interesting application of thiocarbonyl chloride is in the synthesis of bis(trifluoromethyl)thioketene, the first thioketene to be handled, distilled and stored[86]. Diethyl sodiomalonate and thiocarbonyl chloride yielded tetraethyl 1,3-dithietane-$\Delta^{2,\,\alpha:4,\,\alpha^1}$-dimalonate (116)

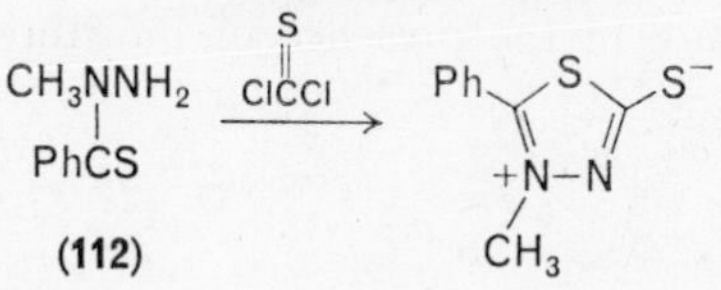

(112)   (113)

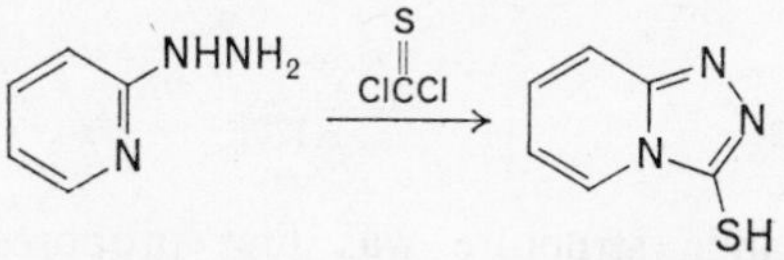

(114)   (115)

which, on heating with sulphur tetrafluoride in the presence of hydrogen fluoride, gave 2,4-bis[2,2,2-trifluoro-1-(trifluoromethyl)ethylidene]-1,3-dithietane (117). This is the dimer of bis(trifluoromethyl)thioketene (118) and when it was pyrolysed at 750° in a platinum tube packed with quartz rings, the monomer was obtained in 70% yield.

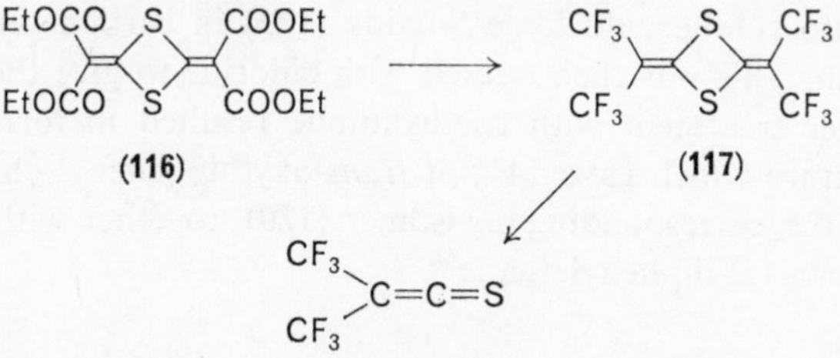

(116)   (117)

(118)

A recent report[87] of the utilization of N,N-dimethylcarbamyl chloride and N,N-disubstituted carbamates with organolithium reagents to form ketones indicates a potentially useful application of the corresponding thiocarbonyl compounds.

V. THIOAROYL CHLORIDE S-OXIDES

There are three main routes for entry into this interesting group of thiocarbonyl compounds which may be regarded as chlorosulphoxides. Reaction of phenylmethanesulphonyl chloride (119) and triethylamine in benzene or cyclohexane gave a mixture of trans-stilbene, cis-diphenylethylene sulphone, triethylammonium chloride, triethylammonium methanesulphonate and cis- and trans-oxythiobenzoyl chloride, (120) and (121), respectively[88]. These compounds are stable at −5° for several weeks but at room temperature in the presence of moisture or light, rapidly decompose.

The thioacyl halide structure was first proposed for the product obtained[89] by treating camphor-10-sulphonyl chloride with triethylamine but this is the first instance in which both geometrical isomers have been isolated. It is also the first demonstration of *cis–trans* isomerization about a double bond in which one of the atoms of the double bond is a second (or higher) period element.

The same mixture of *cis-* and *trans-*isomers was prepared by a second general procedure. Phenylmethanesulphinyl chloride (122) was first converted into thiobenzaldehyde-*S*-oxide (123) in carbon tetrachloride–ether medium. This was then treated with chlorine to give the dichloride (124). Further treatment with triethylamine resulted in formation of a reaction mixture which gave 14% of *trans*-oxythiobenzoyl chloride (121) and 28% of the corresponding *cis*-isomer (120), together with *meso-* and *d,l*-1,2-dichloro-1,2-diphenylethane[90].

A similar reaction sequence applied[90] to 2-methoxynaphthalene-1-thial-*S*-oxide (125) gave 2-methoxy-1-thionaphthoyl chloride-*S*-oxide (126).

Thiocarbonyl chloride *S*-oxide (127) is another interesting member of this family of chlorosulphoxides. Prepared[91] in 32% yield by oxidation of thiocarbonyl chloride with *m*-chloroperbenzoic acid, or by hydrolysis of trichloromethanesulphenyl chloride[92], it undergoes an interesting cyclo-addition[91] with cyclopentadiene at $-40°$ to give a relatively unstable

(125)   (126)

product (128) which was stabilized by oxidation to the sulphone (129).

This procedure illustrates the third general method for the preparation of thioacyl chloride S-oxides. It has also been applied successfully to the synthesis of 120 and 121 by oxidation of thiobenzoyl chloride with monoperphthalic acid[93].

(127)   (128)   (129)

VI. REFERENCES

1. H. Staudinger and J. Siegwart, *Helv. Chim. Acta*, **3**, 824 (1920).
2. H. Houben, *Ber.*, **39**, 3224 (1906).
3. (a) R. Mayer and S. Scheithauer, *J. Prakt. Chem.*, [4] **21**, 214 (1963); (b) F. Block, *Compt Rend.*, 1342 (1937).
4. R. Mayer and S. Scheithauer, *Chem. Ber.*, **98**, 829 (1965).
5. C. V. Jorgensen, *J. Prakt. Chem.*, [2] **66**, 43 (1902).
6. W. J. Middleton, E. G. Howard and W. H. Sharkey, *J. Org. Chem.*, **30**, 1375 (1965).
7. C. G. Krespan, *J. Org. Chem.*, **25**, 105 (1960); P. E. Aldrich, E. G. Howard, W. J. Linn, W. J. Middleton and W. H. Sharkey, *J. Org. Chem.*, **28**, 184 (1963); W. T. Miller, M. B. Freedman, J. H. Freid and H. F. Koch, *J. Am. Chem. Soc.*, **83**, 4105 (1961).
8. D. J. Park, R. Seffl and J. R. Lacher, *J. Am. Chem. Soc.*, **78**, 59 (1956).
9. J. F. Harris and F. W. Stacey, *J. Am. Chem. Soc.*, **85**, 749 (1963).
10. I. L. Knunyants and A. V. Fokin, *Bull. Acad. Sci. USSR, Div. Chem. Soc.*, 705 (1955).
11. N. N. Yarovenko, S. P. Motornyi, L. I. Kirenskaya and S. S. Vasilyeva, *J. Gen. Chem. USSR*, **27**, 2301 (1957).

12. B. R. O'Connor and F. N. Jones, *J. Org. Chem.*, **35**, 2002 (1970).
13. A compilation of references relation to these methods may be found in: H. Tilles, *The Chemistry of Organic Sulfur Compounds* (Eds. N. Kharasch and C. Y. Meyers) Vol. 2, Pergamon Press, New York, 1966, pp. 311–336.
14. E. Tietze and S. Peterson, *Ger. Pat.*, 853,162 (1952); Houben-Weyl, *Methoden der Organischen Chemie*, Vol. 9, Thieme Verlag, Stuttgart, 1955, p. 789.
15. E. F. Orwoll, *U.S. Pat.*, 2,668,853 (1954); *Chem. Abstr.*, **49**, 2496d (1955).
16. P. F. Frankland, F. H. Garner, F. Challenger and D. Webster, *J. Soc. Chem. Ind.*, **39**, 313T (1920).
17. O. B. Helfrich and E. E. Reid, *J. Am. Chem. Soc.*, **43**, 591 (1921).
18. H. Jonas, *Ger. Pat.*, 910,297 (1952); *Chem. Abstr.*, **49**, 3239e (1955).
19. E. Tietze and F. Schmidt, *Ger. Pat.*, 873,836 (1953).
20. A. Schonberg and A. Stephenson, *Ber.*, **66B**, 567 (1933); ref. 13, pp. 333–335.
21. K. V. Martin, *U.S. Pat.*, 3,048,629 (1962).
22. W. Sundermeyer and W. Meise, *Z. Anorg. Allgem. Chem.*, **317**, 334 (1962).
23. W. Walter and R. F. Becker, *Justus Liebigs Ann. Chem.*, **733**, 195 (1970).
24. W. Walter and K. D. Bode, *Angew. Chem.*, **78**, 517 (1966).
25. U. Hasserodt, *Chem. Ber.*, **101**, 113 (1968).
26. E. Lieber, C. N. R. Rao, C. B. Lawyer and J. P. Trivedi, *Can. J. Chem.*, **41**, 1643 (1963); see also E. Lieber and J. P. Trivedi, *J. Org. Chem.*, **25**, 650 (1960).
27. L. Birkofer and K. Krebs, *Tetrahedron Letters*, 885 (1968).
28. M. Zbirovsky and V. Ettel, *Chem. Listy*, **52**, 95 (1958); *Chem. Abstr.*, **52**, 16335i (1958).
29. G. M. Dyson and R. F. Hunter, *J. Soc. Chem. Ind.*, **45**, 81T (1926); G. M. Dyson and H. J. George, *J. Chem. Soc.*, 1702 (1924); G. M. Dyson, H. J. George and R. F. Hunter, *J. Chem. Soc.*, 3041 (1926); G. M. Dyson, H. J. George and R. F. Hunter, *J. Chem. Soc.*, 436 (1927).
30. O. Billeter and H. Rivier, *Ber.*, **37**, 4317 (1904).
31. O. Billeter and A. Strohl, *Ber.*, **21**, 102 (1888); O. Billeter, *Ber.*, **20**, 1629 (1887).
32. H. Rivier and J. Schalch, *Helv. Chim. Acta*, **6**, 612 (1923).
33. H. Rivier, *Bull. Soc. Chim. France*, [3] **35**, 837 (1906); H. Rivier and P. Richard, *Helv. Chim. Acta*, **8**, 490 (1925); A. Kaji and K. Miyazaki, *Nippon Kagaku Zasshi*, **87**, 727 (1966); *Chem. Abstr.*, **65**, 15255h (1966).
34. P. Klason, *Ber.*, **20**, 2376 (1887); P. Reich and D. Martin, *Chem. Ber.*, **98**, 2063 (1965).
35. H. Bergreen, *Ber.*, **21**, 337 (1888); W. Autenrieth and H. Hefner, *Ber.*, **58**, 2151 (1925).
36. R. P. Mull, *U.S. Pat.*, 2,711,421 (1955); *Chem. Abstr.*, **50**, 6507h (1956); B. K. Wasson and J. M. Parker, *U.S. Pat.*, 2,901,501 (1959); *Chem. Abstr.*, **54**, 1324i (1960).
37. F. Arndt, E. Milde and G. Eckert, *Ber.*, **56B**, 1976 (1923).
38. H. Rivier, *Bull. Soc. Chim. France*, [4] **1**, 737 (1907).
39. W. Ried, H. Hillenbrand and G. Oertel, *Justus Liebigs Ann. Chem.*, **590**, 123 (1954).
40. J. Fabian, H. Viola and R. Mayer, *Tetrahedron*, **23**, 4323 (1967).
41. J. I. Jones, W. Kynaston and J. L. Hales, *J. Chem. Soc.*, 614 (1957); V. R. Mecke, R. Mecke and A. Luttringhaus, *Z. Naturforsch.*, **10B**, 367 (1955).

42. A. J. Downs and E. A. V. Ebsworth, *J. Chem. Soc.*, 3516 (1960); D. C. Moule, *Can. J. Chem.*, **48**, 2623 (1970); A. J. Downs, *Spectrochim. Acta*, **19**, 1165 (1963).

43. F. Seel and J. Langer, *Chem. Ber.*, **91**, 2553 (1958).

44. L. Bournelle, *Acad. roy. Belgique, Cl. Sci., Mem. Collect.*, 8; S. G. Smith, *J. Am. Chem. Soc.*, **83**, 4285 (1961).

45. (a) R. Hull, *J. Chem. Soc.*, 1777 (1968); (b) P. L. Southwick, J. A. Fitzgerald and G. E. Milliman, *Tetrahedron Letters*, 1247 (1965).

46. J. C. Crawhall and D. F. Elliott, *J. Chem. Soc.*, 2071 (1951).

47. Y. Iwakura, A. Nabeya, T. Nishiguchi and K.-H. Ohkawa, *J. Org. Chem.*, **31**, 3352 (1966); Y. Iwakura, A. Nayeba and T. Nishiguchi, *J. Org. Chem.*, **32**, 2362 (1967).

48. W. Walter and K.-D. Bode, *Angew. Chem. internat. Ed.*, **5**, 447 (1966); W. Walter and K. J. Reubke in *The Chemistry of Amides*, (Ed. J. Zabicky), Interscience, London, 1970, pp. 481–484.

49. R. R. Crenshaw, J. M. Essery and A. T. Jeffries, *J. Org. Chem.*, **32**, 3132 (1967).

50. S. Scheithauer and R. Mayer, *Chem. Ber.*, **98**, 838 (1965).

51. H. Eckenroth and K. Kock, *Ber.*, **27**, 1368 (1894); M. Delepine, *Ann. Chim. (Paris)*, **[8] 25**, 546 (1912).

52. A. Cohours, *Ann. Chim. (Paris)*, **[3] 19**, 159 (1847); W. C. Zeise, *Justus Liebigs Ann. Chem.*, **55**, 304 (1845); W. Fomin and N. Sochanski, *Ber.*, **46**, 245 (1913).

53. L. T. Tschirgaeff, *Ber.*, **32**, 3334 (1899); S. Wametkin and D. Kursanoff, *J. Prakt. Chem.*, **[2] 112**, 164 (1926).

54. R. Leuckart, *J. Prakt. Chem.*, **[2] 41**, 179 (1890).

55. M. S. Newman and H. A. Karnes, *J. Org. Chem.*, **31**, 3980 (1966); H. Kwart and E. R. Evans, *J. Org. Chem.*, **31**, 410 (1966); H. M. Relles and G. Pizzolato, *J. Org. Chem.*, **33**, 2249 (1968).

56. O. C. Billeter, *Ber.*, **20**, 1630 (1887); O. C. Billeter and A. Strohl, *Ber.*, **21**, 102 (1888); O. C. Billeter and H. Rivier, *Ber.*, **37**, 4319 (1904); H. Bergreen, *Ber.*, **21**, 340 (1888); T. S. Warunis, *Ber.*, **43**, 2974 (1910).

57. O. C. Billeter, *Ber.*, **43**, 1856 (1910); M. Battegay and E. Hegazi, *Helv. Chim. Acta*, **16**, 999 (1933).

58. R. P. Mull, *J. Am. Chem. Soc.*, **77**, 581 (1955).

59. O. C. Billeter and A. Strohl, *Ber.*, **21**, 102 (1888).

60. J. V. Braun, *Ber.*, **35**, 3377 (1902).

61. A. W. Hofmann, *Ber.*, **2**, 116 (1869); E. Fromm and M. Bloch, *Ber.*, **32**, 2212 (1899).

62. M. Delepine and P. Schving, *Bull. Soc. Chim. France*, **[4] 7**, 894 (1910).

63. E. Leiber, C. B. Lawyer and J. P. Trivedi, *J. Org. Chem.*, **26**, 1644 (1961).

64. E. Leiber and C. B. Lawyer, *U.S. Dept. Comm., Office Tech. Serv.*, PB Report 154,269 (1962); *Chem. Abstr.*, **58**, 4543g (1963).

65. T. Bacchetti and A. Alemagna, *Rend. Ist. Lombardo Sci. Pt. I. Classe Sci. Mat. e Nat.*, **91**, 617 (1957).

66. F. D. Popp and C. W. Klinowski, *J. Chem. Soc. (C)*, 741 (1969).

67. T. Bacchetti, A. Alemagna and B. Danieli, *Tetrahedron Letters*, 3569 (1964).

68. H. Staudinger and I. Siegwart, *Helv. Chim. Acta*, **3**, 840 (1920).

69. W. Ried and B. M. Beck, *Justus Liebigs Ann. Chem.*, **673**, 124 (1964).

70. R. Huisgen, H. J. Sturn and M. Seidel, *Chem. Ber.*, **94**, 1555 (1961).
71. H. Reimlinger, J. J. M. Vandewalle, G. S. D. King, W. R. F. Linger and R. Merenyi, *Chem. Ber.*, **103**, 1918 (1970).
72. R. Huisgen, *Angew. Chem.*, **72**, 359 (1960).
73. H. Viola, S. Scheithauer and R. Mayer, *Chem. Ber.*, **101**, 3517 (1968).
74. H. Begreen, *Ber.*, **21**, 337 (1888); L. Gattermann, *Ber.*, **28**, 2869 (1895).
75. H. Rivier and P. Richard, *Helv. Chim. Acta*, **8**, 490 (1925).
76. J. R. Durig and R. C. Lord, *Spectrochim. Acta*, **19**, 769 (1963).
77. W. J. Middleton, *J. Org. Chem.*, **30**, 1390 (1965).
78. G. Laban and R. Mayer, *Z. Chem.*, **7**, 227 (1967).
79. E. Lindner, H. Weber and H.-G. Karmann, *J. Organometal. Chem.*, **17**, 303 (1969).
80. V. V. Alekseev and M. S. Malinovskii, *Zh. Obshch. Khim.*, **21**, 3437 (1961).
81. T. Beckett and G. M. Dyson, *J. Chem. Soc.*, 1358 (1937).
82. E. Bergmann, *U.S. Pat.*, 2,525,200 (1950); *Chem. Abstr.*, **45**, 3424 (1951).
83. K. T. Potts and C. Sapino, Jr., *Chem. Commun.*, 672 (1968).
84. K. T. Potts, S. K. Roy, S. W. Schneller and R. M. Huseby, *J. Org. Chem.*, **33**, 2559 (1968).
85. D. S. Tarbell, C. W. Todd, M. C. Paulson, E. G. Linstrom and V. P. Wystrack, *J. Am. Chem. Soc.*, **70**, 1381 (1948).
86. M. S. Raasch, *J. Org. Chem.*, **35**, 3470 (1970).
87. U. Michael and A.-B. Hornfeldt, *Tetrahedron Letters*, 5219 (1970).
88. J. F. King and T. Durst, *Can. J. Chem.*, **44**, 819 (1966); E. Wedekind and D. Schenk, *Ber.*, **44**, 198 (1911).
89. E. Wedekind, D. Shenk and R. Stusser, *Ber.*, **56**, 633 (1923).
90. J. Strating, L. Thijs and B. Zwaneburg, *Rec. Trav. Chim. Pays-Bas*, **86**, 641 (1967).
91. B. Zwanenburg, L. Thijs and J. Strating, *Tetrahedron Letters*, 4461 (1969).
92. J. Silhanek and M. Zbirovsky, *Chem. Comm.*, 878 (1969).
93. B. Zwanenburg, L. Thijs and J. Strating, *Rec. Trav. Chim. Pays-Bas*, **89**, 687 (1970).

CHAPTER 12

Chloroformate esters and related compounds

DENNIS N. KEVILL

Department of Chemistry, Northern Illinois University,
DeKalb, Illinois, 60115

I. INTRODUCTION

The compounds to be given major coverage in this chapter are described
in the literature as chloroformates, chlorocarbonates or (mainly in peptide
synthesis applications) as alkoxy-, aryloxy- or arylalkyloxy-substituted

carbonyl chlorides. In *Chemical Abstracts* they are found under the general heading of formic acid and are listed under 'formic acid, chloro-, esters of'. The alternative names arise from the possibility of considering them as named from carbonic acid (H_2CO_3) by replacement of a hydroxyl group by chlorine, from formic acid (H_2CO_2) by replacement of a hydrogen by chlorine, or from carbonyl chloride ($COCl_2$) by replacement of a chlorine by an alkyloxy, aryloxy or arylalkyloxy group.

The parent acid (HOCOCl) is extremely unstable and attempts to prepare it invariably lead to carbon dioxide and hydrogen chloride. The esters are of varying stability, dependent upon the nature of the organic group introduced. Aryl, methyl and primary alkyl chloroformates are fairly stable. Secondary alkyl and benzyl chloroformates are somewhat less stable but are commercially available. Tertiary alkyl chloroformates can be prepared but they are quite unstable and are not commercially available; they can be considerably stabilized by placing the tertiary carbon at a bridgehead.

Due to their inherent instability and their lachrymator properties, handling and storage of chloroformates should be with suitable caution. Storage should be at cool temperatures. This was impressed upon the author when a bottle of benzyl chloroformate was carelessly stored upon an open shelf. After several months, the bottle exploded violently.

Methyl, ethyl and isopropyl chloroformates are produced in bulk and they have suggested applications as intermediates for dyes, plastics, chemicals, pharmaceuticals and flotation agents[1]. In the laboratory, benzyl chloroformate (benzyloxycarbonyl chloride) is often used, under relatively mild alkaline conditions, to introduce a protective group during peptide synthesis[2, 3, 4]:

$$OH^- + RNH_2 + C_6H_5CH_2OCOCl \longrightarrow H_2O + Cl^- + C_6H_5CH_2OCONHR$$

The benzyl–ester link will subsequently undergo fairly rapid cleavage, either by treatment with 2N hydrobromic acid:

$$C_6H_5CH_2OCONHR + HBr \longrightarrow C_6H_5CH_2Br + CO_2 + RNH_2$$

or by hydrogenolysis:

$$C_6H_5CH_2OCONHR + 2[H] \longrightarrow C_6H_5CH_3 + CO_2 + RNH_2$$

Modifications include use of *p*-nitrobenzyl chloroformate, which gives more readily crystallized derivatives, and azo-substituted benzyl chloroformates, which give coloured derivatives and simplify column chromatography. The bridgehead chloroformate, 1-adamantyl chloroformate, has recently been found to be useful[5, 6]. Since the use of chloroformate esters to introduce these urethane-type protecting groups into amino compounds has already been discussed within this series[4], it will not be given further consideration during this chapter.

A review of the chemistry of chloroformates[7] appeared in 1964. This review attempted a complete coverage through 1962 and it included 750 references. While not completely neglecting reaction mechanism, consistent with the affiliation of the authors with the research and development department of the plastics division of a large industrial corporation, the emphasis was largely upon preparation, physical properties and reactions of chloroformates. It concluded with a survey of the utility of chloroformates in polymer synthesis. The present chapter is intended to supplement and not to replace this review. Emphasis will be upon reaction mechanism and especially upon work since 1962. Also, there will be coverage of other haloformates and of the related chloroglyoxalate esters ($ROCOCOCl$).

II. SYNTHESIS

Alkyl chloroformates are nearly always synthesized[1, 8] by interaction of alcohols with carbonyl chloride (phosgene) at or below room temperature (to minimize dialkyl carbonate formation), preferably in the presence of a tertiary amine (essential when acid-sensitive groups are present and also used to increase the rate of reaction). The reaction probably proceeds through an intermediate which is a very effective acylating agent[9, 10]:

$$\left[\begin{array}{c} O \\ \| \\ -N-C-Cl \\ | \end{array} \right]^{+} \xrightarrow{\text{ROH}} \begin{array}{c} O \\ \| \\ RO-C-Cl \end{array} + \begin{array}{c} | \\ -N^{+}-H \\ | \end{array}$$

Aryl chloroformates are synthesized from the less reactive phenols at temperatures above 75° or, more readily, by first converting the phenol to an alkali metal phenoxide and reacting with phosgene in the presence of an organic solvent[11]. Again, tertiary amines, preferably of the *N,N*-dimethylaniline type, are frequently used as catalysts.

III. THERMAL DECOMPOSITION

A. In the Liquid Phase

The thermal decomposition of a chloroformate ester to yield the corresponding chloride and carbon dioxide is one example of a fairly widespread type of molecular decomposition. These decompositions are accompanied by the ejection of small highly stable molecules. Some examples are given below:

$$ROCOCl \longrightarrow RCl + CO_2$$

$$ROCOCOCl \longrightarrow RCl + CO + CO_2$$

$$ROSOCl \longrightarrow RCl + SO_2 \text{[12,13]}$$

$$R-N=N-O_2CR' \longrightarrow RO_2CR' + N_2 \text{[14,15]}$$

$$R-N=N-O_2CR' \longrightarrow RO_2CR' + N_2O \text{[16,17]}$$
$$\qquad\quad \downarrow$$
$$\qquad\quad O$$

The latter two reactions have been discussed, together with closely related reactions, within a previous volume of this series[18].

One interesting feature of these decompositions is that they frequently (but not always) proceed with retention of configuration[19]. The reactions can formally be considered as a nucleophilic replacement of a relatively large anion by a smaller anion, generated by loss of a stable molecule from the larger anion. Within the Hughes–Ingold terminology, they are designated as $S_N i$ reactions (internal nucleophilic substitution). As recognized by Ingold[19], the mechanism is less well defined than the $S_N 1$ and $S_N 2$ mechanisms and several detailed mechanisms have been proposed, dependent upon the substrate and the reaction conditions. The diazonium type compounds have been considered to favour initial ionization to give a diazonium salt ion pair, followed by loss of nitrogen and subsequent collapse of the new ion pair[16]:

$$R-N=N-X \underset{\longleftarrow}{\overset{\longrightarrow}{\rule{1cm}{0pt}}} R-N\overset{+}{=}NX^- \longrightarrow R^+\cdot N_2\cdot X^- \longrightarrow RX + N_2$$

This mechanism is not, however, favoured for chloroformate or chlorosulphinate decompositions, where the decomposition has usually been expressed in terms of either cyclic or tight ion-pair processes, with the possibility of a continuous spectrum of mechanisms intermediate between the two extremes[20–27].

Using chloroformates as an example, the covalent process can be expressed:

and the ionic process:

There is also evidence for some chlorosulphinate decompositions favouring an S_N2 mode of decomposition[19].

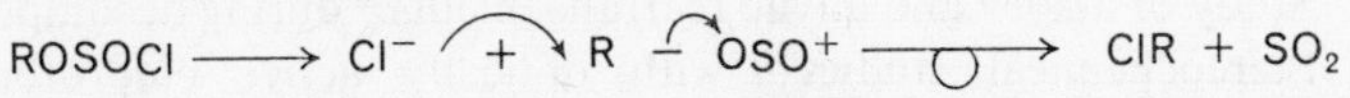

$$\text{ROSOCl} \longrightarrow \text{Cl}^- \overset{\frown}{(} + \overset{\curvearrowright}{R} \overset{\frown}{\text{OSO}^+} \xrightarrow{\quad} \text{ClR} + \text{SO}_2$$

n-Butyl chlorosulphinate yields 1-chlorobutane without any 2-chlorobutane and (+)-1-methylheptyl chlorosulphinate gives little or no rearrangement and a preponderance of inversion[28]. However, chlorosulphinates with phenyl substituents[20] or branched alkyl groups[29] have been found to rearrange.

One feature of nucleophilic substitutions, including S_Ni reactions, is for elimination reaction to compete with substitution, provided β-hydrogens are present. In terms of covalent processes, this will involve competition between the four-centred transition state for substitution and a six-centred transition state for elimination:

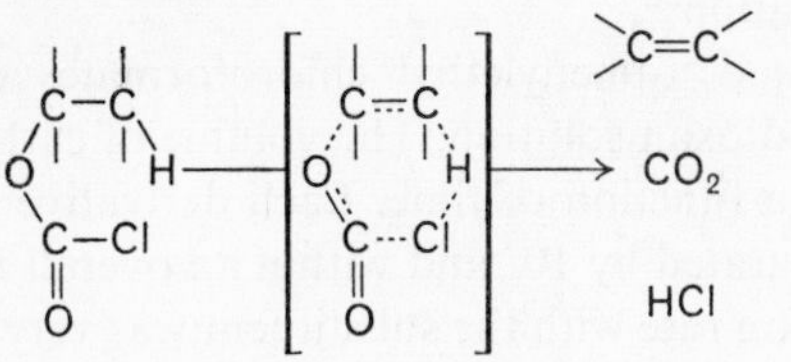

One feature of such a six-centred transition state is that it requires a *cis*-elimination process.

Alternatively, if ionic processes operate, there will be a common step involving formation of an ion pair which either collapses to substitution product or the counter-ion abstracts a proton from the carbonium ion (a process which has been studied for elimination from *t*-butyl derivatives[30]). While an ionic process does not demand a *cis*-elimination, it is possible to visualize how a rapid proton extraction could find the chloride ion ideally situated for this types of process:

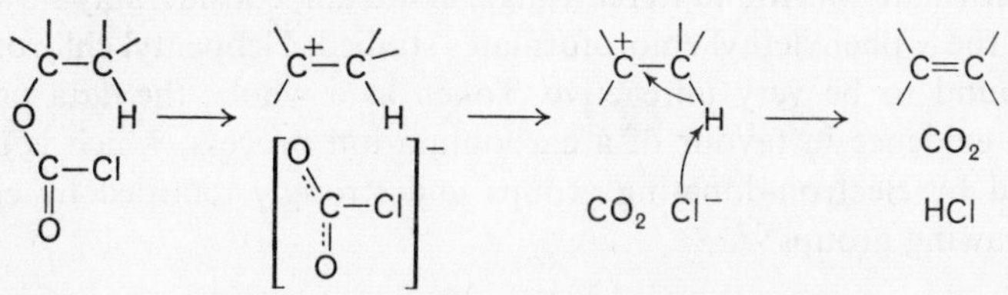

At least five standard tools for investigation of reaction mechanism have been applied to the study of chloroformate decompositions:

(a) Effect on the reaction rate of variation within the R group, including Hammett $\rho\sigma$ correlations[31] for α-phenylethyl derivatives.

(b) Response of reaction rate to solvent ionizing power[32] variations.

(c) Isotopic labelling, especially a search for [18]O equilibration[33] within 'unreacted' chloroformate ester.

(d) Study of alkyl[34] and allylic rearrangements[35] during decomposition.

(e) Stereochemical studies[36] with optically active chloroformates.

Contrary to earlier reports, it has been found that pure aryl chloroformates are very stable thermally[37]. Either in solution or in the gas phase, the relative stabilities are $C_6H_5 \gg 1° > 2° > 3°$. Benzyl chloroformates decompose quite readily to the respective chlorides. While not demanding a completely ionic process, these results do suggest that the favoured transition states incorporate considerable charge development on the α-carbon, such that the same factors which stabilize or destabilize carbonium ions influence this charge development.

In solution, a quantitative study of the electronic requirements at the α-carbon of the transition state has been carried out in terms of the Hammett $\rho\sigma$ equation[38].

Five p-substituted α-phenylethyl chloroformates were investigated in both toluene and dioxan solution. The volume of carbon dioxide liberated was measured as a function of time. Each derivative was studied at three temperatures, separated by 10° and within an overall range of 30–90°. The variation in reaction rate with the substituent was very large. The m-bromo was very slow and the p-methyl was very fast in terms of the analytical technique employed. The Hammett plot[31] gave, in each solvent, a good linear plot for four compounds but the unsubstituted parent compound reacted slower than required to fall on this plot. Such behaviour has also been observed in other systems[39, 40]. The slope of the line (ρ value) was $-3\cdot86$ in dioxan and $-3\cdot56$ in toluene. Although the enthalpies and entropies of activation varied in a seemingly random fashion, they were in general lower in toluene than in dioxan. This effect is commonly observed in reactions with relatively ionic transition states[41]. Benzyl chloroformate was found to rearrange, in dioxan, considerably slower than any of the α-phenylethyl chloroformates studied. Neopentyl chloroformate was found to be very unreactive. Taken as a whole, the data presented strong evidence in favour of a carbonium ion process, which is strongly assisted by electron-donating groups and strongly retarded by electron-withdrawing groups:

$$R-O-\underset{\underset{O}{\|}}{C}-Cl \; \underset{\longleftarrow}{\dashrightarrow} \; R^+ \; -\underset{O}{\overset{O}{C}}-Cl \; \longrightarrow \; RCl + CO_2$$

One means of obtaining evidence for carboxylate containing ion-pair intermediates, such as the one outlined above, is to demonstrate oxygen equilibration in unreacted substrate, presumably due to ion-pair return[33]. This approach has been elegantly incorporated into a study of the closely related decompositions of arylalkyl thiocarbonates:

$$Ar-CHO-C-SR \longrightarrow ArCHSR + CO_2$$
$$\hspace{1.1cm} | \hspace{0.4cm} \| \hspace{2.3cm} |$$
$$\hspace{1.1cm} Ph \hspace{0.4cm} O \hspace{2.3cm} Ph$$

This reaction has been shown to involve ion-pair intermediates[42] and the observation that ether–oxygen ^{18}O-labelled p-chlorobenzhydryl thiocarbonates undergo equilibration faster than decomposition[43], supported by the observation of a faster racemization than decomposition[43, 44], indicates the following mechanism, with k_b less than k_{-a}:

$$R'-O-C-SR \underset{k_{-a}}{\overset{k_a}{\rightleftharpoons}} [R^{+'}O_2^-CSR] \overset{k_b}{\longrightarrow} CO_2 + [R^{+'}S^-R]$$
$$\hspace{2.2cm} \|$$
$$\hspace{2.2cm} O \hspace{6cm} \downarrow k_c$$

$$\text{where } R' = p\text{-}ClC_6H_4CH \hspace{3cm} R'SR$$
$$\hspace{4.6cm} |$$
$$\hspace{4.6cm} C_6H_5$$

A corresponding scheme for the analogous chloroformate decomposition would be as follows[23, 45, 46]:

$$R-O-C-Cl \underset{k_{-a}}{\overset{k_a}{\rightleftharpoons}} [R^{+-}O_2CCl] \overset{k_b}{\longrightarrow} CO_2 + [R^+Cl^-]$$
$$\hspace{1.8cm} \|$$
$$\hspace{1.8cm} O \hspace{6.2cm} \searrow k_c$$
$$\hspace{9.5cm} RCl$$

However, since Cl^- is a much better leaving group than SR^-, it is reasonable to suppose that k_b will be larger than k_{-a} and ion-pair return (and equilibration) will be unimportant. This has been supported for the partial decomposition of α-phenylethyl chloroformate, with ^{18}O label in the ether–oxygen, in dioxan at 60°. Unreacted chloroformate showed no equilibration between ether and carbonyl oxygens[46]. While this observation would also be consistent with a covalent four-centred process[47] for decomposition, comparison with the analogous thiocarbonate decompositions strongly indicates an ionic process in which internal return of chloroformate anion is unable to compete with its rapid loss of carbon dioxide to generate chloride ion. Internal return of the chloride ion leads to product. It should, of course, be pointed out that, with experiments of this type, there is always the possibility of 'hidden return' (a rapid internal return which occurs without equilibration or racemization).

The decomposition of optically active α-phenylethyl chloroformate, in either dioxan or toluene, proceeded with retention of configuration[38]. This is consistent with the above ionic scheme in which steps k_b and k_c are sufficiently rapid for the anion to rarely, if ever, take up a position (either by migration or rotation of the cation) such as to lead to return, either to chloroformate or chloride, with inversion of configuration. A four-centred covalent transition state would, of course, also lead to a retention of configuration[47]. Retention of configuration in the resultant chloride was also observed in an earlier study of the reaction of α-phenylethanol with phosgene in the presence of a small excess of quinoline[48]; larger amounts of quinoline led to an inverted product. Similarly, the decomposition of neat optically active 2-octyl chloroformate at 130° proceeded with retention of configuration[49], but the pyridine-catalysed reaction proceeded with inversion[50].

Inversion in the presence of tertiary amines has also been observed for reactions proceeding through chlorosulphinate ester intermediates[51]. The rationalization given for these catalysed decompositions[48, 52] involves a nucleophilic substitution at the acyl carbon, followed by S_N2 inversion in an attack by the displaced chloride ion at the α-carbon:

$$ROC\!-\!Cl \; + \; :\!N\!- \; \longrightarrow \; ROC\!-\!\overset{+}{N}\!- \; + \; Cl^-$$
(with carbonyl O on ROC groups)

$$Cl^- + R\!-\!O\!-\!C\!-\!\overset{+}{N}\!- \; \longrightarrow \; ClR \; + \; CO_2 \; + \; :\!N\!-$$

Support for this mechanism comes from the observation of a chloride ion catalysis to the decomposition of chloroformate esters in acetonitrile[53] or in the absence of solvent[54]. The S_N2 characteristics of this process are supported by the relative second-order rate coefficients for the reaction

$$ROCOCl + Cl^- \longrightarrow ClR + CO_2 + Cl^-$$

in acetonitrile at 25·0°, as the nature of the R group is varied[53]:

Me (26) > Et (1) > n-Pr (0·94) > isoPr (0·12) > isoBu (0·042)

and by the suppression of otherwise observed rearrangement on addition of pyridine hydrochloride to the neat decomposition of n-butyl, isobutyl, n-pentyl and isopentyl chloroformates[54].

The study of alkyl group rearrangements as a tool in investigating reaction mechanisms has been greatly refined by the advent of gas chromatography[34]. Recently, a study of this type was followed by a report[54] of extensive alkyl rearrangement during the thermal decomposition of alkyl chloroformates, in the liquid phase and under reflux at temperatures within the range 125–180°. Several primary and secondary butyl and pentyl chloroformates were examined. In all cases, both elimination and substitution were observed and the rearrangement patterns were similar to those observed during reactions of aliphatic amines with nitrous acid[55]. The findings were consistent with an ionic process for decomposition, within which carbonium ions are free to rearrange. An alternative process involving concerted displacement–rearrangement was also mentioned as a possibility. Such a process would proceed by way of a five-centred transition state:

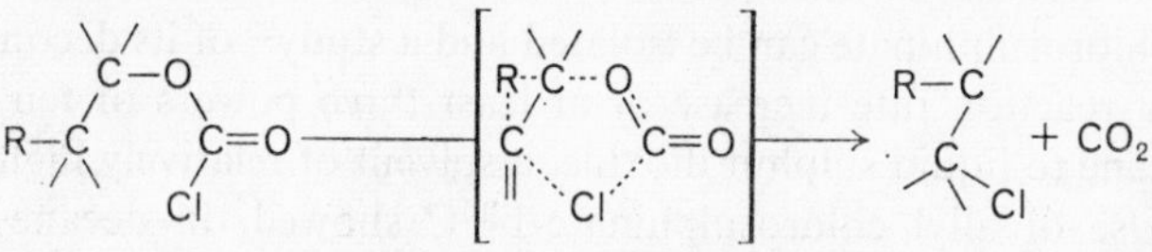

Olefin by-products may arise, in part, by elimination from a carbonium ion, possibly after rearrangement. One important feature, however, is the formation of terminal olefins as the main by-product from secondary chloroformate esters, in violation of Saytzeff's rule. The explanation put forward involved a *cis*-elimination, with chlorine preferentially removing hydrogen from the least sterically hindered carbon atom. This elimination process was formulated as passing through a six-membered cyclic transition state but, as outlined earlier in this chapter, a *cis*-elimination could also result from an ionic process and, if it was preferentially generated from the least crowded conformer, the chloride would be ideally situated for extraction of a terminal hydrogen from the counter-ion.

Consistent with a previous observation[38], decomposition of neopentyl chloroformate was very slow and the products (formed in small yield even after 32 hours at 180°) were completely rearranged, with 20 parts of *t*-pentyl chloride for each part of alkene (19% 2-methyl-1-butene; 81% 2-methyl-2-butene).

Studies of allylic rearrangements during $S_N i$ decomposition, what is termed $S_N i'$ decomposition, were first made for the decomposition reactions of chlorosulphinates. Meisenheimer and Link[56], studying the reactions of 1- and 3-ethylallyl alcohol with thionyl chloride, found the main products to be with rearrangement, the primary alcohol giving secondary chloride and vice versa. Similar results[57] were obtained for 1- and 3-methylallyl alcohols with thionyl chloride in the absence of solvent. However, in dilute ether solution *complete* rearrangement was observed. With optically active *trans*-1,3-dimethylallyl alcohol, optical purity was preserved.

Roberts, Young and Winstein[58] proposed a cyclic $S_N i'$ process for the conversion of the intermediate chlorosulphinate to product:

$$\text{(structure)} \quad \longrightarrow \quad \text{(structure)} \quad + \ SO_2$$

Later it was pointed out[20, 59] that rigidly oriented ion-pair intermediates could also account for the observed rearrangements.

Allyl chlorosulphinate can be isolated and a study[22] of its decomposition showed a reaction rate increase of at least three powers of ten in going from decane to liquid sulphur dioxide, a solvent of relatively high ionizing power. Use of allyl chlorosulphinate-1-^{14}C showed, in decane at 140°, virtually complete rearrangement during decomposition with the allyl chloride being 99·1% allyl chloride-3-[14]C. In the absence of solvent at 80°, the allyl chloride was 88–89% allyl chloride-3-[14]C. In other solvents substantial, but reduced, amounts of rearrangement were observed.

A study of 1-trifluoromethylallyl chlorosulphinate[60] gave rather different results, the rate increase in going from decane to nitrobenzene at 100° being approximately seven. In the absence of solvent, in decane and in nitrobenzene, the only product observed was 3-trifluoromethylallyl chloride. While rearrangement can be explained by a suitably oriented intimate ion pair, the small variation in reaction rate with solvent ionizing power suggested a mechanism with little charge development in the transition state and closely akin to a classical cyclic $S_N i'$ process[58]. Especially revealing was the observation that introduction of the *electron-withdrawing* trifluoromethyl substituent *increased* the rate of decomposition in decane at 100° by a factor of about 10³. It was concluded that the substituted allyl chlorosulphinate decomposed by an essentially covalent process and the parent allyl chlorosulphinate decomposed (at least in the more polar solvents) by an essentially ionic process.

Studies with allyl chloroformates have been carried out by Olivier and Young[23]. The kinetics and products of the thermal decomposition of 1- and 3-methylallyl chloroformates in a series of solvents were studied.

Studies with allyl chloroformates, formally analogous to those with chlorosulphinates, are simplified by the comparitive ease of isolation. The decomposition had quite different product characteristics to those observed for corresponding chlorosulphinates. Even in *n*-butyl ether, appreciable amounts of both 1- and 3-methylallyl chlorides are obtained and, in all the solvents studies, product ratios were quite similar irrespective of whether 1-methylallyl or 3-methylallyl (crotyl) chloroformate was decomposed. It was suggested[23] that both reactions proceed by alkyl–oxygen fission and the carbonium–chlorosulphinate ion pair, but not the carbonium–chloroformate ion pair, is sufficiently stable for the structural orientation required for collapse to the $S_N i'$ product to be attained. This argument is rather difficult to follow since one would picture after orientation an intimate ion pair of the type[22, 61]

$$
\begin{array}{c}
\text{CH} \\
\text{CH}_3\text{CH} \overset{\diagup}{} + \overset{\diagdown}{} \text{CH}_2 \\
\\
\text{O} - \text{O} \\
\diagdown \, \text{S} \, \diagup \\
| \\
\text{Cl}
\end{array}
$$

and, after loss of sulphur dioxide, no special preference for $S_N i'$ relative to $S_N i$. If the ion-pair mechanism is accepted for these decompositions, and as we shall see there is other evidence for it, a reverse interpretation would appear to be required. The results can be rationalized if the chloroformate passes through the oriented intimate ion pair, of the type indicated above, but the chlorosulphinate anion decomposes, after carbon–oxygen heterolysis, prior to orientation and so as to leave the chloride ion favourably disposed for $S_N i'$ attack. This chlorosulphinate mechanism could be considered as an extremely polar version of the classical six-centred mechanism[58]. Indeed, the results from the subsequent[60] study of the 1-trifluoromethylallyl derivative suggest, especially in solvents of low ionizing power, the possibility of an essentially non-polar decomposition for the chlorosulphinate.

Decomposition of (+)-1-methylallyl chloroformate in four solvents gave the following amounts of excess retention in the 1-methylallyl chloride portion of the product, the remaining percentage of 1-methylallyl chloride being racemic: toluene (15%), methylene chloride (25%), n-butyl ether (28%) and dioxan (49%). 1-Methylallyl chloroformate decomposed slightly faster than 3-methylallyl chloroformate in both n-decane and methylene chloride.

Striking evidence in favour of an ionic process came from a comparison of the rates of decomposition of 1-methylallyl chloroformate in a variety of solvents. The dielectric constants ranged from n-decane (2·0 at 20°) to nitrobenzene (34·8 at 25·0°) and, at 75·0°, the rate in nitrobenzene was about 10,000 times that in decane. The intermediate rates did not correlate perfectly with dielectric constant but it is well recognized that dielectric constant is only one of several factors controlling ionizing power. The general trend of increased reaction rate with increase in dielectric constant of the solvent can be considered as strong support for an ionization mechanism. In agreement with the conveniently measurable rates of allyl chloroformate decompositions resulting from formation of resonance stabilized carbonium ions, ethyl chloroformate was found to be extremely stable in methylene chloride at 75·0°.

Olivier and Young[23], bringing together qualitative data from the literature, noted a similarity between the relative rates of liquid-phase decomposition of ROCOCl and the stability of R^+ (as indicated by the behaviour of RCl in solvolysis[62]). Updating this sequence to include more recent data, we can express it: triphenylmethyl[63] > benzhydryl[63] > t-alkyl[64, 65, 66], α-phenylethyl[38] > 1-adamantyl[45] > 1-methylallyl[23] > 3-methyl-allyl[23] > benzyl[38], secondary alkyl[54] > primary alkyl[54] ≫ phenyl[37].

In a study of ionic decomposition processes, it would be advantageous to study t-alkyl chloroformates. t-Butyl chloroformate was found to decompose rapidly above 10° and to yield, in addition to carbon dioxide, isobutylene and hydrogen chloride[64], rather than t-butyl chloride. t-Pentyl chloroformate has been reported[66] to be slightly more stable, although it has not been isolated in pure form, but again decomposition would be expected to proceed with extensive, if not complete, elimination.

The bridgehead tertiary chloroformate, 1-adamantyl chloroformate, has desirable characteristics for a study of ionic decomposition in that it has a tertiary structure and, due to substitution being at a bridgehead involving reasonably small (six-membered) rings, the decomposition product will be exclusively 1-adamantyl chloride. Also, rearside nucleophilic participation by external chloride ion[53, 54] (an especially troublesome side-reaction in chlorosulphinate decompositions[60, 67]) is sterically excluded. The interpolated reaction rate[45], in nitrobenzene at 50°, was approximately 20 times that measured for 1-methylallyl chloroformate[23] and, consistent with a slightly more polar transition state, the rate variation in going from decane to nitrobenzene, at 54°, was 205,000 for 0·06M solutions. Indicating considerable ionic character, even in a low polarity solvent, the rate in decane was quite sensitive to the concentration of the more polar 1-adamantyl chloroformate. The solvent influence ratio can be compared with a value[23] of approximately 10,000 for 0·1M solutions of 1-methylallyl chloroformate, at 75°. By interpolation, the corresponding (0·1M) 1-adamantyl chloroformate[45] ratio can be estimated as about 140,000, at 54°.

In benzene a very small amount of acid formation (0·5% at 54°) accompanied the decomposition[45]. Increased, but still small, amounts of acid formation accompanied decomposition in nitrobenzene (3·0% at 15°) and it appears that alkylation of the solvent can accompany the decomposition. Addition of silver hexafluoroantimonate, to remove chloride ion and prevent its return, led to the isolation of 1-(*m*-nitrophenyl)-adamantane. This experiment was based upon the preparation of 1-(*m*-nitrophenyl)-apocamphane by Beak and Trancik[68].

Addition of chloride ion as its tetra-*n*-butylammonium salt led to a modest increase in decomposition rate and the extent of solvolysis accompanying decomposition was reduced. The reduction was in accord with a competition between chloride ion and solvent for dissociated 1-adamantyl carbonium ions. It required, at 15°, a competition factor of 550 in favour of chloride ions relative to nitrobenzene molecules. In the absence of external chloride ion, it is reasonable to assume that the production of dissociated carbonium ions can be equated to solvolysis:

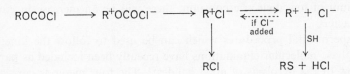

To summarize, the mechanisms proposed for decomposition in the liquid phase have been predominantly ionic in character. There does not appear to be any good evidence for four-centred covalent processes for either chloroformate or chlorosulphinate decompositions and $S_{N}i$ reactions appear to involve heterolysis of the carbon–oxygen bond followed by decomposition of the anion within a tight ion pair and internal return of chloride ion. For allylic systems, chloroformates appear to follow a similar pathway and internal return of chloride to the mesomeric carbonium ion yields appreciable amounts of both $S_{N}i$ and $S_{N}i'$ products. In contrast, allyl chlorosulphinates can follow a mechanism of this type in relatively polar solvents but in solvents of low polarity, and for 1-trifluoromethylallyl chlorosulphinate even in quite polar solvents, a six-membered cyclic covalent transition state appears to be favoured. Due to the greater strain involved in forming four-membered rings, the observation of covalent processes for $S_{N}i'$ but not $S_{N}i$ reactions is not surprising. It has been suggested that elimination reaction from secondary alkyl chlorosulphinates is ionic in character[69]. However, the recent[54] observation of a predominance of Hofmann (anti-Saytzeff) elimination in the decomposition of secondary alkyl chloroformates was explained in terms of a six-membered cyclic transition state resulting in a *cis*-elimination from the least sterically hindered carbon atom. It is, however, possible that the dominance of

Hofmann elimination could be a feature of rapid proton abstraction within an ion pair.

B. In the Gas Phase

Gas-phase reactions of chloroformate esters fit into a general pattern of 'molecular elimination reactions'. Maccoll and Thomas[70] have proposed a spectrum of polarities in the transition state, running from quasi-heterolytic in the case of alkyl halides to essentially symmetrical processes without charge development in the cases of substituted cyclobutanes. The concept of varying polarity has not yet been put on a quantitative basis, but one would predict that chloroformate ester decompositions would lie towards the quasi-heterolytic extreme. However, since there is evidence for a spectrum between covalent multicentred transition states and ionic processes for decompositions in the liquid phase, it is natural to suppose that in the absence of solvent, the ultimate condition for providing a medium of low dielectric constant, there will be an increase in covalent character.

Experimental procedures which can be used to follow the kinetics of gas-phase molecular eliminations have recently been included as part of a review of the pyrolysis of alkyl halides[71]. The four methods currently in use can be classified as static, flow, shock tube and chemical activation; the first two of these have been applied to chloroformate ester decompositions.

The decomposition of chloroformate esters (molecular eliminations) can proceed, as in the liquid phase, to both substitution and elimination products.

$$
-\overset{\displaystyle |}{\underset{\displaystyle H}{C}}-\overset{\displaystyle |}{\underset{\displaystyle |}{C}}-O-\overset{\displaystyle O}{\underset{\displaystyle \|}{C}}-Cl
\quad
\begin{cases}
\nearrow & -\overset{\displaystyle |}{\underset{\displaystyle H}{C}}-\overset{\displaystyle |}{\underset{\displaystyle |}{C}}-Cl \; + \; CO_2 \\
\\
\searrow & {>}C{=}C{<} \; + \; HCl \; + \; CO_2
\end{cases}
$$

Since elimination reactions usually have higher activation energies than the corresponding substitutions[72] and gas-phase decompositions are carried out at higher temperatures than in the liquid phase, elimination in the gas-phase constitutes a greater proportion of the overall decomposition. The formation of three molecules per chloroformate molecule in elimination but only two in substitution is a factor which has to be taken into account if pressure measurements are used to follow the kinetics.

In a discussion of measurements which have been carried out, it may be advantageous first to mention some recent observations by Lewis and Witte[73]. They used a diffusively stirred flow system[24], with nitrogen carrier gas, and analysed the steady-state effluent. They studied benzyl and allylic chloroformates. Radical chain processes[74] were ruled out for allyl chloroformate by showing an insensitivity to added nitric oxide[75]. Only allyl chloroformate decomposed with exhibition of a linear Arrhenius plot, and even this was curved at the lower temperatures. Allyl chloroformate was also the only compound insensitive to seasoning effects. Benzyl, 1-methylallyl and 3-methylallyl all gave markedly curved Arrhenius plots, indicating a mixture of a low-activation energy heterogeneous reaction and a high-activation energy homogeneous reaction. The worst behaviour was for 1- and 3-methylallyl chlorides which interconverted and decomposed to butadiene with rate coefficients almost independent of temperature; these were believed to be entirely heterogeneous reactions.

Lewis and Witte concluded that their results 'show the importance of heterogeneous processes and the severe limitations inherent in these studies of very polar materials in the gas phase, which have convinced us that further work on chloroformates is unpromising'.

This warning is to be borne in mind in the following review of earlier literature. However, the pessimism is probably not completely justified. The heterogeneous component was greatest for the reactions of 1- and 3-methylallyl chlorides and yet 1-methylallyl chloride has been found to decompose essentially homogeneously, and with a linear Arrhenius plot, when static pressure increase measurements were used to follow the extent of decomposition in a vessel coated with a thin carbonaceous layer[76]. The specific rate at 346° was 1.7×10^{-4} s^{-1}, as compared to an essentially identical value[73] for heterogeneous reaction at 245°. Also, 3-methylallyl chloride[77], 3,3-dimethylallyl chloride[78] and 1,1-dimethylallyl chloride[78] dehydrohalogenations have been successfully studied by the static method. The disubstituted compounds decompose much faster than they interconvert and, at 294°, the 3,3-disubstituted compound decomposed twice as fast as the 1,1-disubstituted. This is contrary to the relative solvolysis rates and the transition state was believed to have more multi-centred (six-centred) covalent character than is usual for dehydrohalogenations.

Since the static method can be applied to at least one homogeneous decomposition of a polar substrate for which the flow method failed miserably, there is every reason to believe that homogeneous reactions of chloroformates, which made at least some contribution during the flow reactions, could be successfully examined in seasoned vessels under static conditions. There would, of course, be some loss of versatility as regards simultaneous analysis for several products.

Study of one alkyl chloroformate, isobutyl chloroformate, accompanied the first systematic investigation of the pyrolysis kinetics of alkyl halides[79].

The static method, with clean glass vessels, was employed. In terms of the Arrhenius equation, $k = Ae^{-E/RT}$, a frequency factor (A) of 10^{13} s^{-1} and an activation energy (E) of 40 kcal/mole has been estimated[80] from the data reported at 267° and 302°.

It was found later that t-butyl chloride decompositions became reproducible only after decomposition products had placed a carbonaceous layer on the walls of the vessel[81]. However, Choppin, Frediani and Kirby reported[82] ethyl chloroformate decomposition to proceed quite smoothly at 150–195° by a homogeneous unimolecular mechanism. They reported only a slight effect on either packing the vessel or coating the walls. Their rates were considerably faster than those reported for isobutyl chloroformate; the rate at 175° was almost identical to that reported for isobutyl chloroformate at 302°. At low pressure, the rates fell off and an approach to second-order kinetics was observed. They reported a frequency factor of 5.5×10^{10} s^{-1} and an activation energy of 29.4 kcal/mole. The only products were ethyl chloride and carbon dioxide. Foreign gases, other than in maintaining the high pressure rate, had little effect[83].

Isopropyl chloroformate was similarly investigated[80] and was found to decompose in clean glass vessels, at least in part, by a heterogeneous reaction. Both substitution and elimination reaction were observed in the temperature range of 220–232°, with 45–52% of overall reaction being elimination. A flow method was developed, with nitrogen carrier gas, and used in the range 180–220°. In the flow experiments the reactor walls eventually became coated with decomposition products. However, unless there is an order of magnitude error in the reported data, it appears that at 220° the flow decomposition rate is some 30 times that of the static (which was already assumed to be in part heterogeneous)! Concurrent four- and six-membered ring decomposition processes were postulated for substitution and elimination and the more negative entropy of activation for elimination was considered to arise from a freezing of more degrees of freedom in six- rather than four-membered rings. The overall decomposition, as measured in the flow system, had a frequency factor of 3.1×10^9 s^{-1} and an activation energy of 26.4 kcal/mole.

A reinvestigation[84] of ethyl chloroformate pyrolysis showed that, at slightly higher temperatures, elimination accompanied substitution. A sensitivity to surface conditions was noted and 27–59% elimination was indicated. Although rate data were not quoted, the ability to follow the reaction kinetically at 330° was quite inconsistent with the Arrhenius parameters of the earlier work[82], which indicate a half-life of about 1 s at this temperature[24].

Prior to the next gas-phase chloroformate ester investigation[24], Maccoll and Thomas[85] pointed out the similarity between unimolecular solvolysis rates and pyrolysis rates for alkyl halides and, on this basis, they postulated considerable dipole development in the pyrolysis transition states. Indeed, Ingold[86] had taken this argument to the extreme and proposed the actual development of an intimate ion-pair intermediate. The relative rates observed[24] for decomposition of ROCOCl at 240°, as R is varied, were $CH_3 = 1$, $C_2H_5 = 2.2$, $i\text{-}C_3H_7 = 220$, $sec\text{-}C_4H_9 = 640$. The acceleration by α-substituents was considered to indicate a rather polar transition state with positive charge development on the α-carbon. The relatively slow decomposition[87] of trichloromethyl chloroformate, coupled with the formation of two molecules of phosgene rather than carbon dioxide plus carbon tetrachloride, was also rationalized on this basis. The experimental procedure involved a nitrogen carrier gas-flow technique. The rate for the ethyl compound was similar to that of Lessing[79] for the isobutyl but only 10^{-3} to 10^{-4} (depending on temperature) that of Choppin and co-workers[80, 82]. There was about a 10 kcal/mole higher activation energy for elimination as compared to substitution, for both isopropyl and sec-butyl chloroformates, coupled with a compensating more favourable entropy of activation of about 20 entropy units (approximately -20 e.u. for substitution and approximately 0 e.u. for elimination). Maccoll[88] noted that the relative rates of pyrolysis for ethyl, isopropyl and sec-butyl chloroformates paralleled very closely those of the corresponding chlorides and bromides and suggested 'an ion-pair transition state' (intermediate?) 'of the type $R^+CO_2Cl^-$ in the case of chloroformates, analogous to R^+X^- in the case of halides'.

Stereochemical studies[25] were carried out in the same flow apparatus as was used for kinetic studies. A slow flow rate was used to ensure extensive decomposition. Optically active sec-butyl chloroformate decomposed with clean retention of configuration. The alkene products from erythro- and threo-sec-butyl-3-d_1 chloroformates had deuterium contents[89] requiring a completely stereospecific cis-elimination. The stereochemistry was consistent with four- and six-membered cyclic transition states for substitution and elimination. However, such transition states do not appear to accord with the entropies of activation[24] of -17 e.u. for substitution and -1 e.u. for elimination. As pointed out earlier[80], more degrees of freedom should be frozen as the ring size of a cyclic transition state is increased. It was subsequently pointed out[73, 74] that the activation parameters could be in error.

It was further shown by Lewis and Herndon[26] that, at 240°, neopentyl chloroformate decomposed at roughly the same rate as ethyl chloroformate to yield unrearranged neopentyl chloride and methylbutenes, the latter requiring a rearrangement. Similarly 1-butyl chloroformate, at 255°, gave only 1-butyl chloride as the substitution product but both 1-butene and 2-butenes, the 2-butenes requiring rearrangement, as the elimination products. The substitution results are in marked contrast to liquid-phase decompositions[54], where neopentyl chloride, at 180°, gave no unrearranged products and decomposed relatively slowly compared to *n*-butyl chloroformate, at 150°, which gave as substitution product mainly (78%) rearranged *sec*-butyl chloride. The formation of unrearranged neopentyl chloride does not necessarily rule out an intimate ion-pair intermediate. Fraser and Hoffmann[90] have produced evidence that for neopentyl tosylate, in solution, the first formed ion pair is unrearranged and product formation from an unrearranged ion pair (for which collapse would be an extremely favoured reaction path in the gas phase) can compete with rearrangements. Recent views of intimate ion-pair intermediates have stressed the retention of some weak covalent binding[91, 92]. Rearrangements would require further separation, a process much more difficult in the gas phase than in solution. A mechanism for carbon dioxide loss to give substitution product with retention of configuration and without rearrangement, and which also incorporates the observed facilitation by alkyl groups could be as follows:

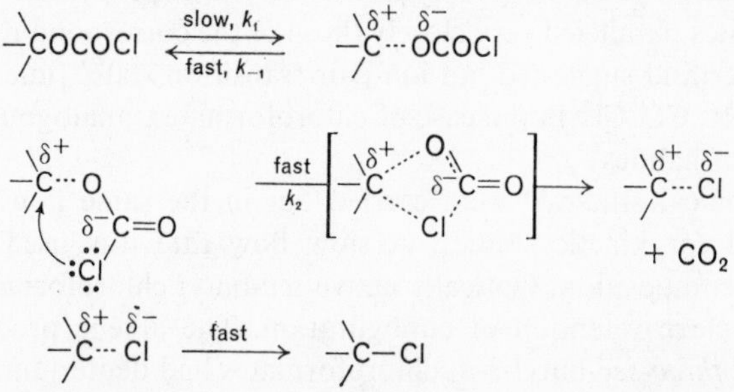

The above scheme treated by steady-state kinetics would give an experimental rate coefficient, $k = k_1 k_2 / (k_{-1} + k_2)$. If ion-pair return is dominant, $k_{-1} \gg k_2$ and $k = k_1 k_2 / k_{-1}$.

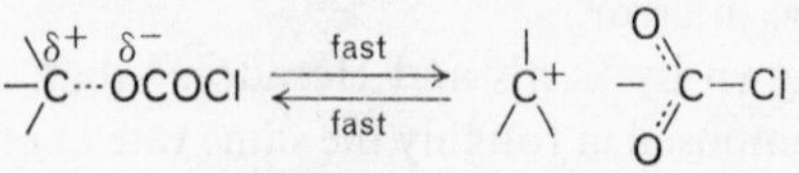

Presumably, when gas-phase elimination is after a rearrangement (and in the liquid phase), further separation of charge can occur with formation of a non-bonded carbonium ion. One driving force, compensating to some degree for increased charge development, will be the increased resonance energy of a fully symmetrical chloroformate anion. In the liquid phase, loss of carbon dioxide can lead to a new ion pair and allow collapse to rearranged chloride. In the case of 1-adamantyl chloroformate[45], there is compelling evidence for even further charge separation in fairly polar solvents.

The gas-phase loss of carbon dioxide from the fully developed anion, while producing a very stable molecule, would also require concentration of the negative charge on one atom, without any compensating solvation effects. A chloride-ion transfer reaction to a suitably situated hydrogen atom, with simultaneous ejection of carbon dioxide, could well be a more favoured pathway in the gas phase. The absence of transfer to the α-carbon, after rearrangement[26], can possibly be correlated with the stereochemically favoured pathway requiring additional charge separation over that required for elimination:

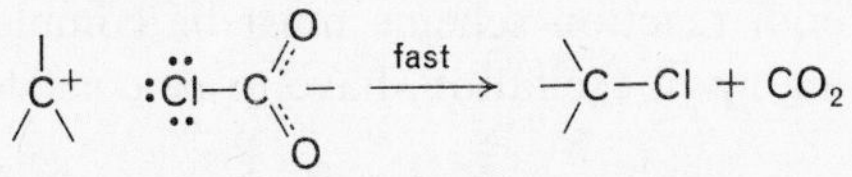

$$
\underset{\diagup \diagdown}{\overset{|}{C^+}} \quad \ddot{:}\overset{..}{\underset{..}{Cl}}-C\overset{O}{\underset{O}{\diagdown}} - \quad \xrightarrow{\text{fast}} \quad \underset{\diagup}{\overset{\diagdown}{C}}-Cl + CO_2
$$

It should be noted that, in the above schemes, intimate ion pair formation is rate-limiting and the products and their stereochemistries are governed by competition between fast processes.

An elimination scheme within the intimate ion pair (competitive with ion-pair return, formation of free ions and with the substitution process) can be visualized as follows:

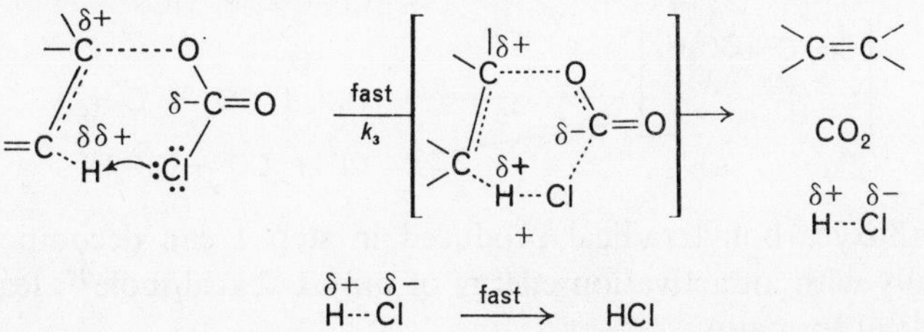

The recent investigation by Lewis and Witte was partially reviewed at the beginning of the section[73]. One additional feature of interest in this work was the observation of reversible allylic rearrangements of the

$$
\underset{\underset{OCOCl}{|}}{CH_3-CH-CH=CH_2} \rightleftharpoons CH_3-CH=CH-CH_2OCOCl
$$

chloroformate esters; these rearrangements are analogous to those observed for other allylic carboxylates[93-95].

IV. PHOTOLYSIS

There has been little work on this aspect of chloroformate ester chemistry. There is, however, a study[96] of the photolysis of various substituted ethyl formates ($RCO_2C_2H_5$) in hydrogen-donating solvents such as cyclohexane. This study included ethyl chloroformate (R = Cl). Electron-withdrawing substituents (R = CN or Cl) gave 15–20% yields of the ethyl carboxylate.

In a typical experiment, 38 g of ethyl chloroformate and 147 g of cyclohexane were mixed and irradiated, for 25 h, with a 350 W high-pressure mercury arc lamp. Gas (450 ml) was evolved, which was found to be a mixture of hydrogen chloride, carbon dioxide, ethane and carbon monoxide, plus small amounts of methane and ethylene. The liquid products were ethyl cyclohexanecarboxylate (15%, based on ethyl chloroformate consumed), cyclohexyl chloride (4·5%) and bicyclohexyl, plus small amounts of methyl cyclohexyl carbinol, diethyl oxalate and ethyl cyclohexane.

Clearly, the overall reaction scheme must be complex. The following four steps, following upon excitation, have to be considered:

$$Cl-\underset{\underset{O}{\|}}{C}-OC_2H_5 \xrightarrow[n \to \pi^*]{h\nu} \left[Cl-\underset{\underset{O}{\|}}{C}-OC_2H_5 \right]^*$$

$$\left[Cl-\underset{\underset{O}{\|}}{C}-OC_2H_5 \right]^* $$

$$Cl^{\cdot} + {}^{\cdot}\underset{\underset{O}{\|}}{C}-OC_2H_5 \quad (1)$$

$$Cl-\underset{\underset{O}{\|}}{C}^{\cdot} + {}^{\cdot}OC_2H_5 \quad (2)$$

$$HCl + CO_2 + C_2H_4 \quad (3)$$

$$Cl^{\cdot} + CO + {}^{\cdot}OC_2H_5 \quad (4)$$

The ethoxycarbonyl radical produced in step 1 can decompose exothermically with an activation energy of only 1–2 kcal/mole[97], leading to ethyl radical formation.

The route to ethyl cyclohexanecarboxylate would appear to involve step 1, followed by abstraction of a cyclohexane hydrogen by a chlorine atom to give a solvent radical which can combine with the ethoxycarbonyl radical. Alternatively, subsequent to step 1, two solvent radicals can dimerize, the solvent radical can combine with a chlorine atom, or the solvent radical can combine with an ethyl radical (formed by loss of carbon dioxide from the ethoxycarbonyl radical). Minor products may arise following steps 2, 3 or 4.

A study[96] was also made of phenyl chloroformate photolysis. Conditions were similar to those just described for ethyl chloroformate, but a 1 kW high-pressure mercury arc lamp was used for 7·2 h. Gaseous products were hydrogen chloride and carbon monoxide. The main 'liquid product' was phenol, and only traces of phenyl cyclohexanecarboxylate were obtained. The main homolysis step is believed to be step 4 (possibly step 1 followed by loss of carbon monoxide), to give a resonance-stabilized phenoxy radical.

In a different type of study, photolysis has been used to form the parent chloroformic acid as an unstable intermediate, which then decomposes to carbon dioxide and hydrogen chloride. The original technique[98] employed was photolysis of chlorine and formic acid mixtures by use of a 500 W tungsten lamp. The high quantum yields indicated a chain mechanism which most logically proceeded through chloroformic acid.

$$Cl_2 \xrightarrow{\ h\nu\ } 2Cl^{\cdot}$$

$$Cl^{\cdot} + HCOOH \longrightarrow HCl + {}^{\cdot}COOH$$

$${}^{\cdot}COOH + Cl_2 \longrightarrow ClCOOH + Cl^{\cdot}$$

$$ClCOOH \longrightarrow HCl + CO_2$$

also

$$COOH \longrightarrow CO_2 + H^{\cdot}$$

$$H^{\cdot} + Cl_2 \longrightarrow HCl + Cl^{\cdot}$$

$$2Cl^{\cdot} \longrightarrow Cl_2$$

The first direct evidence for chloroformic acid formation is due to Herr and Pimentel[99, 100]. Flash photolysis with a unit capable of dissipating 1500 J in about 30 μ s produced about 300 μ mole of quanta in the spectral range 5000–2000 Å. The flash was followed by rapid-scan i.r. spectroscopy. A new absorption at 768 cm^{-1} was observed in the first spectral trace (70 μ s after the flash) and this disappeared within 3·6 ms.

A study of the 768 cm^{-1} band under higher resolution allowed a determination of its kinetic behaviour as a function of temperature[101]. The decomposition was unimolecular over a temperature range of 15–70° and the frequency factor was 5×10^{13} s^{-1} and the activation energy was 14 kcal/mole.

A mechanism with a rate-determining *cis–trans* conformation change was proposed. The activation energy of 14 kcal/mole is not too far removed from the 17 kcal/mole proposed as a barrier to rotation in formic acid[102].

Support for the assignment of the 768 cm^{-1} band as a C—Cl stretching motion came from the band width and, more directly, from the absence of frequency shift on deuteration.

V. PHYSICAL PROPERTIES AND CONFORMATION

References to reports of basic physical properties of chloroformate esters (such as density, refractive index, molar refractions and parachors) have been given in an earlier review[7]. This review also included three references to studies of i.r. spectra and four references to Raman spectra studies. The references to spectral studies need updating in the light of more recent work, mostly carried out with instrumentation of increased sophistication.

As part of a broadly based study of carbonyl group vibration frequencies and intensities in various compounds, methyl, ethyl, n-butyl and isobutyl chloroformates were investigated in carbon tetrachloride solution[103]. During a study of donor characteristics of the carbonyl group, an i.r. study was made of methyl chloroformate in both the gas phase and in dilute carbon tetrachloride solution[104]. As part of a study of frequency correlations for C—H and C=O bands, concentrating mainly on thiolesters, a very accurate study was made of methyl chloroformate, vinyl chloroformate and methyl thiochloroformate in dilute carbon tetrachloride solutions[105]. The ethyl chloroformate absorption spectrum in the 2000–700 cm^{-1} range has been presented in detail[106].

The most detailed i.r. study that has been made for a chloroformate ester appears to be that of Collingwood, Lee and Wilmshurst[107] for methyl chloroformate (and also for dimethyl carbonate). The absorption spectrum was studied in detail for the 4000–400 cm^{-1} range, both in the vapour and in solution. Carbon tetrachloride was used as the solvent for the 4000–1400 cm^{-1} range and carbon disulphide for the 1400–400 cm^{-1} range. The Raman spectrum of the liquid was also obtained. It is generally believed that there is a barrier to rotation around the carbonyl carbon–ether oxygen bond in chloroformate esters and there have been several attempts to determine whether the molecule normally exists in the *cis* or the *trans* conformation.

cis conformation   *trans* conformation

It was predicted that both *cis* and *trans* conformations should give rise to three types of bands in the i.r. However, these bands should have distinguishable characteristics for the different conformations. The vapour spectrum showed the three types of bands expected from molecules in the *cis* conformation and none of the three expected for the *trans* conformation. Assuming the molecule to be in the *cis* conformation, it was then possible completely to assign the fundamental modes, other than torsional modes.

The *cis* conformation had previously been indicated by an electron diffraction study[108] of methyl chloroformate. This gas-phase investigation, using vapour from liquid samples at or below room temperature, was part of a wider study which also included methyl formate and methyl acetate. All three esters were found to have approximately planar heavy-atom skeletons with the ester methyl group *cis* to the carbonyl oxygen atom. Some rotatory oscillation of the methoxy group relative to the carbonyl group was indicated. For methyl chloroformate, the average dihedral angle of rotation from the planar conformation was 20° and limiting values for the average were 0° and 30°. Bond distances and bond angles were reported for all three molecules. For methyl chloroformate, bond distances were as follows (the error values are only rough estimates): $C=O$, 1.19 ± 0.03 Å; $C(carbonyl)-O(ether)$, 1.36 ± 0.04 Å; $C(methyl)-O$, 1.47 ± 0.04 Å; $C-Cl$, 1.75 ± 0.02 Å. Bond angles were as follows: $O=C-O$, $126 \pm 4°$; $C-O-C$, $111 \pm 4°$; $Cl-C-O$, $112 \pm 3°$. The belief that the *cis* conformation arises, at least in part, due to resonance was supported by the marked shortening of the $C(carbonyl)-O(ether)$ bond and the apparent widening of the $C-O-C$ angle. Similar effects were observed for methyl formate and methyl acetate.

Electron diffraction has also been used to investigate the structure of monochloromethyl chloroformate[109]. The bond distances and angles (and error estimates) were essentially identical to those for methyl chloroformate with the exception of the $C(carbonyl)-O(ether)$ and $C(chloromethyl)-O$ distances which were both 1.40 ± 0.04 Å, as opposed to 1.36 ± 0.04 and 1.47 ± 0.04 for the corresponding distances in methyl

14

chloroformate[108]. A *trans* arrangement of the chlorine atom of the chloro-methyl group and the carbonyl carbon atom was observed. Monochloro-methyl chloroformate was thought to have less C(carbonyl)—O(ether) double-bond character and less hindrance to libration than methyl chloro-formate. The increased libration and the shortening of the C(chloro-methyl)—O bond were tentatively explained by a contribution to the resonance hybrid of a structure of type

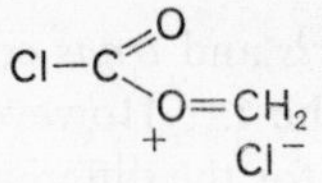

$$Cl-C{\overset{\displaystyle O}{\underset{\underset{Cl^-}{+}}{\diagdown O=CH_2}}}$$

Contrary to a statement in an earlier review[110], this paper does not report a *trans* arrangement of the chloromethyl group and the carbonyl oxygen. The inclusion of structures such as the one drawn above as contributors to the resonance hybrid confirms that a *cis* conformation, similar to that reported for methyl chloroformate[108], is assumed.

Dipole moments can be a powerful tool in investigating molecular structure, including conformational effects. There have been several reviews of the theory, experimental procedures and applications of this technique[111]. However, as LeFèvre has pointed out[112], the best success has been obtained when quantitative interpretation is not required, and a choice can be made between two alternatives with dipole moments which differ greatly and in an obvious manner. At the other extreme, the complex conformations of many organic molecules make an interpretation of their dipole moments extremely difficult. Chloroformate esters appear to lie between these two extremes and, although accurate measurements are now available, different schools have interpreted the dipole moments in terms of both of the two basic conformations (*cis* or *trans*).

Following early and rather inaccurate reports[113, 114, 115], the first detailed study was by Mizushima and Kubo[116]. They investigated several esters, including methyl and ethyl chloroformates, over a fairly wide temperature range (about 180°) in the gas phase and recorded the variations of dipole moment. For methyl acetate they obtained a constant value of 1·696 D within the range 35–210°, in good agreement with a previous determination by Zahn[117]. They concurred with Zahn[117, 118] that this implied a structure of the type

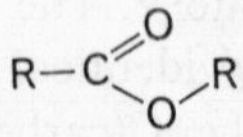

$$R-C{\overset{\displaystyle O}{\diagdown O}}R$$

with a *cis* arrangement of methoxyl methyl and carbonyl oxygen. The carboxylic resonance in methyl acetate was assumed to be sufficiently pronounced for the double-bond character of the C—O bond severely to restrict rotation. With dimethyl carbonate the dipole moment increased slightly with temperature and this was interpreted in terms of a lower barrier to C—O rotation. Due to the lower barrier, at elevated temperatures the population of other conformations, with higher dipole moments, was increased.

Substituting chlorine, the dipole moments of both methyl and ethyl chloroformates were examined as a function of temperature. In both cases, the dipole moment fell as the temperature increased. At 35°, the values were 2·38 and 2·56 D for methyl and ethyl chloroformates and these values fell to 1·68 and 1·43 D at 207°. This was interpreted in terms of the molecule being, at lower temperatures, primarily in a *trans* conformation. Increased amounts of rotation were assumed to result as the temperature was raised. The relative dipole moments of *cis* and *trans* isomers were obtained by vector addition of bond moments and, for methyl chloroformate, estimates were given of 2·15 D for the *trans* and 1·58 D for the *cis* conformations. These estimates were admitted to be crude, since they neglect mutual effects between bond moments as well as (possibly large) resonance effects. The conclusion that the molecule exists at low temperatures in the *trans* conformation is, of course, contrary to the later i.r.[107] and electron diffraction[108] evidence.

LeFèvre and Sundaram[119] measured the dipole moments and molar Kerr constants for several methyl esters, including methyl chloroformate, in benzene. They concluded that there was nothing especially unusual as regards the methyl chloroformate structure. Both polarity and polarizability considerations indicated in all cases that, while the effective conformation was not exactly *cis* (as regards methyl group and carbonyl oxygen), a 30° rotation from the *cis* structure could put *a priori* calculations in accord with observations. They assumed the conformations to be very similar in both the vapour phase and in solution. The dipole moment for methyl chloroformate in benzene at 25° was reported to be 2·37 D.

In 1967, the dipole moments were reported[120] for methyl, phenyl and the C_2–C_5 primary alkyl chloroformates in dilute benzene solutions at 25°. The value for methyl chloroformate of 2·38 D was in excellent agreement with the earlier value[119] of 2·37 D. The C_2 through C_5 primary alkyl chloroformates had values of 2·66, 2·70, 2·71 and 2·71 D respectively. Phenyl chloroformate had a value of 2·39 D. The small value for phenyl chloroformate, relative to the primary alkyl chloroformates, was believed to be due to the electron-withdrawing properties of the phenyl group.

An attempt was made to determine the conformation for methyl chloroformate on the basis of average bond dipole moments reported by Smith[121] and the structure data reported for methyl formate[122]. Vector addition gave values of 1·6 D for the *cis* conformation and 2·0 D for the *trans* conformation. This approach parallels the one carried out thirty years earlier[116] when corresponding values of 1·58 D and 2·15 D were obtained. Not surprisingly, the Canadian workers concluded, as the Japanese workers had earlier, that the most probable conformation is with the methyl group *trans* to the carbonyl oxygen.

The Australian school, led by LeFèvre, has recently attacked[123] the conclusions reached by Bock and Iwacha[120] (who overlooked the 1962 paper[119] of the Australian school) with much vigour. They express their belief that underlying uncertainties in the bond moments used in vector addition render the Bock and Iwacha procedure almost useless. They re-examined the results in benzene and obtained new data with carbon tetrachloride as solvent. The dipole moments (μ) and molar Kerr constants of solutes at infinite dilution, $_\infty(mK_2)$, are given in Table 1.

TABLE 1. Dipole moments (μ) and molar Kerr
constants at infinite dilutions[123]

Solute	Solvent	μ (D)	$10^{12} \, _\infty(mK_2)$
MeOCOCl	CCl_4	2·32	71·4
MeOCOCl	C_6H_6	2·35	104
EtOCOCl	CCl_4	2·58	117
EtOCOCl	C_6H_6	2·62	157
PhOCOCl	CCl_4	2·28	169
PhOCOCl	C_6H_6	2·33	184

Solute–benzene interactions do not have any appreciable effect upon the dipole moment measurements which are virtually identical in benzene and the relatively inert carbon tetrachloride. The molar Kerr constants at infinite dilution are, however, consistently greater in benzene than in carbon tetrachloride and this can most plausibly be attributed to the formation of transient, stereospecific, solute–benzene complexes with Kerr constants larger than for uncomplexed solute molecules.

In addition to uncertainties as regards the relevant bond moments to be subjected to vector addition, the inductive and electromeric effects were also supposed to render this simplified addition procedure invalid. An alternative approach was used. Microwave spectroscopy studies[122] of methyl formate have indicated a dipole moment of 1·77 D and have shown the location of the dipole moment relative to a required *cis* conformation. This dipole moment was combined[123] with a C—Cl bond moment of 1·25 D (from experimental moments of acetyl chloride[124] and phosgene[125]) and, assuming a Me—O bond moment of 1·14 D, consideration of rotation around the C(carbonyl)—O(ether) bond gave calculated dipole moments of 2·29 D for the *cis* and 2·60 D for the *trans* conformations. This is consistent with a *cis* or near *cis* disposition of methyl group and carbonyl oxygen in methyl chloroformate (2·32 D in CCl_4 and 2·35 D in C_6H_6). However, the variation of dipole moment with angle of rotation is quite small and a detailed structure could not be assigned. One experiment which does not seem to fit the *cis* conformation proposal is the *reduction* in dipole moment of both methyl and ethyl chloroformates with increase in temperature, which was observed by the Japanese workers[116]. It seems generally agreed[116, 120, 123] that the *trans* conformation will have the higher dipole moment and any change from a *cis* conformation should lead to an *increase* in dipole moment.

In summary, for methyl chloroformate, i.r.[107] and electron diffraction studies[108] indicate a basically *cis* arrangement of methyl group and carbonyl oxygen. This is consistent with the structures proposed for many other esters, where the assignment of conformation is on a sounder footing, including microwave spectroscopy evidence[122]. Dipole moment results, although now well established experimentally for dilute solutions, have been interpreted by different groups of workers as supporting both *trans* and *cis* conformations. Difficulties in interpretation arise due to inherent uncertainties in predicting dipole moments as a function of structure, coupled with a fairly low sensitivity of the dipole moment to rotation about the C(carbonyl)—O(ether) bond[123]. Since there is general agreement[116, 120, 123] that the *trans* conformation will have the higher dipole moment, the *reduction* in dipole moment observed in the gas phase on increasing the temperature[116] does seem to present some evidence in favour of a stable (low temperature) *trans* conformation. On balance, at the present time, the structure of methyl chloroformate, in both the vapour phase and solution[119], appears best represented[108] by a *cis* conformation, with an approximately 20° effective rotation around the C(carbonyl)—O (ether) bond and with some rotatory oscillation.

VI. NUCLEOPHILIC SUBSTITUTION REACTIONS

A. General Considerations

Nucleophilic substitution reactions of chloroformate esters (ROCOCl) formally parallel those of other types of carboxylic acid esters. However, acid-catalysed substitutions are not of any major importance and mechanisms to be given consideration are S_N1 or S_N2 ($B_{AL}1$ or $B_{AL}2$[126])

attack at the α-carbon of the R group and unimolecular or bimolecular substitution at the acyl carbon ($B_{AC}1$ or $B_{AC}2^{126}$). These substitutions can be complicated by concurrent or subsequent loss of carbon dioxide.

As is generally the case for carboxylic acid esters[126], bimolecular attack occurs most readily at the acyl carbon and special conditions are needed to see $B_{AL}2$ attack. For methyl benzoate, a $B_{AL}2$ (S_N2) attack was observed by use of methoxide ion as the nucleophilic reagent[127]. The concurrent, and faster, $B_{AC}2$ attack merely regenerated substrate and dimethyl ether was isolated in 74% yield. Observation of $B_{AL}2$ attack with chloroformate esters would require chloride ion as the nucleophile. An experiment of this type has been carried out in acetonitrile using tetra-ethylammonium chloride[53]. A parallel study was carried out using tetra-ethylammonium nitrate, a salt containing a considerably weaker nucleophile. The second-order rate coefficients, reported in Table 2, show clearly the $B_{AL}2$ characteristics of attack by chloride ion and the $B_{AC}2$ characteristics of attack by nitrate ion.

TABLE 2. Second-order rate coefficients (l/mole–s) for nucleophilic substitution reactions[53] of chloroformate esters (ROCOCl) with Cl^- and with NO_3^- in acetonitrile, at 25·0°

R:	Me	Et	n-Pr	isoPr	isoBu	Ph
$10^5 k_2^{Cl^-}$	164	6·2	5·8	0·75	0·26	a
$10^5 k_2^{NO_3^-}$	20	10·6		3·5	11·7	670

a No detectable reaction.

The pathway for unimolecular reaction appears to be more delicately balanced and it is, of course, a function only of the structure of the chloroformate ester and its environment. An additional complication is that the formation of an R^+ carbonium ion can be either direct or by formation and subsequent decomposition of a carboxylium ion ($ROCO^+$)*. Possible schemes leading to R^+ are formulated below:

$$ROCOCl \xrightarrow{\text{slow}} ROCO^+ + Cl^- \xrightarrow{\text{fast}} R^+ + CO_2 + Cl^-$$
$$ROCOCl \xrightarrow{\text{slow}} R^+ + CO_2 + Cl^-$$
$$ROCOCl \xrightarrow{\text{slow}} R^+ + OCOCl^- \xrightarrow{\text{fast}} R^+ + CO_2 + Cl^-$$

* This is the nomenclature used by Beak (references 196, 198, 199). While 'alkoxycarbonyl ion' would be more precise in the present instance, 'carboxylium ion' can be used in a general sense for alkoxy-, aryloxy- or arylalkyloxy-substituents.

On the other hand, reactions which show unimolecular characteristics and which, also, give products with carbon dioxide retention must have proceeded through a carboxylium ion ($B_{AC}1$ mechanism). Unfortunately,

$$ROCOCl \xrightarrow{\text{slow}} ROCO^+ + Cl^- \xrightarrow[\text{fast}]{Y^-} ROCOY + Cl^-$$

it appears that studies of this type have usually been carried out under conditions where the ROCOY product is itself unstable. There is, however, stereochemical evidence for carboxylium ion formation in the hydrolysis of secondary alkyl chloroformates[128]. Solvolysis of the tertiary alkyl chloroformate, 1-adamantyl chloroformate, is conveniently explained in terms of a $B_{AL}1$ (S_N1) type mechanism[45], since no ROCOY type product is formed under conditions where this was shown to be stable. Any carboxylium ion formed would be required to decompose appreciably faster than it reacted with the alcohol solvent, to be able to invoke it as an intermediate.

Chloroformates have been extensively used to study substitution reactions involving replacement of chlorine attached to an acyl carbon. The reason for this is that acid chlorides (RCOCl) tend to react too fast for the kinetics to be followed by conventional techniques. Insertion of an oxygen atom between the R group and the acyl carbon slows down the reaction considerably and frequently places it within a velocity range convenient for study. For example[129], in 89.1% by volume acetone (10.9% water) at $-20°$, acetyl chloride (CH_3COCl) solvolyses with a specific rate of $10.9 \times 10^{-4}\ s^{-1}$ and methyl chloroformate with a specific rate of $0.0012 \times 10^{-4}\ s^{-1}$, a rate difference by a factor of about 10^4. At least two factors appear to be responsible for the slow nucleophilic substitution rates of chloroformate esters.

Bimolecular reaction rates for attack at acyl carbon are sensitive functions of the electron deficiency at the acyl carbon. In the presence of an attached ether linkage the electronegativity of the carbonyl oxygen can be countered by conjugative release from the ether oxygen. This can be

regarded as giving initial state stabilization of a type not possible for related carboxylic acid chlorides.

Unimolecular reaction rates are favoured by electron release. The conjugative (mesomeric) stabilization of a carboxylium ion will, to some extent, be counterbalanced by the initial state conjugation mentioned above. There will also be a very powerful electron-withdrawing (destabilizing) inductive effect of the attached alkoxy group. This electron-withdrawing inductive effect will be moderated by electron release from the R group.

In discussing nucleophilic substitution reactions of chloroformate esters, it must be borne in mind that in many respects the mechanisms have features common to substitution reactions of carbonyl halides in general. These general features are discussed in another chapter of this volume. Emphasis will be placed upon aspects of chloroformate ester substitutions which differ in character from those for related replacements within carboxylic acid chlorides.

B. Solvolytic Reactions

It was realized quite early in the nineteenth century that chloroformate esters hydrolyse to give the corresponding alcohol or phenol and hydrogen chloride, with liberation of carbon dioxide. Complications can arise with alkyl chloroformates because the alcohol formed in the hydrolysis is also able to react with further chloroformate to give as a side-product the disubstituted carbonate. It was also discovered, very early, that alkenes can be formed during the hydrolysis.

Aliphatic alcohols were found to react readily with chloroformate esters to yield disubstituted carbonates and hydrogen chloride. Thiols gave corresponding monothiocarbonates. Phenols were found to be relatively unreactive, even at elevated temperatures. Addition of base to the phenol and chloroformate ester led smoothly to high yields of the aryl alkyl or diaryl carbonate. From reactions with carboxylic acids, esters were isolated.

References to this early work can be found in the review of Matzner, Kurkjy and Cotter[7].

In 1937, Leimu reported[130] a kinetic study of the methanolysis of several chloroformate esters. The relative rates obtained as R was varied

$$ROCOCl + MeOH \longrightarrow ROCOOMe + HCl$$

($ClC_2H_4 > CH_3 > C_2H_5 > i\text{-}C_3H_7$) were consistent with a bimolecular mechanism. A study[131] of the hydrolysis of ethyl chloroformate in 100% water gave a specific rate of $2\cdot1 \times 10^{-4}\ s^{-1}$. Investigations in three mixed aqueous-organic solvents, of varying composition, were also reported.

A study of the hydrolysis of several acid chlorides, including ethyl chloroformate, was made by Hall[132]. He obtained an order of relative rates: $CH_3COCl > C_6H_5COCl > C_2H_5OCOCl < (CH_3)_2NCOCl$. He concluded that the carboxylic acid chlorides and ethyl chloroformate hydrolysed by a bimolecular mechanism and the dimethylcarbamyl chloride by rate-determining ionization, with liberation of a chloride ion. The hydrolysis of ethyl chloroformate in pure water had an activation energy of 19·0 kcal/mole and an entropy of activation of $-12\cdot4$ e.u. The entropy of activation contrasted sharply with a value of $+5\cdot6$ e.u. for dimethylcarbamyl chloride hydrolysis in pure water and the values were nicely consistent with bimolecular and unimolecular processes. Confirmation of the duality in mechanism came from a study with 2-substituted piperidines in the low-ionizing power solvent benzene[133]. Under these bimolecular conditions, ethyl chloroformate reacted considerably faster than dimethylcarbamyl chloride.

Bimolecular hydrolysis of ethyl chloroformate would be expected to lead to ethyl hydrogen carbonate (C_2H_5OCOOH) and this compound has been shown[134] to decompose rapidly to yield ethanol and carbon dioxide. This decomposition can be pictured[135] as occurring by the following pathway:

$$C_2H_5OCOOH \rightleftharpoons C_2H_5-\overset{\oplus}{\underset{H}{O}}-CO_2^- \longrightarrow C_2H_5OH + CO_2$$

Crunden and Hudson[135] studied the solvolysis of methyl, ethyl and isopropyl chloroformates in both 65% aqueous acetone (35% water) and in formic acid containing 1% water. In the aqueous acetone, they observed the rate sequence: $CH_3 > C_2H_5 < i\text{-}C_3H_7$. In moist formic acid they obtained the rate sequence: $CH_3 < C_2H_5 \ll i\text{-}C_3H_7$. Addition of sodium formate to the solvolysis in formic acid increased the rate by 27% for ethyl chloroformate and by about 7% for isopropyl chloroformate, as compared to a 1000-fold increase for addition to p-nitrobenzoyl chloride solvolysis[136]. Similarly, ethyl and isopropyl chloroformates were insensitive to a variation in the water content within the range 1–5%. Solvolysis rates for p-nitrobenzoyl chloride were proportional to the water content in this range[136]. It was concluded that, in the aqueous acetone, methyl and ethyl chloroformates solvolyse by bimolecular mechanisms but the isopropyl chloroformate predominantly by a unimolecular mechanism. In moist formic acid, it was concluded that the unimolecular mechanism was dominant. However, the small rate decrease from ethyl to methyl would seem to indicate an appreciable bimolecular contribution to the solvolysis of methyl chloroformate, even in moist formic acid.

A stereochemical study[135] of the solvolysis of optically active 2-octyl chloroformate in 98% formic acid at 21° led to the formation of 2-octyl formate with almost complete inversion of configuration. While the

solvolysis of secondary alkyl chloroformates appears to proceed primarily
by a unimolecular mechanism the evidence produced by Crunden and
Hudson did not allow a distinction to be made between the following
mechanisms:

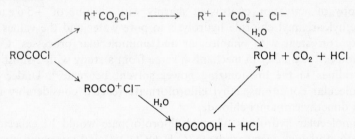

The path involving nucleophilic capture of the carboxylium ion appeared
unlikely in 98% formic acid since inversion and not retention was observed.
However, to give the formate (rather than the alcohol) the reaction would
have to go through a mixed anhydride, in place of the hydrogen carbonate,
and inversion could possibly be an unusual characteristic of the decomposition of mixed anhydrides of formic acid. Mixed anhydrides of carboxylic acids usually decompose with retention of configuration[137]. Two
plausible pathways not considered would be a concerted ionization–
decomposition to give directly the carbonium ion, the chloride ion and
carbon dioxide and an S_N2 type nucleophilic attack on the carboxylium
ion to lead to inverted products.

To clarify the mechanism of solvolysis for secondary alkyl chloroformates, further work was carried out by Green and Hudson[138]. They studied
the rates of solvolysis of cyclopentyl, cyclohexyl, cycloheptyl and cyclooctyl
chloroformates and found each of them to react at a similar rate in formic
acid and in 65% aqueous acetone. They all reacted slightly faster than
isopropyl chloroformate and there was a gradual increase in rate from C_5
to C_8. This behaviour contrasted sharply with the relative rates of S_N1
reactions of cycloalkyl sulphonates[139, 140] and halides[141, 142]. For both of
these types of compounds, the rate order $C_5 > C_6 < C_7 < C_8$ is well established and, also, the variations in rate are considerably larger than for
cycloalkyl chloroformate solvolyses. The differences in behaviour were
taken to indicate that, for chloroformate esters, the unimolecular reaction
involves a rate-determining ionization to give a carboxylium ion $(ROCO)^+$.
This would parallel the rate-determining step proposed by Hall for
dimethylcarbamyl chloride[132]. The previous stereochemical study of
2-octyl chloroformate in 98% formic acid was extended to 85% aqueous
acetone and it was found that the reaction led to 2-octanol which was

formed with 90% *retention* of configuration, as opposed to 90% *inversion* in the formolysis. Boiling the secondary chloroformates in a 1M phosphate buffer led to 30–60% alkene formation, except for *trans*-4-*t*-butyl-cyclo-hexyl chloroformate. For this latter compound the chloroformate group is in an equatorial position and $E2$ (but not $E1$) elimination will be greatly retarded[143].

The observations were correlated by the following scheme:

$$ROCOCl \rightleftharpoons ROCO^+ + Cl^- \tag{1}$$

$$ROCO^+ + H_2O \xrightarrow{E2} alkene + H_3O^+ + CO_2 \tag{2a}$$

$$H_2O + ROCO^+ \xrightarrow{S_N2} H_2\overset{+}{O}R + CO_2 \tag{2b}$$

$$ROCO^+ + H_2O \longrightarrow ROCO\overset{+}{O}H_2 \xrightarrow{H_2O} ROH + CO_2 + H_3O^+ \tag{2c}$$

$$H-C\overset{\displaystyle O}{\underset{\displaystyle O}{\big\langle}} + ROCO^+ \xrightarrow{S_N2} HCO\overset{+}{O}R + CO_2 \tag{2d}$$

$$ROCO^+ + HCOOH \longrightarrow ROCO-\overset{+}{\underset{H}{O}}COH \xrightarrow{:B} ROCOH + CO_2 + BH^+ \tag{2e}$$

The stereochemical results require that in formic acid S_N2 attack on the carboxylium ion (2d) dominates over nucleophilic addition to the carboxylium ion (2e) but in aqueous acetone the reverse is true and (2c) dominates over (2b). The kinetic results would also be consistent with a one-step ionization–decomposition for which, in the transition state, carbon–chlorine heterolysis was running ahead of carbon–oxygen heterolysis. However, such a scheme could not as conveniently explain the observed stereochemistries and the lack of elimination in the presence of a 4-*t*-butyl substituent within cyclohexyl chloroformate.

A study has been made[144] of the effect of a *para*-*t*-butyl group upon the rate of solvolysis of phenyl chloroformate in an aqueous–dioxan solvent at 25°. The rate of this bimolecular substitution is reduced, as one would predict, but the effect is quite modest. Introduction of the substituent lowers the specific rate by a factor of one-third.

Kivinen[145] has carried out an extremely detailed study of the kinetics of solvolysis of ethyl chloroformate in water, methanol, ethanol and several binary mixtures containing one of these three hydroxylic compounds as one component. The rate measurements in water agree fairly well with those of Hall[132] but rather poorly with those of Bohme and Schurhoff[131]. In water the entropy of activation is -24 e.u. (Hall[132] gives $-12\cdot4$ e.u.), in methanol $-34\cdot4$ e.u. (in agreement with a value calculated from the data of Leimu[130]) and in ethanol $-33\cdot7$ e.u. Kivinen concluded that the macroscopic kinetic behaviour was very similar to that for bimolecular solvolyses at the saturated carbon of alkyl halides and he expressed a belief that the bimolecular solvolysis of ethyl chloroformate and other acid chlorides can best be represented by an S_N2 rather than an addition–elimination mechanism. Acid chloride solvolysis mechanisms are considered elsewhere in this volume.

Kivinen[146] has also studied the specific rate of solvolysis of ethyl chloroformate in water and deuterium oxide and he reports values, at 25°, of $38·9 \times 10^{-5}$ s^{-1} and $20·0 \times 10^{-5}$ s^{-1}, corresponding to a k_{H_2O}/k_{D_2O} value of 1·95. There has also been a report concerning the effect of pressure on the rate of hydrolysis of ethyl chloroformate in water at 0°[147].

Queen[148] has studied, in great detail, the hydrolysis of a series of alkyl chloroformates and phenyl chloroformate (and dimethylcarbamyl chloride) in water. By use of extremely sensitive temperature control and a precise conductometric technique for determining the extent of reaction, he was able to determine not only the heat ($\Delta H^{\ddagger}$) and entropy ($\Delta S^{\ddagger}$) of activation but also the heat capacity of activation ($\Delta Cp^{\ddagger}$). Studies in D_2O as well as in H_2O allowed solvent isotope effects to be determined. The k_{H_2O}/k_{D_2O} value for ethyl chloroformate of 1·82 was in fairly good agreement with the value of 1·95 previously reported by Kivinen[146]. The hydrolysis rates for ethyl chloroformate were in reasonable agreement with those of Hall[132] and Kivinen[145]. Data extracted from Queen's paper are presented in Table 3. Similar to the explanation previously given by Hall[132] for the positive

TABLE 3. Rate coefficients, activation parameters and solvent isotope effects for hydrolysis of chloroformate esters (ROCOCl) in water[148]

R	$10^4 k^a$ (s^{-1})	$\Delta H^{\ddagger}_{298}{}^b$	$\Delta S^{\ddagger}_{298}{}^c$	$\Delta Cp^{\ddagger}_{298}{}^d$	k_{H_2O}/k_{D_2O}
Me	1·23	16·2	−19·1	−36·0	1·89^e
Et	0·76	17·1	−17·0	− 3·5	1·82^e
Pr	0·85	17·1	−16·6	+ 2·2	—
isoPr	2·36	24·1	+10·1	− 9·1	1·25^e
Ph	34·6	14·1	−19·8	−42·2	1·79^f

^{a} Determined very accurately but at slightly varying temperatures, within the range $10 \pm 0·2°$.
^{b} kcal/mole.
^{c} cal/mole-deg.
^{d} cal/mole-deg.
^{e} At 25·0°.
^{f} At 7·5°.

entropy of activation for dimethylcarbamyl chloride solvolysis, Queen assumed the sudden increase in entropy for isopropyl chloroformate solvolysis to follow from a change in mechanism from bimolecular for phenyl, methyl and primary alkyl chloroformates towards unimolecular for secondary alkyl chloroformates. Also, there was a related discontinuity in the solvent isotope effect. This conclusion was consistent with the findings of Hudson and co-workers[135, 138] for solvolyses in aqueous acetone or formic acid. Queen proposed a mechanism proceeding through a tetrahedral intermediate for the bimolecular solvolyses, as opposed to the one-step S_N2 mechanism proposed by Kivinen[145]. For the solvolysis of isopropyl chloroformate, two concurrent mechanisms were proposed. One involves unimolecular fission of the acyl–halogen bond (as proposed by Green and Hudson[138]) and the other either S_N1 or S_N2 displacement at the alkyl group. It was emphasized, however, that other explanations are possible. Also, while the heat capacities of activation may ultimately be useful in assigning mechanisms to these solvolyses, there is need for more data from related systems to be available for comparison before more than very tentative assignments can be made on this basis.

Aryl chloroformates have been shown to be reactive towards dimethylformamide[149]. The reaction, to give aryloxy-substituted immonium salts, is believed to proceed as follows:

$$Ar-O-COCl + O=CH-N(CH_3)_2 \longrightarrow$$

$$Ar-\overset{\frown}{O}-\underset{\underset{O}{\|}}{C}-\overset{\frown}{O}-CH=\overset{\oplus}{N}(CH_3)_2 \longrightarrow ArO^- + CO_2 + Cl-CH=\overset{\oplus}{N}(CH_3)_2$$
$$\overset{\diagdown}{Cl^-}$$

$$Ar-O^- + Cl-CH=\overset{\oplus}{N}(CH_3)_2 \longrightarrow Ar-O-CH=\overset{\oplus}{N}(CH_3)_2 + Cl^-$$

The reaction sequence is initiated by a, presumably bimolecular, replacement of chloride by nucleophilic attack by the solvent.

C. Non-solvolytic Reactions

The products isolated from non-solvolytic nucleophilic attack on chloroformate esters show a similar duality in character to those isolated from solvolytic reactions. For bimolecular substitution reaction $(ROCOCl + Y^- \rightarrow ROCOY + Cl^-)$ the isolated product depends on the stability of ROCOY. With azide and fluoride, for example, stable azidoformates and fluoroformates can be isolated. With iodide or nitrate, for example, the nitratoformates and iodoformates are much less stable than the chloroformates, and loss of CO_2 leads to RY products. With nucleophiles of type OZ^-, loss of carbon dioxide from ROCOOZ can occur in either of two ways:

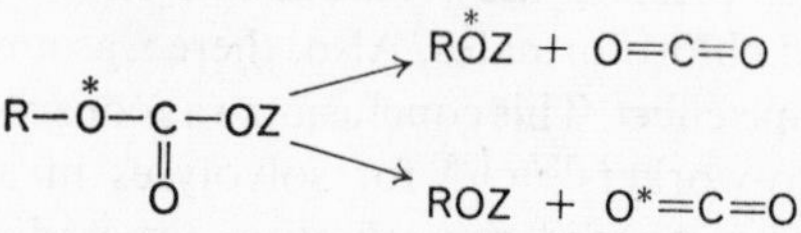

$$R-\overset{*}{O}-\underset{\underset{O}{\parallel}}{C}-OZ \begin{cases} \longrightarrow R\overset{*}{O}Z + O=C=O \\ \\ \longrightarrow ROZ + O^*=C=O \end{cases}$$

The R—O bond retention pathway can formally be considered to involve ionization of Z^+ and attack by Z^+ at the ether oxygen. The R—O bond fission pathway can formally be considered to involve ionization of OZ^- and attack at the α-carbon of the R group. If the α-carbon is asymmetric, optical activity studies can help to distinguish between these pathways.

A possible complication, believed to operate in the tertiary-amine-catalysed decomposition of chloroformate esters[50], is for the initially liberated chloride ion to attack at the α-carbon with ejection of the substituted formate ion (or its fragments).

For unimolecular reaction, ionization to give a carboxylium ion followed by a fast union with Y^- will lead to ROCOY and, ultimately, to the same products as for bimolecular attack. However, loss of carbon dioxide from the carboxylium ion, direct ionization to R^+ and $OCOCl^-$, or a concerted ionization–decomposition (to give R^+ and Cl^- directly) will lead to products derived from the carbonium ion. Also, formation of the carboxylium ion and attack at its α-carbon by Y^- (or Cl^-) will correspond to a fast S_N2 reaction subsequent to a slow unimolecular ionization.

Although formulated as negative, the nucleophile can, of course, be neutral (as in attack by tertiary amines), with corresponding charge differences in the above schemes. Also, in the general case, elimination reaction will accompany substitution.

It has been known for many years that the addition of hydroxide can strongly catalyse hydrolysis reactions[150] and, as one would predict, addition of other powerful nucleophiles greatly speeds up bimolecular reactions, but with diversion to non-solvolytic products.

Amines in aqueous solution react very rapidly with chloroformate esters. Hall[151] has been able to obtain semi-quantitative second-order rate coefficients for reactions of ethyl chloroformate with amines by working in sufficiently acid solutions. The buffers used were phenolsulphonic acids which had bulky groups ortho to the hydroxyl, so as to render the phenolic function inert to acylation by ethyl chloroformate. In these aqueous acidic solutions, the second-order rate coefficients (1/mole–s) for attack by free amine on ethyl chloroformate were approximately 10^4 for piperidine, 10^2 for 2-methylpiperidine and n-butylamine, 10 for sec-butylamine and 1 for t-butylamine. The most obvious characteristic was a very marked steric hindrance effect.

The reactions with phenols are less rapid but, in the absence of ortho-substituents, 10^{-2} to 10^{-1} M concentrations of phenoxide ion will dominate over the solvent water in reaction with ethyl chloroformate[152]. A study, at 0°, of the reaction of ethyl chloroformate with several para-substituted phenols has been made under conditions where the solution was initially adjusted to pH of 8·0 by addition of sodium hydroxide. The background solvolysis rate was $3·9 \times 10^{-5}$ s^{-1} and, as an example, with 0·055 M phenol added at pH 8·0 the specific rate of reaction of the ethyl chloroformate was 75×10^{-5} s^{-1}. Calculating the concentration of phenoxide ion at pH 8·0 and allowing for background solvolysis, second-order rate coefficients were calculated. It was found that a plot of the logarithms of these rate coefficients against the pK_a of the phenol was linear and the slope (α value) of this Brønsted relationship was 0·78.

It might be noted that in reactions involving p-nitrobenzyl chloroformate under quite strongly alkaline conditions significant amounts of di-(p-nitrobenzyl) carbonate were often formed as a by-product[153]. This presumably involves attack upon p-nitrobenzyl chloroformate by the anion of p-nitrobenzyl alcohol.

Reaction of alkyl and aryl chloroformates with catalytic amounts of triphenylphosphine in diethyl ether, in the presence of 5% quinoline, leads to quite good yields of the corresponding chloride; even aryl chlorides can be prepared in this way[154]. Several aliphatic chloroformates gave good yields of aliphatic chlorides even in the absence of quinoline and, in some cases, also in the absence of solvent. Reaction of several aliphatic chloroformates (ROCOCl) with a 10% excess of the phosphine led to phosphonium chlorides (RPPh$_3$)$^+$Cl$^-$, usually in high yields.

Green and Hudson[155] have studied the influence of various anions on the rate of reaction of ethyl chloroformate in 85% aqueous acetone at 25°. The order of reactivities normally observed, in aqueous solvents, for S_N2 attack at a saturated carbon is $S_2O_3^{2-} > SCN^- > I^- > N_3^- > OH^- > Br^- > Cl^- > AcO^-$. This order is similar to the oxidation–reduction potentials[156]. Towards ethyl chloroformate, the order was quite different (Table 4). A detailed explanation could not be given for the extremely marked difference in relative rates from those for S_N2 attack. It appears that there must be large differences in the relative importance of solvation, repulsive and bond-formation energies in the two types of substitutions.

One complication to the study by Green and Hudson was that only the azide and phenoxide attacked to give products which were stable under the experimental conditions. The overall effect of the other anions was to

TABLE 4. Second-order rate coefficients for reactions of ethyl chloroformate with anions in 85% aqueous acetone at 25° [155]

Anion[a]:	$Me_2C{=}N{-}O^-$	OH^-	OPh^-	NO_2^-	N_3^-	F^-	$S_2O_3^{2-}$
k_2[b]:	500×10^3	167	50	31	17·5	0·22	$\sim 0·6$[e]

[a] No detectable reaction with Cl^-, Br^-, CNO^-, SCN^-, ClO_3^- or NO_3^-.

[b] In units of l/mole–min.

[e] Attack at the ethyl group.

catalyse the solvolysis. It was assumed that the catalysis involved nucleophilic substitution at the acyl carbon followed by rapid reaction of the new compound, leading eventually to ethanol. However, when the final product was ethanol, it was difficult to be sure that the catalysis was not, at least in part, a general-base catalysis to a direct solvolysis, followed by decomposition of the hydrogen carbonate.

Green and Hudson did not investigate the rate of chloride exchange. This has been done[157] by reacting ethyl chloroformate with [36]Cl-labelled tetra-n-butylammonium chloride in several water–acetonitrile compositions at 30°. Assuming formation of a symmetrical tetrahedral intermediate, the rate of exchange is one-half the rate of chloride addition to the carbonyl group. Chloride exchange was found to compete very effectively with solvolysis for a range of water concentration from 40 to 85%.

In inert aprotic solvents, not only is competition from the solvent eliminated but, also, nucleophilicities of anions are enhanced (and spread out) in these solvents. Hydrogen-bonding assistance towards breaking the carbon-leaving group bond and towards dissociation of ion pairs, which prevents internal return, is greatly reduced. The overall result of these features is a tendency towards bimolecular and away from unimolecular substitutions. Early measurements of this type were made by Conant, Kirner and Hussey[158, 159] for the reaction of ethyl chloroformate with potassium iodide in acetone. These rates were compared to those for other organic chlorides. Product studies do not appear to have been carried out but the product would be expected to be ethyl iodide formed via the unstable iodoformate. Indeed, this reaction can be a very useful synthetic route to alkyl iodides. For example, cholesteryl iodide can be very conveniently prepared by interaction of cholesteryl chloroformate with sodium iodide in acetone[160]. Under similar conditions, reaction with sodium azide leads to the stable cholesteryl azidoformate.

The order of reactivities for reaction of phenyl, methyl, primary alkyl and isopropyl chloroformates with chloride and nitrate ion in acetonitrile[53] has already been reviewed and data presented in Table 2. Reaction with fluoride ion will be discussed later, in connexion with the synthesis of fluoroformates.

Reaction of optically active 2-octyl chloroformate with potassium acetate, either in the absence of solvent or in ethanol, led to some alkene together with 2-octanol and 2-octyl acetate formed with retention of configuration[50]. Similar results were obtained with potassium benzoate and Tarbell and Longosz[137] have isolated the intermediate benzoic-2-octylcarbonic anhydride and confirmed that it decomposes with retention of configuration. Reaction of 2-octyl chloroformate with sodium p-toluenesulphinate[50] led to 2-octyl p-toluenesulphinate, which on hydrolysis gave 2-octanol of retained configuration.

The reaction of ethyl and methyl chloroformates with an acetonitrile solution of silver p-toluenesulphonate has been investigated[161]. A rather slow precipitation of silver chloride was observed and, on warming the resultant solution, carbon dioxide was evolved and the alkyl p-toluenesulphonate could be isolated. The intermediate carbethoxy- and carbomethoxy-p-toluenesulphonates were isolated as a colourless oil and colourless crystals respectively. Reaction of ethyl chloroformate with ^{18}O-labelled silver p-toluenesulphonate gave ethyl p-toluenesulphonate with no loss of label. The carbon dioxide evolved is derived entirely from the chloroformate ester. Reaction of isobutyl chloroformate led to both isobutyl and sec-butyl p-toluenesulphonate plus isobutene. The tertiary butyl ester would have been unstable under the experimental conditions, even if initially formed[162]. Due to the carbonium ion type rearrangements and the path used for loss of carbon dioxide, the decomposition of the carbalkoxy-p-toluenesulphonate was believed to proceed by ionization to give a carbalkoxy-p-toluenesulphonate ion pair, followed by loss of carbon dioxide from the cation and carbonium ion equilibration. These processes were followed by either collapse to alkyl p-toluenesulphonate or elimination within the ion pair to lead to isobutene and p-toluenesulphonic acid. It is of interest that, in corresponding reactions with alkyl chlorosulphites, the sulphur dioxide is lost in a manner corresponding to carbon dioxide loss from alkyl chloroformates but rearrangement of the isobutyl group is not observed[163]. Cholesteryl chloroformate has been shown to react with silver trifluoroacetate in ether to yield cholesteryl trifluoroacetate[164].

There have been several investigations of the reactions of chloroformate esters with silver nitrate in acetonitrile. This reaction affords a convenient synthetic route to alkyl nitrates. Aryl chloroformates do not react to give phenyl nitrates, unknown compounds, but usually give nitro-substituted phenols.

Boschan[164] showed that n-hexyl chloroformate reacted to give n-hexyl nitrate with 76% retention of an ^{18}O-label introduced adjacent to the hexyl group. Optically active 2-octyl chloroformate gave 2-octyl nitrate with optical activity consistent with 68% retention of configuration accompanied by 32% inversion of configuration. The kinetics were shown to be approximately second-order. Two competing mechanisms were proposed:

$$R-\overset{*}{O}-\underset{\underset{O}{\|}}{C}-Cl + AgNO_3 \longrightarrow R-\overset{*}{O}-\underset{\underset{O}{\|}}{C}-ONO_2 + AgCl \qquad (A)$$

$$R-\overset{*}{O}-\underset{\underset{O}{\|}}{C}-ONO_2 \longrightarrow \left[\begin{array}{c} \text{Four-centred} \\ \text{transition state} \end{array}\right] \longrightarrow \overset{*}{R}ONO_2 + CO_2$$

$$R-\overset{*}{O}-\underset{\underset{O}{\|}}{C}-Cl + Ag^+ \longrightarrow R-\overset{*}{O}-\underset{\underset{O}{\|}}{\overset{\oplus}{C}} + AgCl \qquad (B)$$

$$NO_3^- + R-\overset{*}{O}-\underset{\underset{O}{\|}}{\overset{\oplus}{C}} \longrightarrow RONO_2 + {}^*O{=}C{=}O$$

A more detailed kinetic study for methyl[165], ethyl[166], isobutyl[166] and phenyl[166] chloroformates in acetonitrile showed these chloroformates to be unreactive towards silver perchlorate and to react with dissociated nitrate ions from silver nitrate at approximately the same rate as with tetraethylammonium nitrate. These results require reaction to be a bimolecular displacement by nitrate ion at the acyl carbon, and the silver ion, except for rendering some assistance to chloride loss from the tetra-hedral intermediate[165], is inert and serves only to precipitate chloride ion from solution. In order to explain the 76% retention of the alkyl–oxygen bond observed by Boschan[164], it is necessary to assume the nitratoformate can proceed to products by two pathways. The pathway leading to retention could proceed by the four-centred transition state proposed by Boschan; it is, however, attractive to consider an ionization to give a nitronium ion[165] which can attack at the oxygen adjacent to the alkyl group[165] or at the aromatic ring in the case of aryl chloroformates[167]. The pathway leading to inversion presumably involves attack by nitrate ion at the α-carbon of the alkyl group, this attack could be either upon the nitratoformate by excess nitrate ion or upon the carboxylium ion formed by an ionization which also liberates a nitrate ion.

A kinetic study of the reactions of an acetonitrile solution of isopropyl chloroformate gave quite different results[166]. The dominant mechanism was now the silver-ion assisted mechanism (B) proposed by Boschan (or one of several possible variations of (B)). Both silver perchlorate and tetraethylammonium nitrate react with isopropyl chloroformate at measurable rates. The rate of reaction with silver nitrate can be taken as the sum of two processes. One process can be equated to the rate of reaction with the same concentration of silver perchlorate and a smaller contribution is due to bimolecular attack by dissociated nitrate ions. The overall situation for reactions of alkyl chloroformates with silver nitrate in acetonitrile is rather reminiscent of the solvolysis in aqueous acetone, which has been discussed earlier[135]. Methyl and primary alkyl chloroformates react by a bimolecular nucleophilic substitution. Isopropyl chloroformate reacts mainly by an ionization mechanism but an appreciable bimolecular nucleophilic substitution component is also present.

Mortimer[168] reacted n-propyl chloroformate with an acetonitrile solution of silver nitrate in the presence of a fairly small amount of pyridine as catalyst. At low temperatures (approx. $-17°$) chloroformate was consumed but no carbon dioxide was liberated. Mortimer assumed formation of the nitratoformate ester as an intermediate. The intermediate decomposed on warming to room temperature, with evolution of carbon dioxide and formation of n-propyl nitrate. The pyridine-catalysed reaction was proposed as a quick, mild and quantitative synthesis of alkyl nitrate esters.

The reaction of phenyl chloroformate with silver nitrate in acetonitrile was found to lead to quantitative reaction in terms of silver chloride precipitation and a 64% yield of o-nitrophenol was isolated[169]. It was thought that this may be formed through phenyl nitrate, which would rearrange to the more stable isomer. Subsequently, it was found that this reaction gives, in addition to the o-isomer, substantial amounts of p-nitrophenol[167]. It was shown that the *ortho* and *para* isomers do not interconvert under the reaction conditions. Also, decomposition of phenyl chloroformate in the presence of p-cresol led to nitration of the p-cresol and a 43% yield of 2-nitro-4-methylphenol was isolated. These results were nicely consistent with the previous proposal that the nitratoformate ester ionizes to produce a nitronium ion[165].

In a further investigation[170] of the reactions of several aryl chloroformates with silver nitrate in acetonitrile, the lack of dependence upon the silver-ion concentration[166] was confirmed and for a series of *para*-substituted derivatives a Hammett ρ value of $+1.5$ at $10°$ was obtained, consistent with rate-limiting attack by nitrate ion at the acyl carbon[165, 166]. The fit to second-order kinetics was only approximate, as previously reported for reactions with n-hexyl[164] and methyl[165] chloroformates. With

2,6-disubstituted aryl chloroformates, in addition to the 4-nitro-derivative of the phenol, fairly small amounts of the quinone dimers were also produced as dark red solids.

$$O= \underset{R}{\overset{R}{\bigcirc}} = \underset{R}{\overset{R}{\bigcirc}} =O \quad (R = Me \text{ or } i\text{-Pr})$$

With 2,4,6-trimethylphenyl chloroformate, dark red crystals separated from solution which were identified as 3,5,3′,5′-tetramethyl-(4,4′)-stilbenequinone.

$$O= \underset{CH_3}{\overset{CH_3}{\bigcirc}} =CH-CH= \underset{CH_3}{\overset{CH_3}{\bigcirc}} =O$$

Reaction of phenyl chloroformate with silver trifluoracetate[170] led to isolation of o-hydroxyphenyl trifluoromethyl ketone, conceivably formed by a similar mechanism to that for silver nitrate reaction.

D. Competing Solvolysis–Decomposition

Concurrent with the decomposition of 1-adamantyl chloroformate in nitrobenzene, a small amount of solvolysis, to give 1-(m-nitrophenyl) adamantane, was observed[45]. It has been found[171] that, in hydroxylic solvents, reaction proceeds with appreciable amounts of both solvolysis and decomposition. In alcoholyses, primary and secondary alkyl chloroformates are known to give disubstituted carbonates[7]. Alcoholysis of 1-adamantyl chloroformate gave none of this type of product, which was shown to be stable under the experimental conditions, but gave a mixture of 1-adamantyl alkyl ether and 1-adamantyl chloride. In t-butanol, 1-adamantanol was formed and it was independently shown that 1-adamantyl t-butyl ether decomposed to 1-adamantanol under the experimental conditions.

Additions of tetra-n-butylammonium bromide, chloride, iodide or perchlorate caused moderate rate increases, of similar magnitude, and the product composition was virtually unchanged. It was found that solvolysis was favoured relative to decomposition as either the temperature or the ionizing power of the solvent was increased. Some typical kinetic data are presented in Table 5.

The entropy of activation of +7·6 e.u. for the solvolysis–decomposition in aqueous ethanol is some 16–20 e.u. more positive than values which have been reported[172] for solvolyses of 1-adamantyl halides in this solvent. The bulk of this unusually large difference can be considered to arise from loss of carbon dioxide. This could be a consequence either of a direct

TABLE 5. Specific reaction rates at 25·0° and enthalpies ($\Delta H^{\ddagger}$) and entropies ($\Delta S^{\ddagger}$) of activation for the competing solvolysis–decomposition of 1-adamantyl chloroformate

Solvent	k_1 (s^{-1})	$\Delta H^{\ddagger}$ (kcal/mole)	$\Delta S^{\ddagger}_{298}$ (e.u.)
80% Ethanol	$(1 \cdot 15 \pm 0 \cdot 02) \times 10^{-2}$	$22 \cdot 3 \pm 0 \cdot 6$	$+7 \cdot 6 \pm 2 \cdot 2$
Methanol	$(3 \cdot 36 \pm 0 \cdot 03) \times 10^{-3}$	$21 \cdot 9 \pm 0 \cdot 2$	$+3 \cdot 7 \pm 0 \cdot 7$
Ethanol	$(5 \cdot 53 \pm 0 \cdot 01) \times 10^{-4}$	$23 \cdot 8 \pm 0 \cdot 1$	$+6 \cdot 3 \pm 0 \cdot 4$
Isopropanol	$(1 \cdot 64 \pm 0 \cdot 05) \times 10^{-4}$	$24 \cdot 4 \pm 0 \cdot 2$	$+6 \cdot 0 \pm 0 \cdot 6$
t-Butanol	$(9 \cdot 89 \pm 0 \cdot 04) \times 10^{-5}$	$23 \cdot 8 \pm 0 \cdot 4$	$+3 \cdot 1 \pm 1 \cdot 4$

ionization–decomposition to two ions and a molecule or a consequence of loss of carbon dioxide after ionization to an ion pair, with circumvention of appreciable ion-pair return. Other factors being equal, ion-pair return tends to reduce experimental entropy of activation measurements for the overall process[173].

A reaction scheme is proposed in which loss of carbon dioxide leads to a 1-adamantyl-chloride ion pair. This ion pair then either collapses to 1-adamantyl chloride or it separates, with separation being followed by solvolysis:

$$R^+CO_2Cl^- \longrightarrow R^+Cl^- \nearrow\begin{array}{l} RCl \\ \\ R^+ + Cl^- \xrightarrow{R'OH} ROR' + HCl \end{array}$$

The lack of any increase in RCl formation on adding up to 0·1M tetra-n-butyl-ammonium chloride indicates little, if any, 1-adamantyl chloride formation from dissociated ions. The mechanism leading to formation of the $R^+CO_2Cl^-$ aggregate is not established but initial ionization to give a carboxylium ion $(ROCO)^+$ would require its decomposition to be faster than capture by alcohol molecules.

In order to carry out a treatment of the solvolysis–decomposition in terms of the Grunwald–Winstein equation ($\log k/k_0 = mY$)[174], a study was also made in several aqueous–organic solvents of known Y-value. Both overall kinetics and product composition were determined in each case[175, 176]. In this way, it was possible to carry out parallel treatments of both the solvolysis and the decomposition components of the overall specific rates. These results are shown in Table 6. The quantitative correlation of specific decomposition rates against solvent ionizing power (Y-values) is we believe the first to be reported for an S_Ni type decomposition. The pronounced dependence of both solvolysis and decomposition

TABLE 6. m-Values for reactions of 1-adamantyl chloroformate at 25·0°

Solvents	No. of points	Solvolysis–decomposition	Solvolysis	Decomposition
80–100% Ethanol	3	0·650 ± 0·005	0·684 ± 0·010	0·587 ± 0·003
90–100% Methanol	2	0·658	0·669	0·612
Dry alcohols[a]	4	0·721 ± 0·042	0·906 ± 0·012	0·537 ± 0·055
80–90% Dioxan	2	0·641	0·704	0·551
80–95% Acetone	3	0·424 ± 0·002	0·442 ± 0·003	0·381 ± 0·003
All but aq. acetone[b]	9	0·682 ± 0·030	0·795 ± 0·036	0·546 ± 0·032
All alcoholic[c]	7	0·682 ± 0·028	0·800 ± 0·036	0·543 ± 0·023

[a] Methanol, ethanol, isopropanol and t-butanol. The points for 100% methanol and 100% ethanol were also included in the plot for the appropriate aqueous alcohol system.

[b] Summation of first four entries.

[c] Summation of first three entries.

rates upon solvent ionizing power is consistent with the postulated ionic mechanism. Also, a greater dependence would be expected for solvolysis rates, relative to decomposition rates. Following common development of the R^+Cl^- ion pair, solvolysis requires further separation and an additional dependence upon solvent ionizing power (more strictly, dissociating power) while decomposition features a competing ion-pair collapse.

A delicate balance between attack at the acyl carbon and ionization is indicated by introduction of an appreciable second-order contribution on addition of small concentrations of sodium ethoxide to the reaction in ethanol. This second-order reaction leads to the mixed carbonate and not the ether. For example, in the presence of 0·006M sodium ethoxide, 95% of initial reaction involves bimolecular attack and only 5% is underlying unimolecular solvolysis–decomposition. Similarly, addition of azide has been shown[177] to lead to 1-adamantyl azidoformate.

It is of interest that examples of decomposition under solvolytic conditions which have been abstracted[178] from the early literature relate to alkyl chloroformates with chlorine substituted within the alkyl group. Such substitution would tend to favour bimolecular attack at the acyl carbon and disfavour ionization mechanisms. It would appear that these decompositions must proceed by mechanisms which are different to the one operating for 1-adamantyl chloroformate. The mechanisms of these reactions are currently under investigation within our laboratories.

VII. ELECTROPHILIC ASSISTANCE TO REACTIONS

A. Friedel–Crafts Reactions

Ethyl chloroformate has been reported to be inert towards benzene, even in sealed tubes at 150°[179]. Friedel and Crafts were able to induce a reaction by addition of aluminium chloride. The product was, however, not the aromatic ester but ethylbenzene, formed by loss of carbon dioxide[180]:

$$\text{EtOCOCl} + \text{ArH} \xrightarrow{\text{AlCl}_3} \times \begin{array}{l} \nearrow \text{ArCOOEt} + \text{HCl} \\ \\ \searrow \text{ArEt} + \text{CO}_2 + \text{HCl} \end{array}$$

They proposed that the reaction involved hydrogen chloride saponification of the ester to give ethyl chloride, which then entered into a Friedel–Crafts alkylation[181]. Rennie[182] confirmed the experimental results but proposed a direct decomposition of ethyl chloroformate to ethyl chloride and carbon dioxide, without the need for hydrogen chloride intervention.

The scheme proposed by Rennie was supported by Buu-Hoi and Janicaud[183]. They studied the catalysed reactions of ethyl chloroformate with a variety of aromatic systems and in no instance were they able to detect any ester formation. They investigated the efficiency of other chlorides in promoting decomposition and found $FeCl_3$ to be effective but $ZnCl_2$, $SnCl_4$, $AsCl_3$ and $SbCl_3$ to be ineffective. The claim that zinc chloride is ineffective as a decomposition catalyst is surprising and inconsistent with earlier reports[184–186]. Underwood and Baril[186] found that addition of zinc chloride to ethyl chloroformate produced a spontaneous evolution of carbon dioxide. They isolated 17·7% ethylene and 25·7% ethyl chloride, together with 25·3% unchanged ester. Similarly, n-propyl chloroformate gave 5% propylene and 23% n-propyl chloride. Zinc chloride was also found to be effective in decomposing several other types of esters, in some cases with formation of alkyl chlorides (with chlorine incorporated from the zinc chloride). Bowden[187] subsequently reported that a variety of alkyl carboxylate esters could be used, in conjunction with aluminium chloride, to alkylate benzene. Frequently, carbonium-ion type rearrangements were observed to occur.

Recently, it has been found that antimony pentachloride is effective in promoting decomposition of ethyl chloroformate in ethylene chloride solution[188]. Proprionyl chloride formed an equilibrium concentration of an addition complex but with ethyl chloroformate reaction occurred, with a heat of reaction of $-21\cdot4\pm0\cdot2$ kcal/mole, to give ethyl chloride and carbon dioxide. Only traces of hydrogen chloride, which would have accompanied ethylene formation, were observed. Chloroformates have been found to be useful intermediates for generating carbonium ions from primary and less reactive secondary alcohols in antimony pentafluoride–sulphur dioxide solution[189]:

$$ROH + COCl_2 \longrightarrow ROCOCl \xrightarrow{\text{SbF}_5\text{—SO}_2} R^+(SbF_5Cl)^- + CO_2$$

The basic scheme for Friedel–Crafts alkylation of aromatics can be expressed:

$$ROCOCl + AlCl_3 \longrightarrow (ROCO)^+AlCl_4^- \longrightarrow R^+AlCl_4^- + CO_2$$

$$R^+AlCl_4^- \underset{ArH}{\overset{}{\rightarrow}} \begin{matrix} RCl + AlCl_3 \\ RAr + HCl + AlCl_3 \end{matrix}$$

There is no direct evidence for the carbalkoxylium ion and there could be direct decomposition to R^+, CO_2 and $AlCl_4^-$. Evidence for the $R^+AlCl_4^-$ ion pair comes from work using ethyl chloroformate and aluminium bromide, a mixture which reacts with benzene to give ethylbenzene. In the absence of benzene, and with slight cooling, a brown solution is formed and, on being warmed to $35°$, ethyl halides are evolved. The composition of the evolved gases was found to be 82% ethyl bromide and 18% ethyl chloride. This was rationalized in terms of halide-ion capture from a $C_2H_5^+(AlClBr_3)^-$ ion pair[190].

A detailed study of the reactions of several alkyl chloroformates with benzene and with several alkylbenzenes was made by Kunckell and Ulex[191, 192]. Friedel–Crafts reactions of alkyl chloroformates have been briefly reviewed[193].

Friedel–Crafts reactions of aryl chloroformates do lead to the corresponding esters[194]:

$$Ar'OCOCl + ArH \xrightarrow{\text{AlCl}_3} ArCOOAr' + HCl$$

An aromatic carboxylium ion would not be expected to eject carbon dioxide to give a high-energy aryl carbonium ion. This aromatic entity is, therefore, presented with the opportunity to act as the electrophile in a

Friedel–Crafts acylation. Phenyl, *p*-chlorophenyl and *p*-phenylphenyl chloroformates attacked benzene to give isolated yields of 50–64% of the corresponding benzoates. Attack by *p*-phenylphenyl chloroformate on toluene was primarily at the *para* position and a 30% yield of the *p*-toluate was isolated.

B. Silver-ion Assisted Reactions

Several silver salt–chloroformate ester reactions were reviewed in section VI, C. In most instances the role of the silver ion, other than the obvious one of precipitating chloride ion, was not clear. A rather detailed kinetic investigation of the reactions of silver nitrate with several chloroformate esters in acetonitrile[165, 166] indicated that for phenyl, methyl and primary alkyl chloroformates, the rate-limiting factor was attack by dissociated nitrate ion at the acyl carbon and, at most, the influence of the silver ion upon the kinetics was to give some secondary assistance to chloride-ion loss from a tetrahedral intermediate. With isopropyl chloroformate, the dominant pathway was a silver-ion-assisted ionization but, even here, there was a minor contribution from the pathway operating for primary alkyl chloroformates. Consistent with these observations, the hydrolysis rate of *n*-butyl chloroformate, in water, was found to be independent of the concentration of added mercuric ion[195], a powerful electrophilic reagent towards alkyl halides and, also, towards an aqueous solution of dimethylcarbamyl chloride[195], which is known to favour ionization reactions in water[132]. Similarly, the rates of solvolysis of ethyl chloroformate in acetonitrile–water mixtures were unchanged by addition of silver perchlorate[157]. In this section, we will concentrate on those silver-ion-assisted reactions which take place in the presence of a weakly nucleophilic or non-nucleophilic counter-ion, and which proceed many orders of magnitude faster than the corresponding reactions in the absence of silver ion. The most inert of the counter-ions which have been investigated is the essentially non-nucleophilic hexafluoroantimonate. The fluoborate is quite useful but it is much easier to extract a fluoride ion[196]. This property of fluoborate is quite well known and a standard preparation of fluoroaromatics is by decomposition of the appropriate diazonium fluoborate. Perchlorate is not completely non-nucleophilic. It competes successfully against benzene in silver-ion-assisted reaction of methyl iodide[197], and its effectiveness as a competitor in silver-ion-assisted reactions of alkyl chloroformates in benzene can lead to useful mechanistic information.

These silver-ion-assisted reactions can *formally* be considered as proceeding through carboxylium ions (ROCO$^+$). It has been shown[196, 198, 199] that the reactions have many points of resemblance to diazonium-ion reactions and, in some instances, to Friedel–Crafts reactions. They afford routes to aliphatic carbonium ions of high energy which are often difficult to prepare in other ways. For example[196], 1-apocamphanyl chloroformate will react quite readily with silver fluoborate in chlorobenzene at ambient temperatures. Under identical conditions, 1-apocamphanyl chloride is unreactive. It is, however, possible to deaminate 1-aminoapocamphane[200]. The reaction yields[196, 198] 52% of 1-apocamphanyl fluoride and 24% of a mixture of isomers of (1-apocamphanyl) chlorobenzene (40% *ortho*, 32% *meta* and 28% *para*) and 10% of 1-apocamphanyl chloride. Under Friedel–Crafts conditions, with boron trifluoride in chlorobenzene at 50–60°, 1-apocamphanyl chloroformate will react to give 48% 1-apocamphanyl chloride and 18% of a mixture of isomers of (1-apocamphanyl) chlorobenzene[196, 199].

With silver hexafluoroantimonate substituted for silver fluoborate, the yield of isomers of (1-apocamphanyl) chlorobenzene was raised to 81%[198, 199] and the percentage composition was essentially unchanged; there was also 17% 1-apocamphanyl chloride and less than 5% 1-apocamphanyl fluoride. In the presence of two equivalents of the base tetramethylurea the yield of 1-apocamphanyl fluoride was raised to 29%. The isomer distribution of the (1-apocamphanyl) chlorobenzenes was identical when produced by action of either nitrosyl chloride or nitrosyl hexafluoroantimonate with 1-aminoapocamphane in chlorobenzene or when produced by the action of silver fluoborate with 1-apocamphanyl chlorosulphinate in chlorobenzene. The identical isomer distribution from these five reactions in chlorobenzene strongly suggests that they proceed through a common intermediate: the 1-apocamphanyl carbonium ion.

In nitrobenzene, silver hexafluoroantimonate reacted readily with 1-apocamphanyl chloroformate to give a 24% yield of *m*-(1-apocamphanyl) nitrobenzene and 16% of 1-apocamphanyl chloride[198, 199].

Reaction of 1-aminoapocamphane with nitrosyl chloride in nitrobenzene gave 4% of *m*-(1-apocamphanyl) nitrobenzene. This novel alkylation of nitrobenzene, to give the *meta* isomer, confirmed that attack was by a cationic (and not a radical) species. It has subsequently been found that 1-adamantyl chloroformate reacts with silver hexafluorantimonate in nitrobenzene to give *m*-(1-adamantyl) nitrobenzene[45].

Aryl chloroformates in reaction with silver fluoborate or silver hexafluoroantimonate do not give any products resulting from fission of the aryl–oxygen bond[196, 199]. Reaction of phenyl chloroformate with silver fluoborate in refluxing chlorobenzene gave a 46% isolated yield of the fluoroformate. In the presence of tetramethylurea, an 82% yield of phenyl *N,N*-dimethylcarbamate was isolated. Abstraction of a fluoride ion from fluoborate by acylium ions is well documented[201, 202].

Both cyclohexyl chloroformate and cyclohexyl chloride gave an overall 81% (39% *ortho*, 6% *meta* and 55% *para*) of cyclohexylchlorobenzene on treatment with silver fluoborate in chlorobenzene at room temperature[199]. However, the ring-alkylation products are probably formed via cyclohexene which, in the presence of simultaneously produced fluoboric acid, can alkylate the chlorobenzene. Indeed, in the presence of two equivalents of tetramethylurea, cyclohexyl chloroformate produced only 0·5% cyclohexylchlorobenzenes, together with 27% cyclohexene, 48% cyclohexyl fluoride and 13% cyclohexyl chloride! Reactions of *n*-propyl chloroformate and *n*-propyl chloride show similar behaviour. In the absence of added base, both give the same ratio of *ortho* and *para*-isopropylchlorobenzenes (47 : 53) but, under identical conditions, the chloroformate gave 56% (84% silver chloride) and the chloride 6% (13% silver chloride); less than 1% propene was identified. Again, the presence of two equivalents of tetramethylurea cut down the ring-alkylation products to less than 1%. With silver hexafluoroantimonate, in the presence of two equivalents of tetramethylurea, 47·5% propene was identified. A comparison of these studies with and without added base (towards which the chloroformate esters are unreactive) shows very clearly the need, when acid is concurrently produced, to consider subsequent reactions of initially formed alkenes.

A kinetic and product study[203] of the reactions of alkyl and arylalkyl chloroformates with silver perchlorate in benzene at 25·0° has been carried out. Phenyl chloroformate was inert under these conditions.

$$ROCOCl \xrightarrow[C_6H_6]{AgClO_4} \begin{array}{l} ROClO_3 + R'OClO_3 \\ \uparrow \\ alkene + HClO_4 \longrightarrow polymer \\ \downarrow C_6H_6 \\ RC_6H_5 + R'C_6H_5 (+HClO_4) \end{array}$$

Due to the explosive properties of alkyl perchlorates[204], product isolation was not attempted. The p.m.r. spectra of the aliphatic protons in the product solutions were integrated, after addition of a known concentration of toluene as an internal integration standard. The products identified and the percentage conversion to these products are indicated in Table 7. The only products identified were alkylbenzene and alkyl

TABLE 7. Percentages of products identified in the p.m.r. spectra after reaction of 0·16M chloroformate esters (ROCOCl) with 0·20M silver perchlorate in benzene at 25·0°

R	$ROClO_3$	RC_6H_5	$R'OClO_3$	$R'C_6H_5$
Me	86[a]	14[a]		
Et	47	37		
n-Pr			70[c]	10[c]
isoPr	70	10		
isoBu	0[b]	0[b]		37[d]

[a] No toluene added, these percentages are ratios.
[b] No evidence for methylene doublet of isobutyl group.
[c] Isopropyl derivatives.
[d] t-Butylbenzene.

perchlorate; carbonium-ion rearrangements were observed with appropriate alkyl groups. In some instances the initially colourless solution became brown. The lack of material balance may be due in some instances to polymerization (when unidentified peaks are present) and in others due to loss of alkene vapour from the loosely stoppered reaction mixture.

In general, much of the alkylbenzene (and perchlorate ester?) may well be formed via an acid-promoted alkylation involving initially formed alkene[199]. However, the toluene formation, when methyl chloroformate is the substrate, shows that direct alkylation can compete with collapse to perchlorate ester. Reaction of methyl[197] or ethyl[205] iodide with silver perchlorate in benzene leads only to the perchlorate ester, without any benzene alkylation. This difference in behaviour is nicely consistent with the diazonium-ion analogy for chloroformate–silver ion reactions[196, 199]. It is well established that carbonium ions formed from diazonium ions are 'hot' and less selective than those formed from alkyl halides[206].

Initial second-order rate coefficients were found to increase in value with the initial silver perchlorate concentration (0·01–0·07M). Typical kinetic data are reported in Table 8. Rate-determining attack by perchlorate ion at the acyl carbon, followed by fast precipitation of silver chloride and progress to stable products, would give a fast reaction for phenyl

TABLE 8. Initial second-order rate coefficients[a], k_2, for reaction of 0·1M chloroformate esters (ROCOCl) with 0·035M silver perchlorate in benzene at 25·0°, and the relative rates of reaction

R	$10^4 k_2$ (l/mole–s)	Relative rate[b]
Methyl	0·0113	0·022
Ethyl	0·49	1·00
Allyl	0·76	1·48
n-Propyl	0·48	0·94
Isopropyl	30·0	58·0
Isobutyl	0·41	0·84
Benzyl	2·5	4·8
p-Nitrobenzyl	0·058	0·114
Cholesteryl	13·1	25·0
Cholestanyl	26·0	52·0

[a] $d[AgCl]/dt = k_2[ROCOCl][AgClO_4]$.
[b] Calculated at 0·0165, 0·035 and 0·070M $[AgClO_4]$ and a mean value taken.

chloroformate, which is inert under the experimental conditions. A mechanism with well-developed ionization character is indicated by the considerably faster reaction of isopropyl chloroformate, relative to primary alkyl chloroformates. A silver-ion-assisted reaction with concurrent S_N2 attack at the α-carbon of the R group is ruled out by almost identical rates for ethyl, n-propyl and isobutyl chloroformates, a series of alkyl derivatives for which steric hindrance to S_N2 attack would increase appreciably. The observation of extensive carbonium-ion-type rearrangements with both n-propyl and isobutyl chloroformates would appear to rule out either S_N2 attack of the above type or a fast S_N2 attack upon a preformed carboxylium ion.

The extensive carbonium-ion rearrangements and the absence of any benzoate ester product would appear to require one or the other of two basic reaction paths. In the low polarity solvent, benzene, these reactions almost certainly involve aggregates and the following schemes are intended only to indicate the basic features. The first scheme involves a silver-ion-assisted ionization, to give a carboxylium ion, which can be followed by loss of carbon dioxide:

$$ROCOCl + Ag^+ \xrightarrow{\text{slow}} (ROCO)^+ + Cl^- Ag^+ \xrightarrow{\text{fast}} R^+ + CO_2 + Cl^- Ag^+$$

The second scheme involves a silver-ion-assisted ionization–fragmentation process to yield the carbonium ion directly:

$$\text{ROCOCl} + \text{Ag}^+ \xrightarrow{\text{slow}} \text{R}^+ + \text{CO}_2 + \text{Cl}^-\text{Ag}^+$$

Along the α-methylated series (methyl, ethyl and isopropyl) the rate differences are quite large. However, in the absence of suitable models, it is not possible to decide whether the magnitudes correlate with inductive stabilization of a proximate or a remote positive charge. Benzyl chloroformate is an important member of the series. The phenyl group would be expected to stabilize positive charge development at the α-carbon by a conjugative mesomeric effect[207] but its effect upon a remote positive charge should be one of destabilization, due to its electron-withdrawing inductive effect; the polar substituent constants (δ^*) are $+0\cdot215$ for benzyl, zero for methyl (by definition) and $-0\cdot190$ for isopropyl[208]. The observation that benzyl chloroformate reacts about 200 times faster than methyl chloroformate suggests, at least for benzyl chloroformate, appreciable charge development at the α-carbon. Consistent with this proposal, a p-nitro-group reduces the rate of reaction of benzyl chloroformate by a fairly large factor. On the other hand, allyl chloroformate, despite the mesomeric stabilization of an incipient carbonium ion, reacts only some 50% faster than n-propyl chloroformate. Also, benzyl chloroformate reacts about 12 times slower than isopropyl chloroformate, in contrast to a faster reaction by a factor of about 60 for the corresponding chlorides under the same reaction conditions.

The available evidence can best be accommodated by a transition state of varying structure in which both carbon–oxygen and carbon–chlorine heterolyses can have progressed to varying degrees.

$$\overset{\delta+}{\overbrace{\text{R}\cdots\text{O}=\text{C}}}\overset{\delta-}{\cdots}\overset{}{\text{Cl}} \text{ Ag}^+$$
$$\underset{\text{O}}{\overset{\|}{}}$$

A possible alternative to this mechanism would involve slow reaction to a $(\text{ROCO})^+\text{Cl}^-\text{Ag}^+$ ion triplet, and with a fast internal return of chloride occasionally circumvented by loss of carbon dioxide, followed by reaction to products. The overall rate would again be a function of the ease of both carbon–oxygen and carbon–chlorine heterolyses.

Aryl chloroformates will react with silver ion only at elevated temperature, to give products with the aryl–oxygen bond retained[199]. These reactions can be considered as lying at one extreme of a spectrum of mechanisms with varying amounts of R—O and C—Cl heterolyses in the transition state of the rate-determining step. Benzyl chloroformate, with its high reactivity relative to methyl chloroformate and its large sensitivity to ring substitution, probably has appreciable R—O bond heterolysis at the transition state. However, the low reactivity relative to isopropyl suggests that C—Cl bond heterolysis is also well developed; this heterolysis will be hindered by a remote phenyl group but assisted by a remote isopropyl group. If one is prepared to discount the possibility of fast internal return from an ion triplet, the reaction with benzyl chloroformate is probably best represented as by-passing the carboxylium ion and proceeding directly to the carbonium ion. Indeed, for alkyl and arylalkyl chloroformates, the physical difference between the two pathways may be quite small, with the difficult, and quite common, task of attempting to determine whether or not extremely unstable intermediates exist along the reaction pathway.

Analogous mechanisms can no doubt be applied to Friedel–Crafts reactions of chloroformate esters. The same difference in behaviour between aryl and alkyl chloroformates is observed and carbonium-ion rearrangements are prevalent in both types of reaction.

VIII. OTHER HALOFORMATES

A. Preparation of Fluoroformates

Relative to chloroformates, fluoroformates have increased thermal stability[209] and increased susceptibility to hydrolysis[155]. One major difficulty in preparation is the sensitivity to moisture, acids and bases.

The majority of the fluoroformate syntheses involve an exchange of acyl chloride with inorganic fluoride. The basic difference in the two principal techniques is as to whether the exchange is carried out at the phosgene (carbonyl chloride) stage or at the chloroformate ester stage.

The first reported preparation of fluoroformate esters involved a very early use of thallium in organic synthesis. Treatment of methyl or ethyl chloroformate with powdered anhydrous thallous fluoride led to the corresponding fluoroformates[210, 211]. Subsequent modifications involved treatment with potassium fluoride in an autoclave at elevated temperatures to prepare methyl[212] and ethyl[213] fluoroformates. In both instances the yields were only fair and, after cooling, pressure measurements indicated appreciable decomposition. Olah and Pavlath[214] used potassium fluoride in dry acetylacetone, with exposure to u.v. radiation, to convert ethyl chloroformate to the fluoroformate in 51% yield. Similar yields were obtained by Kitano and Fukui[215] in nitrobenzene and without irradiation. Anhydrous hydrogen fluoride[216] and sodium fluoride in tetramethyl-enesulphone[217] have also been employed. Nakanishi, Myers and Jensen[218] returned to the original use of thallous fluoride and synthesized several primary and secondary alkyl and cycloalkyl fluoroformates in 50–70% yields. In the syntheses of carbohydrate fluoroformates, it was found that lower reaction times and better yields were obtained by refluxing with thallous fluoride in acetonitrile[219]. Green and Hudson[155] prepared ethyl fluoroformate in 80% yield by stirring the chloroformate in carbon tetrachloride with an equivalent of dry ammonium fluoride in the presence of one-half an equivalent of acetamide.

The alternative approach, introduced by Emeleus and Wood[220], has been favoured by industrial research laboratories. It involves treatment of the alcohol or phenol with either carbonyl fluoride (COF_2) or carbonyl chlorofluoride (COClF). Emeleus and Wood prepared both these carbonyl compounds by interaction, in an autoclave, of phosgene with an appropriate amount of antimony trifluoride. Other inorganic fluorides which have been substituted include AsF_3 and SiF_4[221]. A simple procedure for COF_2 involves interaction of phosgene with sodium fluoride in acetonitrile[222].

Emeleus and Wood prepared ethyl and phenyl fluoroformates by use of COF_2 in toluene or benzene, in the presence of a pyridine catalyst for the ethyl derivative and in an autoclave at 100°, in the absence of catalyst, for the phenyl derivative. Reaction of ethanol with carbonyl chlorofluoride at solid carbon dioxide temperature gave ethyl fluoroformate and no ethyl chloroformate; some diethyl carbonate was also formed.

Carbonyl chlorofluoride has been found to be a versatile reagent. It has been used[221, 223] to prepare several aryl fluoroformates from the corresponding phenols. In a typical experiment, the phenol was dissolved in toluene, in the presence of a tributylamine catalyst, and three equivalents of carbonyl chlorofluoride were added. After agitation for 5h at 50°, volatile products were bled off and the product vacuum distilled. The increased thermal stability of *t*-butyl fluoroformate, over that of *t*-butyl chloroformate[64], has allowed it to be made[224] at reduced temperatures by interaction of *t*-butanol with COClF. It is reasonably stable at 0°. The *p*-methoxy- and 3,4,5-trimethoxy-benzyl fluoroformates were also prepared.

With cyclic ethers, in the presence of suitable catalysts, COClF reacts to give ω-chloroalkyl fluoroformates[209].

$$\underset{\underset{O}{\diagdown\diagup}}{(CH_2)_x} + COClF \longrightarrow Cl(CH_2)_x OCOF$$

Tertiary amines are good catalysts for 3- and 4-membered rings. For higher-membered rings, addition of ethylene glycol leads to liberation of hydrogen chloride. The hydrogen chloride also serves as a catalyst, probably by ring opening to the alcohol which then reacts with the COClF (activated by the tertiary amine).

Sheppard[225] has prepared aryl fluoroformates by interaction of phenols with COF_2. The phenol is placed in a pressure vessel, which is evacuated at solid CO_2 temperature and a 25–50% mole excess of COF_2 added. After heating at 100–200° for 1–4 h, the residue is dissolved in methylene chloride and fractionated or recrystallized. The method has been extended to aliphatic fluoroformates with moderate success[226]. Usually, the aliphatic fluoroformates were not isolated but were subjected to further reaction within the pressure vessel.

A related method for the preparation of primary and secondary alkyl and cycloalkyl fluoroformates involves reaction of the alcohols with carbonyl bromofluoride (COBrF). The COBrF was prepared from bromine trifluoride and carbon monoxide:

$$BrF_3 + 2CO \longrightarrow COBrF + COF_2$$

The subsequent reaction was started at $-60°$ and completed at 30 to 40°, no bromoformate was formed and yields were quite good[227].

$$COBrF + ROH \longrightarrow ROCOF + HBr$$

Trifluoromethyl fluoroformate has been prepared by insertion of carbon monoxide into trifluoromethyl hypofluorite, under the influence of u.v. radiation at 35°. The reaction also progressed slowly in the dark at 90°. It was believed to be a chain reaction involving the $F_3COCO\cdot$ radical[228].

B. Physical Properties and Conformation of Fluoroformates

Routine physical properties can be abstracted from appropriate references given in section VIII, A. The boiling points of methyl, ethyl and phenyl fluoroformates are some 30–40° less than for the corresponding chloroformates. Ethyl fluoroformate has a density marginally greater than water[211].

More sophisticated measurements include a study of the microwave spectrum of methyl fluoroformate[229], proton and fluorine n.m.r. and dipole moment studies in benzene solution for several fluoroformate esters[230], and a study of the fluorine n.m.r. in ethyl, methyl and isopropyl fluoroformates[231].

The microwave spectroscopy report[229] was of a preliminary nature. It included structures of type MeOCOX, where X was F, CN, C≡C—H and CH_3. A heavy atom skeleton *cis*-geometry was deduced for X = CN or C≡C—H. This was in accord with related microwave[122] and electron

15

diffraction[108] studies. Although the geometry was probably identical for
X = F or CH_3, the evidence was less clear-cut on account of O, F and CH_3
having similar masses. The fluorine -19 n.m.r. study[231] was as part of a
wider study of acid fluorides. It was concluded that the major contribution
to the shielding of the fluorine nucleus was from conjugation transmitted
through the carbonyl group, with the field effect of the substituent being
of lesser importance.

The dipole moment study in dilute benzene solution at $25 \cdot 0°$ paralleled
that reported earlier for chloroformate esters[120]. The dipole moments
were uniformly higher, by about $0 \cdot 2$ D, than for the corresponding
chloroformates. A similar vector-model bond-moment calculation to that
carried out for methyl chloroformate[120] predicted values of $2 \cdot 33$ D for the
trans (carbonyl oxygen relative to methyl) conformation and $1 \cdot 64$ D for
the *cis* conformation of methyl fluoroformate. The experimental value
($2 \cdot 61$ D) is actually considerably higher than either of these values,
bringing to mind the adverse comments, by LeFèvre and co-workers[123],
about this type of calculation. It was concluded that the fluoroformate is
probably in the *trans* conformation; the value calculated for this conformation is the closest to the experimental value. The results of proton and
fluorine n.m.r. measurements support the *trans* conformation only if a
'through-space' coupling mechanism is assumed for the long-range spin-
coupling (J_{H-F}^{1-5}). However, comparison of methyl group signals in
methyl and ethyl fluoroformates show one broadened signal for the
methyl and spin–spin coupling for the ethyl compound; this suggests that
the number of intervening bonds (4 or 5) is the critical factor. Methyl,
ethyl, *n*-propyl, *n*-butyl, *n*-pentyl and phenyl fluoroformates were examined
in these studies.

The conformation is clearly far from being well established. Possibly,
with further refinement, the microwave spectroscopy study[229] will
eventually give a conclusive answer.

C. Chemical Properties of Fluoroformates

Although it is generally accepted that fluoroformates are thermally
more stable than corresponding chloroformates[209], there does not appear
to have been any quantitative comparison. Aryl fluoroformates are not
decomposed to aryl fluorides by basic catalysts such as pyridine or Lewis
acid catalysts such as boron trifluoride-etherate, which are effective for
alkyl fluoroformates[218, 232]. Some carbon dioxide is evolved but the
decomposition products are high-boiling and apparently consist of polymer
and carbonate[233]. Decomposition can be smoothly effected[221, 223] by passing
the gaseous vapours through a quartz tube containing platinum gauze

maintained at temperatures of 660–790°. The rate and course of the reaction are very sensitive to the nature of the walls of the vessel and the packing material[221]. Phenol and tars can be troublesome side-products. However, under the conditions described above, yields of aryl fluoride can be in excess of 90%. An S_Ni-type mechanism was proposed[223]. However, in view of the obvious heterogeneity of the reaction, any detailed mechanism would have to include the nature of the surface effects.

Alkyl fluoroformates can readily be decomposed to alkyl fluorides. On heating at 70–80° with an equimolar amount of pyridine, ethyl, isopropyl, sec-butyl and cyclohexyl chloroformates decomposed to give 60–75% yields of the corresponding fluorides[218], presumably by the following pathway:

$$ \text{ROCOF} + \text{N} \bigcirc \longrightarrow \text{ROCON}^{\oplus} \bigcirc + \text{F}^- $$

$$ \text{F}^- + \text{ROCON}^{\oplus} \bigcirc \longrightarrow \text{RF} + \text{CO}_2 + \text{N} \bigcirc $$

Cyclopentyl fluoroformate did not decompose to give the fluoride.

Boron trifluoride-etherate was found to be very effective for decomposing these alkyl fluoroformates, including cyclopentyl fluoroformate[232]. Decomposition occurred readily, even at ice-bath temperature. Indeed, these Lewis acid-catalysed decompositions took place considerably more readily than the corresponding reactions of chloroformates, where temperatures of 60–90° were required. Boron trifluoride catalysis has also been used to decompose ω-chloroalkyl fluoroformates to the corresponding ω-chloroalkyl fluorides (on passing the vapour over a metal-oxide catalyst, these can be dehydrochlorinated to give ω-fluoroalkenes)[209]. t-Butyl fluoroformate is sufficiently stable at low temperatures for preparation and use[224], at pH 7–10, in the synthesis of amino-acid derivatives[3, 4].

In 85% aqueous acetone at 0°, ethyl fluoroformate solvolyses[155] with a specific rate of 46×10^{-3} s^{-1}. Under identical conditions, ethyl chloroformate has a specific rate of $1 \cdot 6 \times 10^{-3}$ s^{-1}. A previous study[234], in aqueous acetone, had shown rate ratios favouring the chloro-compound (RCl/RF) of from 10^6 for triphenylmethyl halides, 10^4 for acetyl halides and less than 10^2 for benzoyl halides. However, for reaction with hydroxide ion, benzoyl fluoride reacted some 40% faster than benzoyl chloride. These ratios were rationalized in terms of the relative extents of C—X bond breaking (C—F stronger than C—Cl) and O—C bond making (carbon

more electron-deficient with $C—F$ than with $C—Cl$) at the transition
state. Consistent with this view, nucleophilic aromatic substitution at a
$C—X$ bond, in cases where the transition state of the rate-determining
step is considered to involve bond making to carbon with only a change in
hybridization of the $C—X$ bond, favours RF over RCl by a large factor[235].
With ethyl haloformates, the RCl/RF rate ratio of significantly less than
unity requires a transition state in which $C—O$ bond making is considerably progressed relative to $C—X$ bond breaking and this is nicely
consistent with an addition–elimination mechanism progressing through a
tetrahedral intermediate. Conversely, the RCl/RF rate ratios of appreciably
greater than unity for acetyl and benzoyl halides suggest a fundamental
difference in solvolysis mechanism for these compounds, and they throw
considerable doubt upon the validity of extending conclusions reached
for chloroformate ester solvolyses to other acid-chloride solvolyses.
Consistent with the previously determined ratios[234], dimethylcarbamyl
chloride, which has independently[132, 195] been shown to solvolyse in
water by an ionization mechanism,

$$(CH_3)_2NCOCl \xrightarrow{\text{slow}} (CH_3)_2NCO^+ + Cl^-$$

solvolyses about 10^6 times faster than the corresponding fluoride[236].

Ethyl fluoroformate can be reduced to fluoromethanol by lithium
aluminium hydride in dry diethyl ether[237].

Fluoroformates have been used as intermediates in the syntheses of aryl
and suitably substituted alkyl trifluoromethyl ethers. The synthesis of aryl
trifluoromethyl ethers is a general one[225]. The fluoroformate is reacted
with sulphur tetrafluoride in a pressure vessel at autogeneous pressure and
temperatures of 160–175° for several hours, in the presence of an anhydrous
hydrogen fluoride catalyst. After removal of hydrogen fluoride by a base
wash or by sodium fluoride, yields of 50–80% were isolated in pure form
by distillation. A parallel synthesis can be conducted with the fluorine of
the fluoroformate replaced by a perfluoroalkyl group.

With alkyl fluoroformates[226] the reaction is not a general one but is
restricted to an initial choice of alcohols with electron-withdrawing
substituents on the β-carbon. The intermediate fluoroformate and the
hydrogen fluoride, formed simultaneously in the synthesis from alcohol
and carbonyl fluoride, were immediately reacted with sulphur tetrafluoride.
With one electronegative substituent, overall yields were 30–40% and,
with three, they were increased to 50–60%. Methanol did progress to
methyl trifluoromethyl ether (29%) but other unsubstituted alcohols gave
only tars or carbonaceous products.

D. Bromoformates and Iodoformates

Although bromoformates were first reported[238] prior to fluoroformates[210, 211], very little work has been done with these compounds. The preparation involved treatment of the respective alcohol in dry diethyl ether, or petroleum ether, with carbonyl bromide and subsequent distillation of the ether layer at reduced pressures. In this way ethyl, propyl, isopentyl and benzyl bromoformates were prepared as transparent liquids. Protected from moisture, they were of moderate thermal stability. By comparison with the relative stabilities of chloroformate esters, the ability to prepare and distil (96° boiling point at 9 mm) benzyl bromoformate suggests that many bromoformate esters should be quite stable. Indeed, for ethyl bromoformate an atmospheric pressure boiling point of 116° was measured. Reaction of carbonyl bromide with 8-hydroxyquinoline led, however, not to the bromoformate but to a mixture of 5-bromo- and 5,7-dibromo-8-hydroxyquinoline.

There has been very little subsequent work with these compounds. Bromoformates of polyethylene glycols have been prepared by interaction with carbonyl bromide[239] and ethyl bromoformate has been included in a study of nuclear quadrupole resonance spectra of ^{35}Cl- and ^{79}Br-containing compounds[240].

Iodoformates do not appear to have been characterized. There is one reference in *Chemical Abstracts* to an iodoformate[241]. This is a reference to a patent dealing with aryl haloformate inhibition of the polymerization of vinyl chloride. It is possible that this ester, *p*-nitrophenyl iodoformate, is claimed as a member of a class without having been individually investigated. However, in view of the reasonable stability of alkyl and arylalkyl bromoformates[238], preparation of the more stable members of the iodoformate family may well be feasible. Reaction of cholesteryl chloroformate with sodium iodide in acetone, at 35°, led to cholesteryl iodide[160]. This was anticipated, because the iodoformate of this secondary system, with the additional possibility of neighbouring group participation in carbonium-ion formation, would be expected to be one of the less stable iodoformate esters.

IX. HALOGYLOXALATES

A. Introduction

The haloglyoxalate nomenclature, for compounds of the type ROCOCOX, is consistent with the haloformate nomenclature for compounds of the type ROCOX. It is the nomenclature employed in *Chemical Abstracts*. It visualizes the replacement of the hydrogen of glyoxalic acid by a halogen.

An alternative naming, as halooxalates, is parallel to the naming of compounds of type ROCOX as halocarbonates. This naming visualizes the replacement of a hydroxyl group of oxalic acid by a halogen. Related to the latter nomenclature is the naming as alkyl or aryl oxalyl halides or as the abbreviated form, alkoxalyl or aroxalyl halides.

The presence of a second carbonyl group leads to important modifications of the properties, relative to haloformates. The i.r. spectrum now shows two distinct carbonyl-stretching frequencies, which have been documented for the ethyl, n-octyl and phenyl chloroglyoxalates[242].

Only chloroglyoxalates have been subjected to detailed investigation. Ethyl fluoroglyoxalate has been prepared by reaction of EtOCOCOOH with potassium hydrogen fluoride[243]. Bromoglyoxalates and iodoglyoxalates do not appear to have been reported.

Considerable work has been carried out on derivatives of haloglyoxalate esters. Important derivatives are 2-substituted oximes, semicarbazones and arylhydrazones of the type:

$$R-O-\underset{\underset{O}{\|}}{C}-\underset{\underset{X}{|}}{C}=N-Z$$

where X = Cl or Br and Z = $-OH$, $-NHCONH_2$ or $-NHAr$. Several procedures are available for synthesis of these compounds, but these do not pass through the haloglyoxalate. The 2-oxime of ethyl chloroglyoxalate can be converted to the corresponding iodo-compound by treatment with sodium iodide in acetone[244]. These derivatives have been widely used in the syntheses of a variety of heterocyclic systems. Frequently, the bromine or chlorine is introduced as the final step in their preparation[245]. In other syntheses, it is already present in the reactants; for example[246]:

$$p\text{-}NO_2C_6H_4N_2X + N_2CHCO_2Et \xrightarrow{\text{MeOH, 0°}} p\text{-}NO_2C_6H_4NHN=\underset{\underset{X}{|}}{C}-CO_2Et$$

The extensive chemistry of these derivatives would be out of place in a volume devoted to carbonyl halides and only the preparation and reactions of the 'parent' haloglyoxalates will be discussed.

B. Preparation of Chloroglyoxalates

Three basic techniques exist for the preparation of chloroglyoxalates and all have been used quite extensively. The references given for the uses of the various techniques are not intended to be all inclusive.

Interaction of the alcohol[242, 247] or phenol[242] with commercially available oxalyl chloride is a reasonably good procedure. The main disadvantage is the tendency for further reaction to give the dioxalate[242]. An inert diluent such as diethyl ether[242] or chloroform[247] can be used and the reaction can be accelerated, and hydrogen chloride driven off, by use of a steam bath[242]. A good procedure is to treat the oxalyl chloride with an equivalent of the alcohol, added dropwise[247]. Allyl, cyclohexyl, octyl, menthyl, cetyl and p-nitrobenzyl esters were prepared in this way.

A second method involves treatment of the appropriate dialkyl oxalate with phosphorus pentachloride[248]. For the preparation of ethyl chloroglyoxalate, this was the method favoured by Barré[249]. The reaction proceeds through an alkyl alkoxydichloroacetate[250].

$$ROCOCOOR + PCl_5 \longrightarrow ROCCl_2COOR + POCl_3$$

$$ROCCl_2COOR \longrightarrow ClCOCOOR + RCl$$

The third method involves interaction of the monosubstituted oxalic acid, or the potassium salt of this acid, with thionyl chloride. Ethyl chloroglyoxalate has been made by treatment of diethyl oxalate with excess anhydrous oxalic acid at 130° to give the monosubstituted acid, which was then converted to the chloroglyoxalate by heating with thionyl chloride[251]. This method of preparation of ethyl chloroglyoxalate has also been used by Bergmann and Kalmus[252]. More usually, this procedure has involved preparation, and reaction with thionyl chloride, of the potassium salt. Southwick and Seivard[253] considered this method to be superior to the reaction with phosphorus pentachloride. Dialkyl oxalate is treated with an equivalent amount of potassium acetate, in the appropriate alcohol, to give the potassium salt of the monosubstituted acid:

$$(COOR)_2 + KOAc \xrightarrow[\text{reflux}]{ROH} \begin{array}{c} COOR + ROAc \\ | \\ COO^-K^+ \end{array}$$

Heating with thionyl chloride on a steam bath led to the methyl or ethyl chloroglyoxalates. Overberger and Gainer[254] used basically the same procedure, but with petroleum ether as a diluent in the second step. This reaction has been used in other work for preparation of methyl[255], ethyl[242], n-butyl[256] and n-hexyl[256] chloroglyoxalates. It would appear to be restricted, however, to substrates which allow bimolecular attack of acetate at the α-carbon of the R group. The most general procedure is the interaction of the alcohol or phenol with oxalyl chloride, and this was the method chosen for the synthesis of 1-adamantyl chloroglyoxalate[176].

C. Reactions of Chloroglyoxalates

The carbonyl-bound chlorine is very susceptible to nucleophilic displacement. Hydrolysis leads to the corresponding monosubstituted acids.

Unlike the monosubstituted carbonic acids, these half-esters can be isolated if the calculated amount of water is used. With excess water, further hydrolysis to alcohol or phenol and oxalic acid can occur[257]. With alcohols or phenols, disubstituted oxalate esters are produced[242]. Quinoline was found to be a useful catalyst and hydrogen chloride acceptor, especially for the reactions with phenols. It is of interest that 1-adamantyl chloroglyoxalate solvolyses in methanol to give the mixed oxalate ester[176]. This is in contrast to the solvolysis–decomposition, to give the methyl ether and the chloride, which was observed for 1-adamantyl chloroformate under identical conditions[171]. This is just one example, several more are given below, of the tendency for chloroglyoxalates to retain the ($-OCOCO-$) unit under reaction conditions where chloroformates eject their ($-OCO-$) unit as carbon dioxide. Nucleophilic displacement of chloride occurs readily with amines and their derivatives[253].

Nucleophilic displacement by the potassium salt of the half-ester leads to anhydrides. Methoxalic acid anhydride, prepared in this way is highly active and useful for hydroxyl group determinations[255].

$$CH_3OCOCOCl + K(OOCCOOCH_3) \longrightarrow (CH_3OCOCO)_2O + KCl$$

Friedel–Crafts reactions of alkyl chloroglyoxalates are further examples of reactions proceeding without loss of the ($-OCOCO-$) unit, under conditions where chloroformates react with loss of the ($-OCO-$) unit. Bert[258] showed that, in the presence of aluminium chloride, ethyl chloroglyoxalate reacts with cumene to give the *para*-substituted derivative:

$$EtOCOCOCl + (CH_3)_2CHC_6H_5 \xrightarrow{AlCl_3} p\text{-}(CH_3)_2CHC_6H_4COCO_2Et$$

In cold nitrobenzene, a similar reaction occurs with benzene, to give a 53% yield of ethyl phenylglyoxalate. Other ethyl arylglyoxalates were similarly produced in 70–80% yields. These arylglyoxalates can be catalytically reduced to mandelates $(ArCH(OH)CO_2Et)$[250]. This technique has also been used for *p*-reaction with anisole[254].

Using aluminium chloride in carbon disulphide as catalyst, 1,2,3-trimethoxybenzene was reacted with ethyl chloroglyoxalate for 5 h at water-bath temperature. The CS_2 was removed, the mixture treated with ice-cold HCl and the oil taken up into diethyl ether. The resultant ethyl ester of 2-hydroxy-3,4-dimethoxyphenylglyoxalate was saponified to the free acid, which was ring-closed by dehydration in acetic anhydride, at temperatures below 120°, to give the coumarandione[259].

A simpler procedure to several coumarandiones is the direct reaction of aryl chloroglyoxalates with aluminium chloride in carbon disulphide[260]. These ring closures can be considered as intramolecular Friedel–Crafts reactions. Only the more stable coumarandiones were isolated in good yield. In other systems, products which can be visualized as resulting from a subsequent hydrolytic ring-opening were isolated. Reaction of m-tolyl chloroglyoxalate led to approximately 50% of 6-methylcoumarandione. However, phenyl chloroglyoxalate gave primarily $o\text{-HOC}_6\text{H}_4\text{COCO}_2\text{H}$ and the decarbonylation product, $o\text{-HOC}_6\text{H}_4\text{CO}_2\text{H}$.

The reaction of organocadmium compounds with chloroglyoxalate esters would be expected to provide a convenient route to α-keto-esters. Gilman and Nelson[261] isolated, from the addition of ethyl chloroglyoxalate to an *in situ* preparation of diethylcadmium, the α-hydroxyester, presumably formed by further reaction of the α-keto-ester.

$$\text{EtOCOCOCl} \xrightarrow{\text{CdEt}_2} \text{EtOCOCOEt} \xrightarrow{\text{CdEt}_2} \text{EtOCOC(OH)Et}_2$$

Stacey and McCurdy[262] attempted to limit the second reaction by a slow addition of the organocadmium compound to the ethyl chloroglyoxalate. They isolated only traces of the keto-ester and primarily the ethoxalyl derivative of the α-hydroxyester; further reaction occurs, in this case, due to the presence of excess ethyl chloroglyoxalate. They did manage to increase yields of α-keto-ester by introduction of bulky groups into the organocadmium compound; 7% with diisobutylcadmium, 22% with di-p-tolylcadmium and 50% with di-o-tolylcadmium.

Kollonitsch[263], investigating the reactivity of organocadmium compounds in general, found that the *in situ* preparations, which had always been prepared from the Grignard reagent and a cadmium halide, were unusually active due to the operation of a catalytic salt effect. The catalysis by simultaneously formed magnesium halide was especially powerful. The catalysis by lithium salts was found to be milder in character and sufficient to lead to the α-keto-ester but not for any appreciable further reaction. When n-butyl lithium was reacted with cadmium bromide in ether at $-30°$, the ether removed by distillation, and tetrahydrofuran substituted, addition of this solution to ethyl chloroglyoxalate led to a 70% yield of ethyl α-ketohexanoate. In a similar way, dimethylcadmium, activated by lithium bromide, gave a 51% yield of ethyl pyruvate.

The reaction of ethyl chloroglyoxalate with ketene is best carried out in liquid sulphur dioxide. Treating the resultant acid chloride solution with ethanol gave a fairly good yield of ethyl ethoxalylacetate $(\text{EtO}_2\text{CCOCH}_2\text{CO}_2\text{Et})$[264, 265].

A new class of β-γ-unsaturated α-keto-esters has been prepared by the reaction, for 2 h at 180°, of two equivalents of ethyl chloroglyoxalate with one equivalent of 1,1-diphenylethylene or the p-chloro-derivative[252].

$$(p\text{-ClC}_6\text{H}_4)_2\text{C}=\text{CH}_2 + \text{EtOCOCOCl} \longrightarrow (p\text{-ClC}_6\text{H}_4)_2\text{C}=\text{CHCOCO}_2\text{Et}$$

Irradiation of ethyl chloroglyoxalate in a quartz tube by an unfiltered high-pressure mercury arc (600 W), in an atmosphere of nitrogen, was carried out in the presence of various alkenes[266]. With norbornene, carbon monoxide, carbon dioxide, ethane and ethylene were evolved, and the liquid products were diethyl oxalate (16%), ethyl norbornane-*exo*-2-carboxylate (27%) and a derivative of norbornane-*trans*-2,3-dicarboxylic acid, with one acid function as the ethyl ester and the other as the acid chloride (32%). Similar photochemical reaction with 1-hexene and 1-octene gave 10–15% of the n-alkylsuccinic acid derivatives.

The following mechanism was proposed:

$$\text{ClCOCOOEt} \xrightarrow{h\nu} \cdot\text{COCl} + \cdot\text{COOEt (I)}$$

$$>\text{C}=\text{C}< + \text{I} \longrightarrow \overset{\displaystyle |}{\underset{\displaystyle |}{\cdot\text{C}}}-\overset{\displaystyle |}{\underset{\displaystyle \text{COOEt}}{\text{C}}}- \text{(II)}$$

$$\text{II} + \text{ClCOCOOEt} \longrightarrow -\overset{\displaystyle |}{\underset{\displaystyle \text{ClCO}}{\text{C}}}-\overset{\displaystyle |}{\underset{\displaystyle \text{COOEt}}{\text{C}}}- \; + \text{I}$$

$$\text{II} + \text{a hydrogen donor} \longrightarrow -\overset{\displaystyle |}{\underset{\displaystyle \text{H}}{\text{C}}}-\overset{\displaystyle |}{\underset{\displaystyle \text{COOEt}}{\text{C}}}-$$

The investigation has been extended to attack upon cyclohexane, cyclohexene and toluene[267]. In cyclohexane, a cyclohexyl radical is believed to attack at the carbon of the acid chloride function, to cause a chain reaction. Cyclohexanecarbonyl chloride (51%) and ethyl cyclohexylglyoxalate (2–3%) were formed as the major products. The same major products were obtained, in similar amounts, when the reaction was initiated by dibenzoyl peroxide. The minor products were, however, different. With cyclohexene and toluene the material balance was not good but the products identified suggested that cyclohexen-3-yl and benzyl radicals were unable, under the reaction conditions employed, to attack ethyl chloroglyoxalate.

Rhoads and Michel[27] have investigated the thermal decomposition of several chloroglyoxalate esters. With pyridine present, ethyl, n-butyl, *sec*-octyl and α-phenethyl chloroglyoxalates decomposed readily to give high yields of the corresponding chlorides. With optically active *sec*-octyl and α-phenethyl esters, it was possible to show a clean inversion of configuration. The mechanism was formulated as paralleling the similar pyridine-catalysed decompositions of chloroformate esters[48, 52].

With pyridine absent, ethyl, *n*-butyl and *sec*-octyl chloroglyoxalates were found to be stable, both in the neat phase and in a variety of solvents, at temperatures of up to 250°. The α-phenethyl chloroglyoxalate decomposed in the neat state at 125° to give 40% styrene and 60% α-phenethyl chloride, formed with extensive racemization and a small degree of retention of configuration. In nitrobenzene at 108°, the percentage of chloride was 90%, formed with at least 60% retention of configuration (40% or less racemization). The relative rates in nitrobenzene, *bis*-(2-methoxyethyl) ether and tetralin were as 13·6 : 2·34 : 1.

With 1-adamantyl chloroglyoxalate[176], decomposition of the solid, on standing at room temperature for 2 months, gave a little tar and 1-adamantyl chloride (80%) was isolated. In solution the decomposition was less clean. In nitrobenzene, 94% of the hydrogen chloride assayed after addition to methanol disappeared after 76 h at 95° and only a little tar was produced. In benzene, appreciable amounts of tar were formed and, after several recrystallizations from methanol, a 13% yield of 1-phenyladamantane was isolated. In *n*-decane, much tar was formed and a new carbonyl peak appeared in the i.r. spectrum of the solution. Although the analytical technique employed was strictly applicable only to reactions proceeding to 1-adamantyl chloride, the decompositions were appreciably slower, and the solvent polarity effect was much less, than for 1-adamantyl chloroformate[45].

Rhoads and Michel[27] pointed out that rate-determining alkyl (or aralkyl)–oxygen bond fission would require a chloroglyoxalate to have a decomposition rate intermediate between those of the corresponding chloroformate and chlorosulphinate. To explain the much slower rates, they suggested that carbon–chlorine bond fission was also an important factor. They preferred a rather polar version of a mechanism proceeding through a five-centred transition state over an alternative ionic mechanism, incorporating extensive ion pair return. This was supported for decomposition of α-phenethyl chloroglyoxalate by the low solvent polarity response and the large negative entropy of activation (-27 e.u.) in nitrobenzene. However, the reaction would not follow this mechanism if a faster ion-pair mechanism was available and extensive ion-pair return ($k_{-1} \gg k_2$) is required to explain the lower rates relative to chloroformates. Such return would not be too surprising because other reactions, already outlined, have shown the tendency for the (—OCOCO—) unit to be retained under conditions where the (—OCO—) unit is lost.

$$ROCOCOCl \underset{k_{-1}}{\overset{k_1}{\rightleftarrows}} R^+(OCOCOCl)^- \overset{k_2}{\longrightarrow} R^+Cl^- + CO + CO_2$$

$$\downarrow$$

$$RCl$$

The large negative entropy of activation is not easily correlated with the above ion-pair mechanism but could result from a modification in which the carbonium ion extracts a chloride ion from the chloroglyoxalate ion, a process which would require a restricted orientation. Also, extensive internal ion-pair return tends to lower experimental entropy of activation values for the overall process[173]. Another important factor, which should be considered, is that the low sensitivity of the reaction rate to changes in solvent polarity may result from the incursion of free-radical processes, at least in the lower polarity solvents. The observations concerning 1-adamantyl chloroglyoxalate thermolyses in benzene and decane would be consistent with free-radical processes.

X. REFERENCES

1. *Phosgene Booklet*, Chemetron Chemicals, Organic Chemical Department, 386 Park Avenue South, New York, 1963.
2. M. Bergmann and L. Zervas, *Ber.*, **65**, 1192 (1932).
3. K. D. Kopple, *Peptides and Amino Acids*, W. J. Benjamin, New York, 1966, p. 34.
4. Y. Wolman in *The Chemistry of the Amino Group* (Ed. S. Patai), Interscience, New York, 1968, pp. 682–688.
5. W. L. Haas, E. V. Krumkalns and K. Gerzon, *J. Am. Chem. Soc.*, **88**, 1988 (1966).
6. E. Wunsch, A. Zwiek and E. Jaeger, *Chem. Ber.*, **101**, 336 (1968).
7. M. Matzner, R. P. Kurkjy and R. J. Cotter, *Chem. Rev.*, **64**, 645 (1964).
8. Houben-Weyl, *Methoden der Organischen Chemie*, Vol. 8, George Thieme Verlag, Stuttgart, 1952, pp. 101–105.
9. American Cyanamid Co., *British Pat.*, 586,633 (1947); *Chem. Abstr.*, **41**, 6898 (1947).
10. C. Scholtissek, *Chem. Ber.*, **89**, 2562 (1956).
11. E. L. Wittbecker and P. W. Morgan, *J. Polymer Sci.*, **40**, 367 (1959).
12. E. S. Lewis and C. E. Boozer, *J. Am. Chem. Soc.*, **74**, 308 (1952).
13. P. D. Bartlett and H. F. Herbrandson, *J. Am. Chem. Soc.*, **74**, 5971 (1952).
14. E. H. White, *J. Am. Chem. Soc.*, **77**, 6014 (1955).
15. C. J. Collins, J. B. Christie and V. F. Raaen, *J. Am. Chem. Soc.*, **83**, 4267 (1961).

16. E. H. White and C. A. Aufdermarsh, *J. Am. Chem. Soc.*, **83**, 1174, 1179 (1961).
17. E. H. White and D. W. Grisley, Jr., *J. Am. Chem. Soc.*, **83**, 1191 (1961).
18. E. H. White and D. J. Woodcock in *The Chemistry of the Amino Group* (Ed. S. Patai), Interscience, New York, 1968, pp. 407–497.
19. C. K. Ingold, *Structure and Mechanism in Organic Chemistry*, 2nd ed., Cornell University Press, Ithaca, New York, 1969, pp. 534–538.
20. D. J. Cram, *J. Am. Chem. Soc.*, **75**, 332 (1953).
21. S. Winstein and G. C. Robinson, *J. Am. Chem. Soc.*, **80**, 169 (1958).
22. S. H. Sharman, F. F. Caserio, R. F. Nystrom, J. C. Leak and W. G. Young, *J. Am. Chem. Soc.*, **80**, 5965 (1958).
23. K. L. Olivier and W. G. Young, *J. Am. Chem. Soc.*, **81**, 5811 (1959).
24. E. S. Lewis and W. C. Herndon, *J. Am. Chem. Soc.*, **83**, 1955 (1961).
25. E. S. Lewis, W. C. Herndon and D. C. Duffy, *J. Am. Chem. Soc.*, **83**, 1959 (1961).
26. E. S. Lewis and W. C. Herndon, *J. Am. Chem. Soc.*, **83**, 1961 (1961).
27. S. J. Rhoads and R. E. Michel, *J. Am. Chem. Soc.*, **85**, 585 (1963).
28. W. Gerrard and H. R. Hudson, *J. Chem. Soc.*, 1059 (1963).
29. C. C. Lee and A. J. Findlayson, *Can. J. Chem.*, **39**, 260 (1961).
30. M. Cocivera and S. Winstein, *J. Am. Chem. Soc.*, **85**, 1702 (1963).
31. See, for example, L. P. Hammett, *Physical Organic Chemistry*, 2nd ed., McGraw–Hill, New York, 1970, pp. 355–357.
32. S. G. Smith, A. H. Fainberg and S. Winstein, *J. Am. Chem. Soc.*, **83**, 618 (1961).
33. H. L. Goering and J. F. Levy, *Tetrahedron Letters*, 644 (1961); *J. Am. Chem. Soc.*, **84**, 3853 (1962).
34. W. Gerrard and H. R. Hudson, *Chem. Revs.*, **65**, 697 (1965).
35. R. H. deWolfe and W. G. Young, *Chem. Revs.*, **56**, 784 (1956).
36. Reference 19, pp. 515–526.
37. Reference 7, p. 671.
38. K. B. Wiberg and T. M. Shryne, *J. Am. Chem. Soc.*, **77**, 2774 (1955).
39. C. G. Swain and W. P. Langsdorf, Jr., *J. Am. Chem. Soc.*, **73**, 2813 (1951).
40. For discussion, see J. E. Leffler and E. Grunwald, *Rates and Equilibria of Organic Reactions*, John Wiley, New York, 1963, p. 190.
41. See, for example, A. A. Frost and R. G. Pearson, *Kinetics and Mechanism*, 2nd ed., John Wiley, New York, 1961, pp. 137–142.
42. J. L. Kice, R. A. Bartsch, M. A. Dankleff and S. L. Schwartz, *J. Am. Chem. Soc.*, **87**, 1734 (1965).
43. J. L. Kice, R. L. Scriven, E. Koubek and M. Barnes, *J. Am. Chem. Soc.*, **92**, 5608 (1970).
44. J. L. Kice and M. A. Dankleff, *Tetrahedron Letters*, 1783 (1966).
45. D. N. Kevill and F. L. Weitl, *J. Am. Chem. Soc.*, **90**, 6416 (1968).
46. J. L. Kice and G. C. Hanson, *Tetrahedron Letters*, 2927 (1970).
47. W. A. Cowdrey, E. D. Hughes, C. K. Ingold, S. Masterman and A. D. Scott, *J. Chem. Soc.*, 1252 (1937).
48. A. H. J. Houssa and H. Philips, *J. Chem. Soc.*, 1232 (1932).
49. A. H. J. Houssa and H. Philips, *J. Chem. Soc.*, 108 (1932).
50. A. H. J. Houssa and H. Philips, *J. Chem. Soc.*, 2510 (1929).
51. J. Kenyon, A. G. Lipscomb and H. Philips, *J. Chem. Soc.*, 415 (1930).

52. J. Kenyon, A. G. Lipscomb and H. Philips, *J. Chem. Soc.*, 2275 (1931).
53. D. N. Kevill, G. H. Johnson and W. A. Neubert, *Tetrahedron Letters*, 3727 (1966).
54. P. W. Clinch and H. R. Hudson, *Chem. Commun.*, 925 (1968).
55. J. H. Ridd, *Quart. Rev.*, **15**, 418 (1961).
56. J. Meisenheimer and J. Link, *Justus Liebigs Ann. Chem.*, **479**, 211 (1930).
57. F. F. Caserio, G. E. Dennis, R. H. DeWolfe and W. G. Young, *J. Am. Chem. Soc.*, **77**, 4182 (1955).
58. J. D. Roberts, W. G. Young and S. Winstein, *J. Am. Chem. Soc.*, **64**, 2157 (1942).
59. H. L. Goering, T. D. Nevitt and E. F. Silversmith, *J. Am. Chem. Soc.*, **77**, 4042 (1955).
60. J. A. Pegolotti and W. G. Young, *J. Am. Chem. Soc.*, **83**, 3251 (1961).
61. S. Winstein and G. C. Robinson, *J. Am. Chem. Soc.*, **80**, 169 (1958).
62. A. Streitwieser, *Chem. Revs.*, **56**, 571 (1956).
63. S. T. Bowden, *J. Chem. Soc.*, 310 (1939).
64. A. R. Choppin and J. W. Rodgers, *J. Am. Chem. Soc.*, **70**, 2967 (1948).
65. E. Merck, *German Pat.*, 254,471 (1913); *Chem. Abstr.*, **7**, 1082 (1913).
66. S. Sakakibara, M. Shin, M. Fujino, Y. Shimonishi, S. Inove and N. Inukai, *Bull. Chem. Soc. Japan*, **38**, 1522 (1965).
67. E. S. Lewis and G. M. Coppinger, *J. Am. Chem. Soc.*, **76**, 796 (1954).
68. P. Beak and R. J. Trancik, *Abstracts*, 154*th National Am. Chem. Soc. Meeting*, Chicago, Illinois, September, 1967, S164.
69. C. E. Boozer and E. S. Lewis, *J. Am. Chem. Soc.*, **76**, 794 (1954).
70. A. Maccoll and P. J. Thomas, *Progr. Reaction Kinetics*, **4**, 119 (1967).
71. A. Maccoll, *Chem. Revs.*, **69**, 33 (1969).
72. Reference 70, Table 4.
73. E. S. Lewis and K. Witte, *J. Chem. Soc.* (B), 1198 (1968).
74. H. E. O'Neal and S. W. Benson, *J. Phys. Chem.*, **71**, 2903 (1967).
75. L. A. K. Staveley and C. N. Hinshelwood, *Nature*, **137**, 29 (1936).
76. P. J. Thomas, *J. Chem. Soc.* (B), 1238 (1967).
77. P. G. Rodgers, *Ph.D. Thesis*, University of London, 1966; discussed in reference 71.
78. C. J. Harding, A. Maccoll and R. A. Ross, *J. Chem. Soc.* (B), 634 (1969).
79. E. T. Lessing, *J. Phys. Chem.*, **36**, 2325 (1932).
80. A. R. Choppin and E. L. Compere, *J. Am. Chem. Soc.*, **70**, 3797 (1948).
81. D. Brearley, G. B. Kistiakowsky and C. H. Stauffer, *J. Am. Chem. Soc.*, **58**, 43 (1936).
82. A. R. Choppin, H. A. Frediani and G. F. Kirby, Jr., *J. Am. Chem. Soc.*, **61**, 3176 (1939).
83. A. R. Choppin and G. F. Kirby, Jr., *J. Am. Chem. Soc.*, **62**, 1592 (1940).
84. L. E. Roberts, R. Lashbrook, M. J. Treat and W. Yates, *J. Am. Chem. Soc.*, **74**, 5787 (1952).
85. A. Maccoll and P. J. Thomas, *Nature*, **176**, 392 (1955).
86. C. K. Ingold, *Proc. Chem. Soc.*, 279 (1957).
87. H. C. Ramsberger and G. Waddington, *J. Am. Chem. Soc.*, **55**, 214 (1933).
88. A. Maccoll, *Chem. Soc. Special Publication*, **16**, 159 (1962).
89. D. Y. Curtin and D. B. Kellom, *J. Am. Chem. Soc.*, **75**, 6011 (1953).
90. G. M. Fraser and H. M. R. Hoffmann, *Chem. Commun.*, 561 (1967).

91. D. N. Kevill and N. H. Cromwell, *J. Am. Chem. Soc.*, **83**, 3815 (1961).
92. J. L. Fry, C. J. Lancelot, L. K. M. Lam, J. M. Harris, R. C. Bingham, D. J. Raber, R. E. Hall and P. v. R. Schleyer, *J. Am. Chem. Soc.*, **92**, 2538 (1970).
93. C. S. Marvel and N. O. Brace, *J. Am. Chem. Soc.*, **70**, 1775 (1948).
94. W. J. Bailey and R. Barclay, Jr., *J. Org. Chem.*, **21**, 328 (1956).
95. E. S. Lewis, J. T. Hill and E. R. Newman, *J. Am. Chem. Soc.*, **90**, 662 (1968).
96. Y. Shigemitsu, T. Tominaga, T. Shimodaira, Y. Odaira and S. Tsutsumi, *Bull. Chem. Soc. Japan*, **39**, 2463 (1966).
97. P. Gray and J. C. J. Thynne, *Nature*, **191**, 1357 (1961).
98. H. L. West and G. K. Rollefson, *J. Am. Chem. Soc.* **58**, 2140 (1936).
99. G. C. Pimentel and K. C. Herr, *J. Chim. Phys.*, **61**, 1509 (1964).
100. K. C. Herr and G. C. Pimentel, *Appl. Opt.*, **4**, 25 (1965).
101. R. J. Jensen and G. C. Pimentel, *J. Phys. Chem.*, **71**, 1803 (1967).
102. R. S. Lerner, B. P. Dailey and J. P. Friend, *J. Chem. Phys.*, **26**, 680 (1957).
103. H. W. Thompson and D. A. Jameson, *Spectrochim. Acta*, **13**, 236 (1957).
104. D. Cook, *J. Am. Chem. Soc.*, **80**, 49 (1958).
105. A. W. Baker and G. H. Harris, *J. Am. Chem. Soc.*, **82**, 1923 (1960).
106. L. H. Little, G. W. Poling and J. Leja, *Can. J. Chem.*, **39**, 745 (1961).
107. B. Collingwood, H. Lee and J. K. Wilmshurst, *Australian J. Chem.*, **19**, 1637 (1966).
108. J. M. O'Gorman, W. Shand, Jr. and V. Schomaker, *J. Am. Chem. Soc.*, **78**, 4222 (1950).
109. M. Kashima, *Bull. Chem. Soc. Japan*, **25**, 79 (1952).
110. Reference 7, p. 650.
111. See, for example, B. L. Shaw in *Physical Methods in Organic Chemistry* (Ed. J. C. P. Schwarz), Holden-Day, San Francisco, 1964, pp. 323–337.
112. R. J. W. LeFèvre, *Dipole Moments*, 3rd ed., Methuen, London, 1953, p. 109.
113. C. P. Smyth, *J. Am. Chem. Soc.*, **47**, 1894 (1925).
114. H. Mohler, *Helv. Chim. Acta*, **21**, 787 (1938).
115. S. K. K. Jatkar and V. K. Phansalkar, *J. Univ. Poona, Science and Technol.*, **4**, 45 (1953); *Chem. Abstr.*, **48**, 11859 (1954).
116. S. Mizushima and M. Kubo, *Bull. Chem. Soc. Japan*, **13**, 174 (1938).
117. C. T. Zahn, *Physik. Z.*, **33**, 730 (1932).
118. C. T. Zahn, *Trans. Faraday Soc.*, **30**, 804 (1934).
119. R. J. W. LeFèvre and A. Sundaram, *J. Chem. Soc.*, 3904 (1962).
120. E. Bock and D. Iwacha, *Can. J. Chem.*, **45**, 3177 (1967).
121. J. W. Smith, *Electric Dipole Moments*, Butterworths, London, 1955.
122. R. F. Curl, Jr., *J. Chem. Phys.*, **30**, 1529 (1959).
123. M. J. Aroney, R. J. W. LeFèvre, R. K. Pierens and H. L. K. The, *Australian J. Chem.*, **22**, 1599 (1969).
124. C. G. LeFèvre and R. J. W. LeFèvre, *J. Chem. Soc.*, 1696 (1935).
125. M. J. Aroney, R. J. W. LeFèvre and A. N. Singh, *J. Chem. Soc.*, 564 (1965).
126. Reference 19, pp. 1128–1178.
127. J. F. Bunnett, M. M. Robison and F. C. Pennington, *J. Am. Chem. Soc.* **72**, 2378 (1950).
128. M. Green and R. F. Hudson, *J. Chem. Soc.*, 1076 (1962).
129. I. Ugi and F. Beck, *Chem. Ber.*, **94**, 1839 (1961).

130. R. Leimu, *Chem. Ber.*, **70**, 1040 (1937).
131. H. Bohme and W. Schurhoff, *Chem. Ber.*, **84**, 28 (1951).
132. H. K. Hall, Jr., *J. Am. Chem. Soc.*, **77**, 5993 (1955); see, also, footnote 20 to reference 195.
133. H. K. Hall, Jr. and P. W. Morgan, *J. Org. Chem.*, **21**, 249 (1956).
134. C. Faurholt and J. C. Gjaldbaek, *Dansk. Tids. Farm.*, **19**, 255 (1945); *Chem. Abstr.*, **40**, 513 (1946).
135. E. W. Crunden and R. F. Hudson, *J. Chem. Soc.*, 3748 (1961).
136. E. W. Crunden and R. F. Hudson, *J. Chem. Soc.*, 501 (1956).
137. D. S. Tarbell and E. J. Longosz, *J. Org. Chem.*, **24**, 774 (1959).
138. M. Green and R. F. Hudson, *J. Chem. Soc.*, 1076 (1962).
139. J. D. Roberts and V. C. Chambers, *J. Am. Chem. Soc.*, **73**, 5034 (1951).
140. H. C. Brown and G. Ham, *J. Am. Chem. Soc.*, **78**, 2735 (1956).
141. H. C. Brown, R. S. Fletcher and R. B. Johannsen, *J. Am. Chem. Soc.*, **73**, 212 (1951).
142. H. C. Brown and M. Borkowski, *J. Am. Chem. Soc.*, **74**, 1894 (1952).
143. D. H. R. Barton and R. C. Cookson, *Quart. Rev.*, **10**, 44 (1956).
144. J. Urbanski, *Roczniki Chem.*, **36**, 1441 (1962); *Chem. Abstr.*, **59**, 6222 (1963).
145. A. Kivinen, *Acta Chem. Scand.*, **19**, 845 (1965).
146. A. Kivinen, *Suomen Kemistilehti*, **38B**, 205 (1965); *Chem. Abstr.*, **64**, 5812 (1966).
147. A. Kivinen and A. Viitala, *Suomen Kemistilehti*, **40B**, 19 (1967); *Chem. Abstr.*, **67**, 21212 (1967).
148. A. Queen, *Can. J. Chem.*, **45**, 1619 (1967).
149. V. A. Pattison, J. G. Colson and R. L. K. Carr, *J. Org. Chem.*, **33**, 1084 (1968).
150. J. Nemirovsky, *J. Prakt. Chem.*, **31**, 173 (1885).
151. H. K. Hall, Jr., *J. Am. Chem. Soc.*, **79**, 5439 (1957).
152. R. F. Hudson and G. Loveday, *J. Chem. Soc.*, 1068 (1962).
153. F. H. Carpenter and D. T. Gish, *J. Am. Chem. Soc.*, **74**, 3818 (1952).
154. H. J. Bestmann and K. H. Schnabel, *Justus Liebigs Ann. Chem.*, **698**, 106 (1966).
155. M. Green and R. F. Hudson, *Proc. Chem. Soc.*, 149 (1959); *J. Chem. Soc.*, 1055 (1962).
156. J. O. Edwards, *J. Am. Chem. Soc.*, **78**, 1819 (1956).
157. L. B. Jones and J. P. Foster, *J. Org. Chem.*, **32**, 2900 (1967).
158. J. B. Conant and W. R. Kirner, *J. Am. Chem. Soc.*, **46**, 232 (1924).
159. J. B. Conant, W. R. Kirner and R. E. Hussey, *J. Am. Chem. Soc.*, **47**, 488 (1925).
160. D. N. Kevill and F. L. Weitl, *J. Org. Chem.*, **32**, 2633 (1967).
161. A. Yamamoto and M. Kobayashi, *Bull. Chem. Soc. Japan.*, **39**, 1283 (1966).
162. H. M. R. Hoffman, *J. Chem. Soc.*, 6748 (1965).
163. A. Yamamoto and M. Kobayashi, *Bull. Chem. Soc. Japan*, **39**, 1288 (1966).
164. R. Boschan, *J. Am. Chem. Soc.*, **81**, 3341 (1959).
165. D. N. Kevill and G. H. Johnson, *J. Am. Chem. Soc.*, **87**, 928 (1965).
166. D. N. Kevill and G. H. Johnson, *Chem. Commun.*, 235 (1966).
167. V. A. Pattison, *J. Org. Chem.*, **31**, 954 (1966).
168. G. A. Mortimer, *J. Org. Chem.*, **27**, 1876 (1962).

169. A. Chaney and M. L. Wolfrom, *J. Org. Chem.*, **26**, 2998 (1961).
170. M. J. Zabik and R. D. Schuetz, *J. Org. Chem.*, **32**, 300 (1967).
171. D. N. Kevill, F. L. Weitl and Sr. V. M. Horvath, *Preprints, Div. of Petrol. Chem. Am. Chem. Soc.*, **15** (2), B66 (1970).
172. P. v. R. Schleyer and R. D. Nicholas, *J. Am. Chem. Soc.*, **83**, 2700 (1961).
173. D. N. Kevill and R. F. Sutthoff, *J. Chem. Soc.* (B), 366 (1969).
174. P. R. Wells, *Chem. Revs.*, **63**, 171 (1963).
175. D. N. Kevill and F. L. Weitl, unpublished results.
176. F. L. Weitl, *Ph.D. Thesis*, Northern Illinois University, August, 1969.
177. D. N. Kevill and F. T. Weitl, *J. Org. Chem.*, **35**, 2526 (1970).
178. Reference 7, p. 651.
179. T. Wilm and G. Wischin, *Justus Liebigs Ann. Chem.*, **147**, 150 (1868).
180. C. Friedel and J. M. Crafts, *Compt. Rend.*, **84**, 1450 (1877).
181. C. Friedel and J. M. Crafts, *Ann. Chim.*, [6] **1**, 527 (1884).
182. E. H. J. Rennie, *J. Chem. Soc.*, **41**, 33 (1882).
183. N. P. Buu-Hoi and J. Janicaud, *Bull. Soc. Chim. France*, **12**, 640 (1945).
184. A. Butlerow, *Bull. Soc. Chim. France*, 586 (1863).
185. K. Ulsch, *Justus Liebigs Ann. Chem.*, **226**, 281 (1884).
186. H. W. Underwood, Jr. and O. L. Baril, *J. Am. Chem. Soc.*, **53**, 2201 (1931).
187. E. Bowden, *J. Am. Chem. Soc.*, **60**, 645 (1938).
188. G. Olofsson, *Acta Chem. Scand.*, **21**, 1114 (1967).
189. G. A. Olah and J. M. Bollinger, unpublished results; reported by G. A. Olah and J. A. Olah in *Carbonium Ions*, Vol. 2 (Eds. G. A. Olah and P. v. R. Schleyer), Wiley–Interscience, New York, 1970, p. 765.
190. V. V. Korshak and G. S. Kolesnikov, *J. Gen. Chem. USSR*, **14**, 435 (1944); *Chem. Abstr.*, **39**, 4595 (1945).
191. F. Kunckell and G. Ulex, *J. Prakt. Chem.*, **86**, 518 (1912); *Chem. Abstr.*, **7**, 778 (1913).
192. F. Kunckell and G. Ulex., *J. Prakt. Chem.*, **87**, 227 (1913); *Chem. Abstr.*, **7**, 2219 (1913).
193. F. A. Drahowzal in *Friedel–Crafts and Related Reactions*, Vol. 2 (Ed. G. A. Olah), Interscience, New York, 1964, pp. 644–645.
194. W. H. Coppock, *J. Org. Chem.*, **22**, 325 (1957).
195. H. K. Hall, Jr. and C. H. Lueck, *J. Org. Chem.*, **28**, 2818 (1963).
196. P. Beck, R. J. Trancik, J. B. Mooberry and P. Y. Johnson, *J. Am. Chem. Soc.*, **88**, 4288 (1966).
197. D. N. Kevill and H. S. Posselt, *Chem. Commun.*, 438 (1967).
198. P. Beak and R. J. Trancik, *J. Am. Chem. Soc.*, **90**, 2714 (1968).
199. P. Beak, R. J. Trancik and D. A. Simpson, *J. Am. Chem. Soc.*, **91**, 5073 (1969).
200. P. D. Bartlett and L. H. Knox, *J. Am. Chem. Soc.*, **61**, 3184 (1939).
201. F. Seel, *Z. anorg. allgem. Chem.*, **250**, 331 (1943).
202. G. A. Olah, S. J. Kuhn, W. S. Tolgyesi and E. B. Baker, *J. Am. Chem. Soc.*, **84**, 2733 (1962).
203. D. N. Kevill, W. A. Reis and J. B. Kevill, unpublished results.
204. J. Radell, J. W. Connolly and A. J. Raymond, *J. Am. Chem. Soc.*, **83**, 3958 (1961).

205. D. N. Kevill, V. V. Likhite, H. S. Posselt and B. Shen, *Abstracts*, *3rd Great Lakes Regional American Chemical Society Meeting*, DeKalb, Illinois, June, 1969, p. 56.
206. See, for example, D. V. Banthorpe in *The Chemistry of the Amino Group* (Ed. S. Patai), Interscience, New York, 1968, pp. 586–587.
207. Reference 19, pp. 437–438.
208. Reference 40, p. 222.
209. K. O. Christe and A. E. Pavlath, *J. Org. Chem.*, **30**, 1639 (1965).
210. H. C. Goswami and P. B. Sarkar, *J. Indian Chem. Soc.*, **10**, 537 (1933); *Chem. Abstr.*, **28**, 1332 (1934).
211. P. C. Ray, *Nature*, **132**, 173 (1933).
212. E. Gryszkiewicz-Trochimowski, A. Sporzynski and J. Wnuk, *Rec. Trav. Chim.*, **66**, 413 (1947).
213. B. C. Saunders and G. J. Stacey, *J. Chem. Soc.*, 1773 (1948).
214. G. A. Olah and A. Pavlath, *Acta Chim. Acad. Sci. Hung.*, **3**, 191 (1953); *Chem. Abstr.*, **48**, 7533 (1954).
215. H. Kitano and K. Fukui, *J. Chem. Soc. Japan*, **58**, 603 (1955); *Chem. Abstr.*, **50**, 11262 (1956).
216. M. S. Kharasch, S. Weinhouse and E. V. Jensen, *J. Am. Chem. Soc.*, **77**, 3145 (1955).
217. C. W. Tullock and D. D. Coffmann, *J. Org. Chem.*, **25**, 2016 (1960).
218. S. Nakanishi, T. C. Myers and E. V. Jensen, *J. Am. Chem. Soc.*, **77**, 3099 (1955).
219. V. A. Welch and P. W. Kent, *J. Chem. Soc.*, 2266 (1962).
220. H. J. Emeleus and J. W. Wood, *J. Chem. Soc.*, 2183 (1948).
221. K. O. Christe and A. E. Pavlath, *J. Org. Chem.*, **30**, 3170 (1965).
222. F. S. Fawcett, C. W. Tullock and D. D. Coffmann, *J. Am. Chem. Soc.*, **84**, 4275 (1962).
223. K. O. Christe and A. E. Pavlath, *J. Org. Chem.*, **30**, 4104 (1965).
224. E. Schnabel, H. Herzog, P. Hoffmann, E. Klauke and I. Ugi, *Justus Liebigs Ann. Chem.*, **716**, 175 (1968).
225. W. A. Sheppard, *J. Org. Chem.*, **29**, 1 (1964).
226. P. E. Aldrich and W. A. Sheppard, *J. Org. Chem.*, **29**, 11 (1964).
227. G. A. Olah and S. J. Kuhn, *J. Org. Chem.*, **21**, 1319 (1956).
228. P. J. Aymonina, *Chem. Commun.*, 241 (1965).
229. G. Williams, N. L. Owen and J. Sheridan, *Chem. Commun.*, 57 (1968).
230. E. Bock, D. Iwacha, H. Hutton and A. Queen, *Can. J. Chem.*, **46**, 1645 (1968).
231. A. A. Neimysheva and I. L. Knunyants, *Dokl. Akad. Nauk SSSR*, **177**, 856 (1967).
232. S. Nakanishi, T. C. Myers and E. V. Jensen, *J. Am. Chem. Soc.*, **77**, 5033 (1955).
233. G. A. Olah and D. Kreienbuhl, *J. Org. Chem.*, **32**, 1614 (1967).
234. C. G. Swain and C. B. Scott, *J. Am. Chem. Soc.*, **75**, 246 (1953).
235. J. Miller, *Aromatic Nucleophilic Substitution*, Elsevier, New York, 1968, p. 19.
236. A. Queen, T. A. Nour and H. Gyulai, unpublished results; reported in A. Queen, T. A. Nour, M. N. Paddon-Row and K. Preston, *Can. J. Chem.*, **48**, 522 (1970).

237. G. A. Olah and A. Pavlath, *Acta Chim. Acad. Sci. Hung.*, **3**, 203 (1953); *Chem. Abstr.*, **48**, 7533 (1954).
238. K. W. Rosenmund and H. Doring, *Arch Pharm.*, **266**, 277 (1928); *Chem. Abstr.*, 2741 (1928).
239. Pittsburgh Plate Glass Co., *British Pat.*, 587,933; *Chem. Abstr.*, **42**, 206 (1948).
240. I. P. Biryukov and M. G. Voronkov, *Latv. PSR Zinat. Akad. Vestis*, 39 (1966); *Chem. Abstr.*, **68**, 44645 (1968).
241. *Chem. Abstr.*, **53**, P1848g (1959).
242. M. S. Simon and H. McC. Seyferth, *J. Org. Chem.*, **23**, 1078 (1958).
243. G. Olah, S. Kuhn and S. Beke, *Chem. Ber.*, **89**, 862 (1956).
244. W. Steinkopf and B. Jurgens, *J. Prakt. Chem.*, **83**, 453 (1911); *Chem. Abstr.*, **5**, 3234 (1911).
245. See, for example, G. Werber, F. Buccheri and F. Maggio, *Ann. Chim. (Rome)*, **56**, 1210 (1966).
246. R. Huisgen and H. J. Koch, *Justus Liebigs Ann. Chem.*, **591**, 200 (1955).
247. G. v. Frank and W. Caro, *Ber.*, **63**, 1532 (1930).
248. R. Anschütz, *Justus Liebigs Ann. Chem.*, **254**, 1 (1889).
249. R. Barré, *Bull. Soc. Chim.*, **41**, 47 (1927); *Chem. Abstr.*, **21**, 1632 (1927).
250. K. Kinder, W. Metzendorf and Dschi-yin-Kwok, *Ber.*, **76**, 308 (1943).
251. E. Fourneau and S. Sabetay, *Bull. Soc. Chim.*, **41**, 537 (1927); *Chem. Abstr.*, **21**, 3890 (1927).
252. F. Bergmann and A. Kalmus, *J. Chem. Soc.*, 4521 (1952).
253. P. L. Southwick and L. L. Seivard, *J. Am. Chem. Soc.*, **71**, 2532 (1949).
254. C. G. Overberger and H. Gainer, *J. Am. Chem. Soc.*, **80**, 4556 (1958).
255. R. V. Oppenauer, *Monatsh. Chem.*, **97**, 67 (1966).
256. F. C. Novello, *U.S. Pat.*, 3,066,157 (1962); *Chem. Abstr.*, **58**, 8987 (1963).
257. R. Stollé, *Ber.*, **47**, 1130 (1914); *Chem. Abstr.*, **8**, 2167 (1914).
258. L. Bert, *Bull. Soc. Chim.*, **37**, 1397 (1925); *Chem. Abstr.*, **20**, 1793 (1926).
259. M. Guia and M. Guerzio, *Gazz. Chim. Ital.*, **92**, 1474 (1962).
260. R. Stollé and E. Knebel, *Ber.*, **54**, 1213 (1921); *Chem. Abstr.*, **15**, 3479 (1921).
261. H. Gilman and J. F. Nelson, *Rec. Trav. Chim.*, **55**, 518 (1927).
262. G. W. Stacy and R. M. McCurdy, *J. Am. Chem. Soc.*, **76**, 1914 (1954).
263. J. Kollonitsch, *Nature*, **188**, 140 (1960).
264. J. Beranek, J. Smrt and F. Sorm, *Chem. Listy*, **48**, 679 (1954).
265. J. Smrt, J. Beranek and F. Sorm, *Chem. Listy*, **49**, 73 (1955); *Collection Czech. Chem. Commun.*, **20**, 285 (1955).
266. C. Pac and S. Tsutsumi, *Tetrahedron Letters*, 2341 (1965).
267. C. Pac and S. Tsutsumi, *Bull. Chem. Soc. Japan*, **39**, 1926 (1966).

CHAPTER **13**

The acyl hypohalites

DENNIS D. TANNER

University of Alberta, Edmonton, Alberta, Canada

and

NIGEL J. BUNCE

University of Guelph, Guelph, Ontario, Canada

I. INTRODUCTION

The acyl hypohalites, which have an oxygen–halogen bond, are closely related in their chemical reactivity to alkyl hypohalites and the halogen oxides and are not strictly carbonyl halides, whose reactions are associated primarily with their carbonyl function. Unlike the well-known acyl halides, the acyl hypohalites are not, in general, stable substances that may be isolated and characterized. Rather, they have been postulated as intermediates in a number of reactions and in many cases their existence has been inferred simply from an examination of the final products of these reactions, based on analogy with related processes and from an expectation of what the chemistry of such substances is likely to be. Only perfluoroacyl hypofluorites and acetyl hypochlorite have been isolated in the pure state.

Because the acyl hypohalites (referred to occasionally as acyl hypohalides or halogen carboxylates) are not stable substances, it has not been possible to discuss them in the conventional categories of methods of preparation, physical properties and chemical reactions. Instead, we have approached the subject by discussing separately the hypohalites of fluorine, those of chlorine and bromine together, and finally those of iodine. The fluorine compounds are properly fluorides because of the electronegativity of fluorine; the hypochlorites and hypobromites are very similar, but are different from the derivatives of iodine, largely on account of the greater propensity of iodine towards polycovalency. The discussions of the chlorides and bromides and of the iodides are further divided according to the reaction type by which the hypohalite was prepared. In each case, we have tried to emphasize the limited direct physical evidence that exists for the intermediacy of these substances. A brief concluding section considers the chemistry of some related substances in which the halogen atom is replaced by a nitrogen-containing group.

This account is not intended to be an encyclopaedic catalogue of all the instances in which an acyl hypohalite intermediate has been proposed, but it is hoped that most important literature references up to mid-1970 have been included.

II. ACYL HYPOFLUORITES

Only four members of this isolable but very reactive class of compounds have been prepared; these are the hypofluorites of the fully fluorinated acids, formic through butyric. Perfluoroacetyl hypofluorite was the first of these to be discovered; it is prepared by the action of fluorine on trifluoroacetic acid at room temperature[1]. Water appears to be a necessary catalyst for the reaction[2], but its role is not understood. It is of interest, however, that the reaction requires water as a catalyst, for the acyl hypohalites of the other halogens are readily hydrolysed in its presence (section III, B). Presumably the strength of the H—F bond of the by-product, hydrogen fluoride, provides the driving force for formation of the hypofluorite from the acid (equation 1). Perfluoropropionyl and perfluorobutyryl hypofluorites have been prepared[3] by analogous reactions, but fluoroformyl hypofluorite has been prepared only by the action of fluorine on fluorocarbonyl peroxide (equation 2). Illumination with a mercury arc source is necessary for this last reaction[4].

$$R_FCO_2H + F_2 \xrightarrow{H_2O} R_FCO_2F + HF \tag{1}$$

$$(FCO_2)_2 + F_2 \xrightarrow{h\nu} FCO_2F + \text{other products} \tag{2}$$

The stability of the acyl hypofluorites appears to decrease with increasing molecular weight and, indeed, only the first three compounds have been completely characterized. These three are all colourless gases at room temperature and have been demonstrated to be monomeric by vapour-density measurements. Thermal stability decreases with increasing molecular weight: only fluoroformyl hypofluorite is not reported as being explosive, the others may be decomposed thermally or with an electric spark. The perfluoroacetyl and perfluoropropionyl compounds give CO_2 and carbon tetrafluoride and hexafluoroethane respectively (equation 3). No report of decomposition of FCO_2F to fluorine and CO_2 has appeared.

$$R_FCO_2F \longrightarrow R_FF + CO_2 \tag{3}$$

In the case of trifluoroacetyl hypofluorite, the course of the decomposition has been studied in more detail[5], and a free-radical chain mechanism is proposed to accommodate the observed kinetics: (equations 4–6). An alternative sequence of reactions, replacing (5) with the two-step

$$2\,CF_3CO_2F \longrightarrow CF_3CO_2F + CF_3^{\bullet} + CO_2 + F^{\bullet} \tag{4}$$

$$CF_3^{\bullet} + CF_3CO_2F \longrightarrow CF_4 + CF_3^{\bullet} + CO_2 \tag{5}$$

$$2\,CF_3^{\bullet} \longrightarrow C_2F_6 \tag{6}$$

process (5a, 5b), would make, formally, the decomposition analogous to the Hunsdiecker reaction (section III, B). The observed kinetics are

$$CF_3^{\bullet} + CF_3CO_2F \longrightarrow CF_4 + CF_3CO_2^{\bullet} \tag{5a}$$

$$CF_3CO_2^{\bullet} \longrightarrow CF_3^{\bullet} + CO_2 \tag{5b}$$

equally compatible with a free-radical chain sequence in which fluorine atoms are the chain-carrying species (equations 7–9). The decomposition

$$F^{\bullet} + CF_3CO_2F \longrightarrow F_2 + CF_3CO_2^{\bullet} \tag{7}$$

$$CF_3CO_2^{\bullet} \longrightarrow CF_3^{\bullet} + CO_2 \tag{8}$$

$$CF_3^{\bullet} + F_2 \longrightarrow CF_4 + F^{\bullet} \tag{9}$$

is inhibited in the presence of chlorine, which acts as a free-radical trap and the reaction now assumes the stoicheiometry of (10).

$$CF_3CO_2F + Cl_2 \longrightarrow FCl + CO_2 + CF_3Cl \tag{10}$$

Spectral data are available only for fluoroformyl hypofluorite. The i.r. spectrum shows several strong bands, including those ascribed to the stretching of $C{=}O$ (1930 cm^{-1}), $C{-}F$ (1192 cm^{-1}), $C{-}O$ (993 cm^{-1}) and $O{-}F$ (913 cm^{-1}). The n.m.r. spectrum (40 Mc/s SF$_6$ standard) comprises two doublets due to $C{-}F$ ($+94\cdot4$ p.p.m.) and $O{-}F$ ($-151\cdot8$ p.p.m.), $J = 141$ c/s. The position of the hypofluorite fluorine resonance is close to the values observed for perfluoralkyl hypofluorites[6].

A few reactions of the acyl hypofluorites are known. All are powerful oxidants, liberating iodine from KI solutions (explosively in the case of CF_3CO_2F). Attack on glass yields SiF_4, particularly for FCO_2F, and so the use of polyethylene containers is required when studying these compounds. Fluoroformyl hypofluorite is also known to add fluorine in the presence of a caesium fluoride catalyst, giving difluoromethane dihypofluorite (equation 11). The same compound gives a variety of products

$$FC\overset{\displaystyle{\nearrow O}}{\underset{\displaystyle{\searrow OF}}{}} + F_2 \xrightarrow{\text{CsF}} F_2C(OF)_2 \qquad (11)$$

on photolysis in the presence of sulphur tetrafluoride, including CO_2, COF_2, SF_6, SOF_2, SO_2F_2 and FCO_2SF_5. All the acyl hypofluorites are destroyed by aqueous base; for FCO_2F the reaction follows the stoicheiometry of equation (12).

$$2\ FCO_2F + 8\ NaOH \longrightarrow 2\ Na_2CO_3 + 4\ NaF + 4\ H_2O + O_2 \qquad (12)$$

As already noted, the stability of these compounds decreases as the molecular weight rises and presumably further members of the series will be isolable only with greater difficulty. As in the case of the alkyl hypofluorites, only perfluoro-derivatives are known to date, and probably in both series the fluorinating power of these compounds makes unlikely the isolation of derivatives that are not fully halogenated.

III. ACYL HYPOHALITES OF CHLORINE AND BROMINE

A. Physical Evidence

The only acyl hypochlorite claimed to have been isolated in the pure state is acetyl hypochlorite, which was obtained by Schutzenberger[7] by the interaction of chlorine monoxide and acetic anhydride (equation 13).

$$Cl_2O + (CH_3CO)_2O \longrightarrow 2\ CH_3CO_2Cl \qquad (13)$$

The near-colourless liquid may be distilled under reduced pressure, but it explodes at temperatures approaching $100°$, generating chlorine, oxygen and acetic anhydride. The reversal of equation (13) to form chlorine monoxide and the subsequent decomposition of the halogen oxide to chlorine and oxygen could satisfactorily explain this observation. Water converts it instantly to acetic acid and hypochlorous acid; mercury and zinc give the respective metal chloride and metal acetate. Iodine is reported as producing acetyl hypoiodite, but the evidence for this reaction is quite tenuous.

Bockemüller and Hoffmann[8] obtained colourless solutions by the action of silver carboxylates on solutions of bromine or chlorine in carbon tetrachloride. Filtration yielded solutions which lost their oxidizing power gradually at 0°, or more rapidly on exposure to light. Addition of olefins to the filtered solutions resulted in the formation of 2-haloalkyl carboxylates, and since the silver-containing substances had been filtered off, these may be regarded as derivatives of the non-isolated acyl hypohalites.

Solutions in carbon tetrachloride of acetyl hypochlorite obtained by the action of chlorine on silver acetate, or of chlorine monoxide on acetic acid were found to absorb in the u.v. at λ_{max} 264 mμ (ε 240) in carbon tetrachloride solution[9]. However, since the positions of the chlorine monoxide maxima appear to be identical with those of acetyl hypochlorite[9], and since a complex equilibrium between the halogen oxides and protic solvents exists in non-homogeneous media[10], an evaluation of these results should be made with some reservation. A similar intermediate prepared by the action of chlorine on mercuric acetate had λ_{max} 250 mμ ($\varepsilon \sim$ 240) in acetic acid[11]. The presumed acetyl hypochlorite prepared by this latter method was reported to co-distil with the acetic acid solvent under reduced pressure[12].

Recently a paper appeared which examined the i.r. spectrum of acetyl hypochlorite prepared by the mercuric acetate reaction[13]. Several bands were assigned, including those due to the O—Cl, C—C, C—O and C=O stretching modes. The spectrum was interpreted as showing that the molecule adopts the conformation (1) in which the carbonyl oxygen and

$$CH_3-C\underset{O-Cl}{\overset{O}{\diagup}}$$

(1)

the chlorine lie on the same side of the hypochlorite bond. The Cl—O bond dissociation energy was estimated at $\sim$ 55 kcal/mole. In the same study, the methyl group of acetyl hypochlorite was observed to resonate at δ 2·24 p.p.m. in the n.m.r. (CCl$_4$, $-10°$).

Acetyl hypobromite also absorbs in the u.v. having λ_{max} 320 mμ[9, 14]. In the i.r. a band at 670 cm^{-1} is ascribed to the O—Br stretching frequency by analogy with a similar band in the spectrum of t-butyl hypobromite[14]. This work has been challenged by Ogata and co-workers, who find no evidence of any maximum above 250 mμ in solutions of acetyl hypobromite[15]. An attempt at isolation of acetyl hypobromite was unsuccessful[14].

In similar fashion, some evidence for the intermediacy of benzoyl hypochlorite in the reaction of lithium chloride with benzoyl peroxide (section III, C) comes from observation[16] of a band at 751 cm^{-1} in the i.r. (cf. t-butyl hypochlorite 760 cm^{-1}). An attempt[17] to isolate benzoyl hypochlorite was partly successful. Filtration of the product of reaction of silver benzoate with chlorine in 'Freon 113', followed by removal of the solvent, gave a light-brown solid with retained oxidizing properties; assuming that the oxidizing entity was benzoyl hypochlorite, it had a purity of about 40% by iodometric titration.

B. The Action of Chlorine and Bromine on Metal Carboxylates

This reaction (equation 14) is believed to yield acyl hypochlorites and hypobromites as the first-formed intermediates, and has received very considerable attention, including earlier reviews[18, 19]. Three main reactions

$$RCO_2M + X_2 \longrightarrow RCO_2X + MX \qquad (14)$$

of the acyl hypohalite so formed have been studied: these are (i) electrophilic halogenation of aromatic nuclei, (ii) addition of the elements of the acyl hypohalite to an olefin (Prévost reaction) and (iii) decarboxylation to an organic halide (Hunsdiecker reaction). Although very often one of these processes does not occur to the exclusion of the others, it will nevertheless be convenient to consider them separately.

A problem common to all these discussions, however, is the role that water plays in the reactions of acyl hypohalites, and unfortunately there is considerable lack of agreement about this. For instance, it is well known that the presence of small quantities of water is extremely deleterious to the success of the Hunsdiecker decarboxylation reaction and it is usually supposed that hydrolysis (equation 15) occurs. In support of this,

$$RCO_2X + H_2O \longrightarrow RCO_2H + HOX \qquad (15)$$

if the Hunsdiecker reaction is carried out with improperly dried materials, all that is isolated is the free acid.

The action of water in destroying acyl hypohalites is confirmed by Schutzenberger's experiment[7] in which he showed that acetyl hypochlorite reacts instantly with water. Furthermore, de la Mare and co-workers showed that the equilibrium (16) for hydrolysis lies far to the

$$CH_3CO_2Cl + H_2O \rightleftharpoons CH_3CO_2H + HOCl \qquad (16)$$

side of hypochlorous acid[11]. By contrast, the u.v. spectrum of acetyl hypo-bromite is claimed[14] to be unchanged by the addition of 3% water to the acetic acid solvent (although the u.v. spectrum itself has been questioned[15]). The addition of small amounts of water to the Cristol–Firth modification of the Hunsdiecker reaction, mercuric oxide, carboxylic acid and halogen, is reported not to hinder the formation of the alkyl halide, which arises presumably via the acyl hypohalite (see section III, D). It would thus seem that the part played by water in these reactions is still open to question.

I. Electrophilic halogenation of aromatic compounds

Electrophilic halogenation of aromatic compounds by acyl hypochlorites and hypobromites has been little studied as a preparative procedure, probably because it offers few advantages over more conventional halogenation techniques. Proposals have included the use of silver acetate/halogen[20] and silver trifluoroacetate/halogen[21, 22] as halogenation reagents when presumably the respective acyl hypohalite would be the halogenating entity. From the isomer distributions of the aryl halides obtained, it may be assumed that these are electrophilic rather than free-radical substitutions[22].

More commonly, electrophilic substitution is observed as an unwanted side reaction of the Hunsdiecker decarboxylation of aromatic substances, particularly when an activated aromatic nucleus is present. In the latter reaction, yields of 80–90% of the aryl halide may be achieved if substituted silver benzoates bearing deactivating groups are used; poor yields are obtained with electron-releasing substituents and in addition nuclear halogenation occurs[23-26]. From silver p-anisoate and bromine, 3-bromo-4-methoxybenzoic acid is the major product[23]. In the case of the reaction of bromine with silver 3-(m-methoxyphenyl)-propanoate (2), the product is exclusively 3-(2'-bromo-5'-methoxyphenyl)-propanoic acid (3) rather than m-methoxyphenethyl bromide (4)[20].

$$CH_2CH_2CO_2Ag \qquad\qquad CH_2CH_2CO_2H \qquad\qquad CH_2CH_2Br$$

(2)   (3)   (4)

Several authors have studied the chlorination of aromatic compounds by solutions containing, or potentially containing, acetyl hypochlorite. De la Mare and his school prepared acetyl hypochlorite by the action of chlorine on mercuric acetate/acetic acid followed by distillation[12]. The chlorinations using this reagent have been studied in acetic acid containing various quantities of water[11, 12, 27-29]; under these conditions equilibrium (16) is almost entirely to the right.

In the acidic media normally employed, halogenation by any of the following species must be considered: CH_3CO_2Cl, $CH_3CO_2ClH^+$, HOCl, H_2OCl^+, and Cl^+. Molecular chlorine is not present, for chloride ion is removed by silver perchlorate. For the more reactive substrates, the rate law embodies a component which does not involve the aromatic compound; this is interpreted to mean that the very reactive acetyl hypochlorite is formed in the rate-determining step and that it chlorinates as quickly as it is formed. (Equilibrium (16) has been shown not to be established instantaneously under these conditions[11].) The substance must indeed be a very reactive chlorinating agent, because the reactions are very rapid, yet the equilibrium concentration of acetyl hypochlorite in the aqueous media used must be very small.

Other authors have set up equilibrium (16) by studying the chlorination of aromatic substrates by hypochlorous acid in acetic acid solution[30]. Although hypochlorous acid itself is a relatively unreactive reagent (as might be expected from Ingold's proposal[31] that an effective halogenating reagent Hal—Y should possess a good leaving group Y^- and, at the same time, be easily polarized to $Hal^{\delta+}$—$Y^{\delta-}$), these reactions show a contribution from a very reactive component and again acetyl hypochlorite is implicated by kinetic studies. An alternative explanation for the decreasing reactivity of Cl—Y in electrophilic substitution $Cl—OCOCH_3 > Cl—Cl > Cl—OH$ has been offered[13]; since the Cl—Y bond strengths increase in the order of decreasing reactivity, it is suggested that bond energy factors are responsible.

Mercuric acetate/bromine solutions are also found to be vigorous brominating agents and acetyl hypobromite is thought to be responsible[14]. For the system hypobromous acid/acetic acid/aromatic compounds, however, arguments have been put forward to explain the kinetic results in terms of either acetyl hypobromite[32] or protonated hypobromous acid[33] as the active species.

2. Addition to olefins

The addition of acyl hypohalites to olefins, giving 2-haloalkyl esters, has not been studied very systematically. However, an important observation was made by Bockemüller and Hoffmann[8], who showed that the product of reaction of silver carboxylates and halogen, after removal of the precipitated silver halide, added to olefins and thus provided evidence for the independent existence of acyl hypohalites in silver salt/halogen reaction mixtures. A number of alkanoyl and aroyl hypohalites were added to cyclohexene by Uschakow and Tchistow[34] and by Prévost and Wiemann[35]. Various authors[36-38] have added acyl hypobromites to styrene and p-substituted styrenes. No polymerization of the styrene was reported and in the cases investigated the adducts formed were all of the structure $C_6H_5CH(OCOR)CH_2Br$. This suggests first that free radicals were not involved, and second that the acyl hypobromite added as $Br^+RCO_2^-$. A similar conclusion has been reached for 1-butene[39] and for 1-hexene[38, 40]. The positive nature of the halogen in acyl hypohalites is also seen in their reaction with diethylmercury and diphenylmercury[41], when ethyl bromide and bromobenzene, respectively, result.

$$RCO_2Br + R_2^1Hg \longrightarrow R^1Br + R^1HgOCOR$$

Acetyl hypochlorite has been added to some aryl-substituted unsaturated compounds. The reagent shows lower stereoselectivity than that shown in the addition of chlorine in acetic acid to the same substrates[42]. Unlike the addition of acyl hypoiodites (section IV, B) which has received more attention, no more is known about the mechanism of addition of acyl hypochlorites and hypobromites. Prévost has suggested the intermediacy of a non-isolable complex in solution similar to the Simonini complex $RCO_2Ag \cdot RCO_2I$, but, as noted by Wilson[18], the known facts require no more than addition of the simple hypohalite. Bockemüller and Hoffmann's experiments with the silver-free solutions would seem to exclude a silver-containing reaction intermediate.

From a synthetic point of view, acetyl hypobromite has been used as a reagent capable of adding to steroidal double bonds faster than hypobromous acid and, unlike the latter, under non-acidic conditions[43]. The resulting bromacetate is then easily converted to an epoxide, or, by the use of zinc/copper couple, the original olefin may be regenerated. These results suggest the possibility of the use of acetyl hypobromite as an olefin protecting group.

3. Decarboxylation (Hunsdiecker reaction)

$$RCO_2M + X_2 \longrightarrow RCO_2X + MX \longrightarrow RX + CO_2$$

An enormous number of publications covering both the synthetic and the mechanistic aspects of this reaction has appeared. Two reviews in

1956^{19} and 1957^{18} cover the work prior to those dates, so that this earlier work will be only summarized here.

As a synthetic and degradative technique, the Hunsdiecker reaction is of wide applicability. Secondary and particularly primary aliphatic halides may be prepared in excellent yields, though the reaction fails for tertiary halides. The failure of the reaction with tertiary halides presumably is due to their reaction with the silver salts present in the mixture which assist in the ionization and hence in the destruction of the tertiary halides as they form, (equation 17)[44]. Bridgehead halides in rigid bicyclic systems

$$RX + Ag^+ \rightleftharpoons AgX + R^+ \longrightarrow products \qquad (17)$$

may be prepared in excellent yield; here the tertiary halide is inert to assisted ionization by the silver salts. The reaction is also useful for the formation of aryl halides with the limitation that the aromatic ring should not be substituted by electron-releasing groups, else electrophilic halogenation is preferred over decarboxylation. Great success has been achieved also with a number of perfluorocarboxylic acids[22].

Experimentally, the reaction is usually carried out using the halogen and the dry silver salt with carbon tetrachloride as the reaction medium. Metal salts other than silver may be used, but usually only those of the heavy metals (Pb^{II}, Hg^{I}, Hg^{II}, Tl^{I}) give good yields of alkyl halide and even with these the reaction is often slow[45, 46]. Occasional uses of the alkali and alkaline earth metals have been reported. Carbon tetrachloride is usually the reaction medium of choice, since it is easily rendered anhydrous and is relatively inert towards the free radicals believed to be intermediates in the reaction. In one study[47] the best yield of alkyl bromide from silver palmitate was obtained when carbon tetrachloride was the solvent, out of a field of eleven different reaction media. Other solvents to be used successfully, however, include nitrobenzene, chloroform, trichloroethylene, bromotrichloromethane, ethyl bromide, carbon disulphide, benzene and acetonitrile. In the perfluoro series, the reaction is most often carried out without a solvent, though perfluorotributylamine has been used with success.

Concerning the mechanism of the Hunsdiecker reaction, most workers now favour a free-radical chain pathway for decarboxylation such as that outlined in equations (18)–(21). Proposals involving carbonium ion or

$$RCO_2Ag + X_2 \longrightarrow RCO_2X + AgX \qquad (18)$$

$$RCO_2X \longrightarrow RCO_2^{\cdot} + X^{\cdot} \qquad (19)$$

$$RCO_2^{\cdot} \longrightarrow R^{\cdot} + CO_2 \qquad (20)$$

$$R^{\cdot} + RCO_2X \longrightarrow RX + RCO_2^{\cdot} \qquad (21)$$

carbanion intermediates or an intramolecular pathway (equation 22) have now been largely discarded.

$$R-C\underset{X}{\overset{O}{\lessgtr}}_{O} \longrightarrow RX + CO_2 \tag{22}$$

A mass of evidence now supports the intermediacy of free radicals in the decarboxylation process. The rate of decomposition of the acyl hypohalite may be increased by photoinitiation[8, 17] or by the use of the free-radical initiator azobisisobutyronitrile[17] and in reactions conducted at low temperatures an induction period has been noted[48]. In many cases, typical free-radical processes occur simultaneously with decarboxylation; added alkanes[49] or toluene[22] undergo free-radical halogenation and in one case bibenzyl has been isolated from reaction mixtures to which toluene was added[22]. In another instance a coupled product R—R has been obtained[50] and recently it was reported that the benzoyloxy radical has been observed by e.s.r. spectroscopy in a solution in which the reaction of bromine with silver benzoate was being carried out[51]. When dichloromethane was used as a solvent in the reaction of bromine with silver 2-methylpentanoate, it was noted that small amounts of bromodichloromethane were formed in the reaction[52].

Quite frequently, brominations conducted in carbon tetrachloride[23, 24, 53-55] or tetrachloroethane[56] have yielded the chloride (RCl) in addition to the desired bromide. These products are proposed as arising, in the case of carbon tetrachloride, through the free-radical sequence of equations (23) and (24). Contamination of the product by chloride is most

$$R^{\bullet} + CCl_4 \longrightarrow RCl + {}^{\bullet}CCl_3 \tag{23}$$
$${}^{\bullet}CCl_3 + RCO_2Br \longrightarrow BrCCl_3 + RCO_2^{\bullet} \tag{24}$$

often observed when R is aryl, or more particularly if R is derived from the bridgehead of a rigid bicyclic system. This may be taken as evidence of the unusually high reactivity of bridgehead free radicals in that they are sufficiently indiscriminating to abstract chlorine from carbon tetrachloride in the presence of the (lower) concentration of the acyl hypobromite[55].

By the use of optically active acids either in the biphenyl series[57], or of the constitution $R^1R^2R^3CO_2H$[53, 58-61] it has been demonstrated that the product alkyl bromide is completely racemized and hence that a three-coordinate intermediate is involved. Claims of the production of optically active bromides from silver (+)-α-phenylpropionate[62] and from silver (+)-2-ethylhexanoate[63] could not be substantiated[53, 59, 60, 63-66]. (Optically active 1,2-dibromides from optically active silver 1,2-dicarboxylates may represent a special case and are discussed later.)

That the three-coordinate intermediate is not a carbonium ion is shown by the fact that silver salts RCO_2Ag containing groups R that undergo rearrangements as R^+ yield exclusively unrearranged halides, RX, on treatment with halogens. Thus silver 3,3-dimethylpentanoate is converted to neopentyl bromide[67] and silver cyclobutane carboxylate to cyclobutyl bromide[56, 65, 68]. No rearrangement is observed in the production of halides in the bicyclo-2,2,1-heptyl and the bicyclo-2,2,2-octyl systems, and no change in ring size is observed on treatment of the silver salts (5) and (6) with bromine[68, 69].

(5)   (6)

The effect of temperature on the Hunsdiecker reaction has received little attention. In some cases a low temperature is demanded by the fact that the silver salts are themselves thermally unstable[48]. It has recently been suggested[17] that the yields of primary halides are relatively insensitive to the operating temperature, whereas secondary halides are produced more successfully at lower temperatures, possibly because destruction of the product by assisted ionization (equation 17) is then less important. By contrast, aryl halides are obtained in much higher yields at temperatures above ambient and it is suggested that the relative difficulty of decarboxylation (equation 20) when R is aryl (as opposed to when R is alkyl) results in the radicals $ArCO_2^{\cdot}$ being diverted, especially at lower temperatures, to products in which the carboxyl moiety is retained.

Much interest has recently been displayed in the stereochemistry of the Hunsdiecker reaction. In substituted cycloalkyl systems, it has been noted that the same mixture of *cis*- and *trans*-halides is produced whether the starting carboxylate group lies *cis* or *trans* to the substituent[70-73]. In the 4-*t*-butylcylohexyl system, Eliel and Acharya[71] found that the Hunsdiecker reaction gave *cis*- and *trans*-bromide in a ratio of 34/66, but in a reinvestigation[72] of the reaction, it was reported that the distribution of the two bromides was essentially statistical.

The Hunsdiecker reaction applied to substituted 2-norbornyl systems likewise gives the same mixture of *exo*- and *endo*-bromides whether the *exo*- or the *endo*-carboxylate was the starting material. It was observed[73] that the Hunsdiecker reaction gives *exo/endo*-bromides in the ratio 69/31 whereas the decomposition of the corresponding peroxide in the presence of bromotrichloromethane gives only the *exo*-bromide. It has since been noted[74-76] that in norbornyl systems the ratio of *exo*- to *endo*-products is dependent upon the nature of the transfer species and that the observed differences in substitution reactions are possibly steric in origin.

The action of bromine on some cyclic silver 1,2-dicarboxylates has been studied. Both *cis*- and *trans*-silver cyclohexane-1,2-dicarboxylates are reported to give exclusively the *trans*-dibromide[77]. In the cyclobutane series a similar result has been reported[78], but it has subsequently been claimed that both isomeric dibromides are obtained. The *cis*-diacid was reported[79] to give 14% *cis*- and 78% *trans*-dibromide, while for the *trans*-diacid, the corresponding yields were 26% and 65%. From optically active *trans*-cyclobutane dicarboxylic acid ($\alpha_D = -83\cdot4°$; 46% racemic), Applequist and Fox[80] obtained an all-*trans*-dibromide which retained some optical activity ($\alpha_D = -6°$, optical purity not stated). Similar experiments applied to optically active silver cyclohexane-1,2-dicarboxylate gave an all-*trans*-dibromide which again was optically active, but whose optical activity was highly irreproducible[81].

It may be noted that the stereospecificity of the reaction in tending towards all-*trans*-geometry is not necessarily incompatible with the free-radical pathway of equations (18)–(21). Free-radical additions of hydrogen bromide to bromoolefins[82, 83] and free-radical bromination of bromo-alkanes both yield 1,2-dibromides whose geometry is mainly *trans*[84-86].

The slight tendency of the cyclobutane dicarboxylic acids to yield optically active products and of the optically active cyclohexane dicarboxylic acid to yield optically active, but inverted products, has been alternatively explained as follows[81]. The intermediate diacyl dihypobromite would be unlikely to suffer rupture at both hypobromite centres simultaneously. Fission of one O—Br bond followed by loss of CO_2 yields radical (7). Radical (7) may abstract bromine from another acyl hypobromite molecule (equation 21), but alternatively may abstract bromine

$$
\begin{array}{ccc}
\underset{|}{-}\overset{|}{C}-C\overset{\displaystyle O}{\underset{\displaystyle O-Br}{<}} & \underset{|}{-}\overset{|}{C}-C\overset{\displaystyle O}{\underset{\displaystyle O^{\cdot}}{<}} + Br^{\cdot} & \underset{|}{-}\overset{|}{C}{}^{\cdot} \\
\underset{|}{-}\overset{|}{C}-CO_2Br & \underset{|}{-}\overset{|}{C}-CO_2Br & \underset{|}{-}\overset{|}{C}-CO_2Br
\end{array}
$$

$$(7)$$

intramolecularly (equation 25). When bromine abstraction by (7) is

$$
\begin{array}{cc}
\underset{|}{-}\overset{|}{C}{}^{\cdot} \quad \overset{\displaystyle Br}{\underset{\displaystyle O}{|}} & \underset{|}{-}\overset{|}{C}-Br \\
\underset{|}{-}\overset{|}{C}-C\overset{}{\underset{\displaystyle O}{\diagdown}} & \underset{|}{-}\overset{|}{C}-C\overset{\displaystyle O^{\cdot}}{\underset{\displaystyle O}{<}}
\end{array}
$$

$$(25)$$

intermolecular, racemic, but mainly *trans*-dibromide predominates; when intramolecular abstraction occurs, the optically active system is partly preserved since the second bromine will tend to be directed *trans* to the first. Since the first bromine must be intramolecularly transferred *cis*, net inversion will result.

Also of great interest has been the stereochemistry of the Hunsdiecker reaction for silver α,β-unsaturated carboxylates. The Hunsdiecker reaction does not proceed normally with unsaturated acids and those with α,β-unsaturated acids are difficult to study for they produce mainly polymers. Thus Conly[48] obtained only polymer from the action of bromine on silver methacrylate; in more recent work 3% of CO_2 was evolved, but no 2-bromopropene was obtained[87]. Polymer was also the major product from the similar reaction of silver acrylate[87], though some 1,1,2-tribromoethane (dibromide adduct of vinyl bromide) was also formed.

Aromatic α,β-unsaturated acids have been studied more successfully. Berman and Price[88] obtained *trans*-β-bromostyrene from both *cis*- and *trans*-silver cinnamates, suggesting that a vinyl radical is incapable of maintaining its geometry. Studies involving vinyl radicals generated by other means have led to similar conclusions[89], except in the case of the β-phenylstyryl radicals (**8**), which when generated from the corresponding stilbene acid peroxides (**9**), yielded products in which the stereochemistry

$$\underset{H}{\overset{C_6H_5}{\diagdown}}C{=}\dot{C}{\sim}C_6H_5 \qquad \left(\underset{H}{\overset{C_6H_5}{\diagdown}}C{=}C(C_6H_5){-}CO_2\right)_2$$

(8)   (9)

is largely retained[90]. Berman and Price[88] found that the Hunsdiecker reaction applied to the stilbene acids yielded only polymer, but reinvestigation of the reaction by Sukman[57] afforded the α-bromostilbenes. From the *cis*-acid the ratio of *cis/trans*-bromides was 25/75 while the *trans*-acid gave a ratio of 10/90, implying a modest capability of **8** to maintain its stereochemistry.

In unsaturated acids where the double bond is further removed than the α,β-positions, addition of the acyl hypohalite to the double bond is observed. Thus the silver salts of allylacetic acid[48] (**10**,R = H) and benzylallylacetic acid[91] (**10**,R = $CH_2C_6H_5$) afford bromolactones on treatment with bromine. Likewise[92] a bromolactone has been obtained from the bicyclic silver carboxylate (**11**).

In a number of cases attempted, Hunsdiecker reactions have yielded products other than the expected alkyl halides. For example, whereas the reaction succeeds when halogen[18, 19] or carboxyl[93, 94] substituents are present on the α-carbon, or even when the α-carbon itself is a carbonyl

$$CH_2=CHCH_2\overset{\underset{\displaystyle |}{R}}{C}HCO_2Ag \xrightarrow{Br_2} BrCH_2CHCH_2CHR$$

(with the ring/lactone structure: O—C=O bridging)

(10)

(bicyclic structure with CH_3 CH_3, $C=$, CO_2Ag, CO_2Et)

(11)

group[68], the α-hydroxy[68], amino[68] and acetoxy[95] derivatives yield aldehydes and ketones when their silver salts are brominated (equation 26).

$$R^1R^2\overset{\underset{\displaystyle |}{Y}}{C}CO_2Ag \xrightarrow{Br_2} R^1R^2C=O \tag{26}$$

$$Y = OH, NH_2, OAc$$

N-Acylaminoacids[96], however, yield the isolable, but easily hydrolysed N-acyl-α-bromoamines, the normal product.

$$R^1R^2C(NHAc)CO_2Ag \longrightarrow R^1R^2C(Br)NHAc \xrightarrow{H_2O} R^1R^2C=O$$

Sometimes the radical intermediates undergo rearrangement rather than giving the normal products. This is proposed to be the case for the radicals generated from the cyclic β-hydroxy silver carboxylates **(12)** and **(13)**, which afford open-chain products after the treatment of the original reaction mixture with LiAlH$_4$ (equation 27)[97].

$$(CH_2)_n \overset{\displaystyle OH}{\underset{\displaystyle CH_2}{C}} \begin{matrix} OH \\ CH_2CO_2Ag \end{matrix} \xrightarrow[\text{2. LiAlH}_4]{\text{1. Br}_2} CH_3-\overset{\underset{\displaystyle |}{OH}}{C}H-(CH_2)_n-CH_3 \tag{27}$$

(12) n = 3;
(13) n = 4

Silver β,β,β-triphenylpropionate gives a mixture of esters on treatment with bromine[98] and again it is proposed that the intermediate radical **(14)** rearranges. By studying the rearrangement products of phenyl-substituted

$$(C_6H_5)_3CCH_2CO_2Ag \xrightarrow{Br_2} \longrightarrow (C_6H_5)_3CCH_2CO_2^{\cdot}$$
(14)

$$\longrightarrow (C_6H_5)_2\overset{\cdot}{C}CH_2CO_2C_6H_5$$

$$\longrightarrow (C_6H_5)_2C=CHCO_2C_6H_5 + (C_6H_5)_2C=C(Br)CO_2C_6H_5$$
$$ 9\% 20\%$$

16*

silver β,β,β-triarylpropionates, it has been concluded[99] that the rearrangement possibly involves a transition state (or intermediate) such as 15, but not one such as 16 (R = C$_6$H$_4$X). No rearrangement is observed in the Hunsdiecker reaction of silver δ,δ,δ-triphenylbutyrate[100].

(15)   **(16)**

β,β-Diphenylpropionic acid also rearranges when its silver salt is treated with bromine[101] and yields the lactones (17) and (18). Possible routes to these products involve either 16 (R = H) or 19. Lactone by-products have been obtained in a number of cases, including the reactions of silver δ-p-nitrophenylvalerate[102] and β-phenylbutyrate[61].

$(C_6H_5)_2CHCH_2CO_2Ag \xrightarrow{Br_2}$

(17) (43%)   **(18) (25%)**

(19)

The neophyl radical (20) generated from several sources has been shown[103] to rearrange in part to the more stable tertiary radical (21). It is to be expected that the Hunsdiecker reaction applied to silver β-methyl-β-phenylbutyrate would give a rearranged as well as an unrearranged product. Although both rearranged and unrearranged products are quoted[18] as having been obtained from a Hunsdiecker reaction on β-phenyl-β-methylbutyric acid, a reinvestigation of the reaction indicated that only the unrearranged bromide (from 20) was produced[104]. Possibly in this latter report, due to the reaction conditions, the chain-transfer reaction of **20** with the acyl hypobromite is faster than its rearrangement to **21**.

$$\underset{\overset{|}{CH_3}}{\overset{\overset{CH_3}{|}}{C_6H_5\overset{|}{C}CH_2CO_2Ag}} \longrightarrow \longrightarrow \underset{\overset{|}{CH_3}}{\overset{\overset{CH_3}{|}}{C_6H_5\overset{|}{C}-CH_2^{\textbf{·}}}} \longrightarrow \underset{\overset{|}{CH_3}}{\overset{\overset{CH_3}{|}}{C_6H_5\overset{|}{C}-CH_2Br}}$$

(20)

$$\underset{\overset{|}{CH_3}}{\overset{\overset{CH_3}{|}}{\overset{\textbf{·}}{C}-CH_2C_6H_5}} \longrightarrow \underset{\overset{|}{CH_3}}{\overset{\overset{CH_3}{|}}{Br\overset{|}{C}-CH_2C_6H_5}}$$

(**21**)

C. The Action of Chloride and Bromide Salts on Diacyl Peroxides

Rather little work has been done with these systems; indeed only benzoyl peroxide and valeryl peroxide have received attention.

Bamford and White[105] noted that the rate of decomposition of benzoyl peroxide in *NN*-dimethylformamide (DMF) solution is greatly enhanced in the presence of lithium chloride. Previously Bredereck and co-workers[106] had noticed the increased rate of decomposition of 1-tetralyl hydroperoxide and of benzoyl peroxide in the presence of the hydrochloride or the hydrobromide of butylamine. The reaction between lithium chloride and benzoyl peroxide is first-order both in chloride ion and in benzoyl peroxide, suggesting equation (28) as the rate-determining step. The activation

$$(C_6H_5CO_2)_2 + Cl^- \longrightarrow C_6H_5CO_2Cl + C_6H_5CO_2^- \qquad (28)$$

energy for (28) was estimated at 19 kcal/mole. Unlike the thermal decomposition of benzoyl peroxide (E_a 28 kcal/mole), (28) does not lead to polymerization of added styrene[49, 105], addition to the monomer being preferred[16]. In DMF the final products of the reaction are benzoic acid and the benzoate (**22**), derived from the solvent; in dichloroethane attack on the solvent gives benzoate (**23**).

$$HCON(CH_3)CH_2OCOC_6H_5 \qquad ClCH_2CH_2OCOC_6H_5$$
$$(\textbf{22}) \qquad\qquad\qquad (\textbf{23})$$

In the less reactive solvents, acetic acid[16] or acetonitrile[49], moderate yields of the expected free-radical decomposition product of the hypochlorite, chlorobenzene, were obtained. In acetic acid a transiently observed i.r. band at 757 cm^{-1} was tentatively assigned to benzoyl hypochlorite[16]. The decomposition to chlorobenzene is presumed to take the course described for the silver salt/halogen intermediate (section

III, B) and, as in that reaction, chlorobenzene formation is suppressed by the addition of water to the solvent[49]. Oxidation of hydrogen chloride by benzoyl peroxide gives no more than traces of chlorobenzene[49]. This fact is attributed to destruction of the intermediate acyl hypochlorite by hydrogen chloride, by analogy with the instability of N-chloroamines, N-chloroamides and alkyl hypochlorites towards hydrogen chloride[9, 107].

The addition of alkanes to the chloride ion/benzoyl peroxide reaction mixture largely suppresses chlorobenzene production and, in its stead, alkyl chlorides are formed in moderate to good yield. In the case of toluene[16] both side-chain and nuclear halogenation are observed; the free-radical chlorination is suppressed when cupric chloride is added to the solution (contrast the effect of copper salts with valeryl peroxide below). The mechanism of alkyl chloride formation has been proposed[49] to be that depicted in equations (29)–(31), in which hydrogen abstraction is

$$Cl^{\bullet} + RH \longrightarrow HCl + R^{\bullet} \qquad (29)$$

$$HCl + C_6H_5CO_2Cl \longrightarrow C_6H_5CO_2H + Cl_2 \qquad (30)$$

$$R^{\bullet} + Cl_2 \longrightarrow RCl + Cl^{\bullet} \qquad (31)$$

by way of chlorine atoms. This mechanism is analogous to that observed in the chlorination of substituted toluenes by t-butyl hypochlorite[108] (though alkanes are chlorinated with this reagent by way of t-butoxy radicals abstracting)[109]. The change in reactivity of the reagent, as the ratio of chloride ion to peroxide is varied, has been ascribed to participation in the hydrogen abstraction process by the phenyl radicals that are produced by the decomposition of benzoyl peroxide itself[49].

Alkanes may be halogenated by hydrogen chloride and benzoyl peroxide almost quantitatively[49, 110]. The mechanism of this oxidation is believed[49] to follow the course of equations (29)–(31) with the complication that hydrogen abstraction is, in this case, reversible (equation 32). The reaction

$$R^{\bullet} + HCl \longrightarrow RH + Cl^{\bullet} \qquad (32)$$

of hydrogen bromide and benzoyl peroxide in the presence of cyclohexane yields cyclohexyl bromide and probably follows a similar course[111].

The facile addition of benzoyl hypochlorite to styrene has already been noted and addition of the reagent to cyclohexene is also observed[49]. Both 1,2-dichlorocyclohexane and 2-chlorocyclohexyl benzoate are produced (equations 33–35, X = Cl). In the corresponding reaction of magnesium bromide with benzoyl peroxide and cyclohexene, 1,2-dibromocyclohexane is formed, but the report makes no mention of 2-bromocyclohexyl benzoate[112]. This is consistent, however, with the much more facile attack of bromide ion on acyl hypobromites (equation 34:X = Br) than that of chloride ion on the hypochlorites (see below); consequently this would serve to remove benzoyl hypobromite before it could react with cyclohexene. The greater nucleophilic power of bromide ion compared with that

$$C_6H_5CO_2X + \;\;\bigcirc \!\!\!\!\!\parallel \;\; \longrightarrow \;\; \text{[cyclohexane with OCOC}_6\text{H}_5 \text{ and X]} \qquad (33)$$

$$C_6H_5CO_2X + X^- \;\; \longrightarrow \;\; X_2 + C_6H_5CO_2^- \qquad (34)$$

$$X_2 + \;\; \bigcirc \!\!\!\!\!\parallel \;\; \longrightarrow \;\; \text{[cyclohexane with X and X]} \qquad (35)$$

of chloride ion seems to be general for nucleophilic displacement on divalent oxygen; the attack of nucleophiles on hydrogen peroxide in aqueous solution is fastest for the larger, more polarizable nucleophiles[113].

In the presence of aromatic compounds at least as reactive as chlorobenzene, electrophilic chlorination by the lithium chloride/benzoyl peroxide reagent in acetic acid is observed[16]. This reaction has been studied only in the presence of excess lithium chloride and under these conditions equation (34, X = Cl) appears to be dominant and substitution is by way of the molecular halogen. This may be deduced from the similarity of the resultant o/p ratios and the relative reactivity of benzene and toluene with the reagent compared with those observed for elemental chlorine, and their deviation from the values characteristic of 'positive halogen', i.e. acetyl hypochlorite. The bromination of anisole by lithium bromide/benzoyl peroxide also involves molecular bromine; in this case the colour of bromine is observed immediately on mixing the reagents.

The reaction of benzoyl peroxide with chlorides of metals in a low-valency state leads mainly to oxidation of the metal ion in the cases studied (Cu^{I} [114], Sn^{II} and Sb^{III} [115]). With stannous chloride some stannic chloride is produced; one could speculate that this may have arisen by oxidation of chloride ion to chlorine by benzoyl peroxide, via benzoyl hypochlorite, followed by interaction of the chlorine with more stannous chloride.

Some important differences are noted in the reaction of chloride ion with valeryl peroxide[116]. First, the rate of the reaction in the presence of chloride ion is but little greater than the rate of the thermal decomposition of the peroxide. Second, while valeryl hypochlorite is implicated as an intermediate, its free-radical decomposition appears to be very much more efficient than that of benzoyl hypochlorite, for good yields of the decarboxylation product, butyl chloride, are formed. (This difference between acyl hypochlorites of aliphatic and aromatic acids has been alluded to above in section III, B.) Third, decarboxylation to butyl chloride predominates even when anisole is added to the reaction mixture, though the small amounts of chloroanisoles formed are once again in a ratio characteristic of chlorination by elemental chlorine. It would thus appear that in this system free-radical decarboxylation occurs faster than the attack of chloride ion on the hypochlorite. Butyl chloride formation is reduced, however, by the addition of toluene to the reaction mixture, when benzyl chloride results.

Copper salts have been shown to be very effective catalysts for the decomposition of peroxy compounds[117], including valeryl peroxide[118]. However, in the chloride-ion-assisted decomposition of valeryl peroxide, addition of cupric chloride has very little effect on the reaction rate. The copper-catalysed reaction has been found to involve the oxidation–reduction sequence of equations (36)–(39), and it is believed that the lack of activity of the copper salt in the chloride-ion-promoted reaction is due to removal of the Cu^I species necessary for step (36) by the oxidizing action of either valeryl hypochlorite or molecular chlorine (equation 39, $X = Cl$ or RCO_2).

$$Cu^+ + (RCO_2)_2 \longrightarrow RCO_2^- + RCO_2^{\cdot} + Cu^{2+} \qquad (36)$$

$$RCO_2^{\cdot} \longrightarrow R^{\cdot} + CO_2 \qquad (37)$$

$$R^{\cdot} + Cu^{2+} \longrightarrow R^+ + Cu^+ \qquad (38)$$

$$\text{Oxidation:} \quad 2\,Cu^+ + X - Cl \longrightarrow 2\,Cu^{2+} + X^- + Cl^- \qquad (39)$$

When reactive aromatic compounds or alkenes are added to the reaction mixtures containing copper salts, a large catalytic effect by the copper is once more observed and it is suggested that the aromatic compounds or alkene have removed the oxidizing agents before they are able to oxidize the Cu^I salts and thus interrupt the chain sequence of equations (36)–(38).

Valeryl peroxide also reacts with lithium bromide and, here again, the intermediate hypobromite enjoys only a transitory existence before it is destroyed by the attack of a second bromide ion. This is reflected in the stoicheiometry of the reaction (equation 40), and also by the lack of butyl bromide as a product (presumably (41) is too fast a process to allow

$$2\,Br^- + (C_4H_9CO_2)_2 \longrightarrow Br_2 + 2\,C_4H_9CO_2^- \qquad (40)$$

$$Br^- + C_4H_9CO_2Br \longrightarrow Br_2 + C_4H_9CO_2^- \qquad (41)$$

decarboxylation to compete). The use of cupric bromide as the bromide ion source in contrast gives a high yield of butyl bromide and very little bromine. It is thought that in this case the acyl hypobromite is not involved but rather that the mechanism of equations (36)–(39) is followed.

D. From the Acid by Transhalogenation

The unfavourable enthalpies of these reactions dictate that many transhalogenation equilibria (42) tend to favour the acid rather than the hypohalite.

$$RCO_2H + X-A \rightleftharpoons RCO_2X + HA \qquad (42)$$

$$X = \text{Halogen}$$

In the equilibrium (42) involving chlorine and bromine, the reaction has been shown to lie far to the left and hence the exchange is not a practicable method of making the hypohalites. For example, halogenation of aromatic compounds by solutions of chlorine in acetic acid has been shown[119, 120] to proceed entirely through molecular substitution and not via acetyl hypochlorite. (Electrophilic halogenation of aromatic compounds is discussed in section III, B.)

The action of the halogen monoxide on the acid has received some study (equation 43).

$$X_2O + RCO_2H \rightleftharpoons HOX + RCO_2X \qquad (43)$$

Anbar and Dostrovsky[9] found that for acetic acid the reaction followed the stoicheiometry of (44); on removal of the water formed, they observed

$$Cl_2O + 2\,HOAc \rightleftharpoons H_2O + 2\,ClOAc \qquad (44)$$

the u.v. absorption spectrum of acetyl hypochlorite. This report seems strangely out of accord with the reported hydrolysis equilibrium constant of 400 for acetyl hypochlorite[11] and so far as decomposition products of the hypochlorite are concerned, no more than traces of the desired alkyl halides were formed by the action of chlorine monoxide on several acids[121]. Chlorination at alternative sites predominated. Attempts at transhalogenation to give acyl hypohalites with carboxylic acids and either t-butyl hypochlorite or N-bromosuccinimide were likewise unsuccessful. N-Bromosuccinimide has been used, however, to form olefin bromoacetates in acetic acid solution[122] (equation 45) though the acyl hypobromite is not a necessary intermediate here.

$$(45)$$

In the case of the rather unstable bromine monoxide however, the single case investigated (Br_2O + valeric acid) gave a moderate yield of the free-radical decomposition product of the presumed valeryl hypobromite (i.e., butyl bromide)[123]. This reaction appears to be related to the degradation of acids to alkyl bromides by bromine and mercuric oxide[124]. It is suggested that the Cristol–Firth reaction takes place via the formation of bromine monoxide *in situ* followed by attack on the acid. For example, only those oxides that have oxidizing properties (e.g. PbO, Ag_2O, CdO) may replace mercuric oxide[125]. It is thus likely that this reaction proceeds through the acyl hypobromite and certainly there is at least one common intermediate, for the degradation of a substituted 2-norbornane carboxylic acid gives the same ratio of *exo/endo*-bromide whether the Hunsdiecker method or Br_2/HgO is used[126].

As a synthetic technique, the mercuric oxide method has been claimed as possessing these advantages over the Hunsdiecker degradation: the often hard-to-prepare silver salts are not required and less rigorous exclusion of moisture is necessary. In several instances[56, 100, 127] the method gave a yield of bromoalkane better than that obtained from the silver salt, and for bromocyclopropane[127] without the risk of explosion that had been noted in that case[128].

An apparent instance of transhalogenation by an acyl hypobromite and a different carboxylic acid has been observed by Rottenberg[129] who found Hunsdiecker decomposition products from both the possible acyl hypo-bromites (equations 46–47). The halogen-donating tendency of acyl

$$RCO_2Ag + Br_2 \longrightarrow RCO_2Br + AgBr \qquad (46)$$

$$RCO_2Br + R'CO_2H \longrightarrow RCO_2H + R'CO_2Br \qquad (47)$$
$$\downarrow \qquad\qquad\qquad\qquad \downarrow$$
$$R\dot{B}r + CO_2 \qquad\qquad\qquad R'Br + CO_2$$

hypohalites is also illustrated by the recent report that acetyl hypobromite may be used to effect the *N*-bromination of amides, imides and sulphon-amides[130].

Acyl hypohalites also have been implicated to explain the results of kinetic studies with *N*-haloamides. In the chlorination of *N*-methylace-tamide by hypochlorite ion[131], it is suggested that part of the chlorination is carried out by acetyl hypochlorite if acetic acid is added to the reaction mixture (equation 48). By contrast, the *N*-chlorination of several anilides

$$HOCl + CH_3CO_2H \longrightarrow CH_3CO_2Cl + H_2O \qquad (48)$$

is not catalysed by acetic acid. It has been shown, however, that in the presence of acetic acid, chlorination of aromatic compounds by certain *N*-chloroanilides proceeds at a rate that is dependent upon the concen-tration of acetic acid, implicating acetyl hypochlorite as the reactive chlorinating agent[132]. The same authors suggest that the catalytic action of acetic acid in bringing about the rearrangement of *N*-chloroanilides to *o*- and *p*-chloroanilides also involves acetyl hypochlorite as the chlorinating agent. These results taken together imply that for N-chloroamides equilibrium (49) lies to the right for R = aryl but to the left when R is alkyl. Other examples of chlorination by hypochlorous acid are discussed in section III, B.

$$RN(Cl)COCH_3 + CH_3CO_2H \rightleftharpoons RNHCOCH_3 + CH_3CO_2Cl \qquad (49)$$

Similar studies on the bromination of aromatic substances by bromine in aqueous acetic acid have likewise led to postulation of an acetyl hypobromite intermediate[133].

E. Miscellaneous Reactions

In very few of the reactions described in this section is the intermediacy of the hypohalite established with any degree of security. One such, however, is the report by Schutzenberger[7] that chlorine monoxide and acetic anhydride combine to give acetyl hypochlorite as described in section III, A. Similarly, a group of probably related reactions, described by Rice, most likely involve the generation of bromine monoxide *in situ* by the interaction of bromine and silver oxide. It was observed[134] that the interaction of these two reagents with 1 mole of acetic anhydride led to the production of 2 moles of CO_2; we might postulate equation (50) as

$$Br_2 + Ag_2O \longrightarrow [Br_2O] \xrightarrow{Ac_2O} [2\ AcOBr] \longrightarrow 2\ CO_2 + [2\ CH_3Br] \qquad (50)$$

the sequence involved. Likewise, acetyl chloride under the same conditions yielded 1 mole of CO_2. The reaction has been applied preparatively to a sugar acid chloride[134] and to several aroyl chlorides[135]; in each case the bromide and a quantitative yield of CO_2 were formed. The authors point to the ease of obtaining anhydrous acid chlorides compared with silver salts as an advantage of their technique over the Hunsdiecker method. We may note that this reaction bears the same relationship to Schutzenberger's experiment as does the mercuric oxide technique of Cristol and Firth (section III, D) to the Hunsdiecker reaction.

The intermediacy of acetyl hypobromite has recently been implicated in the bromination of aromatic compounds in acetic acid by a mixture of peracetic acid and bromine[15]. The mechanism proposed (equations 51–53)

$$rate = k[CH_3CO_3H]\,[Br_2]$$

$$CH_3CO_3H + Br_2 \xrightarrow{\text{slow}} Br_2O + CH_3CO_2H \qquad (51)$$

$$Br_2O + CH_3CO_2H \rightleftharpoons HOBr + CH_3CO_2Br \qquad (52)$$

$$CH_3CO_2Br + ArH \longrightarrow CH_3CO_2H + ArBr \qquad (53)$$

is consistent with the kinetics of bromination, which are independent of the concentration of the arene (benzene). Unfortunately, a material balance is lacking and, therefore, the stoicheiometry of this reaction has not been established.

The reaction of chlorine with perbenzoic acid has been studied in aqueous media[136]. The sodium salt of the peracid reacts to give mainly benzoic acid, together with a little CO_2 and some benzoyl peroxide. The reaction is believed to begin as equation (54), but the mechanism is unknown.

$$C_6H_5CO_3^- + Cl_2 \longrightarrow Cl^- + C_6H_5CO_3^\cdot + Cl^\cdot \tag{54}$$

The degradation of carboxylic acids to alkyl halides through the action of lithium chloride and lead tetraacetate[137] is also suggested as proceeding through intermediates other than acyl hypochlorites.

The decomposition of benzoyl peroxide in the presence of several non-metal chlorides has been studied. In the case of thionyl chloride and phosphorus trichloride benzoyl, hypochlorite has been suggested as a necessary intermediate to the products, phosphorus oxychloride, sulphuryl chloride and benzoyl chloride[138]. Since no evidence is put forward to support this proposal and since similar extrusions of oxygen have not been observed in other reactions of acyl hypochlorites, the intermediacy of benzoyl hypochlorite in these reactions must be regarded as speculative.

IV. ACYL HYPOIODITES AND RELATED SUBSTANCES

A. The Reactions of Iodine with Metal Carboxylates[18, 19, 139]

These reactions seem to be somewhat more complex than their chlorine and bromine counterparts, for iodine interacts with silver salts in other ratios besides 1/1. Among the primary products of these reactions, there are proposed to be the acyl hypoiodite (24), the so-called Simonini complex (25) and the iodine tricarboxylates (26).

$$RCO_2Ag + I_2 \longrightarrow RCO_2I + AgI$$
$$\textbf{(24)}$$

$$2\,RCO_2Ag + I_2 \longrightarrow RCO_2Ag \cdot RCO_2I + AgI$$
$$\textbf{(25)}$$

$$3\,RCO_2Ag + 2\,I_2 \longrightarrow I(OCOR)_3 + 3\,AgI$$
$$\textbf{(26)}$$

Little direct physical evidence for acyl hypoiodites (24) has been obtained and their existence is usually inferred by analogy with the behaviour of solutions supposed to contain them with the behaviour of solutions of their chlorine and bromine analogues. Two groups of compounds that may be regarded as derivatives of the acyl hypoiodites have been prepared, the Simonini complexes (25) and the complexes with tertiary amines of the pyridine series. These latter have been obtained by interaction of iodine, the silver salt (usually of an aromatic acid) and the tertiary base, and may be characterized as yellow crystalline solids for which elemental analyses have been obtained[140, 141]. It is thought that they are of the constitution $(Ipy)^+RCO_2^-$ in which an iodine cation has been stabilized by coordination (py = pyridine or substituted pyridine). Such a view is consistent with the great similarity of the electronic absorption spectra of a series of the compounds[142] and the spectra of solutions of INO_3 in pyridine. It has been found by tracer studies that the iodine in these iodopyridinium salts is labile, in that it exchanges rapidly and completely with added elemental iodine[143]. This is not surprising, however, since it has been shown by conductivity measurements that equilibria of the type (55) are set up when iodine itself is dissolved in pyridine[142].

$$2\ I_2 + py \; \overrightarrow{\longleftarrow} \; pyI^+ + I_3^- \qquad (55)$$

The Simonini complexes (25) are also isolable solids in many cases (especially R = aryl) and give good elemental analyses[144-147]. The structure of the Simonini complex has been the subject of some discussion. In the case of iodine silver dibenzoate[146] (25; $R = C_6H_5$), a DMF solution of the complex gives an immediate precipitate of silver iodide with sodium iodide, but no reaction with silver nitrate. This behaviour suggests that actual or potential Ag^+ ions are present in the complex, but that I^- ions are not; this assumes that the complex has the same constitution, on dissolution in DMF, as the solid, and does not dissociate according to equation (56). Such a dissociated complex also would show this behaviour

$$C_6H_5CO_2Ag-C_6H_5CO_2I \; \overrightarrow{\longleftarrow} \; C_6H_5CO_2Ag + C_6H_5CO_2I \qquad (56)$$

with silver ion and iodide ion. Attempts to obtain crystalline material for a crystallographic structure determination and thereby explain this chemical behaviour, have not been successful[146].

Isolation of the Simonini complex followed by heating it in an inert solvent affords mainly the 'Simonini ester', RCO_2R (see below). Other reactions of the complex have been studied, including the oxidation of aromatic amines and the cleavage of 1,2-glycols[148]. Hydrolysis leads to iodate as one product according to equation (57)[145, 147]. In a few cases a

$$3\ (RCO_2)_2AgI + 3\ H_2O \longrightarrow 6\ RCO_2H + 2\ AgI + AgIO_3 \qquad (57)$$

bromine analogue of the Simonini complex has been obtained[147, 149, 150]; in the case of bromine silver dibenzoate[147], hydrolysis takes a more complex course than that of the corresponding iodide and oxygen is evolved.

Simonini complexes have been isolated in the perfluoroalkyl series[151]. The authors considered the possibility of the existence of the bromine analogues in this series but were unable to find evidence for them.

Iodine tricarboxylates are formed when silver salt and iodine are in a molar ratio of 3/2. The colourless solutions may be crystallized at low temperature to give the colourless crystalline, but easily hydrolysed, iodine tricarboxylates, whose stability decreases greatly as the length of the aliphatic chain in R is reduced[152]. Recently[153], perfluorinated iodine tricarboxylates have been prepared by the action of concentrated nitric

$$I_2 + 6\ HNO_3 + 6(R_FCO)_2O \longrightarrow 2\ I(OCOR_F)_3 + 6\ NO_2 + 6\ R_FCO_2H$$

$$(R_F = CF_3, C_2F_5, C_6F_5)$$

acid and iodine on perfluorocarboxylic anhydrides. These latter compounds are not susceptible to hydrolysis.

Hydrolysis of the non-fluorinated iodine tricarboxylates yields iodic acid[152-154] according to equations (58) and (59). Thermal decomposition

$$I(OCOR)_3 + 3\ H_2O \longrightarrow I(OH)_3 + 3\ RCO_2H \tag{58}$$

$$5\ I(OH)_3 \longrightarrow 3\ HIO_3 + 6\ H_2O + I_2 \tag{59}$$

of the long-chain iodine tricarboxylates yields several products of which the alkyl iodide RI and the ester RCO_2R are formed in largest amounts. In the presence of excess iodine, however, up to 80% of the alkyl iodide is formed[152]. Oldham and Ubbelohde propose that the tricarboxylate is converted to the acyl hypoiodite which then decomposes as in a Huns-diecker reaction, equations (60) and (61). Reinvestigation of the reaction

$$I(OCOR)_3 + I_2 \rightleftharpoons 3\ RCO_2I \tag{60}$$

$$RCO_2I \longrightarrow \longrightarrow CO_2 + RI \tag{61}$$

with iodine tripropionate[154] indicated that the course of the reaction is much more complex than equations (60) and (61) would suggest; however, the original authors themselves had noted[152] that the situation was less simple as the length of the alkyl chain in R was reduced.

Electrophilic iodination by acyl hypoiodites has been reported. Acetyl hypoiodite[20] and trifluoroacetyl hypoiodite[21, 22] both give good yields of aryl iodides and the direction of iodination is that which would be expected for an electrophilic substitution. Trifluoroacetyl hypoiodite particularly is reported as a promising reagent, for iodination occurs under relatively mild conditions, and since the decarboxylation of trifluoroacetyl hypoiodite does not proceed rapidly below 100°, electro-philic iodination is observed exclusively.

The interaction of mercuric acetate and iodine in acetic acid has been studied[155]. The reaction is believed to follow the course of equation (62),

$$Hg(OCOCH_3)_2 + I_2 \rightleftharpoons HgIOCOCH_3 + CH_3CO_2I \qquad (62$$

and the equilibrium constant has been obtained by observing the change in the u.v. absorption of the free iodine. At 25° the equilibrium constant is $2 \cdot 43 \pm 0 \cdot 05$ and at 45°, $1 \cdot 99 \pm 0 \cdot 04$, indicating that acetyl hypoiodite formation is a very slightly exothermic process. The iodination of pentamethylbenzene by the reagent was studied and the rate of halogenation by acetyl hypoiodite was compared with the rates of reaction of acetyl hypochlorite and hypobromite with aromatics. The iodo compound was found to be the least reactive, as would befit a compound expected[155] to be more stable.

The addition of acyl hypoiodites to olefins has been studied quite extensively, notably by Prévost in the 1930s and by whose name the reaction is usually called[35, 150, 156-158]. As in the case of the addition of acyl hypochlorites and hypobromites, the addition appears to be an ionic reaction in which addition of (I^+) is followed by (RCO_2^-)[39]. However, depending upon the conditions, several products may be obtained. The simple adducts, 2-iodocarboxylate esters, appear to be first formed. These substances have been proposed as crystalline derivatives of olefins, using iodine and silver 3,5-dinitrobenzoate as the reagent to obtain the iodo-3,5-dinitrobenzoates[39]. When the silver salt is in excess (conditions favouring formation of the Simonini complex) the reaction does not stop at equation (62a), but leads to glycol dicarboxylates, equation (63) (especially for R = aryl)[150, 156].

$$\begin{array}{c} \text{OCOR} \\ | \\ >C=C< + RCO_2I \longrightarrow >C-C< \\ | \\ I \end{array} \qquad (62a)$$

(27)

$$\begin{array}{c} \text{OCOR} \\ | \\ (27) + RCO_2Ag \longrightarrow AgI + >C-C< \\ | \\ \text{OCOR} \end{array} \qquad (63)$$

The reaction has aroused some interest, both as a preparative procedure and on account of its stereochemistry. Most of the mechanistic studies have been concerned with the reaction in the iodine series.

As noted above, the normal reaction in a dry solvent yields initially a *trans*-iodocarboxylate in an ionic sequence. However, in the presence of quite small amounts of water, the product obtained is a *cis*-hydroxycarboxylate[159-161] which may be converted to a *cis*-diol in excellent overall yield, often superior to that obtained by the use of potassium permanganate or osmium tetroxide[149]. Following the work of Winstein[162, 163], the 'wet Prévost' is suggested[149, 160] as involving displacement of the iodine of the *trans*-iodocarboxylate by the neighbouring carboxylate group (equation 64). Hydrolysis of the *ortho*-ester (28) gives two possible *cis*-hydroxyesters,

both of which hydrolyse to the same *cis*-diol. The mechanism has been investigated by Wiberg and Saegebarth[149]; using isotopic labelling, they showed that *both* oxygen atoms of the original acyl hypohalite finish up as the oxygens of the *cis*-diol.

In hindered systems particularly, it has been noted that the *cis*-diol formed is of the opposite stereochemistry to the *cis*-diol obtained by osmium tetroxide oxidation of the olefin[160, 164].

When an acyl hypohalite and a terminal acetylene are allowed to interact, substitution rather than addition occurs (equation 65)[156, 165].

$$RCO_2X + R'C{\equiv}CH \longrightarrow RCO_2H + R'C{\equiv}CX \qquad (65)$$

Almost all the studies of this reaction have been made using iodine as the halogen and usually it is the Simonini complex that is used as the acyl hypoiodite source. Although the reaction has not been extensively investigated, good to excellent yields of iodoacetylenes have been obtained; with acetylene itself both mono- and disubstitution have been reported[165].

Decarboxylation of acyl hypoiodites has been used less frequently as a synthetic route to alkyl iodides than has the corresponding reaction giving alkyl bromides. The reason usually advanced is that ester formation (equation 66) competes with alkyl iodide formation and results in a poor

$$2\,RCO_2Ag + I_2 \longrightarrow 2\,AgI + CO_2 + RCO_2R \qquad (66)$$

yield of the latter. While it is true that ester formation is very frequently observed in the silver salt/iodine reaction, it would seem that one reason has been that many workers have used the 2/1 molar ratio of silver salt to iodine of equation (66), thus favouring Simonini complex formation and ester production[18]. Alkyl iodides have been prepared successfully by the Hunsdiecker reaction, but only by using a 1/1 molar ratio of silver salt and iodine[166] or an excess of iodine[152]. The oxidation–decarboxylation reaction using either silver salts or the salts of alkali metals has found some utility in the production of perfluoroalkyl iodides[19, 167, 168].

Equation (66), however, represents the usual course of decarboxylation in the iodine series and gives the symmetrical ester RCO_2R. Much discussion has centred on the role of the Simonini complex in this reaction, for very often the complex separates from solution and, on heating or isolation followed by heating in the solvent, the ester is produced. Unfortunately, the aryl series, in which the complexes have been best characterized, is useless as a preparative method for esters, a variety of products being formed[145].

The decomposition of iodine silver dibenzoate (25, $R = C_6H_5$) has been studied in several aromatic solvents[147, 169]. Silver iodide is liberated almost quantitatively. Electrophilic iodination of the solvent is observed for benzene and anisole, but not for nitrobenzene or chlorobenzene. Instead, these latter, and to some extent the former, give aryl benzoates and arylbenzenes, both presumably free-radical products. Bromine silver dibenzoate[147] has been subjected to similar reactions giving essentially similar products.

A simple route to the Simonini ester is depicted in equations (67)–(69). Here the alkyl iodide is produced by a Hunsdiecker reaction and reacts

$$RCO_2Ag + I_2 \longrightarrow RCO_2I + AgI \qquad (67)$$

$$RCO_2I \xrightarrow[\text{pathway}]{\text{free-radical}} RI + CO_2 \qquad (68)$$

$$RI + Ag^+RCO_2^- \longrightarrow AgI + RCO_2R \qquad (69)$$

with a further mole of silver carboxylate. This mechanism takes no account of a possible role of the complex and implies that complex formation is an irrelevant side reaction (70) so far as ester production is concerned.

$$RCO_2I + RCO_2Ag \rightleftharpoons RCO_2IRCO_2Ag \qquad (70)$$

Stereochemical studies have shown that when R is of the constitution $R^1R^2R^3C-$ and is optically active, the alkoxy moiety of the ester is optically inactive, while the acyl fragment retains its activity[61], in agreement with equations (67)–(69). It would appear that at least one ionic step is involved whatever the correct mechanism, for unlike the Hunsdiecker reaction, in the Simonini reaction the alkoxy group R is found to have rearranged in cases where R is of such a configuration as to rearrange if it is a carbonium ion[170]. This does not require that the mechanism is ionic throughout as some have suggested[171]. A likely ionic step would be equation (69), assisted ionization of the alkyl iodide followed by reaction with a carboxylate anion. It would be interesting to know in this context whether the same yield of Simonini ester could be prepared under the same reaction conditions by the reaction of equivalent quantities of alkyl iodide and silver salt, but this has not been studied. In the absence of further information, particularly on the possible role of the Simonini complex, the authors are inclined to favour equations (67)–(69) as a working mechanism for the Simonini reaction; in this regard the failure of the reaction for aryl groups R may be held to support step (69), for assisted ionization should be very difficult for such substrates. On this basis, we would not expect ester formation from the action of iodine on the bridgehead silver carboxylates of rigid bicyclic systems, since these also resist assisted ionization to carbonium ions. This point has not been investigated. It may also be noted that the formation of esters in the case of acyl hypoiodites, but not from the hypochlorites or hypobromites may simply reflect the greater ease of assisted ionization of alkyl iodides. It could, however, be related to the production of polyvalent iodine intermediates and a route entirely different from equations (67)–(69), but no information is available on this point.

B. Reaction of Iodide Ions with Diacyl Peroxides

Although the reaction of iodide ions with diacyl peroxides is commonly used as the basis for the iodometric estimation of the peroxides according to equation (71), virtually no reports of the mechanism of the reaction

$$(RCO_2)_2 + 2I^- \longrightarrow 2RCO_2^- + I_2 \qquad (71)$$

have appeared. The single detailed study of the reaction concerns the action of iodide ions on substituted benzoyl peroxides[172]. The kinetics of the reaction were followed spectrophotometrically, and it was found that in ethanol the reaction is first-order in both iodide and the aroyl peroxide. When small quantities of water are added to the solvent the rate is reduced, as has been noted previously[173, 174], and this suggests that changes in polarity are involved in the transition state. This was confirmed both by a study of the ρ–σ correlation for a large number of phenyl-substituted benzoyl peroxides ($\rho = +0.76$, 100% EtOH) as well as by the observation that the rate was significantly affected by the cation associated with the iodide ion.

These data clearly point to a biomolecular reaction between iodide and benzoyl peroxide, but offer no support or lack of support for the intermediacy of benzoyl hypoiodite. Both (72) and (73) are proposed possible pathways for reaction (71), and the ease of attack of bromide ion on acyl hypobromites (section III, C) makes it likely that the attack of iodide ion on a hypoiodite (equation 72a) may be too fast for hypoiodites to be diverted to other products in these systems. To date, however, no information on this point is available. One point, however, is pertinent;

$$(C_6H_5CO_2)_2 + I^- \longrightarrow C_6H_5CO_2I + C_6H_5CO_2^- \tag{72}$$

$$C_6H_5CO_2I + I^- \longrightarrow C_6H_5CO_2^- + I_2 \tag{72a}$$

$$(C_6H_5CO_2)_2 + I^- \longrightarrow C_6H_5CO_2^- + C_6H_5CO_2^{\cdot} + \tfrac{1}{2}I_2 \tag{73}$$

$$C_6H_5CO_2^{\cdot} + I^- \longrightarrow C_6H_5CO_2^- + \tfrac{1}{2}I_2 \tag{73a}$$

if Hammond and Soffer[175] are correct in their suggestion that benzoyloxy radicals are rapidly scavenged in the presence of iodine (section IV, D), it would be anticipated that (72) would be an energetically more favourable reaction pathway than (73), since O—I bond formation is preferred over production of the independent radicals.

C. Acyl Hypoiodites from the Acid

Unlike the failure to produce acyl hypochlorites and hypobromites by the action of positive halogen compounds on carboxylic acids, acyl hypoiodites seem to be prepared quite readily by this reaction. Barton and co-workers[176] found that alkyl iodides and CO_2 were the products of the reaction of t-butyl hypoiodite with carboxylic acids under illumination and suggested equations (74)–(75) as representing the course of the

$$(CH_3)_3COI + RCO_2H \longrightarrow [RCO_2I] + (CH_3)_3COH \tag{74}$$

$$RCO_2I \xrightarrow{h\nu} RI + CO_2 \tag{75}$$

reaction. In benzene solution, fair to good yields of alkyl iodides were obtained, and the acyl hypoiodite was suggested as the first-formed intermediate, which on irradiation cleaves to the acyloxy radical and an iodine atom to start the usual Hunsdiecker decarboxylation sequence. It is of interest that acyl hypohalite formation should apparently be favoured in transhalogenation for iodine but not for chlorine or bromine: one possibility is that because iodine may be polycovalent the reaction may involve trivalent iodine intermediates. The question may be complicated, however, by the structures of the hypoiodites themselves. Hypochlorites, and probably hypobromites of the alkyl and acyl series, are of monomeric structure; while nothing is known about the constitution of acyl hypoiodites: their alkyl analogues are polymeric, the average molecular weight of t-butyl hypoiodite corresponding to about 9–12 monomer units[177].

It is reported that the mercuric oxide/halogen variant of the Hunsdiecker reaction, which uses the acid rather than its metal salt, is successful also with iodine[124, 178, 179]. This observation is of interest in that the mercuric oxide/bromine reaction is thought to involve bromine monoxide as an intermediate (section III, D). Iodine monoxide, however, has never been prepared[180]; hence, the success of the reaction implies either that iodine monoxide exists as an unstable reaction intermediate or that the mercuric oxide/halogen reaction, at least for iodine, proceeds by a pathway other than one involving the halogen monoxide.

Finally, acetyl hypoiodite has been invoked[181] as an intermediate to explain the results of kinetic studies for electrophilic iodination of phenol in the system $C_6H_5OH/I_2/CH_3CO_2H/CH_3CO_2Na(aq.)$.

D. The Interaction of Iodine with Peroxy Compounds and Other Oxidants

The decomposition of benzoyl peroxide in the presence of iodine was studied by Hammond. In benzene or chlorobenzene as solvents[182], the rate of peroxide decomposition is independent of the concentration of iodine, but the product distribution and iodine consumption are changed according to the iodine concentration. With much iodine present considerable quantities of iodobenzene are formed together with (in chlorobenzene) chloroiodobenzenes, mainly the *para*-isomer. Benzoyl hypoiodite was proposed as an intermediate (equation 76) to explain the products.

$$(C_6H_5CO_2)_2 \longrightarrow 2\ C_6H_5CO_2^{\cdot} \xrightarrow{I_2} 2\ C_6H_5CO_2I \qquad (76)$$

In carbon tetrachloride an excellent yield of iodobenzene was obtained[175], and it was discovered that if water were added to the solvent, a nearly quantitative yield of benzoic acid was obtained, presumably because of hydrolysis of the intermediate benzoyl hypoiodite (equation 77). The

$$C_6H_5CO_2I + H_2O \longrightarrow C_6H_5CO_2H + HOI \qquad (77)$$

preparative aspects of this reaction have been studied[183] recently. As reported by Hammond[175, 182], better yields of iodobenzene from benzoyl peroxide result in carbon tetrachloride than in aromatic solvents. Other halogenated aliphatic solvents are also useful. Good yields of *n*-octyl iodide were obtained from dinonanoyl peroxide in both aliphatic and aromatic solvents. It is concluded[183] that the Hunsdiecker method and its modifications are superior for the synthesis of alkyl iodides but that the peroxide/iodine technique is of preparative value for some aryl iodides.

In the presence of olefins the benzoyl peroxide/iodine reagent leads to the production of glycol dibenzoates[184]. Benzoyl hypoiodite was proposed as the reactive intermediate, and this belief was strengthened by the isolation of 2-iodocyclohexyl benzoate from an experiment using cyclohexene as the olefin, and the conversion of the iodoester into 1,2-cyclohexanediol dibenzoate under the reaction conditions (equation 78).

The interaction of peracetic acid and iodine is thought to involve the production of acetyl hypoiodite and several studies with this reagent have

(78)

been carried out. Benzene has been iodinated in yields up to 60% based on the iodine consumed; toluene gave 85% of o- and p-iodotoluenes, but benzoic acid and nitrobenzene were unattacked by the reagent[185]. Iodination of benzene proceeds according to a rate expression (79) which is independent of the benzene concentration and in this respect the

$$\text{rate} = k[\text{I}_2] \, [\text{CH}_3\text{CO}_3\text{H}] \tag{79}$$

electrophilic substitution parallels that of halogenation by acetyl hypochlorite (section III, B). Unlike the latter, however, acid catalysis is relatively unimportant for iodination of benzene by acetyl hypoiodite, although it becomes important for iodination of benzoic acid which is attacked only by the acidified reagent. Possibly here the active entity is protonated acetyl hypoiodite. The simultaneous production of iodic acid suggests that hypoiodous acid is formed at some stage in the reaction, and the sequence (80)–(83) is proposed. Unfortunately, the kinetics do not

$$\text{H}_2\text{O} + \text{CH}_3\text{CO}_3\text{H} + \text{I}_2 \xrightarrow{\text{slow}} \text{CH}_3\text{CO}_2\text{H} + 2\,\text{HOI} \tag{80}$$

$$\text{HOI} + \text{CH}_3\text{CO}_2\text{H} \rightleftharpoons \text{CH}_3\text{CO}_2\text{I} + \text{H}_2\text{O} \tag{81}$$

$$\text{C}_6\text{H}_6 + \text{CH}_3\text{CO}_2\text{I} \text{ (or HOI)} \xrightarrow{\text{fast}} \text{C}_6\text{H}_5\text{I} + \text{CH}_3\text{CO}_2\text{H} \text{ (or H}_2\text{O)} \tag{82}$$

$$\text{HOI} + 2\,\text{CH}_3\text{CO}_3\text{H} \text{ (or 2 HOI)} \longrightarrow \text{HIO}_3 + 2\,\text{CH}_3\text{CO}_2\text{H} \text{ (or 2 HI)} \tag{83}$$

distinguish between HOI and $\text{CH}_3\text{CO}_2\text{I}$ as the electrophilic reagent and the latter is simply the previous authors' own preference. That the reaction is an electrophilic substitution is demonstrated by the relative reactivity of several aromatic substrates towards the reagent, as well as by the direction of iodination by a substituent already present[185, 186].

Subsequent studies[186] by the same authors revealed a more complex situation in that the iodination is autocatalytic, the rate increasing with the concentration of the product aryl iodide, or an added aromatic iodide. This autocatalysis was ascribed to the oxidation of iodoaromatics by peracetic acid, to give iodosobenzene $\text{C}_6\text{H}_5\text{IO}$ or iodobenzene diacetate $\text{C}_6\text{H}_5\text{I(OCOCH}_3)_2$. In acetic acid with added iodine, but in the absence of peracetic acid, iodosobenzene or better, iodobenzene diacetate iodinated m-xylene faster than the iodine/peracetic acid reagent. The reaction had the stoicheiometry of equation (84). It was postulated that acetyl hypoiodite

$$\text{C}_6\text{H}_5\text{I(OAc)}_2 + \text{I}_2 + 2\,\text{ArH} \longrightarrow 2\,\text{ArI} + 2\,\text{CH}_3\text{CO}_2\text{H} + \text{C}_6\text{H}_5\text{I} \tag{84}$$

17

is also the active iodinating agent in the autocatalysed reaction and is produced by equations (85)–(87).

$$ArI + CH_3CO_3H \longrightarrow ArIO + CH_3CO_2H \tag{85}$$

$$ArIO + 2 CH_3CO_2H \rightleftharpoons ArI(OCOCH_3)_2 + H_2O \tag{86}$$

$$ArI(OCOCH_3)_2 + I_2 \longrightarrow ArI + 2 CH_3CO_2I \tag{87}$$

The iodine/peracetic acid reagent has also been studied in combination with olefins[187, 188] and results in the addition of the elements of acetyl hypoiodite to the double bond. A study of the kinetics in this system reveals a situation more complex than that of aromatic iodination, for here addition of the iodine to the olefin is faster than its reaction with peracetic acid (equation 80). The authors conclude that in this system acetyl hypoiodite is not the species responsible for most of the addition.

Subsequent studies on the reaction of propylene with iodine and iodobenzene diacetate or iodosobenzene also gave 1-iodo-2-acetoxypropane as the principal reaction product[189]. Here, by contrast, the authors postulate the initial formation of acetyl hypoiodite (equations 85–87) followed by a Prévost addition of the latter to the double bond.

Acyl hypoiodites have been suggested to arise from aromatic dicarboxylates in another context: the decomposition of aryliodine dibenzoates is proposed to follow a free-radical chain pathway in which benzoyl hypoiodite is a necessary intermediate[190]. Unfortunately, no direct evidence for the intermediacy of the hypoiodite was obtained.

Other oxidizing agents have been proposed to give acyl hypoiodites on reaction with iodine. Oxidation of iodine with lead tetraacetate affords CO_2 and methyl iodide, the following sequence being proposed[176] (equations 88–89). Addition of a carboxylic acid to the reaction mixture

$$Pb(OAc)_4 + I_2 \longrightarrow Pb(OAc)_3I + CH_3CO_2I \tag{88}$$

$$CH_3CO_2I \longrightarrow \longrightarrow CH_3I + CO_2 \tag{89}$$

leads to its preferential decarboxylation, presumably by virtue of equilibria such as (90) and (91) and reactions (92) and (93). Good yields of alkyl

$$Pb(OAc)_4 + RCO_2H \rightleftharpoons Pb(OAc)_3(OCOR) + CH_3CO_2H \tag{90}$$

$$RCO_2H + CH_3CO_2I \rightleftharpoons RCO_2I + CH_3CO_2H \tag{91}$$

$$Pb(OAc)_3(OCOR) + I_2 \longrightarrow RCO_2I + Pb(OAc)_3I \tag{92}$$

$$RCO_2I \longrightarrow \longrightarrow RI + CO_2 \tag{93}$$

iodides have been achieved using this method, although its application is not as wide as the same authors' technique for iodinative decarboxylation using t-butyl hypoiodite (see above, section IV, C).

Finally, the oxidation of iodine by acetyl hypochlorite is reported[7] to give acetyl hypoiodite and chlorine (equation 94). The acetyl hypoiodite

$$2\, ClOAc + I_2 \longrightarrow 2\, IOAc + Cl_2 \qquad (94)$$

so formed is claimed to decompose on heating to iodine, CO_2 and methyl iodide. Hydrolysis gives acetic acid, iodine and possibly iodic acid. Further work is clearly indicated to follow up these very interesting observations.

V. NITROGEN-CONTAINING ANALOGUES OF ACYL HYPOHALITES

This section is limited to a description of several types of compounds of limited stability which can be considered as analogues of acyl hypohalites

$$\overset{O}{\underset{\|}{(R-C-OX)}}.$$

Esters ($X = $ alkyl or aryl) clearly fall outside this classification, as do, for the purposes of this article, peroxy compounds ($X = -OH$, $-OR$, $-OCOR$). In this section the discussion will be limited to a consideration of acyl cyanates ($X = -CN$), acyl nitrites ($X = -NO$) and acyl nitrates ($X = -NO_2$).

A. Acyl Cyanates

No acyl cyanate has ever been isolated, neither is there any spectroscopic or other physical evidence for their existence; rather they have been postulated as plausible reaction intermediates.

A reaction, apparently analogous to the Hunsdiecker process, involves the silver salts of carboxylic acids and cyanogen bromide[191]. From silver benzoate, the products were benzonitrile and benzoic anhydride, the latter normally in much greater amount. The reaction was carried out in several solvents, most successfully in acetronitrile where the yield of benzonitrile amounted to 32%. The following outline of the reaction pathway was proposed. The mechanism of the reaction has not been studied, but in

$$C_6H_5CO_2Ag + BrCN \longrightarrow C_6H_5CO_2CN + AgBr \qquad (95)$$

$$C_6H_5CO_2Ag + C_6H_5CO_2CN \longrightarrow (C_6H_5CO)_2O + AgOCN \qquad (96)$$

$$C_6H_5CO_2CN \longrightarrow C_6H_5CN + CO_2 \qquad (97)$$

view of observations about to be described, it is probable that equation (97) does not take the course analogous to that discussed for the Hunsdiecker reaction (section III, B).

Under more drastic conditions (sealed tube 200–300°), the sodium and potassium salts of carboxylic acids also furnish nitriles on reaction with cyanogen chloride or cyanogen bromide. The salts of acetic, propionic, benzoic and o-chlorobenzoic acids gave moderate yields of the corresponding nitriles[192]. When sodium benzoate–carbonyl–[14]C was used, it was found[193] that the ^{14}C in the products was found mainly in the benzonitrile (equation 98); thus unlike the Hunsdiecker reaction it is not the carboxylate carbon that is lost as CO_2. Similar experiments applied to the reaction

$$C_6H_5{}^{14}CO_2Na + BrCN \longrightarrow C_6H_5{}^{14}CN + CO_2 + NaBr \qquad (98)$$

of sodium propionate with cyanogen bromide gave similar, but less clear-cut results[194]. From this reaction propionyl isocyanate was isolated as a by-product, and the mechanism, equations (99) and (100), was proposed, in which **29**, rather than the acyl cyanate, was the postulated intermediate.

$$R-\overset{O-}{\underset{*}{C}}=O + N\equiv C-Br \longrightarrow \quad R-\overset{O^-}{\underset{*}{\underset{|}{C}}}\overset{|}{\underset{O-C-Br}{-N}}$$

(29)

$$R-\overset{O}{\underset{*}{\overset{||}{C}}}-N=C=O + Br^- \qquad (99)$$

$$\overset{O}{\underset{RC-N}{\overset{||}{\underset{*}{}}}} \longrightarrow \overset{O}{\underset{RC=N}{\overset{||}{\underset{*}{}}}} \longrightarrow RC\underset{*}{\equiv}N + CO_2 \qquad (100)$$

Recently alkyl cyanates have been isolated and are found to isomerize readily to alkyl isocyanates[195]. This suggests, by analogy, that the formation of the acyl cyanate, followed by its rearrangement to the isocyanate (equation 101), is an alternative to (99), direct formation of the isocyanate,

$$RCO_2M + BrCN \longrightarrow MBr + RCO_2CN \longrightarrow (RCO)^+(OCN)^-$$
$$\longrightarrow RCONCO \qquad (101)$$

in that proposed scheme. Whatever the details of the mechanism, it seems clear that the bond from the alkyl or aryl group R to the carbonyl carbon is never broken, for the results of the tracer experiments have been confirmed by the experiments of Barltrop, Day and Bigley[196]. Using terpenoid groups R, they find that the group R changes neither its geometric nor its enantiomeric configuration when the acid salt is converted to the nitrile by means of cyanogen chloride at 250°.

The intermediacy of acyl cyanates may also be implicated in the oxidation of hydrogen cyanide by benzoyl peroxide. In the presence of cyclohexane, cyclohexyl cyanide is produced, and the course of reaction is postulated[111] to be that of equations (102) and (103).

$$HCN + (C_6H_5CO_2)_2 \longrightarrow C_6H_5CO_2CN + C_6H_5CO_2H \tag{102}$$

$$C_6H_5CO_2CN + C_6H_{11}^{\cdot} \longrightarrow C_6H_{11}CN + C_6H_5CO_2^{\cdot} \tag{103}$$

B. Acyl Nitrites

The silver salts of carboxylic acids react with nitrosyl chloride to give acyl nitrites, which may be isolated and characterized. Acetyl nitrite was discovered by Francesconi and co-workers[197], who found that it decomposed on warming, or more violently in sunlight[198], giving mainly acetic anhydride and N_2O_3. Recently, several acyl nitrites have been prepared[199],

$$CH_3CO_2Ag + CINO \longrightarrow AgCl + CH_3CO_2NO \longrightarrow \tfrac{1}{2}N_2O_3 + \tfrac{1}{2}(CH_3CO)_2O \tag{104}$$

and it has been found that anhydride formation is quite general when solutions of the acyl nitrite are heated to about 100°. Pyrolysis of the acyl nitrites in the vapour phase at 200°, however, leads to decarboxylation and the formation of a nitroso compound (equation 105), which under the reaction conditions, usually isomerizes to an oxime (equation 106).

$$RCO_2NO \xrightarrow{200°} RNO + CO_2 \tag{105}$$

$$R^1R^2CHNO \longrightarrow R^1R^2C{=}NOH \tag{106}$$

Carbon dioxide production is essentially quantitative, while the oxime or nitroso compound is obtained in better than 50% yield.

Much more extensive study has been made, however, of the corresponding reaction of salts of perfluorinated acids, especially silver perfluorocarboxylates[200–202], since these salts offer a convenient route to the otherwise rather inaccessible nitrosoperfluoroalkanes. One patented

$$R_FCO_2M + CINO \longrightarrow R_FCO_2NO + MCl \longrightarrow R_FNO \tag{107}$$

report[203] also discusses the formation of the perfluoroacyl nitrites from the perfluorocarboxylic acid and nitrosyl chloride under illumination. These compounds have been studied mainly by Haszeldine and associates. On account of their low volatility, the isolable yellow perfluoroacyl nitrites are suggested[201, 202] as showing molecular association.

On heating the perfluoroacyl nitrites, anhydride formation is not observed, unlike their hydrocarbon analogues[199]; indeed, a convenient route to the perfluoro derivatives is by the action of dinitrogen trioxide (or nitrosyl chloride) on the perfluoroacid anhydride[204–206]. Photolysis, or heating the nitrite vapours to about 200°, causes decomposition to the nitrosoperfluoroalkane and CO_2 to occur in high yield[201]. The original papers stress the necessity for dilution of the nitrite vapours with an inert carrier to minimize the risk of explosion.

It is not known by what route decomposition of the perfluoroacyl nitrite occurs. Haszeldine and co-workers[201] suggest two possibilities; a free-radical pathway (equations 108–111) and a four-centre intramolecular mechanism (equation 112).

$$R_FCF_2CO_2NO \longrightarrow R_FCF_2CO_2^{\bullet} + NO \qquad (108)$$

$$R_FCF_2CO_2^{\bullet} \longrightarrow R_FCF_2^{\bullet} + CO_2 \qquad (109)$$

$$R_FCF_2^{\bullet} + NO \longrightarrow R_FCF_2NO \qquad (110)$$

$$R_FCF_2^{\bullet} + R_FCF_2CO_2NO \longrightarrow R_FCF_2NO + R_FCF_2CO_2^{\bullet} \qquad (111)$$

$$R_FCF_2 \overset{\displaystyle C=O}{\underset{\displaystyle O=N-O}{\vert}} \longrightarrow R_FCF_2NO + CO_2 \qquad (112)$$

Other reactions of perfluoroacyl nitrites to be studied so far include attack on some metal chlorides with the liberation of nitrosyl chloride, hydrolysis and oxidation of many metals to their perfluorocarboxylate salts, with liberation of nitric oxide[202].

Belov and Kozlov have prepared acyl nitrites in aqueous media through the action of sodium nitrite and the acid, or the acid anhydride[207]. They report that the diazotization of aromatic amines by sodium nitrite and an organic acid probably involves an intermediate acyl nitrite[208], in support of which the acyl nitrites, when isolated, act as effective diazotizing agents for the aromatic amines[207].

C. Acyl Nitrates

Acyl nitrates have been prepared, but have not been obtained completely pure. Benzoyl nitrate[209], made by the action of silver nitrate on benzoyl chloride at $-10°$ in carbon tetrachloride solution, is the best known. Essentially no physical characteristics have been reported. Acetyl nitrate is considered to exist in solutions of nitric acid in an excess of acetic anhydride.

Acyl nitrates have been of interest mainly in connexion with aromatic nitration[210, 211]. Aromatic amines and phenols are conveniently nitrated by acetyl nitrate, and with these as well as with aryl alkyl ethers and anilides an unusually high proportion of the o-nitro compound is obtained. Benzoyl nitrate has been studied particularly to test the effect of variation of the leaving group X on nitrations by $NO_2—X$. With benzoyl nitrate, however, the rapid establishment of equilibrium (113) and the production of the more reactive nitrating species thereby makes the substitution

$$2 \ C_6H_5CO_2NO_2 \ \rightleftharpoons \ N_2O_5 + (C_6H_5CO)_2O \qquad (113)$$

proceed entirely through the agency of N_2O_5 and not by benzoyl nitrate. Recently[212] it has been noted that in the case of benzoyl nitrate, dehydrogenation reactions may accompany nitration; such has been observed in the reactions of bibenzyl and of 2,2'-meta-cyclophane.

Nitrations involving acyl nitrates have been discussed elsewhere; it seems probable that in most cases NO_2^+ derived from the N_2O_5 is the reactive nitrating species[213]. Synthetically, acyl nitrates are useful in situations where the usual sulphuric acid/nitric acid nitrating mixture may not be used; nitric acid in excess acetic anhydride has been used very successfully for nitration of acid-sensitive heterocycles[214-217].

With several aromatic substrates, acetyl nitrate is reported to yield acetoxylation as well as nitration products[218, 219]. The mechanism has been interpreted as an electrophilic substitution[218, 219] or an addition–elimination reaction[220, 221]. Addition products of the elements of acetyl nitrate to o-xylene have been isolated[222].

Bordwell and Garbisch[223, 224] have studied the action of acetyl nitrate on a number of olefins. Cyclohexene[223], for example, gives the products **30–33**; subsequently[225], **34** was also found to be a product. The reaction

with aryl-substituted olefins[224] is rather cleaner and it is suggested that a cyclic addition of the acyl nitrate may be involved. In favourable cases up to 70% of nitroacetates may be formed. The addition of acetyl nitrate to cyclohexanone enol acetate has been used as a route to 2-nitrocyclohexanone[225].

The decomposition of benzoyl peroxide in the presence of dinitrogen tetroxide has been investigated; the products of the reaction have been explained as arising through trapping of benzoyloxy radicals by NO_2

followed by breakdown of the intermediate benzoyl nitrate[226]. This proposal is, of course, analogous to the situation in the decomposition of the same peroxide in the presence of iodine (section IV, D).

$$(C_6H_5CO_2)_2 \longrightarrow 2\ C_6H_5CO_2^{\bullet} \xrightarrow{\ NO_2\ } C_6H_5CO_2NO_2 \qquad (114)$$

Acknowledgments

The authors wish to express their thanks to Mrs. Anita Lutzer for her help in the preparation of the manuscript and to Mrs. Betty Tanner and Mr. Alan Ryan for their helpful criticisms of the text.

VI. REFERENCES

1. G. H. Cady and K. B. Kellogg, *J. Am. Chem. Soc.*, **75**, 2501 (1953).
2. G. L. Gard and G. H. Cady, *Inorg. Chem.*, **4**, 594 (1965).
3. A. Menefee and G. H. Cady, *J. Am. Chem. Soc.*, **76**, 2020 (1954).
4. R. L. Cauble and G. H. Cady, *J. Am. Chem. Soc.*, **89**, 5161 (1967).
5. R. D. Stewart and G. H. Cady, *J. Am. Chem. Soc.*, **77**, 6110 (1955).
6. J. H. Prager and P. G. Thompson, *J. Am. Chem. Soc.*, **87**, 230 (1965).
7. P. Schutzenberger, *Compt. Rend.*, **52**, 135 (1861).
8. W. Bockemüller and F. W. Hoffmann, *Ann. Chem.*, **519**, 165 (1935).
9. M. Anbar and I. Dostrovsky, *J. Chem. Soc.*, 1105 (1954).
10. D. D. Tanner and N. Nychka, *J. Am. Chem. Soc.*, **89**, 121 (1967), and references therein.
11. P. B. D. de la Mare, I. C. Hilton and C. A. Vernon, *J. Chem. Soc.*, 4039 (1960).
12. P. B. D. de la Mare, A. D. Ketley and C. A. Vernon, *Research* **6**, 12S (1953).
13. J. C. Evans, G. Y.-S. Lo and Y. L. Chang, *Spectrochim. Acta*, **21**, 973 (1965).
14. Y. Hatanaka, R. M. Keefer and L. J. Andrews, *J. Am. Chem. Soc.*, **87**, 4280 (1965).
15. Y. Ogata, Y. Furuya and K. Okano, *Bull. Chem. Soc. Japan*, **37**, 960 (1964).
16. J. K. Kochi, B. M. Graybill and M. E. Kurz, *J. Am. Chem. Soc.*, **86**, 5257 (1964).
17. N. J. Bunce and L. O. Urban, *Abstracts of the 160th National Meeting of the A.C.S.*, Chicago, September 1970, Abstract ORGN 159.
18. C. V. Wilson, *Org. Reactions*, **9**, 332 (1957).
19. R. G. Johnson and R. K. Ingham, *Chem. Rev.*, **56**, 219 (1956).
20. D. Papa, E. Schwenk and E. Klingsberg, *J. Am. Chem. Soc.*, **72**, 2623 (1950).
21. A. L. Henne and W. F. Zimmer, *J. Am. Chem. Soc.*, **73**, 1352 (1951).
22. R. N. Haszeldine and A. G. Sharpe, *J. Chem. Soc.*, 993 (1952).
23. R. A. Barnes and R. J. Prochaska, *J. Am. Chem. Soc.*, **72**, 3188 (1950).
24. W. G. Dauben and H. Tilles, *J. Am. Chem. Soc.*, **72**, 3185 (1950).
25. F. B. Fisher and A. N. Bourns, *Can. J. Chem.*, **39**, 1736 (1961).
26. K. Birnbaum and H. Reinherz, *Ber.*, **15**, 456 (1882).
27. P. B. D. de la Mare, I. C. Hilton and S. Varma, *J. Chem. Soc.*, 4044 (1960).

28. M. Hassan and G. Yousif, *J. Chem. Soc. (B)*, 459 (1968).
29. P. B. D. de la Mare and J. H. Ridd, *Aromatic Substitution, Nitration and Halogenation*, Butterworths, London, 1959, Chapters 8–10.
30. G. Stanley and J. Shorter, *J. Chem. Soc.*, 246, 256 (1958).
31. C. K. Ingold, *Chem. Rev.*, **15**, 271 (1934).
32. P. B. D. de la Mare and J. L. Maxwell, *J. Chem. Soc.*, 4829 (1962).
33. S. J. Branch and B. Jones, *J. Chem. Soc.*, 2317 (1954); 2921 (1955).
34. M. I. Uschakow and W. O. Tchistow, *Ber.*, **68**, 824 (1935).
35. C. Prévost and J. Wiemann, *Compt. Rend.*, **204**, 989 (1937).
36. D. C. Abbott and C. L. Arcus, *J. Chem. Soc.*, 1515 (1952).
37. W. G. H. Edwards and R. Hodges, *J. Chem. Soc.*, 761 (1954).
38. H. Haubenstock and C. P. VanderWerf, *J. Org. Chem.*, **29**, 2993 (1964).
39. B. I. Halperin, H. B. Donahoe, J. Kleinberg and C. VanderWerf, *J. Org. Chem.*, **17**, 623 (1952).
40. V. L. Hensley, C. L. Frye, G. E. Heasley, K. A. Martin, D. A. Redfield and P. S. Wilday, *Tetrahedron Letters*, 1573 (1970).
41. G. A. Razuvaev and N. S. Vasilenskaya, *Doklady Akad. Nauk S.S.S.R.*, **74**, 279 (1950); *Chem. Abstr.*, **45**, 3800 (1951).
42. P. B. D. de la Mare, C. J. O'Connor, M. J. Rosser and M. J. Wilson, *Chem. Commun.*, 731 (1970).
43. S. G. Levine and M. E. Wall, *J. Am. Chem. Soc.*, **81**, 2826, 2829 (1959).
44. See reference 18, p. 338 and reference 22.
45. R. A. Berry, *Diss Abstr.*, **23**, 4529 (1963).
46. A. McKillop, D. Bromley and E. C. Taylor, *J. Org. Chem.*, **34**, 1172 (1969).
47. A. Adachi and N. Hirao, *Yushi Kagaku Kyokaishi*, **1**, 167 (1952); *Chem. Abstr.*, **48**, 1940 (1954).
48. J. C. Conly, *J. Am. Chem. Soc.*, **75**, 1148 (1953).
49. N. J. Bunce and D. D. Tanner, *J. Am. Chem. Soc.*, **91**, 6096 (1969).
50. F. Kuffner and C. Russo, *Monatsh.*, **85**, 1097 (1954).
51. E. G. Janzen and B. Knauer, *Abstracts of the 158th Meeting of the A.C.S.*, New York, 1969, Abstract ORGN 121.
52. J. Cason and R. H. Mills, *J. Am. Chem. Soc.*, **73**, 1354 (1951).
53. M. Heintzeler, *Ann. Chem.*, **569**, 102 (1950).
54. P. Wilder, Jr. and A. Winston, *J. Am. Chem. Soc.*, **75**, 5370 (1953).
55. F. W. Baker, H. D. Holtz and L. M. Stock, *J. Org. Chem.*, **28**, 514 (1963).
56. J. D. Roberts and V. C. Chambers, *J. Am. Chem. Soc.*, **73**, 3176 (1951).
57. E. L. Sukman, *Diss. Abstr.*, **19**, 1211 (1958).
58. F. Bell and I. F. B. Smyth, *J. Chem. Soc.*, 2372 (1949).
59. R. T. Arnold and P. Morgan, *J. Am. Chem. Soc.*, **70**, 4248 (1948).
60. C. E. Berr, unpublished work quoted in reference 19.
61. S. Oae, T. Kashiwagi and S. Kozuka, *Bull. Chem. Soc. Japan*, **39**, 2441 (1966).
62. C. L. Arcus, A. Campbell and J. Kenyon, *Nature*, **163**, 287 (1949); *J. Chem. Soc.*, 1510 (1949).
63. D. C. Abbott and C. L. Arcus, *J. Chem. Soc.*, 3195 (1952).
64. C. L. Arcus and G. V. Boyd, *J. Chem. Soc.*, 1580 (1951).
65. J. Cason, M. J. Kahm and R. H. Mills, *J. Org. Chem.*, **18**, 1670 (1953).
66. S. Winstein, *Bull. Soc. Chim. France*, **18**, 70c (1951).
67. W. T. Smith, Jr. and R. L. Hull, *J. Am. Chem. Soc.*, **72**, 3309 (1950).

68. C. Hunsdiecker, H. Hunsdiecker and E. Vogt, *Ger. Pat.*, 730,410 (1935); *Chem. Abstr.*, **38**, 374 (1944).
69. W. Parker and R. A. Raphael, *J. Chem. Soc.*, 1723 (1955).
70. D. E. Applequist and A. H. Peterson, *J. Am. Chem. Soc.*, **82**, 2372 (1960).
71. E. L. Eliel and R. V. Acharya, *J. Org. Chem.*, **24**, 151 (1959).
72. F. R. Jensen, L. H. Gale and J. E. Rodgers, *J. Am. Chem. Soc.*, **90**, 5793 (1968).
73. S. J. Cristol, J. R. Douglass, W. C. Firth and R. E. Krall, *J. Am. Chem. Soc.*, **82**, 1829 (1960).
74. C. L. Osborn, T. V. Van Auken and D. J. Trecker, *J. Am. Chem. Soc.*, **90**, 5806 (1968).
75. E. C. Kooyman and G. C. Vegter, *Tetrahedron*, **4**, 382 (1958).
76. P. D. Bartlett, G. N. Fickes, F. C. Haupt and R. H. Helgeson, *Accounts Chem. Res.*, **3**, 177 (1970).
77. P. I. Abell, *J. Org. Chem.*, **22**, 769 (1957).
78. E. R. Buchman, J. C. Conly and D. R. Howton, *Tech. Report Calif. Inst. Tech.*, 90 (1951), quoted in reference 67.
79. M. Avram, E. Marica and C. D. Nenitzescu, *Acad. rep. populare Romine*, *studii carcetari chim.*, **7**, 155 (1959); *Chem. Abstr.*, **54**, 8664 (1960).
80. D. E. Applequist and A. S. Fox, *J. Org. Chem.*, **22**, 1751 (1957).
81. D. E. Applequist and N. D. Werner, *J. Org. Chem.*, **28**, 48 (1963).
82. N. A. Le Bel, *J. Am. Chem. Soc.*, **82**, 623 (1960).
83. H. L. Goering and D. W. Larsen, *J. Am. Chem. Soc.*, **79**, 2653 (1957); **81**, 5937 (1959).
84. W. Thaler, *J. Am. Chem. Soc.*, **85**, 2607 (1963).
85. P. S. Skell and P. D. Readio, *J. Am. Chem. Soc.*, **86**, 3334 (1964).
86. See also D. D. Tanner, D. Darwish, M. W. Mosher and N. J. Bunce, *J. Am. Chem. Soc.*, **91**, 7398 (1969) for other references.
87. S. Furukawa, K. Naruchi, S. Matsui and M. Yuuki, *Bull. Chem. Soc. Japan*, **40**, 594 (1967).
88. J. D. Berman and C. C. Price, *J. Org. Chem.*, **23**, 102 (1958).
89. R. M. Kopchik and J. A. Kampmeier, *J. Am. Chem. Soc.*, **90**, 6733 (1968).
90. M. J. Perkins, *Ann. Reports Chem. Soc.* (*B*), 1968, p. 173.
91. S. Furukawa and B. Kubota, quoted in reference 87.
92. P. Wilder and A. Winston, *J. Am. Chem. Soc.*, **77**, 5598 (1955).
93. J. R. Dice and J. N. Bowen, *J. Am. Chem. Soc.*, **71**, 3107 (1949).
94. A. D. Campbell and D. R. D. Shaw, *J. Chem. Soc.*, 5042 (1952).
95. R. Grewe and E. Vangermain, *Chem. Ber.*, **98**, 104 (1965).
96. K. Heyns and K. Stange, *Z. Naturforsch.*, **7b**, 677 (1952); *Chem. Abstr.*, **48**, 1957 (1954).
97. N. G. Kundu and A. J. Sisti, *J. Org. Chem.*, **34**, 229 (1969).
98. J. W. Wilt and D. D. Oathoudt, *J. Org. Chem.*, **21**, 1550 (1956); **23**, 218 (1958).
99. J. W. Wilt and J. L. Finnerty, *J. Org. Chem.*, **26**, 2173 (1961).
100. J. W. Wilt and J. A. Lundquist, *J. Org. Chem.*, **29**, 921 (1964).
101. U. K. Pandit and I. P. Dirk, *Tetrahedron Letters*, 891 (1963).
102. D. L. Kvalnes, *Diss. Abstr.*, **20**, 90 (1959).

103. S. Winstein and F. Seubold, *J. Am. Chem. Soc.*, **69**, 2916 (1947); W. H. Urry and M. S. Kharasch, *J. Am. Chem. Soc.*, **66**, 1438 (1944); W. H. Urry and N. Nicolaides, *J. Am. Chem. Soc.*, **74**, 5162 (1952).
104. J. W. Wilt, *J. Am. Chem. Soc.*, **77**, 6397 (1955).
105. C. H. Bamford and E. F. T. White, *J. Chem. Soc.*, 4490 (1960).
106. H. Bredereck, A. Wagner, R. Blaschke and G. Demetriades, *Angew. Chem.*, **71**, 340 (1959).
107. D. D. Tanner and M. W. Mosher, *Can. J. Chem.*, **47**, 715 (1969).
108. C. Walling and J. A. McGuinness, *J. Am. Chem. Soc.*, **91**, 2053 (1969).
109. C. Walling and B. B. Jacknow, *J. Am. Chem. Soc.*, **82**, 6108 (1960).
110. D. D. Tanner and N. J. Bunce, *J. Am. Chem. Soc.*, **91**, 3028 (1969).
111. D. D. Tanner, G. Lycan and N. J. Bunce, *Can. J. Chem.*, **48**, 1492 (1970).
112. S. O. Lawesson and N. C. Yang, *J. Am. Chem. Soc.*, **81**, 4230 (1959).
113. J. O. Edwards and R. G. Pearson, *J. Am. Chem. Soc.*, **84**, 16 (1962).
114. D. D. Tanner and N. J. Bunce, unpublished observations.
115. G. A. Razuvaev, B. N. Moryganov, E. P. Dlin and Y. A. Ol'deKop, *Zh. Obshch. Khim.*, **24**, 262 (1954); English translation, *J. Gen. Chem. USSR*, **24**, 265 (1954); *Chem. Abstr.*, **49**, 4575 (1955).
116. J. K. Kochi and R. V. Subramanian, *J. Am. Chem. Soc.*, **87**, 1508 (1965).
117. J. K. Kochi, *Record Chem. Progr.*, **27**, 207 (1966).
118. J. K. Kochi, *J. Am. Chem. Soc.*, **85**, 1958 (1963).
119. P. W. Robertson, *J. Chem. Soc.*, 1267 (1954).
120. P. B. D. de la Mare and M. Hassan, *J. Chem. Soc.*, 1519 (1958).
121. N. J. Bunce and N. G. Murray, unpublished observations.
122. A. Iochev, *12th Inst. Org. Khim. Bulgar. Akad. Nauk*, **2**, 53 (1965); *Chem. Abstr.*, **64**, 11078 (1966).
123. P. W. Jennings and T. D. Ziebarth, *J. Org. Chem.*, **34**, 3216 (1969).
124. S. J. Cristol and W. C. Firth, *J. Org. Chem.*, **26**, 280 (1961).
125. J. A. Davis, J. Herynk, S. Carroll, J. Bunds and D. Johnson, *J. Org. Chem.*, **30**, 415 (1965).
126. S. J. Cristol, J. R. Douglass, W. C. Firth and R. E. Krall, *J. Org. Chem.*, **27**, 2711 (1962).
127. J. S. Meek and D. T. Osuga, *Org. Synth.*, **43**, 9 (1963).
128. J. D. Roberts and V. C. Chambers, *J. Am. Chem. Soc.*, **73**, 3176 (1951).
129. M. Rottenberg, *Helv. Chim. Acta*, **36**, 1115 (1953).
130. T. R. Beebe and J. W. Wolfe, *J. Org. Chem.*, **35**, 2056 (1970).
131. R. P. Mauger and F. G. Soper, *J. Chem. Soc.*, 71 (1946).
132. G. C. Israel, A. W. N. Tuck and F. G. Soper, *J. Chem. Soc.*, 547 (1945).
133. W. J. Wilson and F. G. Soper, *J. Chem. Soc.*, 3376 (1949).
134. F. A. H. Rice, *J. Am. Chem. Soc.*, **78**, 3173 (1956).
135. F. A. H. Rice and W. Morganroth, *J. Org. Chem.*, **21**, 1388 (1956).
136. W. Kirmse and L. Horner, *Chem. Ber.*, **89**, 836 (1956).
137. J. K. Kochi, *J. Am. Chem. Soc.*, **87**, 2500 (1965).
138. M. Karelsky and K. H. Pausacker, *Aust. J. Chem.*, **11**, 336 (1953).
139. J. Kleinberg, *Chem. Rev.*, **40**, 381 (1947).
140. R. A. Zingaro, J. E. Goodrich, J. Kleinberg and C. A. VanderWerf, *J. Am. Chem. Soc.*, **71**, 575 (1949).
141. R. A. Zingaro, C. A. VanderWerf and J. Kleinberg, *J. Am. Chem. Soc.*, **72**, 5341 (1950).

142. R. A. Zingaro, C. A. VanderWerf and J. Kleinberg, *J. Am. Chem. Soc.*, **73**, 88 (1951).
143. J. Kleinberg and J. Sattizahn, *J. Am. Chem. Soc.*, **73**, 1865 (1951).
144. A. Simonini, *Monatsh.*, **13**, 320 (1892); **14**, 81 (1893).
145. H. Wieland and F. G. Fischer, *Ann. Chem.*, **446**, 49 (1926).
146. I. R. Beattie and D. Bryce-Smith, *Nature*, **179**, 577 (1957).
147. D. Bryce-Smith and P. Clarke, *J. Chem. Soc.*, 2264 (1956).
148. P. S. Raman, *Current Science (India)*, **27**, 22 (1958); *Proc. Ind. Acad. Science*, **44A**, 321 (1956).
149. K. B. Wiberg and K. A. Saegebarth, *J. Am. Chem. Soc.*, **79**, 6256 (1957).
150. C. Prévost, *Compt. Rend.*, **197**, 1661 (1933).
151. G. H. Crawford and J. H. Simons, *J. Am. Chem. Soc.*, **75**, 5737 (1953); **77**, 2605 (1955).
152. J. W. H. Oldham and A. R. Ubbelohde, *J. Chem. Soc.*, 368 (1941).
153. M. Schmeisser, K. Dahmen and P. Sartori, *Chem. Ber.*, **100**, 1633 (1967).
154. S.-J. Yeh and R. M. Noyes, *J. Org. Chem.*, **27**, 2978 (1962).
155. E. M. Chen, R. M. Keefer and L. J. Andrews, *J. Am. Chem. Soc.*, **89**, 428 (1967).
156. C. Prévost, *Compt. Rend.*, **196**, 1129 (1933); C. Prévost and J. Weimann, *Compt. Rend.*, **204**, 700 (1937).
157. C. Prévost and R. Lutz, *Compt. Rend.*, **198**, 2264 (1934).
158. L. Birchenbach, J. Goubeau and E. Berninger, *Ber.*, **65**, 1339 (1932).
159. D. Ginsburg, *J. Am. Chem. Soc.*, **75**, 5746 (1953).
160. R. B. Woodward and F. V. Brutcher, *J. Am. Chem. Soc.*, **80**, 209 (1958).
161. L. B. Barkley, M. W. Farrar, W. S. Knowles, H. Raffelson and Q. E. Thompson, *J. Am. Chem. Soc.*, **76**, 5014 (1954); W. S. Knowles and Q. E. Thompson, *J. Am. Chem. Soc.*, **79**, 3212 (1957).
162. S. Winstein and R. E. Buckles, *J. Am. Chem. Soc.*, **64**, 2787 (1942); S. Winstein and R. M. Roberts, *J. Am. Chem. Soc.*, **75**, 2297 (1953).
163. A. Streitweiser, *Chem. Rev.*, **56**, 675 (1956).
164. C. Djerassi, L. B. High, T. T. Grossnickle and R. Erlich, *Chem. Ind.*, 474 (1955).
165. C. Prévost, *Compt. Rend.*, **200**, 942 (1935).
166. T. N. Mehta, V. S. Mehta and V. B. Thosar, *J. Ind. Chem. Soc.*, **3**, 166 (1940).
167. W. T. Miller, Jr., E. Bergman and A. H. Fainberg, *J. Am. Chem. Soc.*, **79**, 4159 (1957).
168. D. Paskovich, P. Gaspar and G. S. Hammond, *J. Org. Chem.*, **32**, 833 (1967).
169. L. Birchenbach and L. Meisenheimer, *Ber.*, **69**, 723 (1936).
170. J. D. Roberts and H. E. Simmons, *J. Am. Chem. Soc.*, **73**, 5487 (1951).
171. See reference 19, p. 259.
172. G. Tsuchihashi, M. Matsushima, S. Miyajima and O. Simamura, *Tetrahedron*, **21**, 1039, 1049 (1965).
173. V. Kokatnur and M. Jelling, *J. Am. Chem. Soc.*, **63**, 1432 (1941).
174. H. A. Liebhafsky and W. H. Sharkey, *J. Am. Chem. Soc.*, **62**, 190 (1940).
175. G. S. Hammond and L. M. Soffer, *J. Am. Chem. Soc.*, **72**, 4711 (1950).
176. D. H. R. Barton, H. P. Faro, E. P. Serebryakov and N. F. Woolsey, *J. Chem. Soc.*, 2438 (1965).

177. D. D. Tanner, G. C. Gidley and N. C. Das, unpublished observations.
178. S. J. Cristol, L. K. Gaston and T. Tiedeman, *J. Org. Chem.*, **29**, 1279 (1964).
179. A. J. Solo and B. Singh, *J. Org. Chem.*, **32**, 567 (1967).
180. T. Moeller, *Inorganic Chemistry*, John Wiley, New York, 1952, p. 436.
181. B. S. Painter and F. G. Soper, *J. Chem. Soc.*, 342 (1947).
182. G. S. Hammond, *J. Am. Chem. Soc.*, **72**, 3737 (1950).
183. L. S. Silbert, D. Swern and T. Asahara, *J. Org. Chem.*, **33**, 3670 (1968).
184. A. Perret and R. Perrot, *Helv. Chim. Acta*, **28**, (1945).
185. Y. Ogata and K. Nakajima, *Tetrahedron*, **20**, 43 (1964), *Tetrahedron*, **20**, 2751 (1964).
186. Y. Ogata and K. Aoki, *J. Am. Chem. Soc.*, **90**, 6187 (1968).
187. Y. Ogata, K. Aoki and Y. Furuya, *Chem. Ind.*, 304 (1965).
188. Y. Ogata and K. Aoki, *J. Org. Chem.*, **31**, 1625 (1966).
189. K. Aoki and Y. Ogata, *Bull. Chem. Soc. Japan*, **41**, 1476 (1968).
190. J. E. Leffler, W. J. M. Mitchell and B. C. Menon, *J. Org. Chem.*, **31**, 1153 (1966).
191. P. H. Payot, W. G. Dauben and L. Replogle, *J. Am. Chem. Soc.*, **79**, 4136 (1957).
192. E. V. Zappi and O. Bouso, *Anales. Asoc. Quim. Argentina*, **35**, 137 (1947); *Chem. Abstr.*, **42**, 7704a, (1948).
193. D. E. Douglas, J. Eccles and A. E. Almond, *Can. J. Chem.*, **31**, 1127 (1953).
194. D. E. Douglas and A. M. Burditt, *Can. J. Chem.*, **36**, 1256 (1958).
195. E. Grigat and R. Putter, *Angew. Chem.* (*Intern. Ed.*), **6**, 206 (1967).
196. J. A. Barltrop, A. C. Day and D. B. Bigley, *J. Chem. Soc.*, 3185 (1961).
197. L. Francesconi and U. Cialder, *Gazz. Chim. Ital.*, **34 I**, 435 (1904).
198. L. Francesconi and G. Bresciani, *Gazz. Chim. Ital.*, **34 II**, 13 (1904).
199. W. Pritzkow and W. Nitzer, *J. Prakt. Chem.*, **25**, 69 (1964).
200. R. N. Haszeldine and J. Jander, *J. Chem. Soc.*, 4172 (1953).
201. M. G. Barlow, R. N. Haszeldine and M. K. McCreath, *J. Chem. Soc.*, (*C*), 1350 (1966), and references cited therein.
202. C. W. Taylor, T. J. Brice and R. L. Wear, *J. Org. Chem.*, **27**, 1064 (1962).
203. W. J. Fraser, *U.S. Pat.*, 3,398,072 (1968); *Chem. Abstr.*, **69**, 76618c (1968).
204. H. A. Brown, N. Kroll and D. E. Rice, U.S. Dept. of Commerce, Office Tech. Service, 4D 418,638; *Chem. Abstr.*, **60**, 14709 (1964).
205. D. E. Rice and G. H. Crawford, *J. Org. Chem.*, **28**, 872 (1963).
206. J. D. Park, R. W. Rosser and J. R. Lacher, *J. Org. Chem.*, **27**, 1462 (1962).
207. V. V. Kozlov and B. I. Belov, *Zh. Obshch. Khim.*, **33**, 1951 (1963); *Chem. Abstr.*, **59**, 8584 (1963).
208. B. I. Belov and V. V. Kozlov, *Zh. Obshch. Khim.*, **32**, 3362 (1962); *Chem. Abstr.*, **58**, 12706 (1963).
209. F. E. Francis, *J. Chem. Soc.*, **89**, 1 (1906).
210. V. Gold, E. D. Hughes and C. K. Ingold, *J. Chem. Soc.*, 2467 (1950).
211. R. O. C. Norman and R. Taylor, *Electrophilic Substitution in Benzenoid Compounds*, Elsevier, Amsterdam, 1965, Chapter 3.
212. M. Fujimoto, T. Sato and K. Hata, *Bull. Soc. Chem. Japan*, **40**, 600 (1967).
213. M. A. Paul, *J. Am. Chem. Soc.*, **80**, 5329, 5332 (1958).
214. V. Migrdichian, *Organic Synthesis*, Reinhold, New York, 1957, p. 1593.
215. K. J. Morgan and D. P. Morrey, *Tetrahedron*, **22**, 57 (1966).
216. I. J. Rinkes, *Rec. Trav. Chim.*, **49**, 1169 (1930); **51**, 1134 (1932).

217. M. J. S. Dewar and D. S. Urch, *J. Chem. Soc.*, 3079 (1958).
218. A. Fischer, J. Packer, J. Vaughan and G. J. Wrights, *Proc. Chem. Soc.*, 369 (1961).
219. A. Fischer, J. Packer, J. Vaughan and G. J. Wrights, *J. Chem. Soc.*, 3687 (1964).
220. P. B. D. de la Mare and R. Koenigsberger, *J. Chem. Soc.*, 5327 (1964).
221. J. H. Ridd, *Studies on Chemical Structure and Reactivity*, Methuen, London, 1966, Chapter 7.
222. D. J. Blackstock, A. Fischer, K. E. Richards, J. Vaughan and G. J. Wrights, *Chem. Commun.*, 641 (1970).
223. F. G. Bordwell and E. W. Garbisch, *J. Am. Chem. Soc.*, **82**, 3588 (1960).
224. F. G. Bordwell and E. W. Garbisch, *J. Org. Chem.*, **27**, 2322, 3049 (1962); **28**, 1765 (1963).
225. A. A. Griswold and P. S. Starcher, *J. Org. Chem.*, **31**, 357 (1966).
226. G. B. Gill and G. H. Williams, *J. Chem. Soc.*, 5756 (1965).

Author Index

This author index is designed to enable the reader to locate an author's name and work with the aid of the reference numbers appearing in the text. The page numbers are printed in normal type in ascending numerical order, followed by the reference numbers in parentheses. The numbers in *italics* refer to the pages on which the references are actually listed. (*) indicates a text reference which does not have a number.

Hale, P.   41 (33), *65*
Hales, J. L.   92 (163), *101*, 361 (41), *378*
Halevi, E. A.   200 (174), *226*
Halford, J. O.   82 (113), *100*
Hall, C.   23 (108), *33*
Hall, Jr., H. K.   121 (60), 123 (71), *135*, 186 (108, 109), 194 (109), 203 (108), 214 (259), 221 (108), *224*, *229*, 322 (27, 28), 339 (28), *344*, 410 (132), 411 (133), 412–414 (132), 416 (151), 427, 438 (132, 195), *450*, *451*
Hall, R. E.   398 (92), *449*
Haller, A.   257 (16), *288*
Halman, M.   338, 339 (103, 104), *347*
Halperin, B. I.   463, 481 (39), *495*
Hals, L. J.   329 (65), *345*
Ham, G.   412 (140), *450*
Hämäläinen, E.   220 (276), *229*
Hambly, A. N.   269 (83, 84), *289*
Hammarberg, G.   75 (56), *98*
Hammes, G. G.   185 (99), *224*
Hammett, L. P.   192 (142), *225*, 385, 386 (31), *447*
Hammond, G. S.   150 (59), *174*, 483 (168), 485 (175), 486 (175, 182), *498*, *499*
Hampel, G.   131, 132 (82), *135*
Hanack, M.   53 (109), *67*
Hancock, J. E.   255, 261 (8), *287*
Hands, G. C.   72 (12), *97*
Hanford, J. B.   123 (70), *135*
Hanford, W. E.   328 (59), *345*
Hann, R. M.   75 (38), *98*
Hanna, J. G.   79 (76), *99*
Hannum, S. E.   10, 19 (55), *31*
Hanson, G. C.   387 (46), *447*
Hantzsch, A.   286 (228), *292*
Hardegger, E.   239 (86), *249*
Hardie, B. A.   109, 114 (27), *134*
Harding, C. J.   395 (78), *448*
Harris, D. R.   329 (67), *345*
Harris, G. H.   80, 82 (107), 92 (162), *99*, *101*, 402 (105), *449*
Harris, Jr., J. F.   93 (170), *101*, 295, 301 (6), 303 (6, 31), *312*, 353, 360–362 (9), *377*
Harris, J. M.   398 (92), *449*

Harris, S. A.   233 (19), *248*
Harrison, M. A.   340 (111a, 112), *347*
Harrison, M. C.   28 (142), *34*
Hart, H.   276 (161, 162), *291*
Hartel, H. V.   245 (140), *251*
Hartley, B. S.   338 (101), *347*
Hartley, R. D.   75 (57), *98*
Harwell, E. J.   80, 82 (88), *99*
Hasek, R. H.   47, 63 (72), *66*
Hasek, W. R.   51 (94), *66*
Hassan, M.   462 (28), 475 (120), *495*, *497*
Hassel, O.   2 (2, 3), 10 (3), *30*
Hasserodt, U.   356, 357, 360 (25), *378*
Haszeldine, R. N.   80 (93, 94), 91 (156), *99*, *101*, 297 (16, 17), *312*, 461 (22), 464 (22, 44), 465, 480 (22), 491 (200, 201), 492 (201), *494*, *495*, *499*
Hata, K.   493 (212), *499*
Hatanaka, Y.   460–462 (14), *494*
Hatch, M.   340 (111), *347*
Hattori, S.   57 (132), *67*
Haubenstock, H.   463 (38), *495*
Haug, A.   73 (16), *97*
Haupt, F. C.   467 (76), *496*
Hauser, T. R.   72 (13), *97*
Haworth, W. N.   255 (2), *287*
Hayamizu, K.   23 (110), 24 (109, 110), *33*, 169 (114), 170 (114, 117), *176*
Hayashi, M.   258 (23, 24), 259 (25), 260 (33), *288*
Hayazu, R.   236 (50), *248*
Haydel, C. H.   70 (1), *97*
Hazlett, R. N.   246 (146), *251*
Heasley, G. E.   463 (40), *495*
Hebbelynck, M. F.   55 (122), *67*
Heck, H. d'A.   198 (163), *226*
Hedberg, K.   3 (4), *30*
Hefner, H.   358 (35), *378*
Hegazi, E.   365 (57), *379*
Heidbuchel, P. W.   193 (144), *225*
Heidelberger, C.   47 (73), *66*, 337 (96), *346*
Heimbach-Jushasz, S.   286 (229), *292*
Heinonen, K.   186 (107), *224*
Heintzeler, M.   465 (53), *495*
Heise, F.   232 (10, 11), *247*
Heitler, A.   273 (120), *290*

18

Subject Index